Lecture Notes in Computer Science 10905

Commenced Publication in 1973
Founding and Former Series Editors:
Gerhard Goos, Juris Hartmanis, and Jan van Leeuwen

More information about this series at http://www.springer.com/series/7409

Sakae Yamamoto · Hirohiko Mori (Eds.)

Human Interface
and the Management
of Information

Information in Applications and Services

20th International Conference, HIMI 2018
Held as Part of HCI International 2018
Las Vegas, NV, USA, July 15–20, 2018
Proceedings, Part II

 Springer

Editors
Sakae Yamamoto
Tokyo University of Science
Tokyo
Japan

Hirohiko Mori
Tokyo City University
Tokyo
Japan

ISSN 0302-9743   ISSN 1611-3349   (electronic)
Lecture Notes in Computer Science
ISBN 978-3-319-92045-0   ISBN 978-3-319-92046-7   (eBook)
https://doi.org/10.1007/978-3-319-92046-7

Library of Congress Control Number: 2018944381

LNCS Sublibrary: SL3 – Information Systems and Applications, incl. Internet/Web, and HCI

Printed on acid-free paper

This Springer imprint is published by the registered company Springer International Publishing AG
part of Springer Nature
The registered company address is: Gewerbestrasse 11, 6330 Cham, Switzerland

Foreword

The 20th International Conference on Human-Computer Interaction, HCI International 2018, was held in Las Vegas, NV, USA, during July 15–20, 2018. The event incorporated the 14 conferences/thematic areas listed on the following page.

A total of 4,373 individuals from academia, research institutes, industry, and governmental agencies from 76 countries submitted contributions, and 1,170 papers and 195 posters have been included in the proceedings. These contributions address the latest research and development efforts and highlight the human aspects of design and use of computing systems. The contributions thoroughly cover the entire field of human-computer interaction, addressing major advances in knowledge and effective use of computers in a variety of application areas. The volumes constituting the full set of the conference proceedings are listed in the following pages.

I would like to thank the program board chairs and the members of the program boards of all thematic areas and affiliated conferences for their contribution to the highest scientific quality and the overall success of the HCI International 2018 conference.

This conference would not have been possible without the continuous and unwavering support and advice of the founder, Conference General Chair Emeritus and Conference Scientific Advisor Prof. Gavriel Salvendy. For his outstanding efforts, I would like to express my appreciation to the communications chair and editor of *HCI International News*, Dr. Abbas Moallem.

July 2018

Constantine Stephanidis

HCI International 2018 Thematic Areas
and Affiliated Conferences

Thematic areas:

- Human-Computer Interaction (HCI 2018)
- Human Interface and the Management of Information (HIMI 2018)

Affiliated conferences:

- 15th International Conference on Engineering Psychology and Cognitive Ergonomics (EPCE 2018)
- 12th International Conference on Universal Access in Human-Computer Interaction (UAHCI 2018)
- 10th International Conference on Virtual, Augmented, and Mixed Reality (VAMR 2018)
- 10th International Conference on Cross-Cultural Design (CCD 2018)
- 10th International Conference on Social Computing and Social Media (SCSM 2018)
- 12th International Conference on Augmented Cognition (AC 2018)
- 9th International Conference on Digital Human Modeling and Applications in Health, Safety, Ergonomics, and Risk Management (DHM 2018)
- 7th International Conference on Design, User Experience, and Usability (DUXU 2018)
- 6th International Conference on Distributed, Ambient, and Pervasive Interactions (DAPI 2018)
- 5th International Conference on HCI in Business, Government, and Organizations (HCIBGO)
- 5th International Conference on Learning and Collaboration Technologies (LCT 2018)
- 4th International Conference on Human Aspects of IT for the Aged Population (ITAP 2018)

Conference Proceedings Volumes Full List

19. LNCS 10919, Design, User Experience, and Usability: Designing Interactions (Part II), edited by Aaron Marcus and Wentao Wang
20. LNCS 10920, Design, User Experience, and Usability: Users, Contexts, and Case Studies (Part III), edited by Aaron Marcus and Wentao Wang
21. LNCS 10921, Distributed, Ambient, and Pervasive Interactions: Understanding Humans (Part I), edited by Norbert Streitz and Shin'ichi Konomi
22. LNCS 10922, Distributed, Ambient, and Pervasive Interactions: Technologies and Contexts (Part II), edited by Norbert Streitz and Shin'ichi Konomi
23. LNCS 10923, HCI in Business, Government, and Organizations, edited by Fiona Fui-Hoon Nah and Bo Sophia Xiao
24. LNCS 10924, Learning and Collaboration Technologies: Design, Development and Technological Innovation (Part I), edited by Panayiotis Zaphiris and Andri Ioannou
25. LNCS 10925, Learning and Collaboration Technologies: Learning and Teaching (Part II), edited by Panayiotis Zaphiris and Andri Ioannou
26. LNCS 10926, Human Aspects of IT for the Aged Population: Acceptance, Communication, and Participation (Part I), edited by Jia Zhou and Gavriel Salvendy
27. LNCS 10927, Human Aspects of IT for the Aged Population: Applications in Health, Assistance, and Entertainment (Part II), edited by Jia Zhou and Gavriel Salvendy
28. CCIS 850, HCI International 2018 Posters Extended Abstracts (Part I), edited by Constantine Stephanidis
29. CCIS 851, HCI International 2018 Posters Extended Abstracts (Part II), edited by Constantine Stephanidis
30. CCIS 852, HCI International 2018 Posters Extended Abstracts (Part III), edited by Constantine Stephanidis

http://2018.hci.international/proceedings

Human Interface and the Management of Information

Program Board Chair(s): **Sakae Yamamoto, Japan and Hirohiko Mori, Japan**

- Yumi Asahi, Japan
- Linda R. Elliott, USA
- Shin'ichi Fukuzumi, Japan
- Michitaka Hirose, Japan
- Yasushi Ikei, Japan
- Yen-Yu Kang, Taiwan
- Keiko Kasamatsu, Japan
- Daiji Kobayashi, Japan
- Kentaro Kotani, Japan
- Hiroyuki Miki, Japan
- Ryosuke Saga, Japan
- Katsunori Shimohara, Japan
- Takahito Tomoto, Japan
- Kim-Phuong L. Vu, USA
- Marcelo Wanderley, Canada
- Tomio Watanabe, Japan
- Takehiko Yamaguchi, Japan

The full list with the Program Board Chairs and the members of the Program Boards of all thematic areas and affiliated conferences is available online at:

http://www.hci.international/board-members-2018.php

HCI International 2019

The 21st International Conference on Human-Computer Interaction, HCI International 2019, will be held jointly with the affiliated conferences in Orlando, FL, USA, at Walt Disney World Swan and Dolphin Resort, July 26–31, 2019. It will cover a broad spectrum of themes related to Human-Computer Interaction, including theoretical issues, methods, tools, processes, and case studies in HCI design, as well as novel interaction techniques, interfaces, and applications. The proceedings will be published by Springer. More information will be available on the conference website: http://2019.hci.international/.

General Chair
Prof. Constantine Stephanidis
University of Crete and ICS-FORTH
Heraklion, Crete, Greece
E-mail: general_chair@hcii2019.org

http://2019.hci.international/

Contents – Part II

Information and Learning

Information in Aviation and Transport

Intelligent Systems

Service Management

Contents – Part I

Information in Virtual and Augmented Reality

Information and Vision

Interacting with Information

The Divergency Model: UX Research for and with Stigmatized and Idiosyncratic Populations

Troy D. Abel[1](✉) and Debra Satterfield[2]

[1] University of North Texas, Denton, TX 76203, USA
troy.abel@unt.edu
[2] Long Beach State University, Long Beach, CA 90840, USA
debra.satterfield@cslub.edu

Abstract. In UX research, people with idiosyncratic or stigmatized conditions may go unrecognized, be dismissed as outliers, or be lumped within the larger normative group. This most directly affects persons with disabilities, medical conditions, or differing lifestyle or belief systems. This tendency may over represent the homogenous group while under representing other idiosyncratic groups based on complex factors such as stigma or perceived risk associated with revealing more accurate aspects of conditions, beliefs or lifestyle. This research bias negatively impacts the outcomes of the UX data and impacts the quality of designed products or services. Therefore, there is a critical need to identify the roles of ethnographic research, cultural programming, and cultural relativism with regard to stigmatized and idiosyncratic populations. The Divergency Model is a UX research and design methodology to identify and measure proximal distances in stigma/conformity and apathy/motivation between individual or target groups.

Keywords: UX · Research · Disabilities · Stigma · The Divergency Model
The Connectivity Model · Ethnographic research

1 Introduction a Subsection Sample

The ability to adequately understand the cognitive, social, emotional, motivational and behavioral tendencies of a target audience are the important underpinnings of user experience (UX) research and design. However, not all members of a target audience are equally willing or able to participate in all forms of UX research. People with disabilities and mental illness are often subject to high levels of stigma by society. In healthcare, stigma is considered to be one of the biggest obstacles in their care and quality of life (Sartorius 2007). The target audience may believe that they either are not able to adequately perform the necessary tasks because of idiosyncratic skills or beliefs or because they believe that their participation would put them in a position of significant personal risk due to disabilities or stigmatizing beliefs or conditions. In addition, failing to include all members of the target audience because of disability or stigma can introduce significant bias into the research due to oversampling of normative populations and under-sampling of idiosyncratic, atypical, or outlier populations. Therefore,

© Springer International Publishing AG, part of Springer Nature 2018
S. Yamamoto and H. Mori (Eds.): HIMI 2018, LNCS 10905, pp. 3–11, 2018.
https://doi.org/10.1007/978-3-319-92046-7_1

there is a critical need to develop an assessment tool for UX researchers to evaluate levels of stigma or proximal idiosyncratic responses in target audiences for the purpose of more robust risk management and mitigation strategies to improve inclusion and accurate representation in their research.

2 UX Research

User participatory design research tools such as usability tests, surveys, focus groups, biofeedback, ethnographic observation, eye tracking, and other types of user testing are widely accepted strategies for informing research-based design. However, typical user testing strategies and procedures often require the participant to have at least a basic command of spoken language, written language, and cognitive skills. There is also the basic assumption that the participant can freely offer their input in unbiased and uninhibited ways and consistent ways without fear of retribution, judgement, or a negative impact. However, many members of society may not fit this typical, homogenous user profile either temporarily or a continuous basis. These nonhomogeneous, atypical and marginalized user groups are a critical consideration in the design of spaces, products and communities designed specifically to serve their unique needs. Factors such as levels of conformity or stigma and motivation or apathy can impact a target audience's willingness and enthusiasm for meaningful participation in user experience design research. Therefore, there is a critical need to understand the roles of conformity and motivation with regard to target audiences.

Good design assumes that the designer and the user have a common language and a common set of expectations that are achieved through the designer's conceptual model and interpreted through the user's mental model. According to Donald Norman, a good conceptual model allows the designer to predict the actions of the user. Mental models are the cognitive constructs that people have of themselves, others, the environment, and things with which they interact (Norman 2016). Therefore, when a target audience has unique, divergent or idiosyncratic ways in they experience and interact with the world through social, emotional, physical, behavioral, or motivational channels, the designer must have a way of interpreting and measuring the ability of their conceptual models to meet those needs.

3 Cultural Relativism

Dealing with differences in the way individuals think, feel, and act is based in our mental programming. According to Hofstede and Hofstede, this cultural programming is learned in childhood through the social environment and through a person's collected life experiences. When these experiences vary in how they are collected, understood or interpreted, one person's mental programming can be vastly different from another person's even when they are members of the same society or family. This is also influenced by personality which is the unique set of mental programs possessed by an individual that can't be shared with any other human being (Hofstede and Hofstede 2005). Therefore, there is a critical need for a decision-making model that accommodates idiosyncratic populations in UX and UI research and design.

4 Ethnographic Research in UX

Ethnographic research can be an important strategy for working with target audiences where alternate methods of data collection are needed. For persons with language or cognitive impairment or persons with a high level of stigma, a nonverbal strategy for observation and data collection is useful because it can be used without causing stress to the audience. The Connectivity Model is a method for observing, collecting, and analyzing ethnographic data based on social, emotional, physical and behavioral (SEPB) categories. The Connectivity Model is particularly effective for use with populations where direct contact, focus groups, or interviews are not possible (Satterfield et al. 2016). According to a 2015 survey, persons with autism showed preference for lower interpersonal interaction than typical peers (Satterfield et al. 2016). Therefore, by using a non-verbal strategy, persons with language disabilities can still provide valuable data to inform the research with causing them any unnecessary duress.

5 Defining Disabilities

With regard to disabilities, our language and ability to understand and quantify the personal experiences of ourselves or someone else can defy our ability to coalesce them into a normative and homogenous population group as is often done in user experience design. The concept of disabilities is inherently vague and incomplete in terms of how it informs another person about the relevance, significance and impact of the idiosyncratic experience of the disabled person.

Disabilities tend to be defined by society based on a perceived comparison to the abilities of a healthy person. However, in many instances, the disabled person may find their condition either to be quite normal or quite impairing depending on the context of a situation. This may be further complicated by what are considered to be invisible disabilities such as language or cognitive conditions or highly visible physical conditions that are considered to be unacceptable or uncomfortable by individuals or society. Barriers such as a stigmatizing physical appearance, atypical language modalities, or unconventional cognitive abilities have traditionally prevented participation in usability and experience design research. This is in part due to complications in how to involve these user groups and the additional legal requirements as indicated by ethical IRB practices.

6 Designing for Disabilities

According to Pullin, "The priority for design for disability has traditionally been to enable, while attracting as little attention as possible... the approach has been less about projecting a positive image than about trying not to project an image at all" (Pullin and Higginbotham 2010). This emphasis on fixing or improving a person rather than meeting a desirability expectation radically changes the nature of both design research and the designed product. In addition, because of disability related differences in physiological and psychological experiences, the learned cultural aspects of persons with disabilities may diverge greatly from normative expectations. In addition, design for disabilities has focused more on drawing attention away from the disability than on creating value or desire for the person using the design. According to Pullin, "Design for disability has traditionally sought to avoid drawing any further unwelcome attention to the disabilities it addresses by trying to be discreet and uncontroversial, unseen or at least not remarked on. Disability can still be a source of discrimination and stigma for many disabled people, whereas a minority of medical engineers and designers are disabled themselves. Designing for and with people whose experiences they will probably never share can heighten sensitivity toward inadvertently causing offence" (Pullin and Higginbotham 2010).

7 Usability Studies and Population/Participant Selection Criteria

An initial literature search for research relating to usability studies participant selection, resulted in very little research and writing on the process of, or guidelines for, the selection process. Currently, normative practitioners rely on the business unit to define the target audience and/or developed personas to include in the study as research participants. These methods can introduce bias into the testing results by eliminating or under representing specific user groups or over representing specific user groups. These biases may be based on each group's ability to perform the test, their ability to comply with the research requirements; a group's lack of willingness to participate based on anxiety, a group's predisposition to lie or please the researcher, or based on the researcher's convenient access to one group over another group. The target audience's own perception of themselves and their situation may differ from the perception of the same person and situation as identified from a researcher's point of view.

Very little literature exists with regard to selecting participants, especially from marginalized populations; however, Rubin states "the selection and acquisition of participants whose background and abilities are representative of your products' intended user is a crucial element of the testing process" (Rubin and Chisnell 2008). Moreover, "Selecting participants involves identifying and describing the relevant behavior, skills, and knowledge of the person(s) who will use your product. This description is known as the user profile..." (Rubin and Chisnell 2008). Finally, "... your test results will only be valid if the people who participate are typical users of the product, or as close to that criterion as possible" (Rubin and Chisnell 2008). We, as practitioners, must strive to be as inclusive as possible when selecting participants for usability testing in any given system. We also must consider guidelines for inclusion and selection of marginalized populations to ensure our system is of greater use to more members of our initially identified audience, as well as marginalized audience members who may otherwise go unnoticed in persona development, especially by the business and marketing units.

8   The Role of Culture and Shared Experience

According to Hofstede and Hofstede, refers to mental programming which is divided into three levels: human nature which is the universal and inherited component; culture which is the learned component shared by a group or category of people; and the idiosyncratic level of personality which is both learned and inherited. Culture is a collective experience or phenomenon that is partly shared with people from the same social environment. Human nature is what all human share. Cultural differences manifest themselves in the areas of what Hofstede calls symbols, heroes, rituals, and values (Hofstede and Hofstede 2005). Therefore, a person may share traits or be a member of multiple cultures based on the person's ability to learn cultural traits. While at the same time they may differ greatly in other areas based on physical, cognitive, social, emotional or behavioral characteristics based on their personality and their unique physiological and psychological makeup.

Therefore, people who align closely with their cultural and with the common traits of human nature are most easily accounted for in user experience design research. Those people who exhibit strong personality traits or traits that are not closely aligned with their expected culture or human nature need to be addressed specifically in user experience design when they will be part of the target user group. Cultural relativism says that one culture has no absolute criteria for judging the activities of another culture or assigning them norms or mores (Hofstede and Hofstede 2005).

9   Measuring Stigma and Motivation Using the Divergency Model

In order to better understand the roles of stigma and motivation, The Divergency Model was created to evaluate the relative similarity or difference from the cultural norm. The center of the diagram represents a neutral position between unmotivated and motivated by a stimuli on the x axis and conformative and stigmatized on the y axis. The farther a target audience or individual moves away from center the more polarized their situation becomes toward motivation and desire versus stigma and apathy. The Divergency Model can be used to quantify an individual respondent within a context or the proximity between multiple respondents with regard their relative levels of conformity and motivation.

The Divergency model combines the motivation to conformity levels of a target audience with regard to a specific situation or stimuli. It identifies stigma and apathy as opposed to comfort and desire. Two or more groups or individuals can be plotted for the purpose of comparing their proximal distance from each other or from the normative population. High Motivation and conformity indicate high desirability and pleasure. Low motivation and stigma indicate low desirability and aversion. The center is neutral in both motivation and conformity (See Fig. 1, and Table 1).

The research survey tool that maps into this chart is based on The Connectivity Model areas of social, emotional, physical, behavioral, and motivational. The horizontal x-axis is the value derived from motivational questions and the survey answers

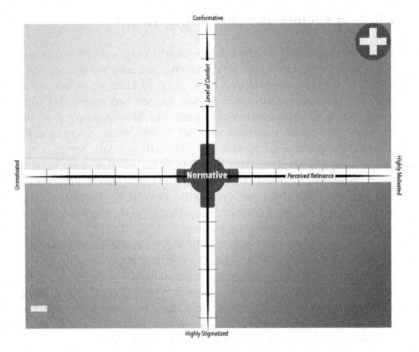

Fig. 1. The Divergency Model

Table 1. The Divergency Model key

ORANGE (Upper right quadrant)	*High conformity + High motivation* – likely to be pleasurable or desirable, most desirable
PURPLE (Lower right quadrant)	*High motivation + High stigma* – likely to be acceptable if conformity can exceed stigma or stigma can be reduced or neutralized, desirable but with anxiety barriers
BLUE (Lower left quadrant)	*Unmotivated + High stigma* – likely to be undesirable or avoided, least desirable
YELLOW (Upper left quadrant)	*Unmotivated + Confirmative* – likely to be acceptable if motivation can be raised, lacks desire or motivation but resonates with the user

to questions of social, emotional, physical and behavioral (SEPB) questions plot into the y-axis indicating levels of comfort or stigma associated with SEPB questions relevant or descriptive of the situational context of the UX/UI artifact under evaluation.

Low stigmatization is associated with conformity with the typical user group associated with this situation and low level social, emotional, or physical barriers. High Stigmatization is associated with a high lack of conformity with the typical user group associated with this situation and a significant combination of social, emotional, or physical barriers.

Low Motivation is associated with a lack of interest in the situation being evaluated without regard to the suitability of the situation in terms of social, emotional or physical barriers. Low motivation increases the likelihood of that a situation will be mastered or accepted. High Motivation is associated with a high level of interest in the situation without regard to the suitability of the situation in terms of social, emotional, or physical barriers. High motivation can increase the likelihood that a situation will be mastered or accepted.

10   Conclusion

Understanding the roles of stigma and motivation can help UX researchers include more people in user experience design research. By identifying which participants are experiencing high stigma or low motivation can help researchers mitigate the impact of these factors. By including idiosyncratic and stigmatized target audiences effectively into UX research, a better balance of under-represented audiences can be achieved.

The Divergency Model is used to identify people with idiosyncratic or stigmatized conditions that might otherwise go unrecognized, be dismissed as outliers, or be lumped within the larger normative group. It specifically addresses persons disabilities, medical conditions, or differing lifestyle or belief systems. This purpose of The Divergency Model is to address the tendency to over represent homogenous user groups while under representing or dismissing other idiosyncratic groups based on factors such as stigma or perceived risk associated with revealing more accurate aspects of conditions, beliefs or lifestyles. The Divergency Model also addresses inherent bias that is introduced when significantly different user groups are not represented. This in turn negatively impacts the outcomes of the UX data and the quality of designed products or services. Therefore, identifying the roles of ethnographic research, cultural programming, and cultural relativism with regard to stigmatized and idiosyncratic populations is critical to UX research. The Divergency Model also measures proximal distances in stigma/conformity and apathy/motivation between individual or target groups therefore representing the inherent diversity in user groups for any product or service.

11   Areas for Future Research

Future research will include testing The Divergency Model with idiosyncratic and stigmatized populations by developing a set of survey questions to plot into the conformity versus motivation quadrants. The following will be researched:

Idiosyncratic Populations in Usability Studies
Non-homogenous and Idiosyncratic Populations (Identifying them)
- Minority/Cultural/Ethnographic/Racial
- Cognitive/Brain Injury/Neurologically atypical
- ASD

Stigmatized Populations in Usability Studies
- Stigmatized/Marginalized populations
- IRB Protected Populations (prisoners, minors (younger than 18), experiencing diminished capacity, mentally or physically challenged, pregnant (particularly for those projects where physical procedures, exercises, etc., will be performed)
- LGBTQ+
- Mental Illness (Depression/anxiety/bipolar)
- Anxiety that is induced by contexts such as a test

Medical Populations with Disabilities and Stigma in Usability Studies
- Epilepsy
- Diabetes
- MR
- Alzheimer's Disease
- Down's Syndrome
- Dyslexia
- HIV/AIDS

User Participatory Strategies: (Categorize These Based on Required Skills)
Linguistic-Based Taxonomies
- Card Sorts
- Tree testing
- Interviews,
- Wizard of Oz
- Cognitive walk thru,
- Talk/speak aloud
- Focus Groups
- Surveys

Non-verbal Communication and Biofeedback
- Eye tracking
- Body sensors
- RFID
- Spatial and non-verbal communication strategies and devices

Ethnographic Observation
- Video modeling
- Ethnographic observations
- Design as Theater
- YouTube Libraries

Design UX Research Strategies or Models/Methodologies:
- Grounded Theory Model
- Definition of Emotions (EQ)
- Definition of Cognition (IQ)
- Kansei Engineering
- Connectivity Model
- Coolabilities (Narratives and Projects)
- Activity Theory

Risk Mitigation in UX Research
- Identifying Risk or Perceived Risk
- Identifying Risk Contexts
- Risk Mitigation Strategies
- Inclusion Criteria for Risk Mitigation

Accessibility in UX Research
- Identifying Accessibility in SEPB Categories
- Accessibility in UX Participation
- Accessibility and Risk Mitigation

References

Hofstede, G.H., Hofstede, G.J.: Cultures and organizations software of the mind. Cultures Organ. **23**, 362–365 (2005). https://doi.org/10.1057/jibs.1992.23

Norman, D.: The Design of Everyday Things (2016). https://doi.org/10.15358/9783800648108

Pullin, G., Higginbotham, J.: Design Meets Disability. Augmentative and Alternative Communication **27**, 123–139 (2010). https://doi.org/10.3109/07434618.2010.532926

Rubin, J., Chisnell, D.: Handbook of Usability Testing [Electronic Resource]: How to Plan, Design, and Conduct Effective Tests, 2nd edn. Wiley, Indianapolis (2008). https://doi.org/10.1007/s13398-014-0173-7.2

Sartorius, N.: Paths of medicine. Croatian Med. J. **48**(3), 396–397 (2007)

Satterfield, D., Kang, S., Lepage, C., Ladjahasan, N.: An analysis of data collection methods for user participatory design for and with people with autism spectrum disorders. In: Marcus, A. (ed.) DUXU 2016. LNCS, vol. 9746, pp. 509–516. Springer, Cham (2016). https://doi.org/10.1007/978-3-319-40409-7_48

Characteristic Analysis of Each Store in Japanese Hair Salon

Nanase Amemiya[✉], Remi Terada, and Yumi Asahi

Department of Management Systems Engineering,
School of Information and Telecommunication Engineering,
Tokai University, Tokyo, Japan
{5bjm2210, 5bjm2117}@mail.u-tokai.ac.jp,
asahi@tsc.u-tokai.ac.jp

Abstract. There are 223,645 Hair salons in Japan in 2011. Those same about quadruple amount of Japanese convenience store. From these things, we can know that there are many Hair salon in Japan. Hair salon moved and there are close stores about 9,000 a year. However new open chain stores about 12,000 therefore Hair salon increasing about 3,000 stores a year of Hair salon in Japan. Hair salon were group at Kanto region and Kinki region because they are high population density in Japan. The struggle for existence Hair salon are store excess also a lot of hair salon are small scale and they are micro enterprises. In that sense, Hair salon are faces severe competition. Customer's hair salon usages frequency is woman 4.5 times a year, men 5.38 times a year. The amount of usage per one times is women 6429 yen, men 4067 yen. Men are more frequently used, and the usage amount is increasing. However, the usage women's rate frequency of Hair salon more than men's rate frequency of hair salon. We use the data is all over Japan of a certain hair salon chain stores of this study. This was provided by Joint Association Study group of Managements Science (JASMAC) 2017 Data Analysis Competition. According to basic statistics, there are many customers with one visit to the store in this hair salon thus high customer rates. A certain hair salon have many people who are 30 to 60 years old. One of 12 stores are a male salon. We used Quantification category 3 that infer the characteristics of customers also, we predict store characteristics from the characteristics of customers. However, it was mixed Mathematical data and Qualitative data thus we had to unify the scale. Therefore, we converted Mathematical data and Qualitative data. As a result, we used Quantification category 3. We Interpreted compound variable answer1 is "Neighboring a working woman and office worker who want to quietness". Answer2 is a customer who emphasize of high temperature and high-quality care. Cluster analysis has Hierarchical approach and Nonhierarchical approach. Nonhierarchical approach be able used for small data on the other hands Hierarchical approach can be used to big data. Therefor we adopted Hierarchical approach. We needed to decide on a group when we used Hierarchical approach. However, it was objectively lacking, and it had disadvantages of reduced reliability if you have decided the number of groups before this analysis. Therefore, we did cluster analysis in several times in addition we decided the number of groups by squared residual sum of squares in cluster. We created an elbow diagram. The difference in the residual sum of squares within the group is It was shrinking sharply smaller than Groups 4 to 5 thus we decided the number of groups to 5 in

S. Yamamoto and H. Mori (Eds.): HIMI 2018, LNCS 10905, pp. 12–30, 2018.
https://doi.org/10.1007/978-3-319-92046-7_2

addition we did k-means method. When initial value of each result, big differences occur in size of group and convergence value. K-means method needs the best solution in multiple analyzes. In this research, we had the purpose to stores characteristic. We developed one-way analysis of variance and multiple comparison by compound variable, answer1 and answer2. We had the purpose to develop that analysis. The purpose was if we did not have significant difference, we would not classify every characteristic. One-way analysis of variance used compound variable, answer1 and answer2. Therefore, we used nonparametric method because those were not normal distribution. Nonparametric method has multiple test method. We used Games-Howell method in this research. Games-Howell method was matching in this research because it did not assumption homoscedastic. We used result hierarchical approach cluster analysis from First time to fourth time. We developed one-way analysis of variance and multiple comparison. The result, we adopted 2'nd time because it is classified definite characteristic. In addition, first time, 3'rd time and fourth time had significant difference one-way analysis of variance. However, there were combination in multiple comparison that did not have significant difference. We developed one-way analysis of variance with result of 2'nd time and compound variable. If answer1 shows a large value to minus, we infer that the customer want glamorous and has enough time to coming from afar. If answer2 shows a large value to minus, we infer that customer important care for low temperature and low humidity. We founded by one-way analysis of variance that there was a difference each group. We divided for each store the group. Then we were find difference in store characteristics. In addition, we suggested optimal marketing strategy based on result of analysis against each store. We will increase the synthesis and make finer interpretations.

After that, we will make suggestions for winning potential customers.

Keywords: Information presentation · Marketing · Beauty · Hair salon Multivariate analysis

1 Introduction

1.1 Background

Hair salon serves us such as perm, cut and make up.

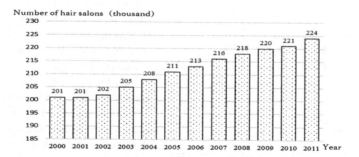

Fig. 1. The trends in the number of stores in the beauty industry [1]

Figure 1 shows the trends in the number of stores in the beauty industry by year. The number of hair salon is flat from 2000 to 2002 but since 2003, this number has continued to increase. Tokyo is the number with the largest number of hair salons. Next is Osaka prefecture, then Saitama prefecture. In addition, hair salon moved and there are close stores about 9,000 a year. However new open chain stores about 12,000 therefore Hair salon increasing about 3,000 stores a year of Hair salon in Japan. There are 223,645 Hair salons in Japan in 2011. Those same about quadruple amount of Japanese convenience store. From these things, we can know that there are many Hair salons in Japan.

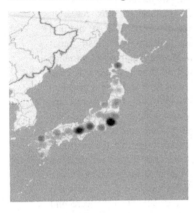

Fig. 2. Heat map of hair salon [1]

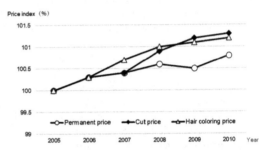

Fig. 3. Service fee consumer price index [1]

Figure 2 shows the number of Japanese hair salons by a heat map. Hair salon were group at Kanto region and Kinki region because they are high population density in Japan.

The average number of employees per facility in the hair salon nationwide is 2.03. Figure 4 shows the percentage of employees. One person is the highest at 29.9%, accounting for about 30% of the total. Next, 2 people followed 21.0%, 5–9 people followed 14.6%. About 70% of the facilities account for less than 5 people. From these facts, it can be inferred that many hair salons are small-scale and microscopic facilities. Figure 3 shows the trend of service fee provided by hair salon indexed as 100 in 2005. Prices offered services are increasing little by little every year. Haircut is the highest price increase, but it is kept at 1 point. Therefore, competition of hair salon is felt.

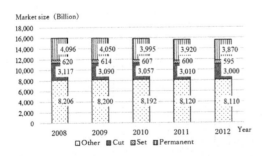

Fig. 4. Trends in services by category of beauty markets as a whole [2]

Figure 4 is a graph showing the changes in market size of beauty services as a whole from surgery for 2008 to 2012. The service scale is decreasing in any service. From this, it is expected that this trend will continue in the future.

Customer's hair salon usages frequency is woman 4.5 times a year, men 5.38 times a year. Men are more frequently used than females. In addition, the amount of usage per one times is women 6429 yen, men 4067 yen.

1.2   Purpose

In recent years, in the retail industry, customer purchasing behavior is gathered by POS data combined with customer ID and POS system and membership number and they use it to manage the store. By utilizing POS data with customer ID, we were able to approach each customer, and we were able to establish a strong relationship with customers. The trend of maintaining or increasing existing customers by strengthening relationships is used in many retail industries. Nakahara and Morita [3], Okuno and Nakamura [4] and Hisamatsu et al. [5] are mentioned as prior studies conducting customer purchase analysis from POS data with customer ID and purchase history. However, in the previous research, the approach to customers is mainly, and there seems to be few approaches to shops selling. In addition, there are many papers on beauty care, but among them there are few papers using statistical methods. Therefore, this research aims to analyze the features of stores using statistical methods and to help them in their management.

2   Data

The data we use is data from the hair salon chain store in Japan. We show below the overview. This data is targeted at 12 hair salons in Kanto area.

2.1   Data Overview

Provide: Joint Association Study group of Managements Science (JASMAC)
        2017 Data Analysis Competition
Period: July 1, 2015–Jun 30, 2016 (For 2 years).

2.2   Basic Aggregate

2.2.1   Customers Master Data

There are more female customers than men in certain hair salon (Fig. 5).

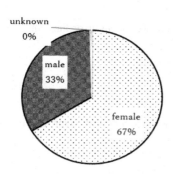

Fig. 5. Percentage of gender

Fig. 6. Heat map of customer's address

Customers live in various parts of Japan (Fig. 7).
They do not change the address after moving (Fig. 6).

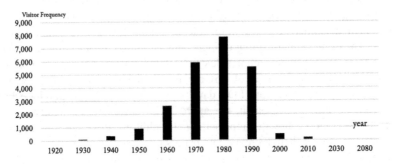

Fig. 7. Percentage by birth year

The 2030's and the 2080's were clearly abnormal.
There were 7915 customers who were no answer.
Most often born in 1980.

"Store I" has a high ratio of men because this store is for men only (Fig. 8).

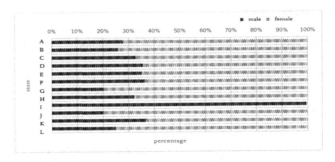

Fig. 8. Gender ratio by store

2.2.2 Accounting Master Data

Before we do basic aggregate of accounting master data, we do an error handling.

- Total of accounting master data: 166,922
  - Free customer: 2,425 (1.5%)
  - Return: 5,757 (3.4%)
  - Sale: 5,148 (3.1%).

These three were abnormal values. So, the total number is 153592 after error handling
(Figs. 9, 10 and 11).

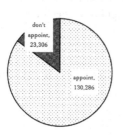
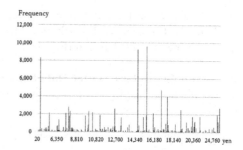

Fig. 9. Nominate ratio (Most customers choose the Hair dresser.)

Fig. 10. Percentage of payment amount (People with one payment 5,400 JPY–8,640 JPY are the most frequent. The next highest percentage is 10,800 JPY–17,280 JPY.)

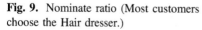
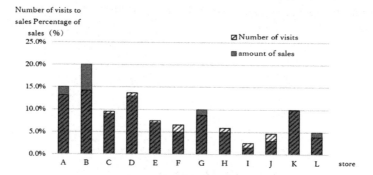

Fig. 11. Number of visitors and sales by store

Most of the customers are not using the point. We thought it would be difficult to save points. A simple calculation means that 1% to 2% of the amount is targeted. Therefore, it is estimated that customers using 1,000 points are excellent customers.

B is the highest in both the number of guests and sales. A, B, G, L the ratio of the number of visitors is below the ratio of sales. The ratio of visitors to C, D, E, F, H, I, J exceeds the ratio of sales. Based on the above, it can be estimated that A, B, G, j is a high percentage of sales per visit.

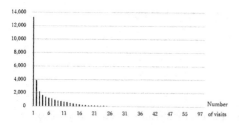
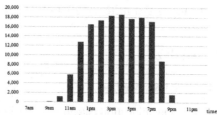

Fig. 12. The number of visits

Fig. 13. Number of accounting by time

There are a lot of people who visit only once. This can be said that the rate of churn is high (Fig. 12). The most used is 4:00 pm. Also, the opening was estimated at about 10:00 and closing at around 10 pm (Fig. 13).

2.2.3   Accounting Details Master Data Sample

Before we do basic aggregate of accounting details master data, we do an error handling.

- Total of accounting master data: 409,347
  - Return: 16,043(3.9%)
  - Sale: 14,846 (3.6%).

These two were abnormal values.
So, the total number is 378,458 after error handling.

2.2.4   Product Master

There are many products of about 3,000 JPY to 5,000 JPY (Figs. 14 and 15).

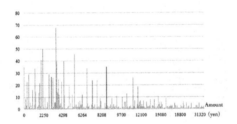

Fig. 14. Price setting

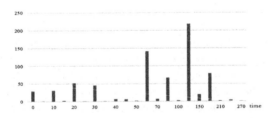

Fig. 15. Treatment time

I can see that there are a lot of treatments that take 120 min.

3   Analyze

3.1   Analysis Method

We used Quantification category 3 that infer the characteristics of customers also, we predict store characteristics from the characteristics of customers. However, it was mixed Mathematical data and Qualitative data thus we had to unify the scale. Therefore, we converted Mathematical data and Qualitative data. As a result, we used Quantification category 3.

For the Quantification category 3, we aimed to conduct statistical analysis on the Internet. We used statistical analysis software provided by Shigenobu Aoki.

3.1.1 Analysis Target

The number of customers to be analyzed is 31,776 out of 31,862 people who are registering customers. 86 people who do not have purchasing behavior between January 2015 and February 2016 (2 years), which is the analysis period, were excluded from the analysis as the reason for the decrease in the number of people. The reason for

Table 1. Result of quantification category 3 ①

Solution	Eigenvalue	Correlation coefficient	Effectiveness %	Cumulative contribution ratio
1	0.18043	0.42477	16.9%	16.9%
2	0.10708	0.32723	10.1%	27.0%
3	0.07461	0.27315	7.0%	34.0%
4	0.06324	0.25149	5.9%	39.9%
5	0.05635	0.23739	5.3%	45.2%
6	0.05217	0.22841	4.9%	50.1%
7	0.04989	0.22336	4.7%	54.8%
8	0.04633	0.21525	4.4%	59.2%
9	0.04429	0.21044	4.2%	63.3%
10	0.04157	0.20389	3.9%	67.2%
11	0.03616	0.19017	3.4%	70.6%
12	0.03481	0.18657	3.3%	73.9%
13	0.03343	0.18285	3.1%	77.0%
14	0.03066	0.17511	2.9%	79.9%
15	0.03021	0.17382	2.8%	82.8%
16	0.02769	0.16641	2.6%	85.4%
17	0.02584	0.16075	2.4%	87.8%
18	0.0208	0.14423	2.0%	89.7%
19	0.01859	0.13633	1.7%	91.5%
20	0.01663	0.12895	1.6%	93.0%
21	0.01549	0.12448	1.5%	94.5%
22	0.01477	0.12155	1.4%	95.9%
23	0.01201	0.10957	1.1%	97.0%
24	0.00919	0.09586	0.9%	97.9%
25	0.00646	0.08039	0.6%	98.5%
26	0.00582	0.07629	0.5%	99.0%
27	0.00574	0.07575	0.5%	99.6%
28	0.00466	0.06824	0.4%	100.0%

narrowing down the analysis target to the customers who had the purchase history is that there is data indicating the characteristics of the customer with the purchase history customer. On the other hand, for customers without purchase history, there are few data showing characteristic. In addition, because it can be considered that the ID of the free customer has multiple customer information, it was necessary to think separately from other customer IDs. Therefore, it is excluded in this research.

3.1.2   Result of Analysis

Table 1 shows the Quantification category 3 using 29 variables.

From Table 1, the contribution ratio of the components is low, and the eigenvalues are also low. Therefore, from this fact, it can be seen that useful results are not obtained with 29 variables.

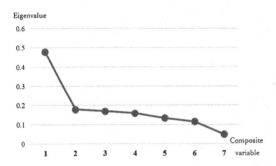

Fig. 16. Scree plot of Quantification category 3 analysis result

We analyzed again except for those with low commonality. We used 9 variables used for analysis, Kanto, Spring use, Summer use, Autumn use, Winter use, permanent use, nomination, morning use and average expenditure 6,429 yen or more. The results of the analysis are shown in Table 1.

A Scree plot of eigenvalues of synthetic variables obtained by analysis is shown in Fig. 16. As shown in Fig. 16, it can be seen that the inclination of the line of the Scree plot suddenly changes from the second line. Therefore, we adopt a second number of synthetic variables according to Scree plot criteria. When adopting up to two synthetic variables, the cumulative contribution ratio of the synthetic variable is 51.1% (Table 2).

Table 2. Quantification category 3 analysis result item category

Item category	Solution 1	Solution 2	Solution 3	Solution 4	Solution 5	Solution 6	Solution 7	Solution 8
Living in Kanto (4 prefectures)	2.98833	0.07177	−0.05667	0.38643	0.25241	0.00711	0.35915	0.68688
Supring use	0.0807	−0.24217	0.51465	−1.48727	−1.15005	0.97211	0.14064	−0.56344
Summer use	−0.12971	0.81873	−0.37256	0.83877	−1.19846	−0.9429	−0.90485	−0.55644
Winter use	−0.18769	−0.32505	0.32275	−0.9023	1.24052	−1.63124	0.3524	−0.69432
Autumn use	−0.274	0.27031	−0.35234	0.50697	1.38412	1.3941	−0.86583	−0.67665
Permanent use	−0.42786	−3.952	−0.55034	1.77861	−0.6795	0.03519	0.46551	−0.22156
Morning use	−0.71703	−0.20928	0.13985	−0.52069	0.11674	−0.15123	−1.05682	2.50122
Nomination	−0.82311	0.91012	−0.62656	0.52194	−0.06895	0.30941	2.22225	0.54494
Average expenditure 6,429 yen or more	−0.87902	0.90648	11.15988	4.55093	0.18613	0.49465	0.83765	0.1226

3.1.3   Consideration

3.1.3.1 Interpretation of Composition Variable · Solution 1

Table 3 shows coloring of variables belonging to each composition variable. Elements living in Kanto (4 prefectures) have the greatest influence on composition variable · solution 1. Next, factors with a high degree of influence are morning use and autumn use. Composite variable · solution 1 is composed of the above three. What can be thought of as a factor that reduces the use in the morning and the use of autumn when living in Kanto (4 prefectures) is the ease of accessing the store from home. The positional relation of the hair salon chain store used in this research is relatively close. Therefore, if you live in Kanto (4 prefectures) it can be said that it is easy to access. On the contrary, it is speculated that customers coming from outside of Kanto (4 prefectures) with poor access are not due to hair salons, but another main schedule. Those who do not live in Kanto (4 prefectures) have long hair trip round trip time thus it is difficult to think of going to a hair salon at the end of work. In other words, I guess that we will use a hair salon when visiting the city center on a holiday etc. or before the schedule from the afternoon. We think that autumn use is not a big factor because of its low value. However, the characteristics of autumn are strong wind, strong temperature and temperature difference, heavy rain due to typhoon. Therefore, the influence of access is considered to be related. From the above, the composition variable · solution 1 was interpreted as "Ease of access".

Table 3. Coloring compound variable scores

	Solution1	Solution2
Living in Kanto (4 prefectures)	2.98833	0.07177
Morning use	-0.71703	-0.20928
Autumn use	-0.274	0.27031
Permanent use	-0.42786	-3.952
Nomination	-0.82311	0.91012
Average expenditure over 6429 yen	-0.87902	0.90648
Summer use	-0.12971	0.81873
Winter use	-0.18769	-0.32505
Spring use	0.0807	-0.24217

3.1.3.2 Interpretation of Composition Variable · Solution 2

Synthetic variable · solution 2 is the one most influenced by permanent use. Next, nomination, Average expenditure 6,429 yen or more, summer use, winter use, spring use. The composition variable · solution 2 consists of the above six elements. The reason why summer use, nomination, Average expenditure 6,429 yen or more decreases when perming is used is the hair length. Because people sweat in summer, I think that perm is not suitable. In addition, permanent and color are correcting hair and discoloration with dyes thus It does not directly affect the change in hair length. It is possible to redo if the dye is washed as soon as possible. On the other hand, cuts cannot be redone to cut hair thus we speculate that customers tend to appoint stylists that will fulfill their wishes. Moreover, cutting tends to be cheaper compared to perm. As the reason that winter use and use of spring have a negative value, the temperature in spring and winter tends to be low, thus I speculated that it is suitable for the season when using perm. In addition, we think that those who use permanence will give preference to gorgeousness rather than quietness. From the above, the composition variable • solution 2 was interpreted as "styling to produce glamorous".

3.2   Cluster Analysis

We got two solutions of the synthetic variables. Perform cluster analysis from the sample score of it.

3.2.1   Analysis Method
Used version: SPSS Statistics 22 from IBM Company.

In this research, we used hierarchical cluster analysis to used large data. When performing hierarchical cluster analysis, it is necessary to determine the number of groups first. Then, we do cluster analysis multiple times. Using the residual sum of squares in the cluster obtained at that time, perform the elbow method. Also, the hierarchical cluster analysis used in this research used the k-means method.

3.2.2   Result of Analysis

3.2.2.1 Determining the Number of Groups
The residual sum of squares within the group used in shown in Table 4.
Also, an elbow diagram is shown in Fig. 17.

Table 4. The residual sum of squares corresponding to the number of groups

Number of group	2	3	4	5	6	7	8	9	10	11	12
Sum of squares residual within group	38,755	21,713	14,501	11,578	10,155	8,975	7,766	7,237	6,386	5,809	5,799

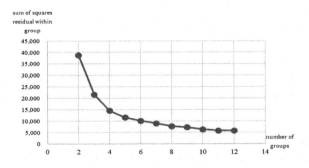

Fig. 17. Elbow diagram

Table 5. Hierarchical cluster analysis

number of group	2	3	4	5	6	7	8	9	10	11	12
sum of squares residual within group	38,755	21,713	14,501	11,578	10,155	8,975	7,766	7,237	6,386	5,809	5,799
comparison with the former		17,041	**7,212**	**2,924**	**1,423**	1,180	1,209	529	851	577	10

As can be seen from this figure, the numerical value sum of squares residual within group decreases toward the group number 5 (Table 5).

The result was changed by riding in the order of data. Therefore, rearrange the data several times and proceed with the analysis.

First time, when the customer number is in ascending order.
Second time, when the store number is in ascending order.
Third time, when the compound variable solution 1 is in ascending order.
Forth time, when the compound variable solution 2 is in ascending order.

3.2.3 Consideration

There was a big difference in each initial value, group size, group convergence value from 1st time to 4th time.

Indicate the positional relationship for each group of solutions 1.2 of the first to 4th synthesis variables.

From this figure, it is found that the average value of the composition variable solution 1.2 of each group from the 1st time to the 4th time is different.

In the k-means method, it is necessary to adopt the best solution from multiple analyzes. In this research, the purpose is to calculate the characteristics of the store. Therefore, one-way analysis of variance and multiple comparison are performed based on the composition variable solution 1.2 obtained by Quantify 3 types.

3.3 One-Way Analysis of Variance and Multiple Comparison

3.3.1 Analysis Method

Used version: SPSS Statistics 22 from IBM Company.

In this research, we used the Welch method when did One-way analysis. Also, we used nonparametric method because the data we used did not follow the normal distribution. In the multiple comparison, the Games-Howell method was used.

3.3.2 Result of Analysis

We were able to judge that only the second time has a significant difference in One-way analysis of variance and multiple comparison. On the other hand, In the first, third and fourth rounds, there was a combination which was judged as significant difference in one-way analysis of variance but there was no significant difference in multiple comparison. As a result, we judged that only the second times output useful results in hierarchical method cluster analysis.

Perform the characteristic analysis of each group from Fig. 18.

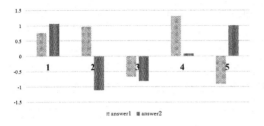

Fig. 18. Composite variable solution of each group average

First, when Synthetic variables answer 1 shows a large positive value, it is considered to be a customer living in the area where access is easy. Conversely, when it shows a large numerical value to a negative, it is speculated that it is a customer other than Kanto 4 prefectures.

When Synthetic variables answer 2 shows a large value to plus, we thought that it is a customer who likes directing and styling. Conversely, when I showed a large value to minus, I thought that it was a customer who likes gorgeousness and styling. When interpreting with only the value of the synthetic variable solution, Customers may live in places where access to stores is reasonably good. Also, we speculated that it is a customer group that likes ornate styling.

3.3.2.1 "Group1" Consideration

Group1's Synthetic variables answer 1 was 0.765 and answer 2 was 1.059. The Synthetic variables answer 1 shows the third highest positive value among the five groups, and the value of the Synthetic variables answer 2 shows the highest positive value among the five groups.

Looking at the results of basic statistics

: The customer group with the highest average age group.
: The highest proportion of female customers.
: Color usage rate is as high as 89%.
: The average usage price of 6,429 yen or more shows the highest value at 100%
: As the proportion of only the first visitor is 1%, and the ratio of more than 4 times per year is 93%, we estimate that it is a customer group with a high maintenance rate
: Spring, summer, autumn and winter are equally used.
: Based on the above, I interpreted the group 1 group of customers as "Changing hair color according to the season A good customer group of young people and elderly who live in the neighborhood".

3.3.2.2 "Group2" Consideration

Group1's Synthetic variables answer 1 was 0.971 and answer 2 was −1.109. Synthetic variables answer 1 has the second highest positive value among the five groups and Synthetic variables answer 2 shows the lowest negative value among the five groups

Looking at the results of basic statistics

: The average age is the third among the five groups, and the largest age group is 26–37 years old
: The only customer group whose male and female ratios are reversed in 64% males and 36% females
: The group with the highest use in winter

: A group with the highest percentage of customers visiting on holiday morning or in the morning
: Store with low color utilization ratio

Based on the above, I interpreted the group 2 group of customers as "A group of customers who prefer a quiet style, centering on men living nearby".

3.3.2.3 "Group3" Consideration

Group3's Synthetic variables answer 1 was −0.677 and answer 2 was −0.816. Group 3 is the only group showing the negative values for both synthetic variable answer 1 and 2. It is a customer group that has a high possibility of living in areas where access to stores is bad and tends to prefer quiet styling.

Looking at the results of basic statistics

: The group with the lowest average age
: The proportion of men is relatively high at 45%
: Residents in Kanto, use in the morning, designated use, Average usage amount
  6429 yen or more, use of perm, use of summer/autumn is low percentage
: Many people use inexpensive menus
: There are few people using coupons

Based on the above, I interpreted the group 3 group of customers as "This group is a group of people trying cut menu, young people who are non-regular".

3.3.2.4 "Group4" Consideration

Group4's Synthetic variables answer 1 was 1.305 and answer 2 was 0.084. People living in areas that are easy to access store, people who prefer quiet styling, and people who like luxury styling are mixed in Group 4.

Looking at the results of basic statistics

: The average age is the second highest among the 5 groups
: Kanto resident, nominated use, summer/fall utilization rate is the highest among
  the 5 groups
: Visit uniformly in spring, summer autumn and winter

Based on the above, I interpreted the group 4 group of customers as "Group of young women in the neighborhood who change hairstyles by season".

3.3.2.5 "Group5" Consideration

Group5's Synthetic variables answer 1 was −0.904 and answer 2 was 1.001. There are many people who live in areas with poor access to stores or prefer quiet styling.

Looking at the results of basic statistics

: Coupon usage rate is the highest among 5 groups
: People who have an average usage price of 6429 yen or more are very high, 99%
: Color menu usage rate 60%

Based on the above, I interpreted the group 5 group of customers as "This group is a group of people trying out the color menu, Young people who are save money".

3.3.2.6 Percentage of Customers by Group

We found that there are many customers in groups 3 and group 5. Therefore, we guess that there are many new customers overall (Fig. 19).

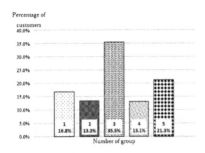

Fig. 19. Number of customers by group for each store

On the other hand, group 1 and group 4, which were presumed to be good customers, are not so many. What the group's customers are seeking is "cut menu", but because customers in Groups 1 and 4 are "color permanent menus", move Group 3 customers to groups 1 and 4 Is difficult. Therefore, Group 3 can also be judged as a customer group that is highly likely to become a departed customer. Because the usage rate of "permanent menu" is high in the group 5 customer, there is a high possibility of moving to group 1 and group 4. Therefore, it is judged that it is a prospective customer highly likely to become a good customer. Therefore, it can be judged that "Styling by color menu is a strength" as the whole 12 stores.

3.3.3   Consideration by Store

From Table 6, it can be seen that Group 3 with the largest number of samples has the largest belonging number in most shops. As Group 3 customers always exist in high proportion at all stores, we can guess that they are basic customer groups. Therefore, it is not considered for interpretation unless it is extremely high or low.

Table 6. Number of group customers by store

group		A		B		C		D		E		F		G		H		I		J		K		L		total
	1	19.2%	658	27.7%	942	11.5%	425	14.0%	662	13.9%	319	17.8%	375	20.1%	514	16.2%	374	2.2%	30	18.8%	317	20.5%	608	19.7%	243	5,467
	2	14.6%	502	12.8%	435	11.7%	431	10.7%	508	11.0%	251	13.6%	288	10.1%	258	8.3%	195	25.4%	347	12.4%	209	15.2%	449	14.3%	177	4,048
	3	27.4%	940	30.7%	1,046	35.6%	1,312	36.8%	1,742	34.8%	796	36.0%	760	26.9%	688	37.6%	869	64.7%	882	29.4%	495	28.3%	838	37.5%	463	10,831
	4	19.5%	669	11.0%	374	13.0%	478	13.3%	631	21.1%	482	14.9%	315	15.2%	389	9.9%	230	4.5%	61	10.9%	183	12.9%	381	10.7%	132	4,325
	5	19.4%	665	17.9%	609	28.1%	1,036	25.1%	1,190	19.2%	439	17.7%	374	27.8%	713	28.0%	648	3.2%	44	28.5%	479	23.2%	687	17.9%	221	7,105
total		100%	3,434	100%	3,406	100%	3,682	100%	4,733	100%	2,287	100%	2,112	100%	2,562	100%	2,314	100%	1,364	100%	1,683	100%	2,963	100%	1,236	31,776

As a result of considering each store based on these, it was possible to divide it into four store patterns.

3.3.3.1 Royal Customer Store

Royal customer stores are "A", "B", "G" and "K" (Fig. 20).

The common thing to the four is that the value of Group 1 is above the average. Group 1 is considered a good customer in a certain hair salon chain. So, it can be said that these four stores are able to acquire good customers and the management is relatively stable.

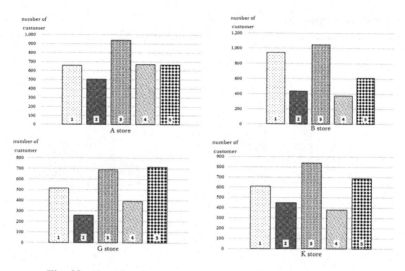

Fig. 20. Number of customers by group by Royal customer store

3.3.3.2 Lead Nurturing Store
Lead Nurturing Stores are "H" and "I" (Fig. 21).

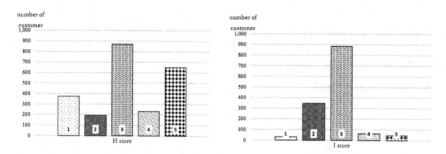

Fig. 21. Number of customers by group by Lead Nurturing Store

H store

The strength of the hair salon I have handled this time is "Color Menu", and the customer of Group 5 can be said to be a promising customer. Because the value of Group 5 is higher than the average, we thought that we could acquire good customers by approaching existing customers.

I store

As the value of group 2.3 with many men is high, what is required in this store is quiet styling (cut). From that we can say that a promising customer of "I" is Group 3. We thought it would be nice to have successful guidance to excellent customers because the value of group 3 is high.

3.3.3.3 Lead Generation Store

Lead Generation Stores are "C", "E", "F" and "L" (Fig. 22).

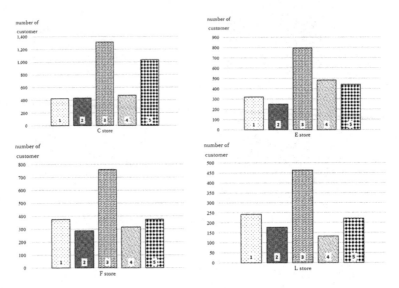

Fig. 22. Number of customers by group by Lead Generation Store

These four stores are stores where the value of the prospect's fifth group is lower than the average or there are few good customers. It is necessary to acquire new customers and promising customers.

3.3.3.4 Miss Match Store

Miss match store is "D" (Fig. 23).

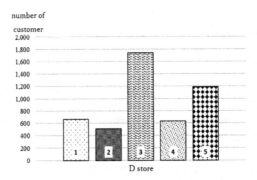

Fig. 23. Number of customers by group by Miss match store

There are few promising customers, and the proportion of groups that may not be repeat customers is very high. In fact, the proportion of repeat customers is very low, and this hair salon may not correspond to the area.

3.3.3.5 Lead Nurturing and Lead Generation Store
Lead Nurturing and Lead Generation Store is "J" (Fig. 24).

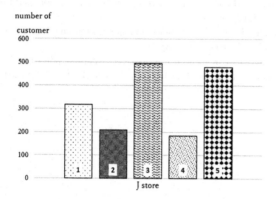

Fig. 24. Number of customers by group by Lead Nurturing and Lead Generation Store

The city of the location is under development. Therefore, it is necessary to simultaneously perform the Lead Nurturing of the current customer and the lead generation of the new customer.

3.3.4   Suggestion by Store

3.3.4.1 Royal Customer Store
Continue to lead nurturing and continue to nurture superior customers. Management will be stable if we maintain current status.

3.3.4.2 Lead Nurturing Store
Because "H store" has many future customers, it is important to promote them to promising customers.
I store is a new store, firstly it is necessary to improve store brand loyalty. In order to secure continuous high-quality customers, it is necessary to develop strengths centering on cuts that tend to favor men.

3.3.4.3 Lead Generation Store
It is urgent to attract prospective customers. Due to the fact that many lead generation shops have many competing stores in the vicinity, differentiation from other stores is necessary. Should appeal about ornate hair style menu which is the strength of certain hair salon.

3.3.4.4 Miss Match Store
J store is not a glamorous hairstyle, which is the strength of a certain hair salon, should enhance the cut menu that this store customer is seeking.

References

1. http://www.mhlw.go.jp/seisakunitsuite/bunya/kenkou_iryou/kenkou/seikatsu-eisei/seikatsu-eisei22/dl/h22/biyou_housaku.pdf. Accessed 02 Dec 2017
2. http://www.beauty-skill.net/wp-content/uploads/2015/01/25kokunai-biyou-shijou-gaiyou.pdf. Accessed 02 Dec 2017
3. Nakahara, T., Morita, H.: Customer purchase analysis for graph partitioning method CD dealers. Oper. Res. **52**(2), 79–86 (2007)
4. Okuno, T., Nakamura, K.: Measurement of personal promotion effect. Inf. Process. Soc. Artic. J. Sci. Model. Appl. **9**(3), 61–74 (2016)
5. Hisamatsu, T., Yamaguchi, K.: Asahi bow not yet: a comparison of FMCG buying patterns using ID-tagged POS data in the drugstore. Oper. Res. **57**(2), 63–69 (2012)

Career that Tend to be Unpaid for Motorcycles Sales Loans

Mari Atsuki[✉] and Yumi Asahi

Department of Management System Engineering,
School of Information and Telecommunication Engineering,
Tokai University, Tokyo, Japan
5bjml120@mail.u-tokai.ac.jp, asahi@tsc.u-tokai.ac.jp

Abstract. Country A is in Latin America and GDP is lower than average. Prices of Japanese products such as motorcycles are on an upward trend. It was influenced by instability of the world economy. The customer selects loan payment. Among them, some customers cannot pay for the specified payment period. If this situation deteriorates, it will be difficult to recover manufacturing costs. Therefore, we analyze the characteristics of customers who loan bankruptcy.

There is a weak positive correlation (0.300) between average income for each province and sales volume. There is a negative correlation (−0.542) to the main income amount and those who go bankrupt. So it can be said that people living in poor countries cannot easily buy a motorbike. Even if they make loans to purchase, they cannot pay. I divided the state data into five. It turned out that there was economic disparity. The southern region is rich and the north region is a poor region. There are many customers who cannot repay the loan to the north region.

In this country, 13% of 10 million young people have not enrolled in school and are not working. In order to see the relationship between academic background and other factors, we quantified the academic qualification. There is a negative correlation between academic records and the proportion of unpaid −0.673. Therefore, I can say that the lower the academic background, the more delayed repayment of the loan. I conducted a multiple regression analysis with the loan's unrepayable party as the dependent variable. The standard partial regression coefficients were −0.543 for educational record number and −0.327 for main income. And I learned that educational background is a factor that has a big influence on judging whether it will be unpaid than main income.

We analyze logistic regression on the probability that that person will be Bad with these elements. We understood what condition customers are likely to be Bad. My goal is to be able to know the percentage of probabilities that your loan can not be repaid from your profile information.

Keywords: Career · Loan · Loan bankruptcy · Educational history
Main revenue · State

1 Introduction

World finances were influenced by Lehman shock in 2008. In the spring of 2009, the economy had bottomed out, and, it had a gradual recovery trend. In August 2011, financial instability was increased by downgrade of US government bonds and impact

© Springer International Publishing AG, part of Springer Nature 2018
S. Yamamoto and H. Mori (Eds.): HIMI 2018, LNCS 10905, pp. 31–40, 2018.
https://doi.org/10.1007/978-3-319-92046-7_3

of the deepening of the European debt crisis. In Latin America, major currencies plunged. And the price of Japanese products in Latin America has been on an upward trend. Hence customers choose installment payment.

Among them, some customers can't pay in the specified payment period. The country's poverty alleviation is the cause. The people's income has increased. And people who were able to afford to some extent purchased loans and purchased products. They don't have loan experience and understanding of contract contents is often inadequate. Therefore, the consciousness to the loan is low, leading to bankruptcy. In addition, some people make new loans even though loan repayment has not been paid. They still make loans even if the down payment gets higher. This is an international issue. If this situation deteriorates, manufacturers have difficulty in recovering manufacturing costs. Thus we analyze the characteristics of customers who will be outstanding.

In this study, we use motorcycles sales data from country A. GDP of country A is lower than average. Motorcycles is used as a practical means of transportation in country A. And it is essential for economic development such as commuting and agriculture.

2 Data Summary

We use motorcycles sales data from country A in September 2010 to June 2012. There are 14,304 data and 359 variables. For example, borrowing amount, birthday, resident state, revenue, career and whether the loan has been repaid or not.

There is not has data of age, and we calculated it using data of application date and birthday. There are some blanks in the main income data. Then supplementing that data with a rating score without missing values. The way is like this. Collect other data with the same value as the rating score of the data whose main income is blank. And we calculate the average value of those main incomes. We put it in the main income amount that was blank.

And we cleaned up the data. We deleted data with blank spaces in key variables such as interest amount and borrowings. With this, the number of data become 13,214.

From now on we will call people who failed to repay the loan as "Bad". People who have once become bad continue to be bad. Therefore, the proportion of data is increasing. Of all the data, 6 months Bad is 9.1%, 12 months Bad is 16.8%, 18 months Bad is 23.0%. As the number of data increases, it leads to more accurate analysis, thus we decided to focus on 18 months Bad. In addition, because it includes "6 months Bad" and "12 months Bad", comprehensive measures can be considered.

There are 4 state variables of state (resident state, resident registration state, working state and first trust survey state.) The same data for all four registered states is 81.8%. The same data for three registered states is 17.1%. Among them, only the resident state is rarely registered in a different state. Therefore, resident state data is the most stable. We use the resident state as state data on that person data.

We made a group of main income. We did it based on the income stratum rank index issued by the economic foundation of country A. The average primary income for Country A is 3,010.48. The A/B layer is wealthy class, the average main income is

145,594.12. The C layer is the middle class, the average main revenue is 2,802.01. The D and E layer are poor and the average of the main income is 1,323.34 and 836.46 respectively. You can see that there is economic disparity.

3   Fundamental Statistics

According to the study of Miyoshi Yusuke (2013), regional features are involved. For example, regional wage rate, security, percentage of young people. Thus we see at the data by state.

3.1   Main Revenue

There is a weak positive correlation (0.300) between the average amount of major revenue by state and sales volume. There is a negative correlation (−0.542) by Bad people. Therefore we can say that people living in poor states can't easily buy motorcycles. Even if they make a loan to purchase, they will not be able to pay.

We classified state data as into division of 5, the North, the Northeast, Midwest, the South and Southeastern area based on a geographic division of the Ministry of Agriculture, Forestry and Fisheries. We corresponded it to the state data arranged in order of the main income. It is shown in Fig. 1. We understand the following. There is an economic disparity that the poor increase from the South to the North. Southeast sales are particularly high. The number of units sold varies by state. The sales volume of North-A is small, however the proportion of Bad is very high.

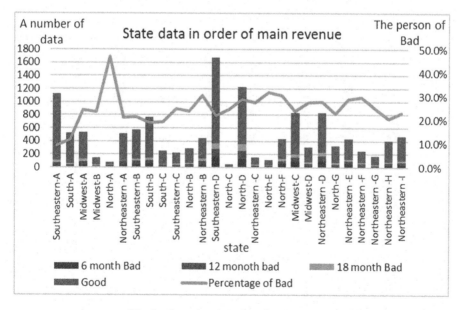

Fig. 1.   State data in order of main revenue

We compared the highest region with the lowest region in average of major revenue. The region with the highest main income is the southeastern part, which is 1784.807. The region with the lowest main income is the north part, which is 1483.431. As expected, there are many bad customers in the north. They are 18.3% of the southeastern population, 28.5% in the north population. We think that the cause of the difference is in occupation. Self-employed is a job with little income and stability. There are more self-employed in the northeastern part than in the southeastern part. Hence even though the northern part has fewer borrowings than the southeastern part, there are many customers who cannot repay the loan.

3.2 Educational History

Another cause of the economic disparity is that 13% of 10 million young people are not enrolled in school, and they are not working yet. In this data of this study, 2.3% is "data without educational history". Considering that people who cannot afford to receive education can think that they cannot afford to purchase motorcycles, it is a convincing data.

In country A, ages 6 to 11 go to Basic school, then until 12 years old is Secondary school, then until 18 years old is High Education. To see the relationship between educational background to other elements, we quantified my academic background. We did it in the following way. Number each academic record in turn. (Basic school: 1, Secondary education: 2, Higher education: 3, Correspondence training: 4). In case of dropout, subtract 0.5 from each number. There is a negative correlation of −0.651 between the academic record and the ratio of unpaid. Therefore we can say that the

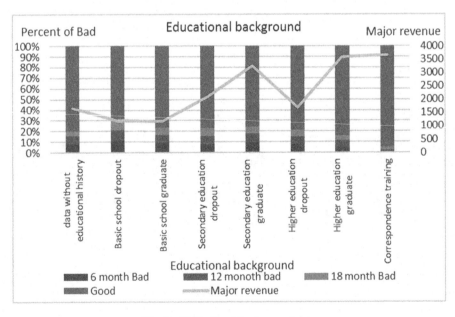

Fig. 2. Educational background data

lower the academic background, the more the repayment of the loan is delayed. However, the correlation between educational background and main revenue is 0.402. The reason is probably the difference between dropout and graduation. The main income amount is higher for secondary education graduation than for higher education dropout. Even if you have a high academic background, you probably cannot get a good job if they dropout. In addition, primary income of people without educational history is higher than those who were in the basic school and the secondary education dropout. This is because someone who has not attended school has been working for a long time. It is because there are advantages in terms of time (Fig. 2).

4 Multiple Regression Analysis

We conducted multiple regression analysis with the proportion of Bad persons by state as the dependent variable. Independent variables are the average of educational background and main income. Because of multiple regression analysis, the weight determination coefficient was 0.542, which was a significant value at 1% level. The standard partial regression coefficients were −0.543 for educational record number and −0.327 for main income. And I learned that educational background is a factor that has a big influence on judging whether it will be unpaid than main income.

5 Logistic Regression Analysis

We used logistic regression. The difference between logistic analysis and multiple regression analysis is the data type of the objective variable. The objective variable of multiple regression analysis is numeric. Logistic regression analysis is nominal scale data. In the multiple regression analysis described above, "how much percentage of people in the state would be Bad" was found. Logistic regression is seen more concretely than that. We know the probability that the person himself will be Bad. Therefore, it meets the purpose of this research. The judgment criterion of this analysis result is that the significance probability is less than 0.05. When the odds ratio (Exp (B)) is larger than 1, it can be said that the larger the explanatory variable, the more it affects the target variable. Conversely, when the odds ratio (Exp (B)) is smaller than 1, it can be said that the larger the explanatory variable, the less influence the target variable.

The objective variable is Bad. The dependent variable is a career. We analyze by main revenue, education history, resident state in order. Main income data has low regression coefficient. In addition, the main income level is four stages, so you can narrow down the target roughly. There are 8 levels of educational history and 26 states. Therefore we can see in detail in order.

5.1 Main Revenue

First, the main revenue analysis the dependent variable. Logistic regression requires that the dependent variable be a binary variable. We decided to use the data was grouped main revenue revel. We made dummy variables of main income revel. These are dependent variables. However, the data of the AB layer is 1.11% (147data) of the total. Furthermore, only 19 data are bad in the AB layer. We excluded it because the number of data is too small.

Table 1. Analysis result of main revenue dependent variable

	B	Stand- ard error	Wald	df	Signifi- cance prob- ability	Exp(B)
E layer	0.651	0.248	6.884	1.000	0.009	1.918
D layer	0.935	0.248	14.258	1.000	0.000	2.548
C layer	0.209	0.252	0.686	1.000	0.407	1.232
constant	- 1.908	0.246	60.203	1.000	0.000	0.148

The results are shown in Table 1. Among them, the significance probability is 0.05 or less only in layers D and E, or the odds of these two layers can be said to be reliable. Odds exceed 1 in all layers. In other words, it can be said that it tends to be Bad if it is this layer. The odds of the D layer is higher than that of the E layer. In other words, it can be said that E layer is less likely to be Bad than D layer. The reason is that motorcycle makers are taking measures against E customers. Therefore, among the poor, the D layer with more income than the E layer tends to be Bad. The C layer has a high significance probability and odds are also relatively low. To see more about Bad, we focus on the D and E layers and proceed with the analysis.

5.2   Educational History

We made dummy variables for each academic background too. These are dependent variables. The data of correspondence education is 0.36% (38 data) of all data. Only 4 cases are Bad in that. So exclude this variable.

The results are shown in Table 2. The significance probability was less than 0.05 and the significant variables were those related to "Basic School" and "Secondary School". All odds are over 1. The lower the academic background is, the more it tends to be Bad. We find that the lower the educational background is, the more likely it is Bad. In each degree, it turns out that the person who is dropping out tends to be Bad more than the person who is Graduate.

The chapter of fundamental statistics shown that the main income of those who do not have an educational history is high. As can be inferred from there, the odds of data without educational history were lower than basic school education and secondary school education. It is a special case that there is no history of education. In addition, the odds of dropped people are higher than those who graduated. It turned out that those who dropped out from this result were likely to be Bad.

Table 2. Analysis result of educational history dependent variable

	B	Stand-ard error	Wald	df	Sig-nificance probability	Exp(B)
No edu-cational his-tory	0.714	0.462	2.391	1.000	0.122	2.042
Basic school dropout	1.323	0.473	7.829	1.000	0.005	3.755
Basic school graduate	0.958	0.446	4.613	1.000	0.032	2.606
Second-ary education dropout	0.969	0.444	4.765	1.000	0.029	2.636
Second-ary education graduate	1.048	0.433	5.846	1.000	0.016	2.851
Higher education dropout	0.829	0.444	3.491	1.000	0.062	2.291
Higher education graduate	0.644	0.443	2.110	1.000	0.146	1.903
constant	-2.100	0.433	23.575	1.000	0.000	0.122

As stated above, the enrollment rate in this country is low. Up to the second degree is compulsory education. However, the school truancy rate and voluntary dropout rate are still high. The reason is often in the environment of parents and family. It becomes impossible to attend school with unstable work and accompanying move. Different family problems such as divorce and divorce. In addition to these, parents are also less aware of their education. We think there is poverty in these backgrounds. They cannot get a stable job and cannot prepare the environment. Living expenses are given priority, there is no educational expenses. For these reasons, it is possible to assume that a person who dropped out is a poor household whose parents' educational awareness is low and who is not in a stable position. In addition, educational institution infrastructure is sometimes inadequate. You can see that there is no money in the country or state. Because it is a poor area it is difficult to get a stable job. And it's difficult to solve educational problems more.

Anyway, we can see that basic school and secondary education are factors that tend to be Bad. We focus on these and proceed with the analysis.

5.3   State

Finally, we analyzed the state data as an explanatory variable. We made a dummy variable like educational background.

The results are shown in Table 3. There were 11 states with significant probability less than 0.05.

Table 3. Analysis result of state dependent variable

	B	Standard error	Wald	df	Significance probability	Exp(B)
Southeastern-A	-1.301	0.187	48.579	1.000	0.000	0.272
South-A	-0.944	0.194	23.593	1.000	0.000	0.389
Midwest-A	-0.230	0.161	2.044	1.000	0.153	0.795
Midwest-B	-0.338	0.309	1.191	1.000	0.275	0.713
North-A	1.287	0.338	14.503	1.000	0.000	3.623
Northeastern -A	-0.400	0.169	5.629	1.000	0.018	0.670
Southeastern-B	-0.392	0.179	4.822	1.000	0.028	0.676
South-B	-0.440	0.156	7.965	1.000	0.005	0.644
South-C	-0.463	0.219	4.462	1.000	0.035	0.630
Southeastern-C	-0.068	0.213	0.102	1.000	0.750	0.934
North-B	-0.244	0.191	1.624	1.000	0.203	0.784
Northeastern -B	-0.008	0.160	0.003	1.000	0.959	0.992
Southeastern-D	-0.273	0.135	4.103	1.000	0.043	0.761
North-C	-0.137	0.410	0.111	1.000	0.739	0.872
North-D	-0.023	0.137	0.027	1.000	0.869	0.978
Northeastern -C	-0.041	0.235	0.031	1.000	0.860	0.959
North-E	0.452	0.261	2.986	1.000	0.084	1.571
Midwest-C	-0.329	0.148	4.930	1.000	0.026	0.719
Midwest-D	-0.130	0.186	0.485	1.000	0.486	0.878
Northeastern -D	-0.014	0.142	0.010	1.000	0.921	0.986
North-G	-0.426	0.203	4.381	1.000	0.036	0.653
Northeastern -E	-0.058	0.165	0.124	1.000	0.725	0.944
Northeastern -F	0.051	0.187	0.076	1.000	0.783	1.053
Northeastern -G	-0.288	0.225	1.641	1.000	0.200	0.750
Northeastern -H	-0.409	0.176	5.386	1.000	0.020	0.664
Northeastern -I	-0.371	0.173	4.576	1.000	0.032	0.690
constant	-0.802	0.115	48.365	1.000	0.000	0.449

Among them, only North-A had an odds of 1 or more. The proportion of Bad in North-A is the highest at 46.9%. The overall Bad rate is 23.1%. But the average primary income is fifth highest. Examining the data, there were 2 data on 5-digit income. We understand this is the cause. The average value without these two data was 1662.19. This is a lower value than the average and it can be said to be a poor state.

The significance probability was 0.05 or less, and the odds other than North-A were all less than 1. That is, other significant variables can be said to be " It is hard to become Bad if you live in that state."

This state is located at the northernmost point of country A. It is a region of climate with high temperature and humidity which is directly under the equator. The average temperature is over 27 °C and annual rainfall is 3500 mm. Many natural remains. Many people sell their timber to make a living. However, it causes the flooding of the river. It will be a further burden to get the damage. Therefore, it becomes Bad in a place where households are hard to stabilize.

The lowest odds is Southeastern-A. In other words, it can be said that it is a state that is least likely to be Bad. Actually the proportion of Bad is 9.4%. It is clear that the economy is stable because the average value of main income is remarkably high as 15530.76.

The reason why economic conditions are good is that it is the center of industry, commerce and finance in country A. It is blessed with climate, infrastructure such as roads, railroads and harbors are in place. The industry has developed and has a large population. Therefore, there are many commercial jobs. There are many wealthy people who have stable jobs. So they are easy to pay off when they form a loan.

The average educational number is 17.04 for State North-A and 21.55 for Southeastern-A. In other words, people in North-A are more likely to drop out than to graduate from secondary degree. Many people in Southeastern-A graduate from secondary degree. There are also people who go on to further advance.

We understand again that it relates to educational history and Bad and main income. And there is also an influence of the economic state of society. For the society to develop, it may be affected by weather and others.

6 Discussion/Summary

We analyzed what kind of career affect to become Bad. Therefore level of revenue is D/E level, academic background is graduation of secondary educational background from dropout of basic school and those living in North-A are prone to Bad. If a customer who meets these conditions wishes to form a loan, should be wary. If we do decision tree analysis, it will be easier to understand. We will make this a future subject. In that case, if you also put variables on other backgrounds, accuracy will improve. Then because there is a possibility that the data is insufficient we have to perform statistics.

References

Ministry of Economy, Trade and Industry: Trade White Paper 2012 Chapter 1 World Economic Trends Section 6 Latin America, Russia Economy (2012)

Miyoshi, Y.: Empirical study of inverse selection in consumer finance market. Kagawa Univ. Econ. Rev. **88**, 537–542 (2016)

Ministry of Agriculture, Forestry and Fisheries: Heisei 21 year Country Report. http://www.maff. go.jp/primaff/koho/seika/project/pdf/nikokukan12-3.pdf. Accessed 18 Dec 2017

Ministry of Education, Culture, Sports, Science and Technology (MEXT): Survey on reforms on academic systems in various countries in FY2005 (2). http://www.mext.go.jp/a_menu/ shougai/chousa/1351481.htm. Accessed 17 Dec 2017

Validation of a Sorting Task Implemented in the Virtual Multitasking Task-2 and Effect of Aging

Frédéric Banville[1,2(✉)], Claudia Lussier[2], Edith Massicotte[2],
Eulalie Verhulst[3], Jean-François Couture[4], Philippe Allain[3],
and Paul Richard[3]

[1] Université du Québec à Rimouski, Rimouski, Canada
frederic_banville@uqar.ca
[2] Université de Montréal, Montréal, Canada
[3] Université d'Angers, Angers, France
[4] Centre de développement et de recherche en imagerie numérique,
Matane, Canada

Abstract. Normal aging is characterized by cognitive, functional, and neuroanatomic changes. Executive functions (EF), linked to the autonomy in the realization of instrumental Activity of Daily Living (iADL) are particularly sensitive to the effect of aging. To date, the best way to describe the cognitive profile of an individual is using traditional neuropsychological assessment. However, these tasks showed a week ecological validity that compromise the prediction of iADL functioning. Fortunately, virtual reality (VR) seems to be an interesting alternative to assess the functional capacities by reproducing everyday life situations accurately. Nevertheless, we know that the elderly could have some difficulty when interacting with technology (Banville et al. 2017). The aims of this study were (1) to differentiate performance of older adult compared to young people in a sorting task implemented in a virtual environment (VE) and (2) to analyze psychometric properties of that novel task. To do this, 30 participants were recruited in the general population and were divided in 2 groups based on their age. The results indicated that older subjects take more time than younger to compete the sorting task. Then, the scores obtained correlated significantly with some neuropsychological tasks, particularly with those assessing EF. To conclude, the VMT-2 could offer the opportunity to make a valid assessment of functional capacities during aging. Further work could validate the relevance to use it when the individual facing with mild cognitive impairment or dementia.

Keywords: Neuropsychological assessment · Virtual reality · Normal aging
Psychometry

1 Introduction

1.1 Normal Aging, Cognitive Reserve and Cognitive Decline

Normal aging can affect cognitive functioning such as episodic memory, speed information processing, working memory and executive functioning (Bherer 2015),

© Springer International Publishing AG, part of Springer Nature 2018
S. Yamamoto and H. Mori (Eds.): HIMI 2018, LNCS 10905, pp. 41–54, 2018.
https://doi.org/10.1007/978-3-319-92046-7_4

including mental flexibility, planning and inhibition (Collette and Salmon 2014). Bherer (2015) has highlighted heterogeneity between individual concerning cognitive functioning alteration by aging. For example, semantic memory seems to be more resistant to aging than autobiographical memory (Calso et al. 2016). It seems to exist also an interindividual variability that increase with aging: cognitive changes can be different according to the people (Bherer 2015; Salthouse 2010). The best way to explain this phenomenon is that some individual has a better cognitive reserve.

The reduction of speed processing (SP) with the age can have some impact on the realization of iADL. Edwards et al. (2009) demonstrated that the slowdown of SP is a good predictor for the removal of driving licence. Executive functions (EF) that represent a set of capacities allowing adaptation to the novel situation are also affected. EF support cognitive, emotional and social control and regulation (Lezak et al. 2012). Even if there is no formal consensus about the definition of EF in the literature (Calso et al. 2016), neuropsychological assessment of EF tends to target planning, inhibition, problem solving, mental flexibility, etc. (Calso et al. 2016; Chan et al. 2008; Sorrel and Pannequin 2008; Stuss and Levine 2002). More specifically, sorting tasks regroup some EF such as abstraction, categorization, information processing, planning and mental flexibility. These tasks imply some abilities to deduct tasks principles, to categorize and to update information and strategies when the environmental contingencies change (Strauss et al. 2006). For Stuss and Levine (2002), several cognitive processes are susceptible to influence performance in a sorting task such as generations and identification of concepts, sustained attention, use of feedback to modify actions and to resist to the interference. When several concepts are possible in the task, alternate and inhibition for perseverative behaviour are essential.

Alteration of executive functioning seems not to affect all the frontal system but dorsolateral area that imply inhibition, planning, deduction of rules and working memory (Allain et al. 2007). So, the decline observed in aging seems to be selective and difference can be observed into a same executive process. In fact, some other problem could deteriorate the executive dysfunction such as SP reduction (Salthouse 1991; Salthouse 1996). Ashendorf and McCaffrey (2008) observed that older people have difficulties to use retraction adequately given by the environment, essentially in a sorting task. To plan some actions appears to be more difficult with aging (Collette and Salmon 2014). Sorel and Pennequin (2008) observed that SP reduction and difficulty to alternate between the tasks explained 58,33% of planning problems.

To sum up, normal aging is linked to several cognitive changes. Some authors consider that SP reduction is one of the principal phenomenon that to explain deterioration of cognitive functioning in normal aging. The causes of this slowdown are multifactorial and are not well identified (Kerchner et al. 2012). Some researcher supposed that these changes could be explained by modification of cerebral white matter (Abe et al. 2002; Bennett et al. 2010; Kerchner et al. 2012). The frontal lobe and their EF associates tend to be more affected by aging. EF are very important to conserve functional autonomy, and then let the person doing social, professional and leisure activities (Calso et al. 2016). Cognitive alteration linked to normal aging could have an impact directedly on iADL and affect independence of older people in everyday life (Ball et al. 2010; Collette and Salmon 2014).

1.2   Assessment of Everyday Functioning

Since some year ago, changes are seen in neuropsychology field, particularly in cognitive assessment's area where tasks are reinvented to be nearer to the everyday functioning and thus to have a better description of human behaviour (Marcotte and Grant 2010). Some authors made the distinction between "activity of daily living" (ADL) and "instrumental of daily living" (iADL) (Tarnanas et al. 2013). ADL refer to the activity addressing basic needs such as moving, dressing, feeding, bathing. In the other part. IADL are associated with the functional autonomy such as use correctly means of transportation, finance, medication use and domestic activities.

Traditional neuropsychological assessment tends to trace a cognitive profile in some distinct and specific domain: EF, SP, etc. That can be useful to give an opinion regardless of the integrity of cognitive functioning. Cognitive functions are then evaluated in an isolate manner and some time artificially regardless to the everyday life (Allain et al. 2014). Functional approaches for cognitive assessment are more natural, ecological because the approaches describe performance, success or fail, in a more precise way. The conclusion is helpful to predict everyday functioning, particularly with aging people (Vaughan and Giovanello 2010). To date, no much method, standardized, ecological and objectives are available for the ADL and iADL assessment (Allain et al. 2014) in a naturalistic way. At this point, that become useful to develop tools that allow to do the best description of how older peoples manage their everyday life to prevent or detect pathological functioning.

Virtual reality (VR) is now recognized as a useful tool for cognitive assessment because that permits a better standardization and experimental control of the environment in the same way that reproduce well every day functioning (Banville et al. 2017). Pratt et al. (1995) wrote that VR authorize the use of interfaces connected to a computer that allows an individual to interact, in real time, with objects implanted in a virtual environment (VE). One of the main advantages of VR is to reproduce with reliability real life in a laboratory condition (Wilson et al. 1997). Thank to the standardization and computer, the task realized in the VE offer reliable data concerning time of realization, success or fail (Wild et al. 2008), interaction interface and navigation in the VE. Our expectation is that VR offer an opportunity to be more precise in the assessment of cognitive function and so far, to detect subtitle changes in the cognitive functioning of older people (Wild et al. 2008).

1.3   Virtual Reality and Aging

Ang et al. (2007) demonstrated that VE ca be a generator of cognitive overload, which is more important if the individual is less familiar with the technology. That is very important to know considering that people older than 60 years old are less comfortable with the computer and the technology (Banville et al. 2017; Sayer 2004). Our work with the Virtual Multitask Test (Banville et al. 2017; Verhulst et al. 2017) have observed that younger people are more comfortable with computers and human computer interface (HCI) and can realize tasks more efficacy, taking less time in the VE, making more action. Navigational difficulties have been observed too. All the immersive process raises the cognitive load of older peoples and create a kind of bottleneck effect that generates the realization of several tasks at the end of the game.

Some studies were interested in the use of VR with aging, principally to detect cognitive deficits liked to dementia or mild cognitive impairment (MCI). Executive functions are often targeted for task development in the VE. For example, *Virtual Reality Day-Out Task* (VR-DOT; Tarnanas et al. 2013), a multitasking environment simulating a fire context. Prospective memory and reasoning are assessing. Results obtained showed that VR-DOT make distinction between healthy older subject and cognitive impaired older subjects.

Allain et al. (2014) have used the non-immersive virtual coffee task (NI-VCT), a tool developed for assessing several cognitive functions needed to realize iADL. Three main results are extracted from this study: (a) participants with Alzheimer's disease are less better than healthy older participants; (b) several measurements of NI-VCT (*Time to completion, Accomplishment score, Total errors, Omissions errors, Commissions errors*) are correlated to several neuropsychological score; (c) the success to the NI-VCT can predict the success in a real environment.

The *Virtual Supermarket* (VSM; Zygouris et al. 2015) developed for an executive function assessment implies that individual purchase products wrote on a specific list. VSM showed that healthy adults are better than MCI participants in the tasks realization. The score of VMS (*Duration, Correct Types, Correct quantities, Bought Unlisted*) are also correlated to neuropsychological evaluation.

1.4 Objectives and Hypothesis

The first objective of this pilot study was to differentiate the behaviour of healthy older and younger adults into a sorting task that simulate an iAVQ task with a high cognitive load. So, the participants must store groceries and at the same time paying attention to multiple tasks such as answering the phone. The second objective, in a psychometric perspective, was to analyze the validity of this task developed into the VMT second version.

The research hypotheses are declined as follows:

1. Older people will be slower in the realization of the sorting task and will make more errors in the storage than younger participants.
2. Score obtained in the VMT-2 (total time, errors with storing fruits, vegetables, fresh aliments) will be correlated to neuropsychological assessment, particularly to the mental flexibility, planning and inhibition.

2 Method

2.1 Participants

Thirty participants have been recruited for this quasi-experimental exploratory study. The recruitment was done at the University of Montreal for younger participants and at the Lanaudière Alzheimer Society for healthy older adults, caregivers for the majority.

To be included in this study, participant must be: (a) aged 18–45 years old (for the "young group") and 60–87 years old (for the "older group"); (b) French speakers. Exclusion criteria were: (a) to have a non-corrected problem with viewing or hearing; (b) to have a neurological or a psychiatric diagnosis; (c) to have a chronic disease (e.g. hypothyroidy); (d) to have a score to the Montreal cognitive assessment (MoCA) equal or less than 26 (Nasreddine et al. 2005). At the end of this research, 3 young participants and two old participants have been excluded based on the MoCA and 2 participants due to technical problems with virtual reality.

2.2   Material

Neuropsychological Assessment. The Delis-Kaplan Executive Function System (D-KEFS from Delis et al. 2001) aimed to assess executive functioning. More specifically, Trail Making Test (TMT) was used to describe speed processing, visuo-motor functioning and mental flexibility. For the TMT, time of realization was considered as a dependent variable. Then, the Stroop Task was used to assess inhibition and mental flexibility. Time to complete the task and errors made were the dependent variables. Finally, the Hanoi Tower Test was administered to assess planning, reasoning and rules breaking. For this D-KEFS subtest, we retained total score, precision ratio and rule violation.

The Behavioural Assessment of Dysexecutive Syndrome (BADS from Wilson et al. 1996) was administered. Specifically, the Zoo Map Test (ZMT) aimed to assess planning and organization. The variables retained was the total number of errors and total score. Finally, the modified version of the Six Element Test (SET) was also used to collect our data. That ecological test assessed planning, inhibition and multitasking capacities. The profile score was kept to the analysis.

Post-immersive Questionnaire. Simulator Sickness Questionnaire (SSQ from Kennedy et al. 1993) was targeted to observe cybersickness during or after the tasks realization in the VE. On the other hand, the iPresence questionnaire (iPQ from Schubert et al. 1999) was used to describe a which point the participant feel present in VE during the assessment.

Virtual Multitasking Test – Second Version. The Virtual Multitasking Test (VMT), second version (Banville et al. 2013, 2017) aims, at the beginning of its development, to assess PM and executive functions using a multitasking paradigm. Different scenarios are implanted into a 6½ rooms virtual apartment, each room including at least one task except the bathroom. Figure 1 give an overview of the VE. At the beginning of the immersion, participants are told that they are visiting their best friend. During the day, he is at work and they must live in his apartment. In the evening, they will go to a show with their best friend. However, during the day, they must perform several tasks alone based on daily life. For instance, they must store the groceries on the counter as quickly as possible (even if they are told there is no time limit to complete the activities), answer the phone, and perform other tasks such as faxing a document, searching for show tickets, drying a shirt, feeding a fish. PM tasks require, among other things, to close a door just when exiting the master bedroom to prevent a dog from climbing on the bed. Unforeseen events occur during the execution of the tasks. For instance, the occurrence of a storm which over throw objects in the guest room and let water seep into the dining room. For example, storms that reverse objects in the guest room and that let water seep into the dining room. Every time a person is exposed to the VMT, they start a training phase of the environment. Afterwards, the experimental phase began, and the person had to carry out the tasks proposed by the scenarios planned by researchers. Figure 2 describes the sequence of each task and unforeseen events.

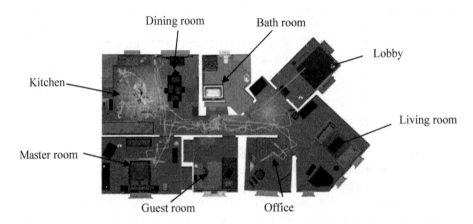

Fig. 1. Navigation landmark for an older participant with head-mounted display (HMD)

Technically, VMT-2 was developed using Unity-3D as a game engine. The interaction with the VE is allowed by an Oculus Development Kit-2 (HMD), keyboard and mouse. The computer using windows 7 as an exploitation system and two nVidia graphic cards (GeForce 550). Data was processes in real time and recorded in a log document (.txt document).

Specifically, for this study, our focus was on the "storing the grocery task'. So, we developed a performance score in order to better describing the behaviour of our participants. The scoring system consists of the following items: (1) total realization time in the VMT-2; (2) total time for the storing groceries task (a natural sorting task); (3) number of errors in the sorting of Fruits & Vegetable products; (4) number of errors in the sorting of Packaged Products (e.g. can, box of pasta, etc.). More specifically, for the sorting of Fruits & Vegetables items and Packaged products, 0 point is given if the product is stored in an inappropriate place or if it is not stored; 1 point is given if the product is stored.

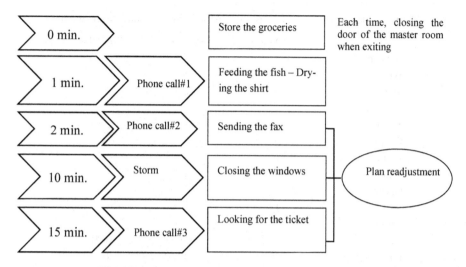

Fig. 2. Temporal sequences of the task into the VMT-2.

Procedure. The experimental procedure was as follows: (1) participant received a phone call for a pre-interview to confirm if he/she was eligible for the study; (2) if yes, participants were then seen in the laboratory for a 2–3 h testing and experimentation session. The meeting is divided in two steps: (a) neuropsychological assessment and (b) immersion in the VMT-2. The order of the two steps was counterbalanced for avoiding tiredness effect on the result.

3   Results

3.1   Demographic Characteristics

Table 1 shows the participant's characteristics. Participants were predominantly women (n = 19). The "young group" was composed of 14 participants, 11 women and 3 men. The participants were aged between 21 and 43 years old (means of age: 26,07 ± 6,23). Number years of education was, in mean 16,14 ± 2,48 years, ranging between 13 and 23 years of education. The "older group" was consisted of 11 participants, 8 women and 3 men. The group was aged between 63 and 72 years old (means of age: 66,80 ± 2,90). The mean years of education was 15,55 ± 3,11 years, ranging between 11 and 20 years. No significant difference was found on the basis of education [$F(1,23) = 0,29$, $p = 0,60$]. A statistic significant difference existed concerning facility and ease for using technology and computers [$F(1,23) = 9,87$, $p = 0,01$], younger were more comfortable with technology.

Table 1. Demographic characteristics of participants according to the group

	Younger (n = 14)	Older (n = 11)	p-value
Gender			
Woman	11	8	–
Man	3	3	
Civil status			
Single	7	1	–
Couple	6	8	
Divorced	1	1	
Widow	0	1	
Occupation			
Job	12	4	
No job	2	7	
Age	26,07 (6,23)	66,80 (2,90)	0,001
Education (years)	16,14 (2,48)	15,55 (3,11)	ns
Ease for using computer	85,71 (12,99)	60,91 (25,77)	0,01

ns: non significant

Table 2. Mean and standard deviation obtained to the post-immersion questionnaires.

	Younger (n = 14)	Older (n = 11)	p-value
Cybersickness	9,86 (5,74)	12 (6,68)	ns
Nauseas	5,86 (3,48)	7,6 (4,45)	ns
Oculomotor	4 (2,86)	4,73 (3,32)	ns
Sens of Presence			
Presence's general factor	2,57 (1,02)	2,73 (1,42)	ns
Spatial	2,11 (0,37)	2,10 (0,27)	ns
Implication	1,80 (0,61)	2,23 (0,93)	ns
Realism	1,93 (0,54)	2,50 (0,32)	0,005

As shown in Table 2, no significant difference was found concerning cybersickness [$F(1,22) = 0,71$, $p = 0,41$]. Younger (mean = 9,86 ± 5,74) and older (mean = 12 ± 6,68) participants showed little or no discomfort when immersion in the VE. When questioning their sense of presence, participants seem to be different on a specific scale, more precisely realism scale [$F(1,23) = 9,63$, $p = 0,01$]. Older participants judged the VE more realistic than younger. This opinion concern particularly the visual and graphic aspects of the VMT-2.

3.2 Performance in the VMT-2

Table 3 shows the main difference based on the mean between younger and older participants during tasks realization into the virtual environment. More specifically, older participants take more time to perform the entire tasks require by the VMT-2 [$F(1,20) = 4,46$, $p = 0,05$]. They were slower than young participants in the realization of the sorting task (i.e. storing groceries [F(1,17) = 5,39, p = 0,03]. However, if we analyze the ratio (in percentage) of time spend in the kitchen compared to the total time for the tasks realization, it exists no statistical difference. The older subjects spend 64% of their time in the kitchen and young subjects 55%. No significant group difference was found [F(1,17) = 1,20, p = ns]. Moreover, no difference was observed between the 2 groups concerning errors score for storing Fruit & Vegetable[F(1,18) = 0,21, p = 0,65] or Packaged Products [F(1,16) = 0,94, p = 0,35].

Table 3. Time of realization and number of errors when storing groceries

	Younger	Older	p-value
Total time into the VMT-2 (min.)	18,84 (8,03)	26,86 (9,69)	0,05
Total time for storing the groceries	10,99 (4,68)	18,55 (8,67)	0,03
Ratio total time for storing groceries/total time	55,47 (7,92)	64,18 (22,63)	ns
Number errors for fruits & vegetables	1,83 (1,40)	2,13 (1,36)	ns
Number error for packaged products	1,64 (3,20)	0,43 (0,79)	ns

3.3 Psychometric Properties of the Storing Groceries Score

First, time required to store the groceries was correlated to MoCA memory sub-test score [r(19) = −0,55, p = 0,02], SET-break of rules score [r(19) = −0,49, p = 0,04] and visual scanning in the TMT condition 1 [r(19) = 0,47, p = 0,04].

Second, the number of errors for storing Fruit & Vegetable was correlated with MoCA abstraction sub-scale [r(20) = −0,55, p = 0,01], ZMT-perseveration score [r(20) = 0,53, p = 0,02], ZMT-errors score [r(20) = 0,53, p = 0,02] and reading speed in the second condition of Stroop Test [r(20) = 0,44, p = 0,05].

4 Discussion

This exploratory study has two main objectives. The first one was, after having created a novel scoring system for the task "storing groceries" in the Virtual Multitasking Task – second edition, to explore the difference in the performance of a group of old subjects compared to a group of young subjects. The second one was to explore the convergent validity of our scores.

For the first objective, as anticipated, older participants take more time to navigate in the VE and to finish the sorting task. Similar studies using technology showed similarities on speed processing (Davison et al. 2017; Raspelli et al. 2012; Sayers 2004). Older participants seem always be slower than young participants. In our perspective, result obtained regarding total time in the VMT could be associated with the ease using technology. Indeed, the older participants said to be less comfortable with computers than younger one.

Moreover, it is very important to consider the time taken for the sorting task in the context where it could represent the capacity of participants to manage unforeseen and additional events during storing groceries. More precisely, for example, unforeseen events, in addition to requiring good capacities to modified plan, required a more important use of Human Machine Interface (HMI). The HCI in order to manipulate objects and realizing other tasks asked by the VE. That also involves navigating toward a place to another. That combination of plan change, and manipulation of the HMI could generate a more important cognitive overload on aging people. If we refer to the correlational analysis, the task seems to solicit mental flexibility, planning, and inhibition, and executive processes that permit the work organization and answering of environmental contingencies. Yet, these functions are typically altered in the course of normal aging (Allain et al. 2007).

SP seems to be an important factor that could contribute to the group difference observed here. Reference to the cognitive slowdown theory, SP plays a role of mediator in the diminution of performance on several cognitive tests during aging (Salthouse 1991; Salthouse 1996). When we reanalyzed time realization, we suppose that it gives a lot of information concerning the management of cognitive overload.

The results showed that the proportion of time devoted by older peoples to the sorting task was like the younger group. But, regarding the standard deviation, we saw that some older participants took very more of time to accomplish tasks. These observations seem linked to the fact that cognitive aging is heterogenous (Bherer 2015). The manipulation of food associated with the multitasking context could create a reduction in speed processing for some people but that not affect the quality of the storing (error score). Indeed, results showed that older participants don't make more errors in the sorting tasks considering Fruits & Vegetables or Packaged products.

The psychometric analysis demonstrated that score for the sorting task correlates with some neuropsychological measures of executive functions. Indeed, total time of realization seems to be associated with the episodic memory and to the respect of rules. In fact, more the participant take time to realize the task, poorer was the memory and more he/she breaks the rules. We suggest that to navigate in the VMT-2 quickly, the participant must remember all instructions. The time of realization of the task is linked to the initial recall of the task after paying attention to an unforeseen event. In addition, participants must organize and to coordinate activities to respect some rules linked to the virtual environment and instructions. The rule braking in the VE could generate perseverative behaviour implied a bad interpretation in relation to the Virtual World. In the other side, we observed a link with speed processing and the time taken for storing groceries. This supposes that visual scanning is important when participants plan to store all the product spread on the island of cooking. See Fig. 3.

Errors made in the storage of Fruits & Vegetables items were negatively correlated to the abstraction score obtained by MoCA. Concrete thinking seems to interfere with the concepts formation that allow the participant to sort the food correctly. Lezak et al. (2012) associated abstraction to the categorization, an ability that could be mobilized in the sorting task extracted to the VMT. So, this kind of error was also correlated to inhibition liked to the break of rules and perseverative behaviours. Lezak et al. (2012) observed that this type of difficulty can be associated with mental inflexibility. That could influence the storage of Fruits & Vegetables items. To finish errors in the storage of Fruits & Vegetable s is correlated to error score to speed processing in the condition 2 of Stroop; this suggests that oculomotor and speed processing or both contributing to the sorting task.

Fig. 3. Storing groceries at the beginning of the task

To conclude, psychometric properties of the sorting task score in the VMT-2 showed correlation with some executive functions: speed processing, mental flexibility, planning and inhibition. The VMT-2 require the accomplishment of multiple tasks that solicit prospective memory, plan reorganization, mental flexibility, attention (Pachoud 2012). These results are compatible with those found by several authors that associate the iADL "natural tasks" executive function and speed processing (Allain et al. 2014; Davison, Deeprose and Terbeck 2017; Zygouris et al. 2015).

4.1 Limits and Future Works

Even if this exploratory study shows again the great potential of virtual reality as a neuropsychological assessment tool, some limits are raised by our work. Firstly, young participants were better in the VMT-2; That was surely favorized by the familiarity with technology compared to older people (Davison et al. 2017; Sekuler et al. 2008). To address this phenomenon, it becomes important to control for the experience and ease with technology. Secondly, the number of participants and the great number of variables limit the statistical power of our analysis and the generalization of the results. Thirdly, 76% of the sampling was women, that let us think that in addition to face some generational effect we could face an effect of gender too. For example, a study demonstrated that school-age girls have a negative attitude and a week's self-efficacy sense with technology than boys (Virtanen et al. 2015). Fourthly, we observe that participants in both groups were relatively schooled. That creates a participant bias; maybe this is not representative of the general population.

Virtual reality is now well known for its relevance for ecological neuropsychological assessment (Allain et al. 2014; Tarnanas et al. 2013). Even if the VMT-2 has been developed to study the everyday functioning, some important criticism should be formulated. Technical problem linked navigational aspect or some malfunction of some applications (for example: the fax) interaction with the object (for example: highlighting) could affect the realism of the VE when compared to the real life. Results obtained with the iPQ presence questionnaire seem to show that older participants were more sensitive to the realism of the VE.

References

Abe, O., Aoki, S., Hayashi, N., Yamada, H., Kunimatsu, A., Mori, H., Yoshikawa, T., Okubo, T., Ohtomo, K.: Normal aging in the central nervous system: quantitative MR diffusion-tensor analysis. Neurobiol. Aging **23**(3), 433–441 (2002). https://doi.org/10.1016/S0197-4580(01)00318-9

Allain, P., Foloppe, D.A., Besnard, J., Yamaguchi, T., Etcharry-Bouyx, F., Le Gall, D., Nolin, P., Richard, P.: Detecting everyday action deficits in Alzheimer's disease using a nonimmersive virtual reality kitchen. J. Int. Neuropsychol. Soc. **20**(5), 468–477 (2014). https://doi.org/10.1017/s1355617714000344

Allain, P., Kauffmann, M., Dubas, F., Berrut, G., Le Gall, D.: Fonctionnement exécutif et vieillissement normal: étude de la résolution de problèmes numériques. Psychologie & NeuroPsychiatrie du vieillissement **5**(4), 315–325 (2007). https://doi.org/10.1684/pnv.2007.0106

Ang, C.S., Zaphiris, P., Mahmood, S.: A model of cognitive loads in massively multiplayer online role-playing games. Interact. Comput. **19**, 167–179 (2007)

Ashendorf, L., Jefferson, A.L., O'Connor, M.K., Chaisson, C., Green, R.C., Stern, R.A.: Trail Making Test errors in normal aging, mild cognitive impairment, and dementia. Arch. Clin. Neuropsychol. **23**(2), 129–137 (2008). https://doi.org/10.1016/j.acn.2007.11.005

Ball, K., Ross, L.A., Viamonte, S.: Normal aging and everyday functioning. In: Marcotte, T.D., Grant, I. (eds.) Neuropsychology of Everyday Functioning, pp. 248–263. The Guilford Press, New York (2010)

Banville, F., Forget, H., Bouchard, S., Page, C., Nolin, P.: Le virtual Multitasking Test, un outil d'évaluation de fonctions exécutives: une étude pilote. Conférence présentée au congrès de la SQRP, Chicoutimi: 22, 23 et 24 mars 2013 (2013)

Banville, F., Couture, J.F., Verhulst, E., Besnard, J., Richard, P., Allain, P.: Using virtual reality to assess the elderly: the impact of human-computer interfaces on cognition. In: International Conference on Human Interface and the Management of Information, pp. 113–123 (2017). https://doi.org/10.1007/978-3-319-58524-6_10

Bennett, I.J., Madden, D.J., Vaidya, C.J., Howard, D.V., Howard, J.H.: Age-related differences in multiple measures of white matter integrity: a diffusion tensor imaging study of healthy aging. Hum. Brain Mapp. **31**(3), 378–390 (2010). https://doi.org/10.1002/hbm.20872

Bherer, L.: Cognitive plasticity in older adults: effects of cognitive training and physical exercise. Ann. N. Y. Acad. Sci. **1337**(1), 1–6 (2015). https://doi.org/10.1111/nyas.12682

Calso, C., Besnard, J., Allain, P.: Le vieillissement normal des fonctions cognitives frontales. Gériatrie et Psychologie Neuropsychiatrie du Vieillissement **14**(1), 77–85 (2016). https://doi.org/10.1684/pnv.2016.0586

Chan, R.C., Shum, D., Toulopoulou, T., Chen, E.Y.: Assessment of executive functions: Review of instruments and identification of critical issues. Arch. Clin. Neuropsychol. **23**(2), 201–216 (2008). https://doi.org/10.1016/j.acn.2007.08.010

Collette, F., Salmon, E.: Les modifications du fonctionnement exécutif dans le vieillissement normal. Psychologie française **59**(1), 41–58 (2014). https://doi.org/10.1016/j.psfr.2013.03. 006

Davison, S.M.C., Deeprose, C., Terbeck, S.: A comparison of immersive virtual reality with traditional neuropsychological measures in the assessment of executive functions. Acta Neuropsychiatrica, 1–11 (2017). https://doi.org/10.1017/neu.2017.14

Delis, D.C., Kaplan, E., Kramer, J.H.: Delis-Kaplan Executive Function System. The Psychological Corporation, San Antonio (2001)

Edwards, J.D., Bart, E., O'Connor, M.L., Cissell, G.: Ten years down the road: predictors of driving cessation. Gerontologist **50**(3), 393–399 (2009). https://doi.org/10.1093/geront/ gnp127

Kennedy, R.S., Lane, N.E., Berbaum, K.S., Lilienthal, M.G.: Simulator sickness questionnaire: an enhanced method for quantifying simulator sickness. Int. J. Aviat. Psychol. **3**(3), 203–220 (1993). https://doi.org/10.1207/s15327108ijap0303_3

Kerchner, G.A., Racine, C.A., Hale, S., Wilheim, R., Laluz, V., Miller, B.L., Kramer, J.H.: Cognitive processing speed in older adults: relationship with white matter integrity. PloS one **7**(11) (2012). https://doi.org/10.1371/journal.pone.0050425

Lezak, M.D., Howieson, D.B., Bigler, E.D., Tranel, D.: Neuropsychological Assessment, 5e edn. Oxford University Press, New York (2012)

Marcotte, T.D., Grant, I.: Neuropsychology of Everyday Functioning. Guilford Press, New York (2010)

Nasreddine, Z.S., Phillips, N.A., Bédirian, V., Charbonneau, S., Whitehead, V., Collin, I., Cummings, J.L., Chertkow, H.: The montreal cognitive assessment, MoCA: a brief screening tool for mild cognitive impairment. J. Am. Geriatr. Soc. **53**(4), 695–699 (2005). https://doi. org/10.1111/j.1532-5415.2005.53221.x

Pachoud, B.: Remédiation cognitive et vie quotidienne. In: Franck, N. (ed.) La remédiation cognitive, pp. 70–90. Masson, Issy-les-Moulineaux (2012)

Pratt, D.R., Zyda, M., Kelleher, K.: Virtual reality: in the mind of the beholder. Computer **7**, 17–19 (1995)

Raspelli, S., Pallavicini, F., Carelli, L., Morganti, F., Pedroli, E., Cipresso, P., Poletti, B., Corra, B., Sangalli, D., Silani, V., Riva, G.: Validating the Neuro VR-based virtual version of the Multiple Errands Test: preliminary results. Presence: Teleoper. Virtual Environ. **21**(1), 31–42 (2012). Repéré à https://www.researchgate.net/profile/Federica_Pallavicini/publication/ 51231110_Validation_of_a_Neuro_Virtual_Reality-based_version_of_the_Multiple_ Errands_Test_for_the_assessment_of_executive_functions/links/ 0912f511e3a4c07378000000/Validation-of-a-Neuro-Virtual-Reality-based-version-of-the-Multiple-Errands-Test-for-the-assessment-of-executive-functions.pd

Salthouse, T.A.: Mediation of adult age differences in cognition by reductions in working memory and speed of processing. Psychol. Sci. **2**(3), 179–183 (1991). Repéré à http:// journals.sagepub.com/doi/pdf/10.1111/j.1467-9280.1991.tb00127.x

Salthouse, T.A.: The processing-speed theory of adult age differences in cognition. Psychol. Rev. **103**(3), 403 (1996). https://doi.org/10.1037/0033-295X.103.3.403

Salthouse, T.A.: Selective review of cognitive aging. J. Int. Neuropsychol. Soc. **16**(5), 754–760 (2010). https://doi.org/10.1017/S1355617710000706

Sayers, H.: Desktop virtual environments: a study of navigation and age. Interact. Comput. **16**(5), 939–956 (2004). https://doi.org/10.1016/j.intcom.2004.05.003

Sekuler, R., McLaughlin, C., Yotsumoto, Y.: Age-related changes in attentional tracking of multiple moving objects. Perception. **37** (6), 867–876 (2008)

Sorel, O., Pennequin, V.: Aging of the Planning process: The role of executive functioning. Brain Cogn. **66** (2), 196–201 (2008)

Schubert, T., Friedman, F., Regenbrecht, H.: Decomposing the sense of presence: factor analytic insights. In: Extended Abstract to the 2nd International Workshop on Presence, pp. 1–5 (1999). Repéré à http://s3.amazonaws.com/academia.edu.documents/31976070/Schubert.pdf? AWSAccessKeyId=AKIAIWOWYYGZ2Y53UL3A&Expires=1500951837&Signature= LV2jB%2FiQd7BmfhtgxswL0un08sM%3D&response-content-disposition=inline%3B% 20filename%3DDecomposing_the_Sense_of_Presence_Factor.pdf

Strauss, E., Sherman, E.M.S., Spreen, O.: A Compendium of Neuropsychological Tests: Administration, Norms, and Commentary, (3e éd. rév). Oxford University Press, New York (2006)

Stuss, D.T., Levine, B.: Adult clinical neuropsychology: lessons from studies of the frontal lobes. Annu. Rev. Psychol. **53**(1), 401–433 (2002). https://doi.org/10.1146/annurev.psych.53. 100901.135220

Tarnanas, I., Schlee, W., Tsolaki, M., Müri, R., Mosimann, U., Nef, T.: Ecological validity of virtual reality daily living activities screening for early dementia: longitudinal study. JMIR Serious Games, **1**(1) (2013). https://doi.org/10.2196/games.2778

Vaughan, L., Giovanello, K.: Executive function in daily life: Age-related influences of executive processes on instrumental activities of daily living. Psychol. Aging **25**(2), 343 (2010). https:// doi.org/10.1037/a001772

Verhulst, E., Banville, F., Richard, P., Tabet, S., Lussier, C., Massicotte, É., Allain, P.: Navigation patterns in ederly during multitasking in virtual environnment. In: Yamamoto, S. (ed.) HIMI 2017. LNCS, vol. 10274, pp. 176–188. Springer, Cham (2017). https://doi.org/10. 1007/978-3-319-58524-6_16

Virtanen, S., Räikkönen, E., Ikonen, P.: Gender-based motivational differences in technology education. Int. J. Technol. Des. Educ. **25**(2), 197–211 (2015). https://doi.org/10.1007/s10798-014-9278-8

Wild, K., Howieson, D., Webbe, F., Seelye, A., Kaye, J.: Status of computerized cognitive testing in aging: a systematic review. Alzheimer's Dementia **4**(6), 428–437 (2008). https:// doi.org/10.1016/j.jalz.2008.07.003

Wilson, B.A., Alderman, N., Burgess, P.W., Emslie, H., Evans, J.J.: The Behavioural Assessment of the Dysexecutive Syndrome. Thames Valley Company, Bury St Edmunds (1996)

Wilson, P.N., Foreman, N., Stanton, D.: Virtual reality, disability and rehabilitation. Disabil. Rehabil. **19**(6), 213–220 (1997). https://doi.org/10.3109/09638289709166530

Zygouris, S., Giakoumis, D., Votis, K., Doumpoulakis, S., Ntovas, K., Segkouli, S., Karagiannidis, C., Tzovaras, D., Tsolaki, M.: Can a virtual reality cognitive training application fulfill a dual role? Using the virtual supermarket cognitive training application as a screening tool for mild cognitive impairment. J. Alzheimer's Dis. **44**(4), 1333–1347 (2015). https://doi.org/10.3233/jad

Impact of Menu Complexity upon User Behavior and Satisfaction in Information Search

Svetlana S. Bodrunova$^{(\boxtimes)}$ and Alexandr Yakunin

St. Petersburg State University, St. Petersburg 199004, Russia
s.bodrunova@spbu.ru

Abstract. *Background.* The growing complexity of website navigation demands more behavior-oriented research. Today, static and sequential menus have become virtually incompatible with adaptive forms of web layouts; tagging-based menus started to dominate; and these navigational elements more and more co-exist on webpages. Previous research compares the three basic types of menus (static, sequential, and expandable) for two dimensions: reduction of mental load (objective, linked to task complexity and structural complexity) and growth of user satisfaction (subjective, linked to menu type). *Objectives.* We hypothesized that growth of task complexity linked to menu complexity leads to selection of non-productive search strategies and to growth of perceived complexity of the interface. *Research design.* Following Kang et al. (2008), we have divided user search strategies into productive (systemic) and non-productive (chaotic) and have conducted an experimental pre-test. Menu complexity was created by using four different menus in one prototype. Structural complexity was assessed by path depth and menu options diversity. The search tasks were designed to be realizable disregarding the menu complexity. Two homogenous groups of 10 assessors were consecutively conducting tasks on six HTML pages of the prototype. A questionnaire was used to assess user satisfaction. *Results.* For low-complexity tasks, menu diversity has virtually no impact upon navigational behavior. But we have discovered impact of menu complexity for high-complexity tasks for multi-menu navigational schemes. Additional tests with newer types of menus show that they make the assessors drop the sequential principle of search. Also, these pages were perceived as the hardest to use.

Keywords: Visual complexity · Menu · Menu type · Task performance
Navigation · User satisfaction

1 Introduction

User behavior and satisfaction in web 1.0 is crucially linked to the efficacy of website navigation. Research on website menus constitutes a substantial part of online navigation studies. Despite the fact that today most of the studies use data-driven approaches to user reactions to website navigation elements, the growing variability of the latter demands more behavior-oriented research methods.

© Springer International Publishing AG, part of Springer Nature 2018
S. Yamamoto and H. Mori (Eds.): HIMI 2018, LNCS 10905, pp. 55–66, 2018.
https://doi.org/10.1007/978-3-319-92046-7_5

But despite the seemingly universal necessity of knowledge on menu structure in the changing design environment, the research that would address these rapid changes remains rare. Earlier works compare the three basic types of menus (static, sequential, and expandable) for two dimensions of reduction of mental load and growth of user satisfaction. The first one links task complexity to user experience, where task complexity is objectivized via measuring structural complexity (path length; [1]); and semantic relevance of menu indices to the search task [2]. The second approach looks for a perfect model of navigation representation, or menu type. Here, the term 'menu' often includes many other navigational elements, e.g. breadcrumbs.

But so far, two issues have not been raised substantially in the behavior-oriented navigation research. First, static and sequential menus have become virtually incompatible with the newest web design trends, especially with adaptive forms of web layouts. At the same time, new forms of navigation, like tagging-based ones, have taken dominant positions in information architecture. Second, these navigational elements tend to more and more co-exist on webpages, which is expected to highly complicate information search and, thus, affect user behavioral strategies and, ultimately, user satisfaction. Menu type diversity (or menu complexity) needs to be re-addressed today at user's end.

In our research, we would like to address both aforementioned issues. Thus, we ask: Does the level of menu complexity influence the selection of search strategy for fulfilling navigation tasks? And does it influence user satisfaction in realization of information search task?

The remainder of the paper is organized as follows. Section 2 reviews the literature on menu complexity and its impact upon task performance. Section 3 presents our methodology and the conduct of the experiment. Section 4 presents our results and discusses them.

2 Menu Types and Task Complexity in User Navigational Behavior

2.1 Current Research on Task Complexity, Navigational Models, and User Performance

In today's research on user navigation on web 1.0 websites, many attempts to establish the factors influencing efficiency of user navigational behavior have been made. The majority of these studies have been using two approaches.

The first approach relates user behavior and the respective user satisfaction to the nature of the task performed by a user. Its main focus is task complexity. In most cases, two dimensions of task complexity are discussed. Starting from the earliest research, the first one is objective and is linked to (and even defined by) the complexity of website structure; the second is subjective and, thus, user-dependent [3].

In more recent works, like the one by authors [1], scholars research upon the inter-relations between operational parameters of user behavior in the process of information search, the objective task complexity, and the subjectively perceived complexity of the task performance.

The authors claim that objective complexity is first and foremost defined by the length of the path that leads to the target information bit. This parameter is also studied as the level of location of target information within the website hierarchy, and the path length can be measured by the number of the levels in the website structure. The authors have also demonstrated that the objective complexity correlates with subjective perception of its complexity.

The study [2] adds another parameter, namely path relevance, to the factors of objective complexity. Path relevance is the extent to which the very description of the task hints to the way to target information. Path optimization is possible via using semantic linkages between the task definition and the menu indices (the names of menu rubrics).

Thus, the task complexity may be linked to a combination of structural (path length) and semantic (path relevancy) parameters of the path. In this case, low task complexity is characterized by a short path length and its high relevance, while high task complexity is, vice versa, characterized by a long path of low relevance. As the study [2] has shown, with the reduction of task complexity, precision and speed of task fulfilment grow, and with the growth of task complexity they both fall. These results correspond to earlier tests of complexity variance [4].

Earlier types of menus have understandably received more scholarly attention than those that appeared later; the former were compared for their efficiency. Thus, the authors [2] have compared the impact of expandable and sequential menus upon user productivity in information search. Expandable menus represent the whole context of possible user selections inside the information architecture, as well as the tree of reviewed pages; sequential menus feature only one level of website architecture and the current position of the user within it. The study has supported the idea of impact of task complexity upon the efficiency of a certain menu type: thus, for high complexity tasks, expandable menus were shown as more efficient. Also, expandable menus were found to be reducing subjective user disorientation (as evaluated by the assessors after the session). In a whole range of other works [5–7] the preference over sequential menus is also given to simultaneous menus that overview the content of a web portal and show all the levels of content location at the same time, even proving that task complexity is the determinant for the efficacy of sequential vs. simultaneous menus. According to another study [8], sequential menus work faster for simple tasks where comparing several sets of search results is not necessary. But with the growth of task complexity, e.g. when the users need to compare data on the screen, the efficiency of simultaneous (or expandable) menus grows. Thus, a linkage between cognitive load, the type of information search task, and the efficiency of a certain type of a navigational model was established.

Similar results have been received in a later study [9] that involved design aspects – namely, on vertical and dynamic menus. Here, the links between task complexity, menu design, and user performance and subjective experience evaluation.

The second approach focuses on finding an optimal model of representation of the website structure in the navigational instruments – that is, in most cases, on the menu type. A range of early studies of 2000s [10–12] have demonstrated that consecutive website navigation based on use of the so-called 'breadcrumbs' was preferential to the traditional menus of the early stages of web 1.0 development. This was especially true for the large web spaces with diversified architecture and strictly hierarchical content distribution between the levels of hierarchy.

This approach has been expanded by the works that take into account the cultural aspects of user perceptions of different menu types. Thus, the study [13] arguments the selection of an optimal menu model via the culturally defined peculiarities of user behavior in the process of information search; the authors call them cognitive styles. Depending on user belonging to either Western or Eastern cultures, the authors differentiate two navigational models – the 'wide' one (oriented to 'overview' perception of the website content) and the 'deep' one (oriented to hierarchical connections within the narrow search task).

Summarizing the main research approaches on the parameters of efficient navigation within a web interface, we may state the following. First, we need to take into account the task complexity – or, in other words, the cognitive load embedded into the structure of the task, and the ways to minimize it. Second, we need to remember that the axes along which the search of an optimal menu model was going were overview vs. linearity and context vs. hierarchy.

2.2   Today's Trends in Web Design and Lack of Research on Menu Complexity

But today's research still does not take into consideration three crucial changes that have been around on the Web in the recent years. This happens due both to general lack of navigation studies and also the rapid development of media design for portable, mobile, and wearable devices.

The first trend is proliferation of adaptive strategies of web design instead of those of responsive design preferred by web developers and designers in the 2000s. Due to adaptation of designed interfaced to portable and mobile devices, webpage design has re-oriented to vertical scrolling, representation via visual blocks, and growth of identical (and often dynamic) webpage elements instead of imitating pages of print media out of which the 'classic' landing pages grew. New web projects have also re-oriented to user-generated content and have integrated user-oriented popularity tracking mechanisms.

These trends, in its turn, have called for appearance and rapid proliferation of new types of menus, such as tag clouds, hub pages, and selective menus. They are dominating mobile and user-generated design of today, but the research on their efficiency is scarce enough.

Third, in the current situation, earlier types of menus co-exist with the newest ones, substantially raising the menu complexity and, allegedly, user frustration in the process of information search – despite the fact that the menus are developed to raise the speed of task fulfilment and user satisfaction. But, till today, we lack studies that would answer to the following questions: What happens with user search behavior if several types of menus are co-functioning on one webpage? How the diversity of the navigational system affects both information search efficiency and subjective user satisfaction?

We try to address these gaps in the current research on user navigation. In it, we employ the findings in the area of cognitive strategies, especially in that of error correction. Thus, a study by the authors [14] shows that error correction strategies are applicable to possible scenarios of information search behavior. This research area is especially important for the web spaces with big amount of textual data and low navigation relevance – e.g. for university web spaces.

As a rule, to reduce the content complexity, such web spaces offer to their users at least several alternative navigation instruments. But it is still unclear whether the efforts of web designers bring more trouble or more satisfaction to the users.

3   Research Design, Method, and Conduct of the Experiment on User Behavior

3.1   The Research Questions and Hypotheses

Following the research on error correction [14], we have divided the possible strategies of information search into productive and non-productive. The former strategy may be described as systemic information search based on use of one menu only or a certain evident principle in combined use of several menu types. The latter strategy may be described as 'the method of trials and errors' when search demonstrates a chaotic, sporadic character revealing the user's disorientation.

We have chosen regressions – repetitions and recursions in referring to menu rubrics of the same menu, as well as returns to previously used menus – as the main parameter revealing a non-productive strategy of information search. Special attention should be given to the cases of repeated search in the menu rubrics where the target information is patently absent. According to the authors, such 'rigid exploration' [14: 427] signifies loss of user control upon the search process.

For this study, we have posed two formal research questions:

RQ1: With the growth of task complexity, does navigation (menu) complexity enhance or diminish efficiency of user search behavior?
RQ2: With the growth of task complexity, does navigation (meny) complexity enhance or diminish subjective user satisfaction?

Accordingly, our research hypotheses look as follows:

H1. With the growth of task complexity, navigation (menu) complexity grows the users' inclination towards non-productive search strategies.
H2. With the growth of task complexity, perceived interface complexity also grows.

3.2   The Research Design

Based on our research premises, we have founded our experiment on two parameters: navigational (menu) complexity/variability and structural task complexity. For the clarity of the experiment, we have omitted the parameters of cultural proximity and semantic relevance from our analysis.

We have used one of the old layouts of the website of St. Petersburg State University, Russia (SPbU). Earlier, we have shown that, of three universities in Russia and the USA, this website demonstrated mid-range results [15, 16]; but, according to our preliminary research, it was the menu complexity that raised the biggest number of claims among the scholarship and studentship of this university who dealt with the university website. Thus, we have elaborated the experimental prototype based on a real-world design of the main hub of a large university web space.

Measuring Navigational (Menu) Complexity. Menu complexity was measured by the number and nature of the navigational instruments (types of menus) offered to a user simultaneously within the search task fulfilment procedure. Thus, on one page, combinations of up to four different types of menus could exist:

(1) an expandable hierarchic menu (see Fig. 1);
(2) an expandable bar menu with dropdown second-level menu (all-encompassing, with low path depth) (see Fig. 2);
(3) a hub page with content-based navigation (see Fig. 1);
(4) a popularity-based selective tag menu (see Fig. 3).

On the real-world SPbU web portal, several varying categorization principles were used. Thus, in cases (1) and (3), classification was based on activity types, in case (2) – on document types, in case (4) – on popularity of certain topics and individual pages among the website users. We kept this scheme in our prototype.

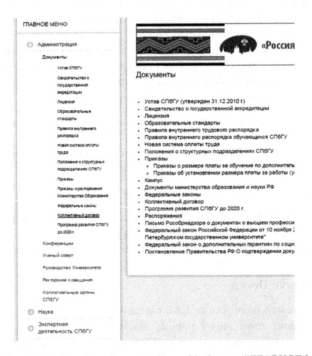

Fig. 1. The real-world combination of case (1): hierarchical menu ('ГЛАВНОЕ МЕНЮ', *main menu*) – and case (3) – content-based hub ('ДОКУМЕНТЫ', *documents*)

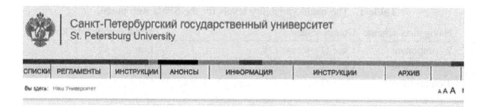

Fig. 2. Case (2): expandable bar menu

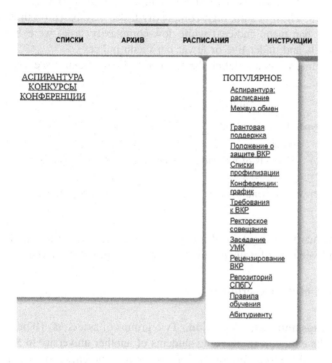

Fig. 3. A combination of case (2), case (3), and case (4) – selective popularity-based tag menu

We have defined three combinations of menu variability – from earlier to earlier + newer menu types. In this way, we could assign the level of menu complexity to each of these combinations (see Table 1). For the real-world SPbU web portal, a three-menu style was typical for most pages, but many sub-nodes had the four-component one.

We have ensured the possibility of alternative search to a full extent via selecting the target information the way that it could be found by any navigation scheme independently of each other.

Measuring Structural Task Complexity. The second parameter in our experiment was structural task complexity. Structural complexity was assessed by two dimensions. The first one, relevant for cases (1) and (3), was path depth (choice between 2 and 3

Table 1. The menu variability levels for the SPbU web portal.

Navigation scheme	Components	Menu complexity level
2-component	Case (1) + case (2)	Low
3-component	Case (1) + case (2) + case (3)	Middle
4-component	Case (1) + case (2) + case (3) + case (4)	High

immersion levels). The second one, relevant for case (2), was menu options diversity: low-complexity bar menu had five or fewer indices, high-complexity bar menu had over five indices. Thus, in sum, high-complexity tasks were based on three immersion levels and 5 + bar menu indices, while low-complexity tasks were based on two immersion levels and under five or fewer bar menu indices.

Thanks to this, we could define the level of task complexity (see Table 2).

Table 2. Levels of task complexity

Menu type	Task complexity	
	High	Low
Case (1)	3 levels of website architecture	2 levels of website architecture
Case (3)		
Case (2)	Bar menu indices $\geq$ 5	Bar menu indices < 5

In accordance with the defined navigational schemes and task complexity levels, six local html-versions of a university website prototype were created.

3.3   Conduct of the Experiment

Assessor Groups and Data Collection. Two groups of assessors, 10 persons in each, were part of the experiment, all being students of another university in St. Petersburg, Russia; this was done to ensure that the assessors have no cause for regular use of the SPbU web portal and are ignorant of its inner features.

During the experiment, the two groups fulfilled the search tasks of varying complexity upon the six html-versions of the SPbU web portal. Each of the tasks (of low and high task complexity) was performed on the three versions with varying navigation schemes.

We used a JavaScript scenario for registration of the website sections reached by a given assessor. The scenario created a descriptor note on each menu index: with each click of a user on a menu section, the respective note would appear in the log journal of the given user. These notes were later used as the research sample.

Description of the tasks. As stated above, the tasks were created the way that they could be fulfilled disregarding the chosen navigational scheme.

Additionally, we loaded the assessors with at least two categorization regimes. The first task was about finding the target information on types of university activities. This categorization scheme corresponded with the tree of the expanded hierarchical menu of the case (1) obviously linked on the real-world web portal to the administrative, teaching, scientific, and expert activities in the university. The second categorization scheme was linked to the case (2) of the upper one-level expandable menu and is oriented to the target document search, as the menu indices mention normative documents, lists, announcements, archives etc. The case (3) of content-oriented navigation actually combines these two categorization schemes, while the case (4) of the tag-based menu breaks them. This is why the task description had to contain the indications of both the document type and the type of university activity (see Table 3).

Table 3. Examples of the task descriptions in the experiment.

Task complexity level	Task #	Task description	University activity referral – case (1)	Document type referral – case (2)
High	1	'To find the Rector's order on the conducting scientific events in the university'	The categorization schemes are indicated by 'Rector's order' and 'scientific events' Optimal path: 'conferences/organization/regulations on scientific events'	The categorization scheme is indicated by 'order' (refers to the categories 'instructions' and 'normative acts/regulations') Optimal path: 'instructions/regulations on scientific events'
Low	2	'To find the information on the dates of theses defenses'	The categorization scheme is indicated by 'theses defenses' Optimal path: 'final state attestation/schedule of defenses'	The categorization scheme is indicated by 'the dates' Optimal path: 'timetables/schedule of defenses'

Additionally, two crucial conditions for task description had to be fulfilled:

1. The task description may refer to several indices of different menus on one page at the same time (see Table 3, comments on Task 1 description), which rises the user's cognitive load.
2. The task must not coincide literally with the menu indices but may contain direct referrals to the name of the target document.

The Conduct of the Pre-test Experiment. The experiment that we have designed consisted of two sessions. Throughout the test, both groups were taking part in it; this is why, for each new session, the task content changed. The changes were related to the menu indices /referrals under scrutiny while the menu structure was preserved. The tasks of high complexity the referrals were related to scientific activities; those of low complexity were linked to studying (see Table 4). Also, to make the process traceable for us, each assessor place had a pen-and-paper set, where the assessors could mark their trajectory if they occurred to have regressions.

Table 4. Examples of the task descriptions in the experiment.

Session	Task complexity	Topic	Group rotation	Navigational scheme
1	Low	Studies	Group 1	(1) + (2) + (3)
			Group 2	(1) + (2); (1) + (2) + (3) + (4)
2	High	Science	Group 2	(1) + (2) + (3)
			Group 1	(1) + (2); (1) + (2) + (3) + (4)

In the immediate aftermath of the sessions, the assessors were asked to evaluate their user experience in terms of perceived navigation complexity. They answered the question 'Please evaluate how easy it was for you to navigate this page'; the available answer options were structured according to a modified Likert scale and included the following answers:

1 – 'extremely uneasy and complicated, to the extent that I find it annoying';
2 – 'not that easy, to the extent that I felt discomfort';
3 – 'usual level of complexity, typical for the majority of websites';
4 – 'easy enough';
5 – 'way too easy, I did it intuitively and with no effort from my side'.

The results of the experiment and presented and discussed below.

4   Results and Discussion

4.1   The Research Results

The results of our experiment are presented in Table 5.

Table 5. Examples of the task descriptions in the experiment.

Task complexity	Menu complexity	Perceived complexity, mean	% of clicks to case (1), of all	% of clicks to case (2), of all	% of regressions
High	(1) + (2)	4.9	9.09	90.9	–*
	(1) + (2) + (3)	2.6	34.4	65.51	5
	(1) + (2) + (3) + (4)	2.9	37.5	62.5	8
Low	(1) + (2)	5	11.3	88.6	–
	(1) + (2) + (3)	4.7	11.86	88.1	–
	(1) + (2) + (3) + (4)	5	15.6	84.3	–

Note. * - the percentage of regressions is less than 1%.

We did not calculate any correlations between the menu complexity levels and the user strategy, as our data were too scarce to be statistically significant, but the mean figures that we have calculated are also telling.

Thus, Table 5 shows that, for the low complexity tasks, menu complexity has virtually no impact upon navigational behavior, as well as to the subjective evaluations of easiness at information search. In this case, all the three navigational schemes saw the assessors search for the target information confidently and consistently; the absence of regressions and productive (mostly linear) patterns of finding the target information clearly tell of this. For the low-complexity tasks, the majority of assessors used the upper expandable menu and viewed the menu indices from left to right.

The impact of menu variability upon user search behavior is found only for the high-complexity tasks, and only in the cases of three and four menus represented on one page. Here, the level of subjective easiness of search drops almost twice and the percentage of regressions rises to 5% for three-menu pages and 8% for four-menu pages. Also, user preferences change radically: if two-menu pages invoked clicks on the upper one-level menu (84% to 90%), the appearance of additional navigation instruments fosters the growth to hierarchical left-side menu (from 9–15% to 37% of moves). Surprisingly, the assessors never used the contextual and tag-based menus, which suggests that, in case of growing complexity, the older, perhaps more reliable and usual navigational instruments are perceived as all-encompassing and become a natural starting point for information seeking. It is on three- and four-menu pages where the assessors' activities lost its consistency: the users allow repetitive study of the rubrics they had assessed earlier and are switching from one menu type to another.

Thus, our H1 and H2 on the growth of non-productive search strategies with the growth of task complexity have proved partially right – they are true for the cases of middle and high menu complexity. Our pre-test results show that, unlike common-vesical expectations, user satisfaction does not grow linearly with the growth of menu complexity: the three-menu pages invoked the results even worse than the highly complicated four-menu ones. Thus, saturating web pages with the maximal amount of navigational instruments shows up as clearly counterproductive, and finding an optimal scheme is trickier than one could expect. Our results suggest that we can recommend a two-menu navigational scheme; we consider all other navigational instruments on the same page excessive, disorienting, and annoying.

Our results are, of course, subject to substantial limitations. Thus, we did not involve such parameters as time of task performance and user metadata into the current report on our data; this was due to the pre-test nature of our experiment. Also, the questionnaire has to be expanded in future to more precisely track the users' attitudes towards the navigation diversity. Also, the assessors' absolute disregard of the newer navigation instruments seems odd and definitely deserves a further investigation. But we can show that, in the online world where the web space complexity grows as well as the search tasks do, excessive complexity seems to be counter-productive.

References

1. Gwizdka, J., Spence, I.: What can searching behavior tell us about the difficulty of information tasks? A study of web navigation. In: Grove, A. (ed.) Proceedings of the 69th Annual Meeting of the American Society for Information Science and Technology (ASIS&T), vol. 43, pp. 1–22. Information Today Inc., Medford (2006)
2. Van Oostendorp, H., Ignacio Madridb, R., Carmen Puerta Melguizo, M.: The effect of menu type and task complexity on information retrieval performance. Ergon. Open J. **2**, 64–71 (2009)
3. Campbell, D.J.: Task complexity: a review and analysis. Acad. Manag. Rev. **13**, 40–52 (1988)
4. Byström, K.: Information and information sources in tasks of varying complexity. J. Am. Soc. Inf. Sci. Technol. **53**, 581–591 (2002)
5. Maldonado, C.A., Jlesnick, M.L.: Do common user interface design patterns improve navigation? In: Proceedings of the Human Factors and Ergonomics Society 46th Annual Meeting, vol. 46, no. 14, pp. 1315–1319. SAGE Publications, Los Angeles (2002)
6. Galitz, W.O.: The Essential Guide to User Interface Design: An Introduction to Gui Design Principles And Techniques. Wiley, Hoboken (2007)
7. Rogers, B.L., Chaparro, B.: Breadcrumb navigation: further investigation of usage. Usability News **5**(2), 1–7 (2003)
8. Hochheiser, H., Shneiderman, B.: Performance benefits of simultaneous over sequential menus as task complexity increases. Int. J. Hum.-Comput. Interact. **12**, 173–192 (2000)
9. Leuthold, S., Schmutz, P., Bargas-Avila, J.A., Tuch, A.N., Opwis, K.: Vertical versus dynamic menus on the world wide web: eye tracking study measuring the influence of menu design and task complexity on user performance and subjective preference. Comput. Hum. Behav. **27**, 459–472 (2011)
10. Instone, K.: Location, path and attribute breadcrumbs. In: Proceedings of the 3rd Annual Information Architecture Summit, pp. 15–17
11. Ahmed, I., Blustein, J.: Influence of spatial ability in navigation: using look-ahead breadcrumbs on the web. Int. J. Web Based Communities **2**(2), 183–196 (2006)
12. Blustein, J., Ahmed, I., Instone, K.: An evaluation of menu breadcrumbs for the WWW. In: Williamson, C.L., Zurko, M.E., Patel-Schneider, P.F., Shenoy, P.J. (eds.) Proceedings of the Sixteenth ACM Conference on Hypertext and Hypermedia (HT 2005), pp. 202–204. ACM Press, New York (2005)
13. Tingru, C., Xinwei, W., Hock-Hai, T.: Building a culturally-competent web site: a crosscultural analysis of web site structure. University of Wollongong Research Online database (2015). http://ro.uow.edu.au/cgi/viewcontent.cgi?article=6013&context=eispapers. Accessed 09 Feb 2018
14. Kang, N.E., Yoon, W.C.: Age- and experience-related user behavior differences in the use of complicated electronic devices. Int. J. Hum. Comput. Stud. **6**(6), 425–437 (2008)
15. Bodrunova, S.S., Yakunin, A.V., Smolin, A.A.: Comparing efficacy of web design of university websites: mixed methodology and first results for Russia and the USA. In: Chugunov, A., et al. (eds.) Proceedings of the International Conference on Electronic Governance and Open Society: Challenges in Eurasia (EGOSE), pp. 237–241. ACM (2016)
16. Bodrunova, S.S., Yakunin, A.V.: U-index: an eye-tracking-tested checklist on webpage aesthetics for university web spaces in Russia and the USA. In: Marcus, A., Wang, W. (eds.) International Conference of Design, User Experience, and Usability, pp. 219–233. Springer, Cham (2017). https://doi.org/10.1007/978-3-319-58634-2_17

Study on Process for Product Design Applying User Experience

Luya Chen[✉], Keiko Kasamatsu, and Takeo Ainoya

Tokyo Metropolitan University, 6-6 Asahigaoka, Hino-shi, Tokyo, Japan
chinn_roga@yahoo.co.jp

Abstract. The purpose of this paper is to propose a product modeling design process centering on user experience for product designers. It can be flexibly substituting in various product design processes, and guide designers to a reasonable design direction instead of simply modeling according to their own aesthetics and design experience.

Keywords: User experience · Human-centered · Product design
Design process

1 Introduction

In the field of product design, when considering product modeling from a concept, there are countless possibilities. It will also be influenced by factors such as the designer's own aesthetics and design experience.

However, if the product design is not considered from the user's viewpoint, the designed product may have problems such as the user does not know how to use or the user experience is not good.

Today, user experience is getting more and more attention. The design process called User Experience Design (UXD) is widely used in major design fields. Among them, the most widely used is in the user interface (UI) design, the web design field. In the field of product design, designers are constantly trying to incorporate user experience design into the product design process. Various user-centered product design process are presented, but they did not specifically consider the part of the modeling.

2 Background and Related Work

For the user experience design process, there is no single answer suitable for all cases. Among them, one approach to improve user experience is Human-Centered Design (HCD). From the viewpoint of usability, it is also called User-Centered Design (UCD).

It was originally proposed for interactive systems to improve system usability.

As the content, mainly model user experience from survey analysis, create concept to realize this experience, make prototype and evaluate. And if necessary depending on the result of the evaluation, return to the previous step and repeat the work (Fig. 1).

© Springer International Publishing AG, part of Springer Nature 2018
S. Yamamoto and H. Mori (Eds.): HIMI 2018, LNCS 10905, pp. 67–75, 2018.
https://doi.org/10.1007/978-3-319-92046-7_6

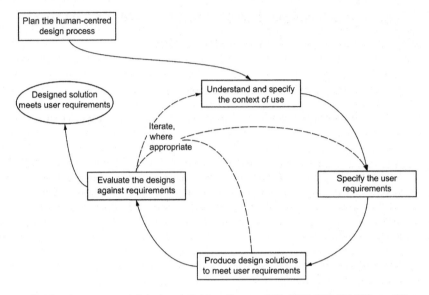

Fig. 1. Human-centered design activities (Source: ISO 9241-210:2010(E), p. 11)

Today many design companies are seeking their own unique user experience design process. But most of them are similar to the idea of this human-centered design process: from the user survey to concept design and do design, receive user feedback and improve the design. The following example is TOSHIBA's user experience design process (Fig. 2):

Fig. 2. TOSHIBA UX design process (Source: TOSHIBA DESIGN CENTER) Exploring needs from users' current situation survey, extracting essential issues, researching, then thinking, realizing and verifying based on essential tasks.

However, the solution derived from these user experience design process is often a system or a service, but not a specific product.

For example, in order to be able to pay quickly when shopping and avoid queuing, it is not to make an improvement design of the cash register, but even losing the action of accounting, and it leads to a new lifestyle of the future and it must be considered in the system (Figs. 3 and 4):

Fig. 3. Amazon go: a new kind of store with no checkout required use the Amazon go app to enter the store, just take the products you want then go out. Payments will be automatically completed in the app

Fig. 4. Alibaba Buy+: VR shopping without going out, you can buy global products at home. Payment will also be completed in an instant

Although it is said that in order to solve a problem, it should not be confined to design a specific product, but for a product designer, such a user experience design process still lacks in the final product modeling design part.

3   General Method for Designing Modeling

When designing product modeling, product designers sometimes brainstorming to draw a lot of different styles of design, then choose the best one. Sometimes find inspiration around a theme. Different product designers have different approaches based on their own design logic. According to the survey and analysis, I have summarized the following major sources of inspiration:

3.1   Inspiration from Psychology

The inspiration that psychology brings is to consider what kind of impression products leave in the minds of users.

For example: black reflects mature, stable, scientific and technological sense and stability. The same coffee poured into cups of different colors, the coffee in yellow cup tastes lighter, the coffee in green cup tastes sour, and the coffee in red cup tastes the most fragrant (Figs. 5 and 6).

Fig. 5. DELL VOS

Fig. 6. Nestle mug (Color figure online)

3.2 Expressing the Beauty of Materials

Different materials can also give people different feelings, and different designs can be derived by expressing the beauty of the material (Fig. 7).

Fig. 7. The peacock chair. The same chair with different materials, wood shows a soft affinity, velvet feels warm

3.3 Expressing the Beauty of Geometry

Mathematics has a rigorous beauty. It is often used by designers in product design.

The geometric principles often used by designers are: Golden ratio, Root rectangles, and so on (Figs. 8 and 9).

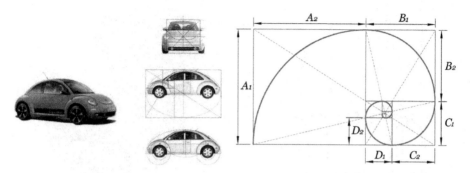

Fig. 8. Volkswagen beetle & golden ratio (Source: Geometry of Design: Studies in Proportion and Composition)

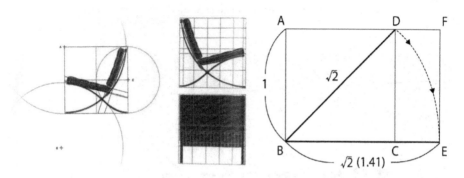

Fig. 9. Barcelona chair & root rectangles (Source: Geometry of Design: Studies in Proportion and Composition)

3.4 Expressing the Beauty of Culture

Different cultures breed different people. Each region has its own unique design style that is unmatched by other regions. Tradition represents the indelible beauty (Fig. 10):

3.5 Inspiration from Nature

There is a design method called bionic design. Designers draw inspiration from the nature or shape of creatures and have designed many excellent products (Fig. 11).

3.6 Details Determined by Commercial Factors

Products for the company are for profit, so they have to consider commercial factors.
    For example: Why milk is sold in square boxes and cola is sold in round bottles?
    Because soft drinks are mostly drunk directly, cylindrical container is more convenient to hold in hand to drink. Milk must be considered for storage, square box saves storage space (Fig. 12):

Fig. 10. Ming style chair from China & egg chair from Denmark

Fig. 11. Sydney opera house inspired by eggshells

Fig. 12. Cola in round bottles & milk in square box

3.7   Details Determined by the Function of the Product

In order to achieve a certain function, the shape of the product needs to be designed in some way, then the design will be more convenient to use, etc. The function of the product limits the shape of the product to some extent. From another perspective, it also points out the direction for product design.

For example: Why the ticket gate where reads the Suica card tilts 13°?

The designer, Shunji Yamanaka, learned from the observation and analysis of pedestrians who passed the ticket gate then found that there are often mistakes in reading cards. Through constant attempts, finally reached the conclusion that the card reading error is the smallest when tilting the card reading place by 13° (Fig. 13).

Fig. 13. The Suica ticket gate

4   The Process of Product Modeling Design Applying UX

It can be seen from the above that product designers actually consider the usability and user experience when designing the modeling of a product, but there is not a uniform systematic process, so the result of the design is often influenced by the designer's own aesthetics and design experience. However, each designer has his own design logic and a design process cannot fit into all designers and design projects. The design process should not only be designed to design a product with a better user experience, but also should be designed to serve the designer to work more smoothly.

It is more important to systematically organize the factors that determine the design of modeling, analyze the relationship between the primary and secondary according to each situation, and find their balance in the design.

4.1   The Factors that Determine the Design of Modeling

According to the analysis, I have preliminarily sorted out several major factors that determine the design of modeling. They are (Fig. 14):

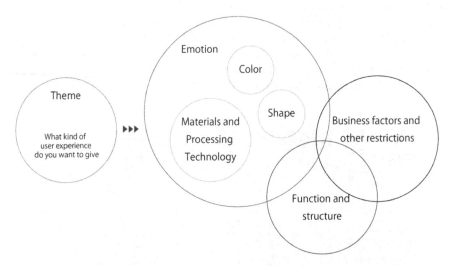

Fig. 14. Factors that determine the design of modeling. The "Theme" comes from the previous user surveys and analysis, to ensure that this part can be combined with a variety of different user experience design processes

In a variety of different product design processes, combined with the results of previous user surveys and analysis, and taking into account the above factors, the product design will be more reasonable.

4.2   How to Design Product Modeling Applying UX

In order to incorporate the user experience when considering factors that determine the design of modeling, maybe we can consider the modeling of the product from the following aspects:

1. **From users impact on modeling style.**
   For example: understand the lifestyle, preferences, etc. from user analysis, make collages and decide the style of the whole modeling.
2. **Structure for problem solving.**
   While the structure limited the modeling, It shows the direction of modeling.
3. **Shape for easy use.**
   In terms of ergonomics, present size and shape.
4. **Shape for presenting operation action.**
   With the goal that users can operate naturally.
5. **Designer's aesthetic sensation and design experience.**

4.3   Future Works

The final molding design process needs farther exploration.

1. The factors determine the design of modeling are still not perfect. In order to obtain more information that is closer to the real situation of the designer, an interview or a questionnaire survey will be required in the future.
2. After the final modeling program is proposed, how to verify is also a problem that needs further discussion.

References

1. TOSHIBA UX Design Process. http://www.toshiba.co.jp/design/ux/en/process.htm
2. Amazon Go. https://www.amazon.com/b?ie=UTF8&node=16008589011
3. Buy+. https://www.youtube.com/watch?v=4RDq0jtjaWc
4. Elam, K.: Geometry of Design: Studies in Proportion and Composition (2001). ISBN 978-1-56898-249-6
5. Yamanaka, S.: Skeletal Structure of Design (2011). ISBN 978-4-8222-6470-3

Issues of Indexing User Experience

Shin'ichi Fukuzumi[1(✉)] and Yukiko Tanikawa[2]

[1] Research Planning Division, NEC Corporation, Minato, Japan
s-fukuzumi@aj.jp.nec.com
[2] Corporate Business Development Division, NEC Corporation, Minato, Japan

Abstract. Usability and User experience are also important factors for not only suppliers but also users when product, system and service are developed and operated. Though these are often confused, concepts are different. Also about measurement and evaluation, the former targets product, system and service themselves, the latter targets experience using them which includes users. This paper proposes that UX shall be evaluate whether it is "good" or it is "bad" not whether it is "high" or it is "low". Expected value by before using and real experience value by during/after use are made these a function respectively. And index of UX is a difference between them. From this, to reduce a gap between high expected value and real experience value becomes to provide good UX for developers and providers.

Keywords: Usability · User experience · Quality · Summative test
Satisfaction

1 Introduction

In 2010, ISO9241-210 "Human-centred design for interactive system" which is an ergonomic related standard about human centered design was published [1], discussion related human centered design (HCD) and user experience (UX) which is newly defined in this standard becomes active in IT business field.

HCD activities are applied to each development process shown in Fig. 1. HCD is a method to give better UX to stakeholders and to provide system and product with high usability for users and stakeholders [2].

On the other hand, traditionally, usability is one of the most important factors for products, system or service. Usability and User experience are also important factors for not only suppliers but also users when product, system and service are developed and operated. Though these are often confused, concepts are totally different.

The former is an index for "easy to use" when user uses product, system or service [3]. So, targets are product, system or service themselves. The latter is to experience whether user is able to do which he/she would like to really do or not. So, this targets to experience using them which includes users [4].

In case of realize "ease of use" or provide experience what user really want to do, some quantitative purpose have to be set in planning and designing phase of product, system or service. To set some concrete purpose, they have to be indices and decided as objective value.

© Springer International Publishing AG, part of Springer Nature 2018
S. Yamamoto and H. Mori (Eds.): HIMI 2018, LNCS 10905, pp. 76–82, 2018.
https://doi.org/10.1007/978-3-319-92046-7_7

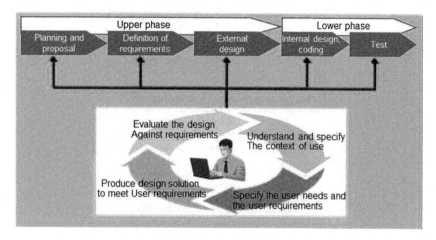

Fig. 1. The relationship among HCD activities and development process

When system and product with high usability could be developed, it is easy to verify their usability by usability test [5]. However, it is difficult to check whether these products or system achieve that a user wants to really do it.

This paper describes a trend of indexing, measurement and evaluation of usability and issues when UX is indexed and possibility of measuring UX and proposed that evaluation index of UX shall be "good"/"bad", not "high"/"low".

2 Usability

As described Sect. 1, usability is an index for "easy to use" when user uses product, system or service, and an index of product, system or service themselves. In this section, usability is describe from a view point of "human" and "development".

2.1 From a View Point of Human

"Easy to use" when using product, system or service are also classified by "efficient operation", "easy to learn", "easy to memorize" and "less error" [3]. There is a research to quantify them directly [6]. In this case, "easy to use" is an evaluation of interaction about UI operation. In a point of index to achieve an aim, usability is defined as "extent to which a system, product or service can be used by specified users to achieve specified goals with effectiveness, efficiency and satisfaction in a specified context of use" in ISO9241-11 [7]. Finally, to realize high "extent", "efficient operation" or "easy to learn" have to be achieved. ISO decided these elements to set easily as objective value.

Recently, there is a trend that usability is a part of quality. "Quality in Use" is defined and software quality model [8]. In this model, "freedom from risk" and "context coverage" are added to three elements in ISO9241-11. Figure 2 shows the quality in use model defined in ISO/IEC25010 [8].

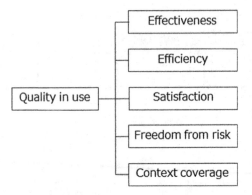

Fig. 2. Quality in use model [8]

2.2 From a View Point of Development

Quality model described before, not only "quality in use" but also "product quality" which has eight elements is defined (Fig. 3).

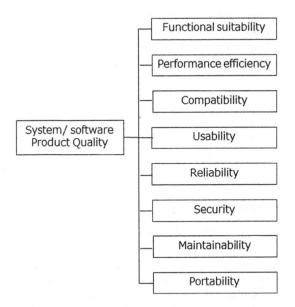

Fig. 3. Product quality model [8]

In this product quality model, "usability" is also listed. This usability is different from the meaning of ISO9241-11, they are, "appropriateness recognizability", "learnability", "operability", "user error protection", and so on. These elements are related to total product quality. This means that to achieve product quality related to usage is to realize quality in use. Figure 4 shows the relationship among usability (correspond to quality in use), dialogue principle (correspond to product quality) and characteristics of information presentation [9].

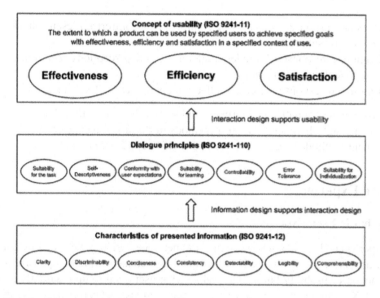

Fig. 4. Relationship among usability, dialogue principle, characteristics of information presentation [9]

2.3   Evaluation and Measurement of Usability

Usability evaluation has been tried to carry out conventionally. However, evaluation and measurement do not distinguish each other. In ISO, two kinds of evaluation formats, they are "summative test" [5] and "Formative test" [10] are defined, respectively. The former is not an evaluation index but a format for measuring "effectiveness", "efficiency" and "satisfaction" defined in ISO9241-11, the latter is a format for identifying usability problem using inspection method, usability test and user observation. Three detail elements of usability are as follows [7]:

- Effectiveness
    - Accuracy (extent to which an actual outcome matches an intended outcome)
    - Completeness (extent to which users of the system, product, or service are able to achieve all intended outcome.
- Efficiency
    - Time used (the time expended in attempting to achieve a goal)
    - Human effort expended (the mental and physical effort expended to complete specified tasks)
    - Financial resources expended (include the costs of using the system, product or service, such as paying wages, or the cost of energy or connectivity

- Materials expended (physical items (e.g. raw materials, water, paper) used as input to the task (including maintenance tasks) and processed by the system, product or service.
- Satisfaction
  - Physical responses (Feelings of comfort or discomfort represent physical components of satisfaction)
  - Cognitive response (Attitudes, preferences and perceptions represent cognitive components of satisfaction)
  - Emotional response (represent affective components of satisfaction).

By using this summative test format, results of evaluation test for these items can be described. Though there are many test methods to do this, the experimenters can refer measurement method of quality in use in ISO/IEC 25022 [11].

3   User Experience

3.1   What is User Experience

User experience (UX) is a general term of experience which is provided by use and a product or service. Sometime, UX is explained as a concept which emphasize not only each function or usability but also to realize what user want to do pleasantly and comfortably. UX is defined that "To have experiences with use in "before use", "during use", "after use" and "through total usage" and "To be able to shape "users want to do" through experience in each step and to achieve it" in User Experience white paper [4]. This definition is concreted of concept, but it is difficult to measure UX by using this definition.

ISO 9241-210 [1] defines UX as follows: "person's perceptions and responses resulting from the use and/or anticipated use of a product, system or service". This definition is difficult to understand for almost engineers though this text is tried to represent that UX could be measured.

3.2   Evaluation and Measurement of UX

As described in Sect. 1, usability and UX are often confused, concepts are totally different. To distinguish them, some trial were carried out. Kurosu et al. corresponded the relationship between product quality and quality in use to UI and UX [12]. This paper explains that quality in use is evaluated by using UI. Quality in use are separated to objective quality in use and subjective quality in use. And total of them defined as UX. These index of UX are similar to usability, so, evaluation results are numeric value, and can be represented by "high" and "low". However, does this index evaluate "experience"? An index of "experience" seems "good" of "bad".

Firstly, "before use experience" is considered. When before use, specification according to user requirements are decided, expected value will be high because of emotional change by imagine for use, environment and promotion. Next, "during use/after use experience" are considered. This is real experience which depends on satisfaction or emotion level which are achievement level of specification or usage feeling correspond to users' original image.

So, each expected value and real experience is represented by "high" and "low". However, user experience is a function of these two elements.

As shown in below, expected value by before using and real experience value by during/after use are made these a function respectively.

- Expected value (before experience)

Before using, expected value for use will be high because specification is decided by user requirements and influence by imaging usage at special environment and promotion.

$$Exp = f(specification, environment, promotion, usage, etc.)$$

- Real experience (experience during use and after use)

By using, user receives high satisfaction because this system can be judged to meet specification originally decided. And high emotion because of enjoyable and comfort by using.

$$Realexp = f(real\ measurements, environments, usability, etc.)$$

And index of UX is a difference between them. That is,

$$UX = f(Exp, Reaexp)$$
$$e.g.\ (Exp - Realexp), (Exp/Realexp)$$

From this, to reduce a gap between high expected value and real experience value becomes to provide good UX for developers and providers.

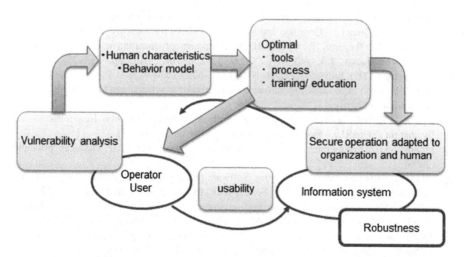

Fig. 5. Secure system operation based on behavior factor analysis of human

4 Conclusion

This paper proposed an indexing method for UX. We apply this method to secure operation system include human Fig. 5.

From this figure, expected value and real experience are able to describe below:

$$Exp = f(\text{specification, usage, environment, possible emergency information, education for counterplan, etc.})$$

$$Realexp = f(\text{real performance, usability, environments, defense against attack, etc.})$$

When system are installed, expected value include emergency information and education for counterplan are raised. When system will attacked, satisfaction by experience are raised if education/provision of information are good. From this, difference between expected value and real experience are minimized.

Like this, when measuring UX, a variety of elements have to be considered.

References

1. ISO9241-210: Human-centred design for interactive systems (2010)
2. Tanikawa, Y., Suzuki, H., Kato, H., Fukuzumi, S.: Problems in usability improvement activity by software engineers. In: Yamamoto, S. (ed.) HCI 2014. LNCS, vol. 8521, pp. 641–651. Springer, Cham (2014). https://doi.org/10.1007/978-3-319-07731-4_63
3. Nielsen, J.: Usability Engineering. Academic Press, London (1993)
4. Roto, V., Law, E., Vermeeren, A., Hoonhout, J. (eds.): Use Experience White Paper - Bringing clarity to the concept of user experience. Result from Dagstuhl Seminar on Demarcating User Experience, 15–18 Sept 2010, pp. 1–12 (2010)
5. ISO/IEC 25062: Software engineering—Software product Quality Requirements and Evaluation (SQuaRE)—Common Industry Format (CIF) for usability Test Reports (2005)
6. Fukuzumi, S., Ikegami, T., Okada, H.: Development of quantitative usability evaluation method. In: Jacko, J.A. (ed.) HCI 2009. LNCS, vol. 5610, pp. 252–258. Springer, Heidelberg (2009). https://doi.org/10.1007/978-3-642-02574-7_28
7. ISO9241-11rev: Ergonomics requirements for interactive system, part11: usability: concept and definition (2018)
8. ISO/IEC 25010: Software engineering, Software product Quality Requirements and Evaluation (SQuaRE) - System and software quality models (2011)
9. ISO9241-110: Ergonomics of human-system interaction - Part 110: Dialogue principles (2006)
10. ISO/IEC 25066: Software engineering, Software product Quality Requirements and Evaluation (SQuaRE), Common Industry Format (CIF) for usability: Evaluation Reports (2015)
11. ISO/IEC25022: Systems and software engineering – Systems and software Quality Requirements and Evaluation (SQuaRE) – Measurement of quality in use, 2016ISO/IEC 25062: Software engineering—Software product Quality Requirements and Evaluation (SQuaRE)—Common Industry Format (CIF) for usability test reports (2005)
12. Kurosu, M., Hashizume, A., Ueno, Y., Tomida, T., Suzuki, H.: UX graph and ERM as tools for measuring kansei experience. In: Kurosu, M. (ed.) HCI 2016. LNCS, vol. 9731, pp. 331–339. Springer, Cham (2016). https://doi.org/10.1007/978-3-319-39510-4_31

The Importance of Online Transaction Textual Labels for Making a Purchasing Decision – An Experimental Study of Consumers' Brainwaves

Pei-Hsuan Hsieh[✉]

National Cheng Kung University, Tainan 701, Taiwan
peihsuan@mail.ncku.edu.tw

Abstract. The purpose of this study was to explore how different transaction risk scenarios on auction platforms can significantly and differently influence online consumers' brainwaves. In this study, four scenarios (a 2×2 experimental design, i.e., one-two transaction products and with-without transaction labels) were developed. Twenty male participants (from engineering and liberal arts programs) assigned to view all four scenarios were recruited. In each scenario, the participants must determine whether they should make a transaction for purchasing products as accurately and quickly as possible. By using a 40-channel digital EEG amplifier with a cap, the participants' purchasing decision-making responses were recorded and timed, and their brainwave data were detected and stored for data analysis afterwards. After the experiment, every participant received an appreciation reward worth about $10 USD. All the brainwave data were analyzed using the software SCAN 4.5. As a result, it was found that participants spent more time and were less accurate without textual transaction labels than with the labels regardless of purchasing one or two items. Specifically, the participants' decision-making response efficiency and accuracy rate performed the worst in the scenario which did not offer any feedback for determining two transaction items. Also, when comparing the with-without textual feedback scenarios, the participants' brainwaves were found to be different.

Keywords: Online transaction · S-O-R (stimulation-organism-response)
Transaction feedback · EEG (electroencephalography)

1 Introduction

Online transactions have become easier for consumers as more applications are available in the marketplace. According to the statistics reported by the comScore MMX Multi-Platform [7], there have been more mobile device users than desktop computer users in Taiwan since May of 2016 in the age group of 25 to 54 years. Using a mobile device to do an online transaction is similar to doing a traditional one on e-commerce websites where consumers still need to go through a series of complex purchasing decision-making processes [5, 10, 24]. From the perspective of the S-O-R (stimulation-organism-response) psychological model, consumers' purchasing decision processes can be systematically presented [16, 31]. Even though consumers save time

© Springer International Publishing AG, part of Springer Nature 2018
S. Yamamoto and H. Mori (Eds.): HIMI 2018, LNCS 10905, pp. 83–97, 2018.
https://doi.org/10.1007/978-3-319-92046-7_8

on completing a transaction by a few clicks via their mobile devices, due to the significant effects of information asymmetry on online transactions, consumers may perceive high risks of financial loss and encounter fraudulent methods, such as low product quality, transaction interception, product price and function misrepresentation, seller evaluation inflation and shill, and delivery failure [8, 12, 21].

Online consumer research on trust behavior has pointed out that online sellers not only should be established as a secure and privacy-oriented website, but also should consider the establishment of credibility, so that consumers feel that the site is trustworthy [6]. However, online fraud has recently become prominent in online transactions, coupled with higher risks involving online auction platforms [12]. On the platforms, continuous or simultaneous types of online auctions are both available to consumers [17]. Sellers can display more than one product item as a combo in one auction, and buyers are able to maintain a degree of anonymity while purchasing products on online auction platforms [19]. Therefore, both the consumers and the sellers often lack information about the counterparty [13]. This anonymous mechanism makes people more opportunistic and increases online market opportunistic behavior [15].

The brain can be divided into different regions, and each region controls behaviors in different ways, including cognition, emotion, decision making, calculation, and memory [23]. Online consumers' purchasing intention and its influential factors, such as purchasing habits, can also be forecasted by analyzing time series brainwave data [22, 25]. When an α wave ranging from 8 to 13 Hz is detected in a person's left frontal lobe, it means that he/she is having a positive emotion and is able to actively take action. Contrarily, a negative emotion in a passive reaction is correlated with brain neural activation in the right frontal lobe [26]. An increasing number of neuromarketing studies has provided empirical evidence that a non-intrusive approach to retrieving consumers' brainwave data is the most objective way to understand consumers' purchasing decision-making processes [33]. In the e-commerce context, the most direct way to best satisfy consumer needs is to detect consumers' active brain neurons, which represent the consumers' thoughts regarding purchasing preferences and product price [2]. For example, electroencephalography (EEG) is one of the practical ways to retrieve consumers' brainwave data.

The present study was framed by the Engel-Kollat-Blackwell (EKB) model and utility theory and adopts the Stimulus-Organism-Response (S-O-R) psychology model, to investigate the effect of different feedback mechanisms on consumer final purchasing decision-making in an online auction purchase process. The purpose of this study was to explore how different transaction risk scenarios on auction platforms can significantly and differently influence online consumers' brainwaves.

2 Literature Review

Consumer purchasing decision-making process refers to all the activities that consumers engage in to meet their needs or wants. Such activities include search, choice, purchase, use, and product evaluation. The process consists of two aspects: the subjective psychological events and objective material-based events. The behaviors displayed during the process are called purchase behaviors [21]. Consumer purchasing decision encompasses the process in which consumers carefully evaluate a product itself as well as its brand value and characteristics, choose, and purchase a product in order to meet certain needs [31]. In other words, to meet their needs, consumers have a purchase motivation, then they analyze and evaluate purchase options, choose one of the many options, and finally decide to execute the optimal purchase option, all the while conducting a systematic decision process from start to finish. Consumer purchase behavior consist of the following steps [9]: problem recognition, search for information, alternative evaluation, choice, and outcome evaluation. The most important of these steps is the choice being made because choices are the end products of complex cognitive activities of human brains, and the accuracy of a choice directly determines whether and how a purchase behavior will occur [31].

2.1 Theory of Consumer Behavior

Decisions made during consumer purchasing process include psychological and cognitive events, as well as the actual final purchase behaviors which are affected by endogenous and exogenous factors [4, 27]. Therefore, the process of purchase behavior – meeting one's needs through evaluating and choosing products – is a process of evaluation and decision [9]. The EKB model assumes that after obtaining sufficient information, consumers may generate several options, evaluate those options, and make a choice. The standards for evaluating option are formed by the consumers' purchase perspectives and outcome expectations, and these standards present themselves as product characteristics preferences. Based on the EKB theoretical model, consumers of digital transactions have a vastly different degree of autonomy and control in their purchase process compared to the purchase behaviors found in traditional stores. Online purchase provides more options in less time, increases purchase speed, and offers innovative post-purchase services [18].

In addition, analyzing consumer behaviors quantitatively enables one to use needs theories to investigate the relationship between desires and utility [34]. Desire is the feeling of not being able to have something, and it has no limits. When a desire is satisfied, another rises. Utility is the sense of satisfaction gained when purchasing a product; it is a subjective feeling that shifts constantly due to time, location, and human factors. According to consumer behavior theories and needs theories, when price is fixed, consumer surplus increases, which determines the price consumers are willing to pay. In other words, as product quantity increases, marginal utility decreases, which lowers the price consumers are willing to pay. Similarly, the larger the utility, the higher the price that consumers are willing to pay. From a psychology perspective, one can apply the S-O-R model to examine consumer purchase behavior. S-O-R stands for stimulation-organism-response. The theory states that consumer purchase behaviors start with motivations formed by the synergy of psychological factors and external stimuli, then proceed to product purchase decisions, actual act of purchase, and finally, post-purchase evaluation [16, 31].

2.2   Fraudulent Online Transactions

Online shopping touts efficient and comfortable shopping experiences, breaking the tradition of physically going out to shop. However, scammers use a variety of methods to conduct illegal transactions online. Especially, when consumers use private communication software to shop, because the platforms do not maintain transaction records, the Q&A tools in such platforms do not count as evidence for completed transactions, making the platform owners unable to assist with handling complaints when conflicts arise or when consumers are scammed as result of pursuing alleged perks [28]. Generally, concerns regarding online transactions are rooted in: (a) the lack of face-to-face interactions between sellers and buyers causes their inability to confirm each other's identities; (b) the steps for completing an online purchase are complex, making it possible for payment information to be leaked and increasing the odds of it being stolen; and (c) online purchases require submitting personal information and credit card numbers at the risk of these being stolen. Kalakota and Whiston [14] pointed out that online transactions have three primary security requirements: privacy, confidentiality, and integrity. Thus, encryption, digital signature, public key infrastructure, key logger, etc., are effective tools to increase online transaction security.

2.3   Online Consumer's Brain Neural Circuits

The human brain has several regions that are associated with decision-making and cognition, including emotions and socialization. These regions' nerves are activated when humans engage in online auctions or e-commerce. For example, the prefrontal cortex and the anterior cingulate gyrus are related to decision-making. If there is uncertainty during a decision-making process, the orbital frontal cortex and parietal cortex become activated. Additionally, the human emotions of happiness and pleasure are managed by three brain regions: nucleus accumbens, anterior cingulate cortex, and putamen. The unpleasant emotions are also managed by three regions: superior temporal gyrus, amygdaloid, and insula cortex. Morin [26] found that detecting the α brainwave (8–13 Hz) produced by the left frontal lobe indicates positive emotions. By inference, such brain activity is a good predictor of being motivated to take action [20]. The right frontal lobe is generally associated with negative emotions, which often makes a person withdraw from taking action. We now briefly describe the electroencephalography (EEG), an instrument for detecting brainwaves, before proceeding to discuss neuralmarketing research findings that involved using this instrument.

2.4   The Application of EEG in Neuromarketing Research

EEG is a non-invasive technology that measures brain activity. It can detect tiny brainwaves under the scalp and does not harm the human body. It converts electrical signal into numbers and sends the data through an amplifier, then displays the time series of the voltage values. It can show the changes that a brain goes through in different conditions or states, so it is a valuable tool for providing data for analysis and research [33]. The greatest advantage of EEG is its time sensitivity (recording up to 2,000 data points per second), which enables it to capture the fast-changing brain activities and processes such as awareness, recognition, and emotion exercises, better than other brain imaging tools like MRI or PET scan [29].

EEG's application in the field of neuromarketing allows economists to examine the brain's processes that drive consumer decisions. In neuromarketing research, often the research participants are invited to observe especially designed experimental materials, and researchers obtain the participants' brainwave's time series graphs in order to understand what factors in the purchase process have impact on consumer's purchase intention [22]. Furthermore, common purchase habits or patterns can be identified, so that key elements can be modified to more closely conform to consumer behaviors [25]. One example is the text configuration on merchandizes and webpages. Because the human brain prefers images over texts, consumers may adjust the proportion and placement of images and texts to improve the brain's ability to process information, to increase the purchase rate of online shoppers and e-commerce users, and to improve consumers' preference of and attention to the merchandizes' characteristics and similar products [26]. Some research studies use mobile lab instruments. They ask participants to explore real or virtual stores and measure the individuals' mental statuses as they purchase products or services, and thus obtain consumers' purchase habits and decision-making process in real-world settings [1]. Neuromarketing can also be applied when investigating consumer preference of brands and advertisement [22]. Price is a major factor in influencing purchase decisions; it can even be used as a tactic to attract buyers. Neuromarketing research can be used to discover areas in the brain that help consumers rationalize their preferences and to induce the most profitable response when stimulated by the highest price [2].

3 Methodologies

The research framework is sketched in the figure below (Fig. 1). The stimulation (S) in the study consisted of the two online transaction feedback mechanisms for displaying product information: one or two products to purchase and with-without textual transaction labels. The organism (O) had two dimensions: consumer background (in this case, college major as either engineering or liberal arts) and consumer purchase needs (one item or a two-item combo purchase). The response (R) was the consumer purchase decision, both in terms of the efficiency in making the decision and in terms of the accuracy of that decision. The entire research process utilized EEG to generate consumer brainwave graphs, and the efficiency and accuracy of purchase decisions were recorded simultaneously.

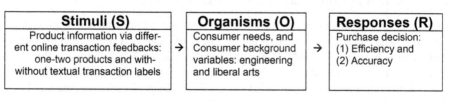

Fig. 1. Research framework.

Based on prior research, five brain regions were the focus of the study: Fz (frontal lobe, focusing on the orbital frontal lobe cortex), FCz (between frontal lobe and parietal lobe, focusing on the anterior cingular gyrus cortex), Pz (occipital lobe cortex), Cz (parietal lobe cortex), and CPz (between occipital lobe and parietal lobe) [3, 29, 30, 35]. Of the five regions, the frontal lobe (Fz and FCz) is located near the forehead and is related to higher cognition (working memory, thinking, judgment, planning, creating), language, personality, and exercise. The parietal lobe (Cz and CPz) is located at the back of the head. It acts as the processing center for sensory information and is responsible for high-level perceptions such as visual-spatial ability, attention, touch, pressure, and pain. The occipital lobe (Pz) is located at the center of the back of the brain and is the simplest region, primarily responsible for vision (shape, size, color) and responds to changes in number size. When purchasing product quantities larger than available items, this area is expected to be stimulated and will show distinctive amplitudes in the waves. In this study, participants were required to observe the images in the experiment that show the number of items available and the number of items desired. When the number to be purchased was greater than the number of items available, the participants must make a purchase decision in a timely manner. That is, the participants' Fz, FCz, Pz, Cz, and CPz regions would experience changes in brainwave voltage, and the EEG graphs would show distinctive amplitudes. It is worth noting that the experiment images were limited to consumer electronics, computers, communication devices, and their peripheral products for bidding. The price level matched the market value, and the price did not include any discounts or promotions.

3.1 Research Participants

This study recruited twenty adult internet users ages 20 or above, with online shopping experiences. Ten of them were from engineering programs, and the rest were from liberal arts programs. Those with experience in online auctions were given priority when recruiting. All participants were right-handed, and none took any medicine chronically. They filled out a personal internet experience questionnaire and signed the informed consent form before proceeding to the EEG experiment. Since the research design involved EEG, which demanded a sustained period of attention, the participants were required to avoid any activities that might disturb their biological indicators (e.g., having caffeinated drinks, taking medicine, or staying up late) for a 24-h period prior to starting the experiment. In addition, to better measure the participants' brainwaves, they were instructed to wash their hair prior to the experiment to remove any hair products like hair spray or wax.

3.2 Research Procedures and Data Collection

The model of the caps procured for the experiment was the 40-channels Quik-Cap for NuAmps. The cap was connected to the NuAmps 40-channel Amplifier, an instrument that receives brainwave signals. The researcher used the software program E-Prime 2.0 to display the experimental materials on the screen in coordination with the EEG machine to collect behavior data, and the program automatically sent the data into Excel statistical software for analysis afterwards. During the experiment, the participants were randomly assigned to four experimental scenarios (one-two transaction products and with-without transaction labels). Each scenario contained 100 sets of images; each set contained three images: ISI (500 ms), target (1500 ms), ITI (1500 ms). The participants put on an electrode cap for capturing brainwaves (Fig. 2), signed the informed consent form, removed all metal materials, all hair accessories, and phones from their bodies. They were asked if they needed to use the restroom to avoid interruptions during the experiment. When putting on the cap, conductive adhesive was inserted between the electrodes and the scalp to aid signal reception. Then the EEG was used to measure the brainwave signal (Fig. 3). The research participants were instructed to observe the screen that showed the number of products available and the number of products desired. There was a break during the 1-h experiment. A compensation of $10 USD was given to participants who completed the entire process, and those who withdrew from the study midway received $3 USD to compensate for their time. After the experiment, the researcher used Excel to organize the behavior data and used SCAN 4.5 to analyze the brainwave data.

Fig. 2. The electrode cap used during the experiment.

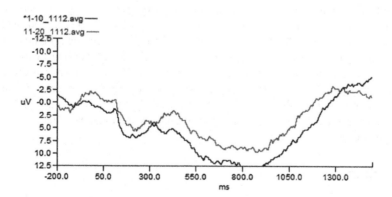

Fig. 3. EEG of Fz in simple scenario and scenario N.

3.3   Scope and Limitations

This research study was designed based on the presumption that consumers (buyers) already had a purchase intention. It investigated how consumers, in the final steps of the purchase process, perceived different transaction mechanisms (product information texts or symbols designed for this study), and how these mechanisms affected the consumers' decision-making efficiency, accuracy, and brainwave activity. The study adopted the EKB model and the final portion of the SOR model – consumers' decision to buy – and excluded the other stages of the decision-making process. We also did not examine if the seller offered any promotions. In addition, the study controlled for similar online shopping contexts regardless of the actual auction platform (e.g., Yahoo, PChome, eBay, Taobao, Ruten.com, and Shopee). We also did not make a distinction of the product categories being considered, such as technology products, used versus new merchandize, etc. All participants had online purchase experiences. All participants were right-handed, and none took medicines chronically.

3.4   Research Data Analysis Methods

Orbital frontal lobe (Fz) and anterior cingular gyrus (FCz) are related to higher level cognition activities, thus they serve as the locations for the electrodes that detect brainwaves related to decisions. These areas are activated when making a purchase decision after considering purchase needs. However, when decision-making involves more uncertainties, the participants will be under pressure. The parietal lobe's activities (CPz) process higher level perceptions, that is, they are sense-related activities, and the orbital frontal lobe cortex (Pz) and parietal lobe cortex (Cz) will be activated. Activities in the occipital lobe cortex and in the core of parietal lobe indicate a focused use of visual receptors.

First, the brainwave data were compiled and calibrated, and eye movements were removed. We then separated the EEG graphs into pre-stimulus (200 ms prior) and post-stimulus (1500 ms after) sections, using the pre-stimulus data as baseline for calibration. Noises (amplitudes that fell outside of the −100 µv–100 µv range) were removed. The brainwave data were filtered via band-pass filtering to keep the frequencies between 0.1 Hz and 40 Hz. Then, we analyzed the data collected using the Fz, FCz, Cz, Pz, and CPz electrodes, from the frontal lobe, to central sulcus, to the parietal lobe cortex which is responsible for visual space and working memory [32]. We also analyzed the average electric potential. Finally, focal analysis was conducted for the coding-phase P3 wave (the largest positive wave occurring between 300 ms and 450 ms after stimulus), the N2 wave (the largest negative wave occurring between 180 ms and 350 ms after stimulus), and the amplitudes and interpeak intervals for these two types of waves [11].

4   Findings

Table 1 organizes the average response time in different scenarios. One can see that regardless of the participants from engineering or liberal arts programs, there was a higher efficiency in terms of response time in simple scenarios (purchasing one item) than complex scenarios (purchasing two product items). However, in simple scenarios, both engineering and liberal arts students had more efficient response time for the scenario with textual transaction feedbacks (Scenario T) than without (Scenario N). While the engineering students had the shortest response time when dealing with the Scenario T, the difference between their response time and that of the liberal arts students was minimal. Similarly, in complex scenarios, both engineering students and liberal arts students had better response efficiency with the Scenario T than the Scenario N, but the engineering students had the longest response time with the Scenario N and the shortest response time with Scenario T.

Table 2 contains the results of analysis for various scenarios in terms of average response accuracy rate. With a 95% confidence interval, regardless of the scenario, the engineering students' average response accuracy rate was greater than that of the liberal arts students, but students in either program had lower accuracy rate in complex scenarios than in simple scenarios. It is worthy to note that the engineering students had the best response time in the Scenario Simple and T; here, they also had the highest accuracy rate.

Table 1. Average response time in various scenarios.

Four scenarios	Engineering students	Liberal arts students
Simple and N	658.5008 ms	667.7183 ms
Simple and T	**642.4801 ms**	643.7842 ms
Complex and N	932.9937 ms	894.1560 ms
Complex and T	883.9645 ms	889.2950 ms

Table 2. Average accuracy rates of various scenarios.

Four scenarios	Engineering students	Liberal arts students
Simple and N	0.991	0.973
Simple and T	0.992	0.979
Complex and N	0.946	0.926
Complex and T	0.956	0.935

Figures 3, 4, 5, and 6 below show the results of the analysis on the EEG data, with the liberal arts students being represented by the red line in the graphs and the engineering students by the blue lines. The response time is the duration between the N2 wave (first negative wave) and the P3 wave (first positive wave). The Fz (cognitive) and Cz (perception) electrode data are presented to demonstrate the results of comparing between the different scenarios; the other electrodes' graphs (FCz, Pz, CPz) are used for supporting evidence. In simple scenarios where one item was to be purchased, we compared the participants' brainwaves in Scenarios N and T. We found that the liberal arts students' Fz (cognitive decision-making) brainwave amplitudes were greater than those of the engineering students; their Cz (perception of stress) amplitudes were also greater that those of the engineering students; the former (cognition) was larger than the latter (perception).

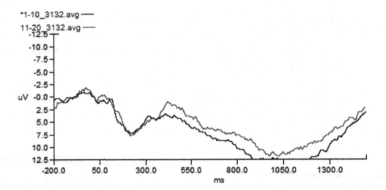

Fig. 4. EEG of Fz in simple scenario and scenario T.

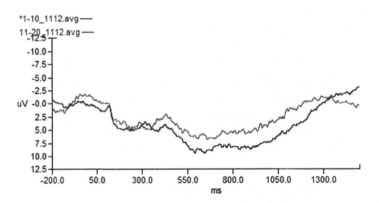

Fig. 5. EEG of Cz in simple scenario and scenario N.

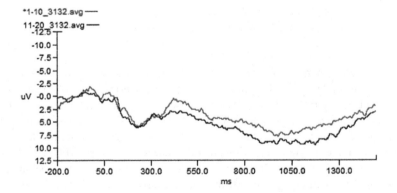

Fig. 6. EEG of Cz in simple scenario and scenario T.

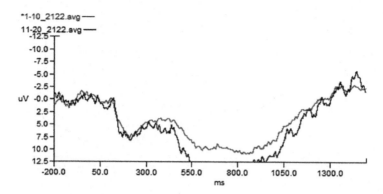

Fig. 7. EEG of Fz in complex scenarios and scenario N.

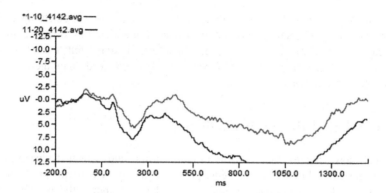

Fig. 8. EEG of Fz in complex scenarios and scenario T.

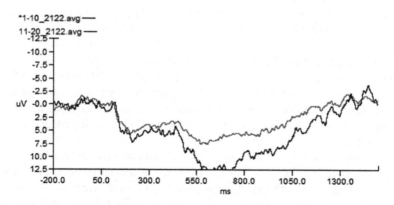

Fig. 9. EEG of Cz in complex scenario and scenario N.

In complex scenarios where two items were to be purchased (Figs. 7, 8, 9, and 10), we compared the participants' brainwaves in Scenarios N and T. We found that the liberal arts students' Fz (cognitive decision-making) as well as their Cz (perception of

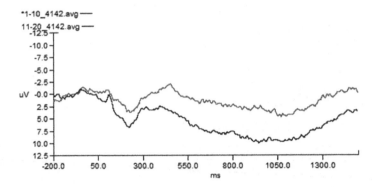

Fig. 10. EEG of Cz in complex scenario and scenario T.

stress) brainwave amplitudes were again greater than those of the engineering students. Furthermore, the latter (perception) had greater amplitude than the former (cognition). This means that in complex scenarios, the liberal arts students' felt text-related stress such that their brainwave amplitudes were more pronounced than their cognitive decision-making brainwave activities.

5   Discussion of Research Findings

Traditional views suggest that students in engineering fields can better process numbers, and students in liberal arts can better process texts. In the experiment, the findings show that regardless of the participants' majors, they had more efficient performance (spending less time) and higher accuracy rates in simple scenarios than complex scenarios. Moreover, the engineering students' average accuracy rates were greater than those of the liberal arts students. However, no visible difference in efficiency was found between the two groups of students. Specifically, both groups had shorter average response time when reacting to Scenario T than Scenario N. It can be inferred that having textual transaction feedbacks allowed participants to perceive more concrete information, thereby increased the response efficiency. Additionally, regarding the accuracy rates, we speculate that in general, engineering students possibly spend more time than liberal arts students in using computers for completing assignments and using labs, thereby demonstrated more composure when presented with either numbers or texts in the brainwave experiment.

The brainwaves change in the two groups of students showed that in simple scenarios, the liberal arts students had greater reaction to text-based stimuli than the engineering students in both cognitive decision-making and perception. Furthermore, the liberal arts students' perceived stress might have affected their decision-making process. However, in complex scenarios, the liberal arts students' brainwave amplitude was again greater than the engineering students'; in other words, in complex scenarios, the liberal arts students' brainwave that perceive text-induced stress had greater amplitude than their brainwaves associated with decision and cognition. This indicates that, the difference between the liberal arts students' and the engineering students' perception of text-based stimuli was greater than the difference between the two groups' cognitive decision-making brain activities. This is possibly because the liberal arts students' perception of stress had more impact than their cognitive decision-making in a complex scenario.

Based on the findings of the study, we recommend the following future research and experimental procedures: (1) Individual differences: Based on the research participants and the research purpose, select appropriate cognitive assignments to better understand the differences among individual participants and treat the differences as variables in the experimental analyses. (2) Increase the sample size to reduce the effect of outliers and to draw more effective conclusions. (3) Attempt to increase the difficulty of the questions to increase the level of differentiation of participant response time, accuracy rate, and of the experiment as a whole. (4) Recruit participants from older age groups. This study involved mostly 20-year-olds, whose mental capacities are at a peak, making it harder to see differences in outcomes.

Finally, this experiment focused on the participants' college major background, inferring that liberal arts students are more text-savvy and engineering students are more number-savvy. This is a traditional Taiwanese view: that students who choose liberal arts or social sciences as their majors generally are weaker in handling numbers and vice versa. However, the participants grew up in an era that emphasizes a well-rounded aptitude development and were all attending top colleges, hence they all had a high level of literacy and numeracy achievement. Simply using their college major categories as a variable was somewhat biased; therefore, it is recommended that future research that aims to differentiate participant characteristics to design the experiment differently to evaluate participant abilities.

Acknowledgment. This research project was supported in part by Prof. Chia-Liang Tsai's EEG Research Lab and the Ministry of Science and Technology (MOST) of Taiwan. I appreciated one of my advisees, Felix Fan Chen, who took a lot of efforts on adopting EEG methods in this research. The students, Li-Chi Hsu, Chia-Chieh Chou, Yu-Yang Chen, Chi-Yu Lin, and Cheng-Han Hsieh, who helped collect data as well as run data analyses to fulfill the graduation requirements were also appreciated.

References

1. Alajmi, N., Kanjo, E., El Mawass, N., Chamberlain, A.: Shopmobia: an emotion-based shop rating system. In: 2013 Humaine Association Conference on Affective Computing and Intelligent Interaction (ACII), pp. 745–750. IEEE, September 2013
2. Aprilianty, F., Purwanegara, M.S., Suprijanto, S.: Using Electroencephalogram (EEG) to understand the effect of price perception on consumer preference. Asian J. Technol. Manag. **9**(1), 58 (2016)
3. Avram, J., Baltes, F.R., Miclea, M., Miu, A.C.: Frontal EEG activation asymmetry reflects cognitive biases in anxiety: evidence from an emotional face Stroop task. Appl. Psychophysiol. Biofeedback **35**(4), 285–292 (2010)
4. Belk, R.W.: An exploratory assessment of situational effects in buyer behavior. J. Mark. Res. **11**, 156–163 (1974)

5. Burnham, T.A., Frels, J.K., Mahajan, V.: Consumer switching costs: a typology, antecedents, and consequences. J. Acad. Mark. Sci. **31**(2), 109–126 (2003)
6. Chen, Y.H., Barnes, S.: Initial trust and online buyer behaviour. Ind. Manag. Data Syst. **107** (1), 21–36 (2007)
7. comScore MMX Multi-Platform. Smartphone usage has doubled in the past three years (2016). https://www.comscore.com/Insights/Blog/Smartphone-Usage-Has-Doubled-in-the-Past-Three-Years
8. Dai, B., Forsythe, S., Kwon, W.S.: The impact of online shopping experience on risk perceptions and online purchase intentions: does product category matter? J. Electron. Commer. Res. **15**(1), 13–24 (2014)
9. Engel, J.F., Kollat, D.T., Blackwell, R.D.: Consumer Behavior. Holt, Rinehart, and Winston, New York (1968)
10. Engel, J.F., Blackwell, R.D., Miniard, P.W.: Consumer Behaviour, 8th edn. Dryden Press, Fort Worth (1995)
11. Farwell, L.A., Donchin, E.: Talking off the top of the head: toward a mental prosthesis utilizing event-related brain potentials. Electroencephalogr. Clin. Neurophysiol. **70**(6), 510–523 (1992)
12. Gavish, B., Tucci, C.L.: Reducing internet auction fraud. Commun. ACM **51**(5), 89–97 (2008)
13. Jones, K., Leonard, L.N.K.: Trust in consumer-to-consumer electronic commerce. J. Inf. Manag. **45**(2), 88–95 (2008)
14. Kalakota, R., Whinston, A.B.: Frontiers of Electronic Commerce. Addison-Wesley, Reading (1996)
15. Katawetaaraks, C., Wang, C.L.: Online shopper behavior: influences of online shopping decision. Asian J. Bus. Res. **1**(2), 66–74 (2011)
16. Kawaf, F., Tagg, S.: Online shopping environments in fashion shopping: an S-O-R based review. Mark. Rev. **12**(2), 161–180 (2012)
17. Kayhan, V.O., McCart, J.A., Bhattacherjee, A.: Cross-bidding in simultaneous online auctions: antecedents and consequences. Inf. Manag. **47**(7), 325–332 (2010)
18. Keeler, L.L.: How to extend your e-mail search. Supervisory Manag. **40**(8), 8–13 (1995)
19. Kim, C., Galliers, R.D., Shin, N., Ryoo, J.H., Kim, J.: Factors influencing internet shopping value and customer repurchase intention. Electron. Commer. Res. Appl. **11**(4), 374–387 (2012)
20. Klimesch, W.: Alpha-band oscillations, attention, and controlled access to stored information. Trends Cogn. Sci. **16**(12), 606–617 (2012)
21. Kotler, P.: Marketing Management: Millennium Edition, 10th edn. Prentice Hall, Upper Saddle River (2000)
22. Lee, N., Broderick, A.J., Chamberlain, L.: What is 'neuromarketing'? A discussion and agenda for future research. Int. J. Psychophysiol. **63**(2), 199–204 (2007)
23. Lieberman, P.: Human Language and Our Reptilian Brain: The Subcortical Bases of Speech, Syntax, and Thought. Harvard University Press, Cambridge (2009)
24. Liebowitz, S.J.: Re-thinking the Network Economy: The True Forces that Drive the Digital Marketplace. Amacom, New York (2002)
25. Madan, C.R.: Neuromarketing: the next step in market research? Eureka **1**(1), 34–42 (2010)
26. Morin, C.: Neuromarketing the new science of consumer behavior. Society **48**(2), 131–135 (2011)
27. Mullen, B., Johnson, C.: The Psychology of Consumer Behavior. Psychology Press, New York (2013)
28. Newman, G.R., Clarke, R.V.: Superhighway Robbery: Preventing e-Commerce Crime. Willan Publishing, Devon (2003)

29. Niedermeyer, E., da Silva, F.L. (eds.): Electroencephalography: Basic Principles, Clinical Applications, and Related Fields. Lippincott Williams & Wilkins, Philadelphia (2005)
30. Nithiya Amirtham, S., Saraladevi, K.: Analysis of attention factors and EEG brain waves of attention deficit and hyperactivity disorder. Int. J. Sci. Res. Publ. 3(3), 1–10 (2013)
31. Peng, C., Kim, Y.G.: Application of the stimuli-organism-response (S-O-R) framework to online shopping behavior. J. Internet Commer. 13(3–4), 159–176 (2014)
32. Scherf, K.S., Sweeney, J.A., Luna, B.: Brain basis of developmental change in visuospatial working memory. J. Cogn. Neurosci. 18(7), 1045–1058 (2006)
33. Teplan, M.: Fundamentals of EEG measurement. Measur. Sci. Rev. 2(2), 1–11 (2002)
34. van Raaij, W.F., Wandwossen, K.: Motivation-need theories and consumer behavior. In: Hung, K. (ed.) NA - Advances in Consumer Research, vol. 05, pp. 590–595. Association for Consumer Research, Ann Abor (1978)
35. Vecchiato, G., Toppi, J., Astolfi, L., Fallani, D.V., Cincotti, F., Mattia, D., Bez, F., Babiloni, F.: Spectral EEG frontal asymmetries correlate with the experienced pleasantness of TV commercial advertisements. Med. Biol. Eng. Comput. 49(5), 579–583 (2011)

A Mobile Augmented Reality Game to Encourage Hydration in the Elderly

Sarah Lehman[1], Jenna Graves[2], Carlene Mcaleer[1], Tania Giovannetti[1], and Chiu C. Tan[1(✉)]

[1] Temple University, Philadelphia, USA
{smlehman,cmcaleer,tgio,cctan}@temple.edu
[2] Centre College, Danville, USA
jenna.graves@centre.edu

Abstract. Dehydration among the elderly is associated with numerous negative health outcomes that result in increase hospitalization, institutionalization, and burden to caregivers. Older adults with dementia are physically capable of fluid intake, but fail to drink due to multiple cognitive deficits, including poor initiation, decreased motivation, and amnesia. We are interested in looking at the use of mobile smartphone technologies to help ensure adequate hydration in older adults with cognitive impairment or dementia. Our approach uses everyday consumer smartphones paired with appropriately designed augmented reality (AR) game that will remind, motivate, guide, and track hydration in older adults with dementia. Our hypothesis is that our system will not only improve hydration, but also that the improvements in hydration will result in improved cognition and mood in community-dwelling older adults with cognitive impairment. This paper describes the design of a feasibility study for an electronic reminder application for older adults.

Keywords: Computer-aided cognition · Reminder system
Older adults · Hydration

1 Introduction

There are currently approximately five million Americans living with dementia today. As the older adult population increases (21.7% of the total United States population is estimated to be older adults by 2040), the number of people suffering from dementia is expected to grow. The exorbitant health care costs associated with this trend are a significant concern for the United States, as well as other countries with similarly changing demographics [3,50]. One of the primary causes of this high cost of care is the difficulty for older adults with dementia in completing everyday tasks. Which leads to increase caregiver burden.

Developments in computing technology have led to the creation of an *Internet of Things* (IoT) [29]. This is an environment where everyday objects such as

© Springer International Publishing AG, part of Springer Nature 2018
S. Yamamoto and H. Mori (Eds.): HIMI 2018, LNCS 10905, pp. 98–107, 2018.
https://doi.org/10.1007/978-3-319-92046-7_9

refrigerators, cabinets, and so on, are embedded with computational intelligence, and networked together. We can harness these new developments to build new systems to help dementia patients live more independent lives. This form of **computer-aided cognition** is especially promising, given the fact that older adults are increasingly computer literate, and will have likely had grown up with these types of technologies.

We approach this computer-aided cognition paradigm by examining a specific everyday activity and how an existing commonplace smartphone device can be used to help older adults. The everyday activity we look at is ensuring adequate hydration. Dehydration is associated with numerous negative health outcomes, including acute confusion, urinary and respiratory infections, medication toxicity, pressure ulcers, muscle weakness, falls [43], constipation, and death [5,7,53]. Dehydration also increase the risk of thrombosis, cardiac arrhythmias, and death. These negative health consequences may result in hospitalization, institutionalization, and increased caregiver burden and/or increased cost of care. In 1996 the estimated medical costs associated with dehydration in the elderly exceeded $1 billion [24]. However, many acute illnesses (e.g., pneumonia) or injuries (i.e., falls) may be caused by dehydration, and hospitalizations for dehydration in older adults have increased by 40% from 1990 to 2000 [56]; therefore, it is likely that the true costs associated with dehydration in older adults is actually much higher in 2016.

In this paper, we will first discuss the challenges faced by older adults with dementia and water intake. Section 3 will discuss our planned study and prior work in this area, and Sect. 4 concludes.

2   Elderly and Hydration

Dehydration in older adults with cognitive impairment is a complex, multidimensional problem. Most older adults are physically capable of fluid intake, but fail to drink due to multiple cognitive deficits, including poor initiation, decreased motivation, and amnesia (i.e., impaired episodic memory) [33]. A review study by [18] suggested that elders in some residential settings who were not fully dependent on caregivers for mobility and activities of daily living and appeared to be capable of obtaining their own fluids were at a somewhat higher risk for dehydration, suggesting that cognitive factors play a meaningful role in dehydration in ambulatory older adults. Based on our review of the extant literature on dehydration and cognitive and functional disability in dementia, we have identified four cognitive targets for improving hydration in the elderly with cognitive impairment.

Poor Initiation. Older adults typically understand that they must drink and know how to attain water, but they may fail to initiate a search for water for at least two reasons. First, the ability to recall an intention (i.e., prospective memory) is a complex cognitive process that involves a network of brain structures, including the prefrontal cortex, parietal cortex, and thalamus [6]. Decline in prospective memory is observed even in healthy older adults and it is impaired in older adults with dementia [15]. In addition to prospective memory impairment, older adults with cognitive impairment may also fail to initiate the search for water, because their experience of thirst, which serves as a salient cue in healthy people, may be blunted due to brain changes [2]. The sensation of thirst is controlled through receptors in the hypothalamus that are sensitive to the fluid balance in the body. Neuropathology affecting the hypothalamus may perturb the natural physiological reminders to drink for older adults with dementia [22].

Lack of Motivation. Because of degraded taste and smell, drinking may not be as pleasurable or rewarding for older adults. Functional neuroimaging studies with healthy participants have demonstrated that the pleasantness and reward associated with drinking when thirsty is linked to activation of the medial orbitofrontal cortex [12,27]. Subtle changes (i.e., thickening) in the orbitofrontal cortex have been reported in healthy older adults and marked dysfunction in this region has been observed in individuals with frontotemporal dementia [46]. Thus, without the natural pleasure and reward that follows drinking, drinking behaviors will not be as strongly reinforced and may decline over time in older adults with dementia.

Unreliable Memory/Poor Monitoring. Older adults may fail to drink because they cannot reliably recall whether or not they have consumed liquid earlier in the day. Episodic memory impairment, a hallmark feature of dementia [32], is associated with degeneration of the hippocampus and surrounding temporal lobe tissue [20]. Episodic memory impairments are associated with problems in a range of everyday tasks (e.g., forgetting to pay bills, forgetting to turn off the stove, etc.) [49]. Even in the early stages of decline episodic memory impairment create difficulties recalling everyday events. Regarding hydration, failure to recall whether or not one has consumed a drink would preclude the accurate monitoring of fluid intake in older adults with dementia and could lead to a reduction in drinking behaviors.

Premature Decay of Intention to Drink. Errors in everyday tasks are sometimes due to the premature decay of an intention [41]. The decay of an intention is more likely if an activity unfolds over an extended period of time or if there is distraction between the time at which the intention was activated and the goal state is achieved. The *doorway effect* or *location updating effect* are terms used to describe the loss of intention that people sometimes experience upon entering a new room [39]. Although this phenomenon has been largely studied in healthy young participants, it is widely noted among clinicians and caregivers that older adults with dementia become quickly distracted and derailed when moving about and searching for objects to achieve even simple everyday goals. For people with dementia, the location of objects (e.g., cup, water) in the home may require more time because of confusion or forgetfulness, thereby increasing the time during which they may become distracted and derailed. Thus, the premature decay of action intentions over time may contribute to the failure to maintain adequate hydration in older adults.

Fig. 1. Illustration of game rewards to motivate participant. Left figure is a prompt to reminder the participant to hydrate. Right figure is the reward after the participant has completed the hydration task.

3   Proposed Study Design

A reminder system is a form of computer-aided cognition that delivers cues to perform daily tasks, such as drink water or take medication. The objective of our proposed study is to understand how older adults, especially older adults with dementia, respond to reminder systems of varying complexity. This is an important preliminary step since it is known that the elderly often face barriers when using technology [25, 34, 38].

A simple reminder application design is a mobile app running on a smartphone that is programmed to set off an alarm at pre-defined intervals to remind the older adult to drink water. However, a simple alarm-type reminder application may not be suitable for older adults with dementia. People with dementia require simple interfaces and more explicit instruction to circumvent the multitude of cognitive challenges that preclude successful task recall and completion. For example, when responding to a timed alert, individuals with dementia may become distracted while moving to a different room to begin the task; they may fail to initiate or complete the task due to apathy, inability to perform the accurate sequence of task steps, and so on.

A complex mobile app can make use of the additional processing capabilities of the smartphone to overcome these difficulties. We have designed a complex mobile app that addresses the four cognitive targets for improving hydration. The hydration reminders are delivered to the user based on the caregiver's predefined schedule, which address the older adults failure to recall the intention to drink due to cognitive decline. The gaming aspect is designed to promote engagement with the system and provide some motivation for older adults, who may not feel inherently thirsty, to hydrate. The intent is to encourage users to care for themselves by placing the emphasis on caring for something else. Figure 1 shows snapshots of the game "reward".

We address the problem of poor monitoring of water consumption, by reminding the user to take a photograph of the cup before and after water consumption as a means of recording water consumption for the user and caregiver. Finally, to address the issue of distraction and premature decay of intention, the application will also help the user to complete the drinking task by prompting and reminding the user to find the drinking vessel (e.g. cup or water bottle) by using the smartphone's camera to scan and locate the appropriate vessel. This recognition is done by having the caregiver affixing QR stickers onto the user's commonly used drinking cups and water bottles. Table 1 summarizes the basic and advanced versions of our reminder application.

Table 1. Summary of different features of the basic and advance versions of the reminder application

Objective	Basic app	Advanced app
Poor initiation	Auditory alarm and text of goal at predefined times	Auditory alarm and text of goal at predefined times. A delayed alarm will be played if participant does not complete hydration task within allocated time
Unreliable memory or poor monitoring	Participant instructed to keep paper logbook	Participant is instructed to take a photo of the cup as part of the game to earn points
Lack of motivation	None	Points are awarded as photos are collected into the photo log; at the end of the day bonus points are acquired for completing both tasks
Premature decay of intention to drink	None	Screen shows customized photo of the objects needed or the scene where the target task must be completed. A written reminder of the task goal is presented along with the photo. The camera can be used to help identify cup affixed with QR stickers

We plan to conduct testing of the basic and advanced versions of the reminder apps in a feasibility study modeled after the Memory for Intentions Screening Test (MIST). The study includes two conditions (Unprompted and Prompted), each lasting one hour.

In the Unprompted Condition participants are instructed to walk to a nearby kitchen to retrieve a very small glass of water every 15 min, such that they must go the kitchen on four different occasions by the end of the Unprompted Condition. Participants will be shown the laboratory kitchen at the start of the Unprompted Condition, and the condition will not begin until the participant is able to repeat the task directions. The experimenter will not provide reminders or cues about the task objective during the Unprompted Condition. All participants will engage in standardized secondary tasks with the examiner during the hour-long, Unprompted Condition.

In the Prompted Condition, participants will undergo a brief tutorial on how to use the reminder application. Participants will be told that they will be prompted by the phone to retrieve a glass of water in the laboratory kitchen on four occasions within one hour. As in the Unprompted Condition, participants will be engaged in standardized secondary tasks with the examiner; however unlike the Unprompted Condition, the participant will be carrying the smartphone in a belt case or lanyard, depending on the participants preference, to receive the system prompts.

All participants will complete both the Prompted and Unprompted Conditions in a two-hour testing session. To control for order effects, the conditions will be counter-balanced across participants. The following dependent variables will be collected separately for each condition: (1) number of drinks consumed (max = 4); (2) time at which drink was consumed = (i.e., minutes before or after each 15-min mark at which a drink was consumed). The participants will be asked to rate their overall experience with the system, using a simplified and modified usability questionnaire to determine the likability of the reminder application.

While participants are completing the tasks, their caregivers will work with a second research assistant who will demonstrate the reminder application using a standardized script. Following the demonstration, caregivers will be asked to rate their impression of the system usability on both a standardized, closed-ended usability questionnaire, and an open-ended interview. Additional information regarding the participants current drinking schedule and preferences also will be obtained during a structured, open-ended interview.

4   Related Work

Simple prompts and environmental adaptations have been successful for increasing fluid intake in older adults in residential care facilities [51]. For example, significant benefits were reported in a sample of 51 nursing home residents following a five-week intervention that incorporated a hydration plan, caregiver education, environmental changes (e.g., adding color to cups and a colorful beverage cart), and increasing beverage options [45]. The goal of the intervention was for participants to drink two additional 8-ounce beverages per day (mid-morning and mid-afternoon). When compared to a two-week baseline prior to the intervention, participants showed a significant increase in total body water as measured by skin impedance testing, reduction in falls, increase in the number of bowel movements, decrease in laxative use, and a decline in the care cost of associated negative outcomes.

Augmented reality technology has been incorporated into many different areas [4], including education [36,44], retail [11], environmental monitoring [54], navigation [19,35,42], security [30], safety [40], and so on. Within the healthcare area, augmented reality applications include systems to promote healthy diet [1], exercise [28], supplement traditional nursing programs [13], and aid in childhood cognitive and motor development [9,23]. Our system falls under this category of AR applications.

Applying virtual or augmented reality specifically for adults with cognitive decline is also an emerging field of research [14,16,31,47,55]. These include training-type applications [17,57], systems to help driving automobiles [26,48], systems that help with physical rehabilitation [21], and so on. These prior work demonstrate the utility of AR systems to help older adults. We take the approach of make full use of commercially available smartphones to make WaterWatcher feasible for wider deployment.

Researchers have also considered usability issues of mobile devices and the elderly [34,52] to identify specific design concepts that would appeal more to older adults [37,58]. Others have also looked at gamification to motivate older adults to use mobile applications [8,10]. While our preliminary design draws upon this prior research, none of these usability studies have targeted older adults with cognitive decline. We will expand the knowledge of this group of little studied users as part of this project.

5   Conclusion

We propose a novel complex smartphone app to promote hydration in older adults with cognitive impairment. The app has many advantages over existing reminder apps. The entire app is designed to target the cognitive impairments that prevent hydration, and incorporates gaming to increase motivation. Our system can be used with commercially-available, and relatively inexpensive, smartphones. Our testing plan includes a feasibility study that will compare the complex app to a more basic app to determine whether the additional features of the complex app are needed to promote hydration in cognitively impaired older adults. Data obtained from this feasibility study will be used to inform a larger, randomized controlled trial in participants' homes to test the effectiveness of the app.

The reminder application described in this paper is designed to increase hydration in older adults, which is expected to have positive effects on cognition and general health. However, the system may be modified to promote a wide range of healthy behaviors and everyday activities with diverse clinical populations (e.g., brain injury, stroke, Schizophrenia, etc.).

References

1. Ahn, J., Williamson, J., Gartrell, M., Han, R., Lv, Q., Mishra, S.: Supporting healthy grocery shopping via mobile augmented reality. ACM Trans. Multimed. Comput. Commun. Appl. 12(1s), 16:1–16:24 (2015)
2. Ainslie, P.N., Campbell, I.T., Frayn, K.N., Humphreys, S.M., MacLaren, D.P.M., Reilly, T., Westerterp, K.R.: Energy balance, metabolism, hydration, and performance during strenuous hill walking: the effect of age. J. Appl. Physiol. 93(2), 714–723 (2002)
3. Anderson, G.F., Hussey, P.S.: Population aging: a comparison among industrialized countries. Health Aff. 19(3), 191–203 (2000)
4. Azuma, R.T.: A survey of augmented reality. Presence: Teleoper. Virtual Environ. 6(4), 355–385 (1997)

5. Bennett, J.A., Thomas, V., Riegel, B.: Unrecognized chronic dehydration in older adults: examining prevalence rate and risk factors. J. Gerontol. Nurs. **30**(11), 22–28 (2004)
6. Burgess, P.W., Quayle, A., Frith, C.D.: Brain regions involved in prospective memory as determined by positron emission tomography. Neuropsychologia **39**(6), 545–555 (2001)
7. Chassagne, P., Druesne, L., Capet, C., Ménard, J.F., Bercoff, E.: Clinical presentation of hypernatremia in elderly patients: a case control study. J. Am. Geriatr. Soc. **54**(8), 1225–1230 (2006)
8. Chou, M.-C., Liu, C.-H.: Mobile instant messengers and middle-aged and elderly adults in Taiwan: uses and gratifications. Int. J. Hum.-Comput. Interact. **32**(11), 835–846 (2016)
9. Correa, A.G.D., De Assis, G.A., Do Nascimento, M., Ficheman, I., De Deus Lopes, R.: GenVirtual: an augmented reality musical game for cognitive and motor rehabilitation. In: Virtual Rehabilitation, pp. 1–6 (2007)
10. Cota, T.T., Ishitani, L., Vieira Jr., N.: Mobile game design for the elderly: a study with focus on the motivation to play. Comput. Hum. Behav. **51**(Part A), 96–105 (2015)
11. Dacko, S.G.: Enabling smart retail settings via mobile augmented reality shopping apps. Technol. Forecast. Soc. Change (2016)
12. De Araujo, I.E.T., Kringelbach, M.L., Rolls, E.T., McGlone, F.: Human cortical responses to water in the mouth, and the effects of thirst. J. Neurophysiol. **90**(3), 1865–1876 (2003)
13. Ferguson, C., Davidson, P.M., Scott, P.J., Jackson, D., Hickman, L.D.: Augmented reality, virtual reality and gaming: an integral part of nursing. Contemp. Nurse **51**(1), 1–4 (2015). PMID: 26678947
14. Foloppe, D.A., Richard, P., Yamaguchi, T., Etcharry-Bouyx, F., Allain, P.: The potential of virtual reality-based training to enhance the functional autonomy of Alzheimer's disease patients in cooking activities: a single case study. Neuropsychol. Rehabil. 1–25 (2015). PMID: 26480838
15. Gao, J.L., Cheung, R.T.F., Chan, Y.S., Chu, L.W., Lee, T.M.C.: Increased prospective memory interference in normal and pathological aging: different roles of motor and verbal processing speed. Aging Neuropsychol. Cognit. **20**(1), 80–100 (2013)
16. Garca-Betances, R.: A succinct overview of virtual reality technology use in Alzheimer's disease. Front. Aging Neurosci. **7**, 80 (2015)
17. Garca-Betances, R.I., Jimnez-Mixco, V., Arredondo, M.T., Cabrera-Umpirrez, M.F.: Using virtual reality for cognitive training of the elderly. Am. J. Alzheimer's Dis. Other Dement. **30**(1), 49–54 (2015)
18. Hodgkinson, B., Evans, D., Wood, J.: Maintaining oral hydration in older adults: a systematic review. Int. J. Nurs. Pract. **9**(3), S19–S28 (2003)
19. Huang, T.-C., Shu, Y., Yeh, T.-C., Zeng, P.-Y.: Get lost in the library? An innovative application of augmented reality and indoor positioning technologies. Electron. Libr. **34**(1), 99–115 (2016)
20. Hyman, B.T., Van Hoesen, G.W., Damasio, A.R., Barnes, C.L.: Alzheimer's disease: cell-specific pathology isolates the hippocampal formation. Science **225**(4667), 1168–1170 (1984)
21. Im, D.J., Ku, J., Kim, Y.J., Cho, S., Cho, Y.K., Lim, T., Lee, H.S., Kim, H.J., Kang, Y.J.: Utility of a three-dimensional interactive augmented reality program for balance and mobility rehabilitation in the elderly: a feasibility study. Ann. Rehabil. Med. **39**(3), 6 (2015)

22. Ishii, T.: Distribution of Alzheimer's neurofibrillary changes in the brain stem and hypothalamus of senile dementia. Acta Neuropathol. **6**(2), 181–187 (1966)
23. Juan, M.-C., Mendez-Lopez, M., Perez-Hernandez, E., Albiol-Perez, S.: Augmented reality for the assessment of children's spatial memory in real settings. PLoS ONE **9**(12), 1–26 (2014)
24. Kayser-Jones, J., Schell, E.S., Porter, C., Barbaccia, J.C., Shaw, H.: Factors contributing to dehydration in nursing homes: inadequate staffing and lack of professional supervision. J. Am. Geriatr. Soc. **47**(10), 1187–1194 (1999)
25. Kenigsberg, P.-A., Aquino, J.-P., Bérard, A., Brémond, F., Charras, K., Dening, T., Droës, R.-M., Gzil, F., Hicks, B., Innes, A., et al.: Assistive technologies to address capabilities of people with dementia: from research to practice. Dementia (2017). https://doi.org/10.1177/1471301217714093
26. Kim, S., Dey, A.K.: Simulated augmented reality windshield display as a cognitive mapping aid for elder driver navigation. In: Proceedings of the SIGCHI Conference on Human Factors in Computing Systems, CHI 2009, pp. 133–142. ACM, New York (2009)
27. Kringelbach, M.L.: The human orbitofrontal cortex: linking reward to hedonic experience. Nat. Rev. Neurosci. **6**(9), 691–702 (2005)
28. Laine, T.H., Suk, H.J.: Designing mobile augmented reality exergames. Games Cult. **11**(5), 548–580 (2016)
29. Li, X., Lu, R., Liang, X., Shen, X., Chen, J., Lin, X.: Smart community: an internet of things application. IEEE Commun. Mag. **49**(11) (2011)
30. Lukosch, S., Lukosch, H., Datcu, D., Cidota, M.: Providing information on the spot: using augmented reality for situational awareness in the security domain. Comput. Support. Coop. Work (CSCW) **24**(6), 613–664 (2015)
31. Manera, V., Chapoulie, E., Bourgeois, J., Guerchouche, R., David, R., Ondrej, J., Drettakis, G., Robert, P.: A feasibility study with image-based rendered virtual reality in patients with mild cognitive impairment and dementia. PLoS ONE **11**(3), 1–14 (2016)
32. McKhann, G., Drachman, D., Folstein, M., Katzman, R., Price, D., Stadlan, E.M.: Clinical diagnosis of Alzheimer's disease report of the NINCDS-ADRDA work group* under the auspices of department of health and human services task force on Alzheimer's disease. Neurology **34**(7), 939 (1984)
33. Mentes, J.C.: A typology of oral hydration: problems exhibitied by frail nursing home residents. J. Gerontol. Nurs. **32**(1), 13–19 (2006)
34. Nikou, S.: Mobile technology and forgotten consumers: the young-elderly. Int. J. Consum. Stud. **39**(4), 294–304 (2015)
35. de Oliveira, L.C., Soares, A.B., Cardoso, A., de Oliveira Andrade, A., Lamounier Jr., E.A.: Mobile augmented reality enhances indoor navigation for wheelchair users. Res. Biomed. Eng. **32**, 111–122 (2016)
36. Petrucco, C., Agostini, D.: Teaching our cultural heritage using mobile augmented reality. J. E-Learn. Knowl. Soc. **12**(3), 115–128 (2016)
37. Pijukkana, K.: Graphical design and functional perception on technology-driven products: case study on mobile usage of the elderly. Proc. Soc. Behav. Sci. **42**, 264–270 (2012)
38. Pijukkana, K., Sahachaisaeree, N.: Graphical design and functional perception on technology-driven products: case study on mobile usage of the elderly. Proc.-Soc. Behav. Sci. **42**, 264–270 (2012)
39. Radvansky, G.A., Tamplin, A.K., Krawietz, S.A.: Walking through doorways causes forgetting: environmental integration. Psychonom. Bull. Rev. **17**(6), 900–904 (2010)

40. Rameau, F., Ha, H., Joo, K., Choi, J., Park, K., Kweon, I.S.: A real-time augmented reality system to see-through cars. IEEE Trans. Vis. Comput. Graph. **22**, 2395–2404 (2016)
41. Reason, J.: Human Error. Cambridge University Press, Cambridge (1990)
42. Rehman, U., Cao, S.: Augmented reality-based indoor navigation using Google glass as a wearable head-mounted display. In: IEEE International Conference on Systems, Man, and Cybernetics, pp. 1452–1457 (2016)
43. Renneboog, B., Musch, W., Vandemergel, X., Manto, M.U., Decaux, G.: Mild chronic hyponatremia is associated with falls, unsteadiness, and attention deficits. Am. J. Med. **119**(1), 71-e1 (2006)
44. Joan, D.R.R.: Enhancing education through mobile augmented reality. J. Educ. Technol. **11**(4), 8–14 (2015)
45. Robinson, S.B., Rosher, R.B.: Can a beverage cart help improve hydration? Geriatr. Nurs. **23**(4), 208–211 (2002)
46. Salat, D.H., Buckner, R.L., Snyder, A.Z., Greve, D.N., Desikan, R.S.R., Busa, E., Morris, J.C., Dale, A.M., Fischl, B.: Thinning of the cerebral cortex in aging. Cereb. Cortex **14**(7), 721–730 (2004)
47. Saracchini, R., Catalina, C., Bordoni, L.: A mobile augmented reality assistive technology for the elderly. Comunicar **23**(45), 65–73 (2015). Copyright - Copyright Grupo Comunicar 2015; Document feature -; Photographs; Diagrams; Tables; Graphs. Accessed 11 Jul 2015
48. Schall, M.C., Rusch, M.L., Lee, J.D., Dawson, J.D., Thomas, G., Aksan, N., Rizzo, M.: Augmented reality cues and elderly driver hazard perception. Hum. Factors: J. Hum. Factors Ergon. Soc. **55**(3), 643–658 (2013)
49. Schmitter-Edgecombe, M., Woo, E., Greeley, D.R.: Characterizing multiple memory deficits and their relation to everyday functioning in individuals with mild cognitive impairment. Neuropsychology **23**(2), 168 (2009)
50. Seshamani, M., Gray, A.: The impact of ageing on expenditures in the national health service. Age Ageing **31**(4), 287–294 (2002)
51. Simmons, S.F., Alessi, C., Schnelle, J.F.: An intervention to increase fluid intake in nursing home residents: prompting and preference compliance. J. Am. Geriatr. Soc. **49**(7), 926–933 (2001)
52. Stamato, C., de Moraes, A.: Mobile phones and elderly people: a noisy communication. Work **41**, 320–327 (2012)
53. Thomas, D.R., Tariq, S.H., Makhdomm, S., Haddad, R., Moinuddin, A.: Physician misdiagnosis of dehydration in older adults. J. Am. Med. Dir. Assoc. **4**(5), 251–254 (2003)
54. Veas, E., Grasset, R., Ferencik, I., Grunewald, T., Schmalstieg, D.: Mobile augmented reality for environmental monitoring. Pers. Ubiquit. Comput. **17**(7), 1515–1531 (2013)
55. White, P.J.F., Moussavi, Z.: Neurocognitive treatment for a patient with Alzheimer's disease using a virtual reality navigational environment. J. Exp. Neurosci. **10**, 129–135 (2016)
56. Xiao, H., Barber, J., Campbell, E.S., et al.: Economic burden of dehydration among hospitalized elderly patients. Am. J. Health Syst. Pharm. **61**(23), 2534–2540 (2004)
57. Yamaguchi, T., Foloppe, D.A., Richard, P., Richard, E., Allain, P.: A dual-modal virtual reality kitchen for (re)learning of everyday cooking activities in Alzheimer's disease. Presence: Teleoper. Virtual Environ. **21**(1), 43–57 (2012)
58. Yusof, M.F.M., Romli, N., Yusof, M.F.M.: Design for elderly friendly: mobile phone application and design that suitable for elderly. Int. J. Comput. Appl. **95**(3), 28–31 (2014). Full text available

MyStudentScope: A Web Portal for Parental Management of Their Children's Educational Information

Theresa Matthews[1(✉)], Jinjuan Heidi Feng[1], Ying Zheng[2], and Zhijiang Chen[1]

[1] Towson University, Towson, MD 21252, USA
tscott2@students.towson.edu
[2] Frostburg State University, Frostburg, MD 21532, USA

Abstract. Research shows that parents and caregivers have challenges optimizing their use of information they receive regarding their children's education. The challenge hinders their ability to effectively participate in their children's educational development. Existing electronic student information systems used by schools are designed from the perspective of the educator or student, not the parent. A prior study aimed at determining areas where challenges were perceived supported the need for a tool tailored to meet the parents' needs. In order to understand current methods of information management using technology, we review literature on Personal Information Management (PIM) and Knowledge Management. Experts in education were interviewed for their opinions regarding the types of information parents should have accessible. The literature review, study results, and recommendations from experts in education were used to inform the design of MyStudentScope (MSS), a web portal for parental management of information regarding their children's education. The portal has four primary functions: monitoring, retrieving, communication and decision making. Parents' use of MSS to archive and retrieve information regarding their children's education is expected to improve parental monitoring of the progress of their children's education over paper-based methods. Studies are underway to evaluate the effectiveness of MSS in improving parental management of information regarding their children's education.

Keywords: Parents · Education · Personal Information Management
PIM · Information organization · Web portal

1 Introduction

Parents and caregivers are inundated with information regarding their children's education received verbally, on paper and digitally via a variety of methods. Parents must be able to optimize their use of the information so that they are able to effectively participate in their children's educational development. In a prior study a survey was conducted to identify areas where challenges are perceived and/or realized for parents managing information regarding their children's education (Matthews and Feng 2017). The results supported the need for a tool to aid parents because existing electronic

© Springer International Publishing AG, part of Springer Nature 2018
S. Yamamoto and H. Mori (Eds.): HIMI 2018, LNCS 10905, pp. 108–121, 2018.
https://doi.org/10.1007/978-3-319-92046-7_10

student information system used by schools are designed from the perspective of the educator or student, not the parent. Subject matter experts in education were also interviewed and literature was reviewed for recommendations regarding the types of information parents should manage regarding their children's education, warning signs for concern and ways to engage educators for maximum benefit for the child's success (Crabtree 1998; Wright and Wright 2008).

Prior research indicated that parents used paper-based methods more often than technology-based methods to archive and retrieve information regarding their children's education; even when an electronic student information system from their child's school is available for their use (Matthews and Feng 2017). Our objective is to bring together the capabilities parents need in one tool with the idea that centralizing the information and key functions will improve their awareness of their children's academic progress as well as make it easier for them to identify challenges that should be addressed.

In this paper we introduce a conceptual framework for information management in the context of children's education, Parental Information Management Model, describe the design and implementation of a web portal based on this framework and influenced by literature reviews, prior study results, and recommendations from experts, and reports of a pilot study during which four users completed tasks using the system and using paper-based methods that simulate situations parents/caregivers may encounter related to their children's education and extracurricular activities.

2   Literature Review

A literature review was conducted in the areas of Personal Information Management (PIM), Knowledge Management (KM) and existing information management systems and models.

2.1   Personal Information Management

Although PIM is generally concerned with an individual's information, the management of information regarding one's child shares the same basic requirements. As described by Buttfield-Addison et al., PIM is the study of the process of information capture, organization and re-finding of information dealt with in daily life (Buttfield-Addison et al. 2012). PIM literature reviewed is documented in a previous paper (Matthews and Feng 2017). Relevant points from the PIM literature review include:

- the concept of organizing information into a data collection, digital library or some other means
- understanding the how users make decisions regarding what they keep
- information retrieval and re-finding methods
- challenges individuals face when attempting to organize information
- challenges individuals face when finding information they had previously organized.

The Sense-Making Approach is the process of seeking out and making sense of information and is based on Brenda Dervin's work in communications. In Spurgin's review of the Sense-Making Approach she offers that in PIM research, "a Sense-Making Approach could help us begin to understand the common types of gaps that people experience that lead them to [attempt to organize their information] and the types of gaps they experience [in the process]" (Spurgin 2006). Per Dervin, sense-making is based on the assumption that (1) it is possible to design and implement systems that are responsive to human needs, (2) it is possible for humans to adapt their behavior to use the systems and (3) achieving these goals requires communication-based methodical approaches (Agarwal 2012).

2.2 Knowledge Management

The term KM is generally applied to information related to an organization or company. However, KM theories may also be applicable to parental information management. As cited by Smith, "knowledge management includes four areas: managing intellectual capital [...]; gathering, organizing and sharing the company's information and knowledge assets; creating work environments to share and transfer knowledge among workers; and leveraging knowledge from all stakeholders to build innovative company strategies" (Smith 2001).

One KM method that is relevant to the topic of parental information management is knowledge codification. The purpose of knowledge codification is capture knowledge, especially tacit knowledge, in a format that is accessible by those who need it. Tacit knowledge is knowledge that is difficult to write down, visualize or transfer from one person to another. According to Smith, nearly two thirds of company information is tacit knowledge that comes from face-to-face interactions (Smith 2001). There must also be a "means by which those in need of the knowledge can discover its existence as reposed, articulated knowledge" (Bakerville and Dulipovici 2006).

The data–information–knowledge–wisdom (DIKW) hierarchy is a method for describing how we move from data to information to knowledge to wisdom and is the basis for much KM research and progress. However, the DIKW hierarchy has been criticized for missing a very important step in data development; progressing actionable data. Mannion presents a representation of the DIKW hierarchy that includes a fifth tier; Decisions. He describes the first three tiers, Data, Information and Knowledge as progressions of data development in the past that reveal patterns and relationships. He describes the Wisdom and Decisions tiers as the future. The future tiers reveal what direction to take (Mannion 2015).

2.3 Information Management Systems

Many school systems use an electronic student information system to allow authorized caregivers to log in from any computer with an Internet connection and view their child's student information. Most of these systems, like SchoolMAX and ParentCONNECTxp, allow the user to view the child's student information, including current attendance records, assignment scores and course grades for all years the student was enrolled in the particular school district. Teachers are usually required to update the information on the sites weekly, at a minimum. The systems are designed such that the school and/or educator is able to provide information regarding the student's academics and attendance to parents on a frequent basis. The systems were not designed to optimize parental use of the provided information. Therefore, parents may not fully benefit from the wealth of information at their disposal because they may be overwhelmed by the volume or unable to view it at the frequency with which it is updated.

Although electronic student information systems provide parents with 24/7 access to their child's grades and attendance records, it lacks other information that according to education experts is critical for parents to truly have a handle on their child's academic progress. We interviewed a student advocate and educator who provides support to families, schools, students of all ages and attorneys in ensuring that student educational needs are met. She recommended that parents document teacher phone calls, keep records of requests for appointments by the parent or teacher, keep copies of school work/assignments especially those with which that parent or teacher has expressed concern, keep copies of any official reports that have been signed and dated and keep children's pre-school portfolios. Her recommendations are in line with recommendations published by other student advocates (Crabtree 1998; Patrikakou 2008; Wright and Wright 2008). In addition to report cards and progress reports, an administrator of a public school system recommends that parents retain major assessment results, benchmarks, suggestions for improvements from teachers and recommendations for screenings from teachers.

A prior literature review revealed one application that was specifically designed to assist parents in organizing and gathering information related to their child's education. The tool's purpose was specific to helping parents participate in their child's Individualized Education Program (IEP) process by documenting and organizing relevant information between IEP meetings in a way that it can be easily accessed during the meeting or reported to teachers and administrators. The tool allows parents to type notes, record audio and takes pictures (Excent 2014; Swanson 2012). Recent inquiries indicate that the application is no longer available.

No other information management app or website designed to assist parents in managing information of their children were identified. Therefore information management systems for other subject areas were also investigated. Medical information management tools were relevant because the needs and challenges of an individual managing his/her healthcare are similar to the needs of a parent managing his/her child's educational development (Fig. 1).

Based on research regarding the use of technology and the self-management of chronic disease, Perry Gee et al. introduced a revised model, eHealth Enhanced Chronic Care Model (eCCM) to show how eHealth tools can be used to improve patient management of their chronic illnesses. Gee used the Theory Derivation process to create eCCM. It is a process used in nursing by which a parent theory or model is chosen to guide the development of a new model. The Chronic Care Model (CCM) is a framework of an all-inclusive approach to caring for chronically ill patients that supports improved clinical outcomes. eHealth is loosely defined as the promotion of positive health outcomes through the use of technology. A critical part of the eCCM is the continued communication between the patient and the provider as depicted in Fig. 2.

Fig. 1. eCCM depiction (Gee et al. 2015)

Parent Information Management Model (PIMM)

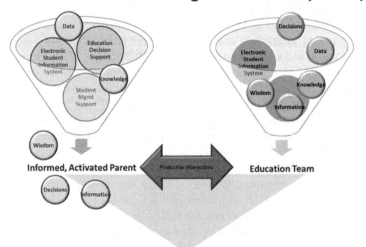

Fig. 2. The parental information management model

One of the two major contributions of this paper to parents' management of information regarding their children's education is a parental information management model based on the eCCM modal. The other contribution is a web portal designed to mitigate challenges identified in prior research and literature (Matthews and Feng 2017; Roshan et al. 2014). Specifically, MyStudentScope (MSS) addresses challenges of (1) monitoring student's academic progress when dealing with large volumes of disparate data; (2) recalling and retrieving relevant data or knowledge over long periods of time when needed; (3) communicating with other partners in education; and (4) making decisions regarding the student's academic career. We introduce the Parental Information Management Model (PIMM) to show how electronic student information system can be used by parents to manage information regarding their children's education. We introduce MSS to demonstrate how new electronic student information system functionality designed from the parents' perspective can be used with PIMM to drive mores positive outcomes with respect to children's academic progress.

3   The Parental Information Management Model

Prior research focused on models to assist patients with chronic illnesses in the management of their healthcare. Based on our literature review and prior research, we conclude that it is also important to focus on models to assist parents in the management of information regarding their children's education. Similar to how CCM was used to guide the development of eCCM, the eCCM model has been chosen to guide the development of a new model that is applicable to parental information management; the Parent Information Management Model (PIMM). Equivalent functional components related to parental management of information regarding their children's education is defined for each functional part of the eCCM (see Table 1).

Table 1. Mapping of eCCM components to PIMM functional components

eCCM	PIMM
Clinical decision support	Education decision support
Delivery system	Electronic student information system
Self-management support	Student-management support
eHealth education	Parental information system education

The goal of the PIMM is to drive parent activation with respect to their involvement in their children's education. Adapting the description of patient activation, parent activation as it relates to the PIMM is the level of skills, knowledge, and confidence that a parent has in managing and influencing his/her child's educational progress.

The CIS element of eCCM provides information to providers to ensure that they are able to provide the right care to patients. An equivalent capability in the area of education is the administrator-facing portion of the existing student information systems. Because that functionality does not directly support the parent, it falls outside of the scope of PIMM and is not depicted in Table 1. A depiction of the PIMM is presented in Fig. 3. Components that support the education team are grayed out because their use by educators is important, but they are not part of the PIMM.

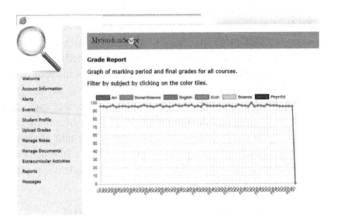

Fig. 3. Sample MSS grade report

Education decision support includes reports, graphs, charts and reminders to assist parents in making decisions regarding the student's education. The reports and graphs are generated based on data regarding the students' grades. Decision support mitigates issues associated with number confusion that parents may encounter when trying to keep track of grades in tabular form.

The electronic student information system is the system via which most school districts provide grade, attendance and other information relevant to a student's academic records to parents. Additional data regarding the student's academic progress is provided in graded assignments that are sent home and other communications with the parent.

Student-management support consists of technologies that enable the parent to prepare for parent-teacher conferences and education program meetings, track grade reports, participate in their child's learning experience and provide input for courses of action to address concerns with their child's academic progress. This information is provided to parents in various forms which is why a system to assist parents is necessary.

There is at a minimum a user guide that accompanies most student information systems. However, most parents do not have time to read a manual to understand how to use systems and tools, therefore this area may continue to be a challenge.

4   MyStudentScope

MSS is a web portal for parental management of information regarding their children's education. MSS functionality falls within the education decision support and student-management support components of the PIMM. It is designed to work in tandem with the current methods and systems via which parents receive information regarding their children's education like existing electronic student information systems. It is designed for personal use and can be accessed via any web browser. The MSS user interface was designed to be simple for parents to navigate with very little training or instruction. After creating an account and logging in, the parent is presented with a welcome screen that provides links to all sub-pages, but highlights the most frequently visited pages by presenting "buttons" for them in the center of the screen.

4.1 Functions

The portal has key pages to support four primary functions: monitoring, retrieving, communication and decision making. The mapping of the web portal functions to the applicable pages is presented in Table 2.

Table 2. MSS function to page mapping

Function	Web portal page(s)
Monitoring	Alerts, Events, Student Profile, Upload Grades, Manage Notes, Manage Documents, Extracurricular Activities, Reports
Retrieving	Manage Documents, Manage Notes
Communication	Messages, Alerts
Decision making	Alerts, Events, Student Profile, Upload Grades, Manage Notes, Manage Documents, Extracurricular Activities, Reports

Monitoring. The monitoring function incorporates viewing information uploaded by the parents to include grade reports like progress reports or report cards, individual assignment grades, information regarding schedules and extracurricular activities as well as work samples and personal notes. The monitoring function also involves the use of an alert capability by which parents can define criteria or events of which they want to be notified. Alerts can be related to grades or scheduled events.

Retrieving. The retrieving function enables parents to retrieve previously saved information regarding their child's education when needed via search mechanisms.

Communication. The communication function allows parents to correspond with educators, coaches, and other providers. It provides a means for parents to attach documents saved in the tool and reports generated by the tool to better facilitate the conversation.

Decision Making. The decision making function enables parents to observe trends and anomalies in educational development by viewing graphs and/or reports of the educational information stored in MSS. Parents are also able to search for and view grades for a specific subject over a specified timeframe. For example, parents can view their child's math grades for the past three years. The view will present the actual grades and/or a graphical representation of them so the parent can identify trends and anomalies.

4.2 User Interface

Upon account creation and initial login, the parent is prompted to add a student for which they would like to manage information. There is no limit to the number of students for which a parent may manage information using MSS. If a parent has added more than one student to his account, he/she must select the student for which they would like to view or manage information.

In order for MSS to generate reports like the graph shown in Fig. 4, the parent must enter grade information. The parent has the option of uploading course or assignment grades manually from a screen like Fig. 5. Because some electronic student information systems allow parents to export grade reports to a file, MSS also allows parents to upload course or assignment grades from a file.

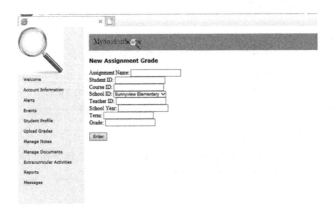

Fig. 4. MSS assignment grade entry form

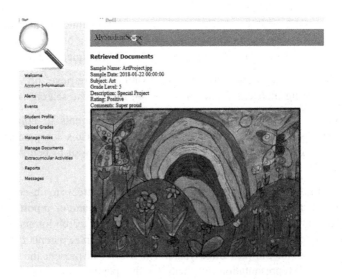

Fig. 5. MSS document retrieval example

On the Manage Documents page, parents are able to upload documents, including but not limited to images, samples of their children's school work or information regarding the student's extracurricular activities. The following metadata may be entered and viewed for each sample: date, subject, grade, description, comments and indication of whether this is a positive or negative entry. Parents may also retrieve previously uploaded documents (see Fig. 6). This will enable parents to review samples of their children's work to observe progress which is particular important from Kindergarten through second grade. Parents are able to retrieve documents by reviewing all uploaded documents or performing keyword searches on the document fields.

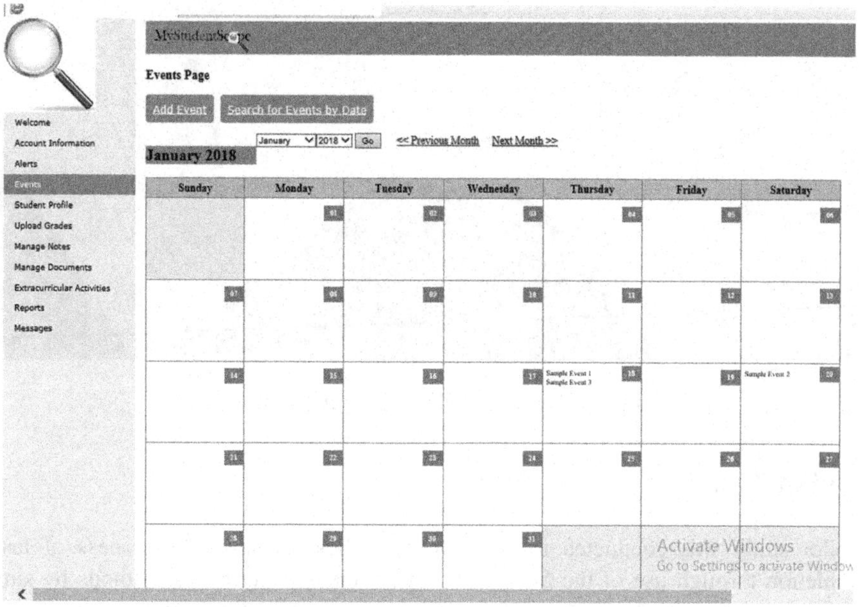

Fig. 6. MSS events page

The Notes Management page is where parents can enter notes or observations regarding an event or activity related to their child's education. The note may be positive, for example to record the receipt of an award or special recognition that is not reflected in the grade reports. Or the note may capture a negative event such as an encounter with another student. The following metadata may be entered and viewed for each entry: date, involved parties, subject area/topic and indication of whether this is a positive, negative or neutral entry. Parents are able to retrieve notes by reviewing all notes or performing keyword searches on the note fields.

By default, parents are presented with a calendar view on the Events page, which displays the current month and events scheduled for the current month (see Fig. 7). Parents can enter school event information, assignment due dates, extracurricular activity dates, etc. in the calendar. Parents are able to scroll to different months to see upcoming or past events. This page allows parents to schedule reminders for upcoming events and detect potential schedule conflicts.

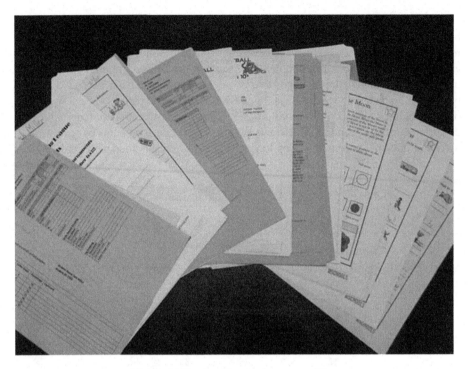

Fig. 7. Folder for pilot test paper condition

5   Pilot Test

A pilot study was conducted to compare the efficiency and effectiveness of task completion through use of the MSS web portal versus paper-based methods by simulating situations parents/caregivers may encounter related to their children's education and extracurricular activities. The goal of the pilot study was to collect preliminary feedback on MSS functionality and to improve the clarity of questions and tasks used to exercise MSS to be used in a later formal user study.

5.1   Experimental Setup

The typical user of MSS would be anyone of various capabilities who is the parent or guardian of or who is responsible for a school-aged child in grades Kindergarten through 12th grade. For the pilot study, target participants were comprised of four parents of students in grades K − 12 who may or may not currently use a school-provided student information system. Each participant completed similar tasks under two conditions; paper and using a MSS prototype. The order of conditions was balanced to control the learning effect. Two users completed the paper condition first and two users completed the MSS condition first. Each user was given a brief demo of MSS prior to starting the MSS condition.

5.2   Method

The participants completed a pre-test questionnaire to provide information regarding their demographics, computer and information management experience. For each condition, participants were presented with a description of a student that included the student's name, school, grade, gender and a summary of the student's extracurricular activities. For the paper condition, participants were asked to complete the tasks listed below using collection of student data including report cards, interim reports, assignment samples, school newsletters, extracurricular activity schedules and other announcements from school. Upon completion of the paper condition, participants completed a questionnaire regarding their experience.

Paper-Based Tasks

1. Determine the student's approximate average grade in a specified subject for the duration of the student's school career.
2. Determine the student's grade in a specified subject for a specified grade-level and term.
3. Determine if the student has any conflicts that will interfere with his/her ability to attend an event on a specified date/time.
4. Determine if a grade received on a current assignment/test is normal for the student.
5. Based on recently received assignment grades, compose a message to one of the student's teachers regarding a concern (positive or negative). Attach or included references any supporting facts.
6. When an incident has been reported by the student, determine if it is the first of its kind or has occurred before.

For the MSS condition, students were asked to complete the same tasks using MSS. Some report card grades, assignment grades and calendar events were pre-loaded into the system. In addition to the Paper-Based Tasks, users were also asked to complete the tasks listed below for the MSS condition. Upon completion of the MSS condition, participants completed a questionnaire regarding their experience.

Additional MyStudentScope Tasks

1. Enter individual assignment grade
2. Upload a file
3. Retrieve uploaded file

Upon completion of all conditions, participants completed a questionnaire to compare their paper condition experience to their MSS experience.

5.3   Results

As previously stated there were three objectives associated with the pilot study; obtain preliminary MSS feedback, improve clarity of questions and improve clarity of tasks. Three of the four pilot users agreed that using MSS was easier than paper for the requested tasks. They also responded that they believe they could be more productive in the management of their children's information if they used MSS. One of the pilot users felt that it was easier to use paper. In her opinion, ease of data retrieval and graphical representation did not outweigh the burden of entering information into MSS.

All participants understood the tasks as written. Most participants were able to navigate to the relevant MSS page to accomplish the requested tasks. One user failed to complete one task, but all other tasks were completed. It was not intuitive to three out of four participants that they should refer to the Notes page to determine whether an incident had occurred previously. For future tests we may consider modifying the Notes page title or description or introducing additional options for completing that task. The three users in question expected to find the incident information on the Messages page.

Most of the users seemed tired after completing the test and were therefore not inclined to add many comments to the post-test questionnaires. In the future, to capture user opinions without burdening them with further writing or forms, post-test interviews will be conducted and the investigator will document user responses.

6 Conclusions and Future Work

Prior research indicated that parents used paper-based methods more often than technology-based methods to archive and retrieve information regarding their children's education (Matthews and Feng 2017). Parents' use of MSS to archive and retrieve this information is expected to improve their effectiveness and efficiency over paper-based methods. Further, we believe that use of MSS will improve parental monitoring of the progress of their children's education, enable parents to more easily identify trends and anomalies in their children's educational development, manage schedule constraints and conflicts, and improve the effectiveness of communications with providers and educators.

For future work, we will complete a formal study where MSS is compared to paper-based methods. Further testing of MSS on common tasks encountered by parents managing their children's educational information and extracurricular schedules is needed to truly measure its effectiveness in improving parental use of this information. Moreover, extensive end user testing is also very important to discover limitations and areas of improvement for MSS.

References

Agarwal, N.K.: Making sense of sense-making: tracing the history and development of Dervin's sense-making methodology. In: International Perspectives on the History of Information Science and Technology: Proceedings of the ASIS&T 2012 Pre-conference on the History of ASIS&T and Information Science and Technology. Information Today, Inc. (2012)

Bakerville, R., Dulipovici, A.: The theoretical foundations of knowledge management. Knowl. Manag. Res. Pract. **4**, 83–105 (2006)

Buttfield-Addison, P., Lueg, C., Ellis, L., Manning, J.: Everything goes into or out of the iPad: the iPad, information scraps and personal information management. In: Proceedings of the 24th Australian Computer-Human Interaction Conference, pp. 61–67. ACM, Adelaide (2012)

Crabtree, R.K.: The Paper Chase: Managing Your Child's Documents (1998). Retrieved 14 February 2013, from Wrightslaw: http://www.wrightslaw.com/info/advo.paperchase.crabtree. htm. Accessed 01 Feb 2018

Excent: MyIEPmeeting Overview (2014). Retrieved 9 July 2015, from supt.excent.com: http:// supt.excent.com/products/myiepmeeting/

Gee, P.M., Greenwood, D.A., Paterniti, D.A., Ward, D., Miller, L.M.S.: The eHealth enhanced chronic care model: a theory derivation approach. J. Med. Internet. Res. **17**(4), e86 (2015). http://www.jmir.org/2015/4/e86/

Mannion, P.: Optimal Analysis Algorithms are IoT's Big Opportunity, 12 January 2015. Retrieved from Electronics 360. http://electronics360.globalspec.com/article/4890/optimal-analysis-algorithms-are-iot-s-big-opportunity. Accessed 01 Feb 2018

Matthews, T., Feng, J.H.: Understanding parental management of information regarding their children. In: Yamamoto, S. (ed.) HIMI 2017. LNCS, vol. 10273, pp. 347–365. Springer, Cham (2017). https://doi.org/10.1007/978-3-319-58521-5_28

Patrikakou, E.N.: The Power of Parent Involvement: Evidence, Ideas, and Tools for Student Success. Center on Innovation & Improvement, Lincoln (2008)

Prince George's County Public Schools: Family Portal for Parents & Guardians Guide (2015). Retrieved 2 September 2015, from PGCPS Student Information System - SchoolMAX: https://docs.google.com/document/d/1w-oBIgUCEsfYAP3uv_ZkyC-txiYk3zn06VVETt3SsK8/edit?pref=2&pli=1

Roshan, P.K., Jacobs, M., Dye, M., DiSalvo, B.: Exploring how parents in economically depressed communities access learning resources, pp. 131–141. ACM (2014)

Smith, E.: The role of tacit and explicit knowledge in the workplace. J. Knowl. Manag. **5**(4), 311–321 (2001)

Spurgin, K.M.: The sense-making approach and the study of personal information management. In: Personal Information Management - A SIGIR 2006 Workshop, pp. 102–104 (2006)

Swanson, G.: Managing Individual Education Programs (IEP) on the iPad, 20 January 2012. Retrieved 8 July 2015, from Apps in Education: http://appsineducation.blogspot.com/2012/01/managing-individual-education-programs.html. Accessed 01 Feb 2018

Wright, P., Wright, P.: The Special Education Survival Guide: Organizing Your Child's Special Education File: Do It Right!, 21 July 2008. Retrieved 14 February 2013, from Emotions to Advocacy: http://www.fetaweb.com/03/organize.file.htm. Accessed 01 Feb 2018

Basic Study on Creating VR Exhibition Content Archived Under Adverse Conditions

Naoya Mizuguchi[1(✉)], Isamu Ohashi[2], Takuji Narumi[1], Tomohiro Tanikawa[1], and Michitaka Hirose[1]

[1] Graduate School of Interdisciplinary Information Studies, The University of Tokyo, 7-3-1 Hongo, Bunkyo-ku, Tokyo 113-8654, Japan
{mizuguchi,narumi,tani,hirose}@cyber.t.u-tokyo.ac.jp
[2] Undergraduate School of Engineering, The University of Tokyo, 7-3-1 Hongo, Bunkyo-ku, Tokyo 113-8654, Japan
isamu@cyber.t.u-tokyo.ac.jp

Abstract. In this research, we propose a VR content creation method which reproduces the original appearance of the cultural properties based spherical image taken under adverse conditions. The proposed method consists of brightness adjustment and spherical image synthesis for improvement of appearance. In the adjustment of the brightness, γ correction is performed so that the contrast is smoothed. In the spherical image synthesis, images are selected according to the user's eye direction, and the image is synthesized after conversion by Moebius transformation. In order to verify the proposed method, the methods were applied to spherical images taken in the real world and evaluated. In the brightness adjustment method, the similarity between feature points is improved by the proposed method. In spherical image synthesis method, it was confirmed that the deviation became small as a result of calculation of the compositional deviation between the actual spherical image taken and the synthesized image.

Keywords: Digital museum · Spherical image
Spherical image synthesis · Brightness adjustment

1 Introduction

In recent years, museums have not only collected and stored cultural property, but also focused on exhibitions and education [1]. In this movement, in addition to simply displaying the cultural property stored in the museum, it is important to pass on the background knowledge such as the history of the cultural property and usage applications. Therefore, VR exhibition using digital archive technology has been advanced, because it is difficult to convey the background knowledge on cultural property in the conventional static exhibition [2]. By using VR technology for exhibitions, it is possible to exhibit a highly immersive exhibition even

© Springer International Publishing AG, part of Springer Nature 2018
S. Yamamoto and H. Mori (Eds.): HIMI 2018, LNCS 10905, pp. 122–131, 2018.
https://doi.org/10.1007/978-3-319-92046-7_11

for cultural properties that cannot be actually touched or entered in traditional exhibitions because of its high cultural value, and understanding of cultural properties deepens [3, 4].

In addition, the digital archive technology has developed rapidly in recent years, and the 360° camera appeared, which can photograph surroundings at once. 360° camera is suitable for archiving cultural properties with large space, and it is also introduced in VR exhibition in museum [5]. However, when offering such a VR exhibition, it is necessary to archive cultural properties, and, the archiving using the 360° camera involves difficulties. This is because many cultural properties cannot be archived as they were at that time as a result of aging and physical restrictions. In the VR exhibit using the archived material under such circumstances, there is a possibility that a misunderstanding may arise in the transmission of the background information. In such a situation, it is necessary to reproduce the original appearance in order to promote correct understanding of the cultural properties.

In this research, we treated the Imperial Train, which is for the Japanese Emperor and Imperial family, as the subject to exhibit using VR technology. The train is of importance because its historical value is very high. Also, the interior of the train is made of silk fabric, resulting in its very high value as a craft.

However, due to the age of the car, it cannot be driven outside. In addition, the interior of the train is very dark because the internal electric circuit is dead, and an indoor light cannot be applied. Moreover, there are various restrictions in the archive for VR exhibition of Imperial trains. First, because there is no light, a light source is required. Furthermore, from the viewpoint of protecting the exhibits, it is impossible to hit a strong light source. Second, the indoor environment is very narrow. The width of the aisle is about shoulder width, and in the observation room there are sofas on the left and right, which cannot be damaged. We need to be able to archive in such constraints In addition, images taken in this way may not be sufficient for VR exhibition. First, it is necessary to ensure uniform brightness. Second, the captured image is the image taken inside the museum, not the archive when actually running. In the images under such conditions, background information may not be transmitted correctly.

Therefore, in this research, we propose VR content creation method to promote understanding of cultural properties based on spherical images photographed under adverse conditions separated from the original context.

2 Related Works

2.1 Creation of Virtual Reality Space

There are several ways to create a Virtual reality space for exhibitions. First of all, it can be thought of a generation method using point clouds like Structure from Motion [6]. Although this method got only sparse point clouds, in recent years, a method of obtaining a dense point clouds called Multi View Stereo was also proposed [7]. However, there are cases in which estimation does not work well in these method, and there is a possibility that a hole may be emptied in the VR space or a form different from the original form may be presumed. This method is inappropriate because the possibility of transmitting wrong background information. In addition, a method using a 3D model such as Image based modeling can be considered [8]. Although these methods are useful for reproducing a single structure, it is difficult to reproduce vast urban spaces and natural landscapes. The VR space using the 360° camera has a limitation on the viewpoint, but because of its photorealistic image, it is possible to preserve the figure as it is. According to Tanaka et al. it is shown that the exhibition using the whole celestial tone has deepened the understanding of the target [5]. In this way, spherical images are considered suitable for the museum exhibition. Therefore, in this research, we will consider the archive using this spherical images.

2.2 Brightness Adjustment and Image Synthesis

In order to correct the captured image, we describe the previous research on the method of synthesizing the image and correcting the brightness.

Lee proposed a method to perform a process for enormous contrast and a filtering process for removing noise [9]. Singh and Kapoor proposed a method to emphasize the contrast of the grayscale image based on the histogram [10]. There is also a method of estimating the light source and changing the brightness of the image as a method of brightening the dark image [11]. In this research, we will consider correcting the darkened portion by correcting the contrast and luminance of the image like these researches.

There are also many studies on image synthesis, and it is known that geometric consistency and optical consistency are important at the time of synthesis. Glassner described the human visual characteristics, signal processing methods such as Fourier transform and wavelet transform, and proposed an image synthesis method using them [12]. Yasuda et al. Proposed a method to cut out regions using temperature information and to do synthesis [13]. These are compositing methods for obtaining geometric consistency, and are often used in today's image synthesis. On the other hand, as a technique for achieving optical consistency, a method of estimating the light source environment from shadows of real objects in an input image was proposed [14]. In this research, we will examine the method for synthesizing, while considering geometric consistency and optical consistency during synthesis.

3   Design of Proposed System

3.1 Overview

We propose VR content creation method to promote understanding of cultural properties based on spherical images photographed under adverse conditions separated from the original context. In Sect. 3.2, we describe the method of correcting the brightness of the spherical image. In Sect. 3.3, we describe the synthesis method of the spherical images. We correct the appearance of the spherical image due to the fact that the location was not originally the place of the cultural property at that time of photographing the spherical image. For that purpose, we propose a method to combine a spherical image with other spherical images taken at the original place.

3.2 Brightness Adjustment

First, the brightness of the spherical image is adjusted. This is because it was thought that in the experience of the spherical image, when the brightness is too low or too high, it adversely affects the viewing of cultural properties. The brightness is adjusted by γ correction. Calculate the histogram after performing γ correction on the original spherical image. The one with the cumulative frequency distribution of the histogram closest to the straight line is adopted as the picture with the largest contrast. The determination as to whether it is close to a straight line is made by the least squares method.

3.3 Spherical Image Synthesis

In this section, we describe a method of reproducing the appearance by synthesizing spherical images. Consider synthesizing spherical images taken at different places. However, if this synthesis is simply performed, there will be deviations in the perspective due to mismatch of the viewpoints, resulting in different appearance. Furthermore, in the synthesis of spherical images, it is difficult to correct all of the deviations over the entire periphery. However, in the viewing of the spherical image, since the part the user is looking at is only limited part of spherical image, we considered to correct only that part. Therefore, in this research, we select dynamically the material to be synthesized from the consecutive spherical images and deform the image before synthesizing to correct that part. The synthesized image is dynamically selected according to the user's eye direction, and the image to be synthesized is deformed using Mobius transformation [15].

The workflow of this method will be explained. First, a spherical image to be a base for combining is selected according to the user's eye direction. Next, we apply the Mobius transformation to the selected image according to the positional relationship between spherical images, and perform the composition (Fig. 1).

First, we will describe a method of selecting a spherical image according to the eye direction. Two spherical image groups to be synthetic materials are prepared. Subsequently, camera parameters are obtained using Structure from Motion, and the positional relationship between all the image of each camera is determined. The positional relationship between the two groups is determined based on parameters in the real world. Then, when improving the appearance from one spherical image, an image closest to the direction of the line of sight is selected and set as a composite material.

Subsequently, the image is transformed so as to synthesize the selected image with geometric consistency. Therefore, in this research, we synthesize after

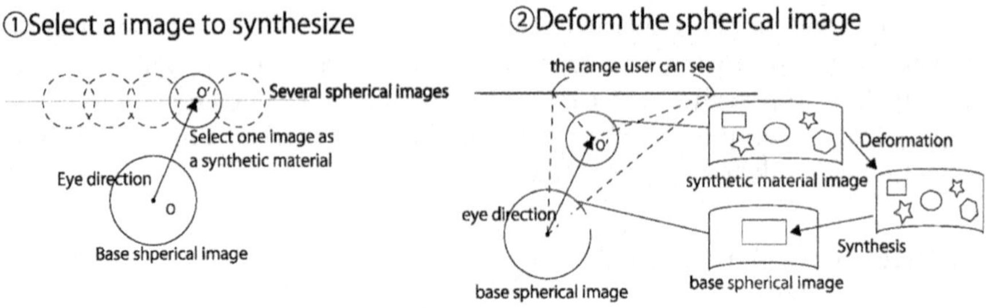

Fig. 1. 1. Select spherical image to synthesize according to the eye direction. 2. Deform the spherical image by mobius transform according to the geometry

deforming the spherical image using Mobius transformation. The coefficients of the Mobius transformation are determined from the position of all spherical images and then from the geometrical positional relationship of the landscape shown in the deformed spherical images. Then, according to the value, deformation of the spherical image is performed, and synthesis is performed.

4   Experiments

In this chapter, we examine the effect of each of the correction methods proposed in Sect. 3. First of all, as to the method of Sect. 3.2 the difference between histogram difference and feature point matching result is checked between the image captured by the single lens reflex camera and the image before correction and the image after correction. The validity of the image composition method of Sect. 3.3 is verified by applying it to an actual image. In order to verify the usefulness of the method, we photograph the train running in Tokyo and the home of Ueno station in Tokyo, synthesized by the proposed method and evaluated.

4.1   Brightness Adjustment

Detailed Procedures. Using the method proposed in Sect. 3.2, confirm whether the brightness is consistent. First of all, we prepared spherical image groups archive inside the Imperial train. Also, as a comparative image, an image of the interior of the train taken with a single lens reflex camera was prepared. First, after executing gamma correction on the spherical images, calculate the histogram to obtain the image with the greatest contrast. To the spherical image obtained in this way, an image which becomes visible from the user when viewing the entire circumference image is cut out so as to be equal to the composition of the single lens reflex. We compare histograms of image taken with single lens reflex camera and video cut out. Also, we measure the similarity of feature points between those images. Comparison of histograms was made to have a size of 300×200 px so as to preserve the aspect ratio, each histogram was calculated, and the degree of coincidence was examined. The degree of similarity between the feature points was calculated by grayscaling the image and after setting the size to 300 × 200 px in the same way, extracting the feature points using the A-KAZE feature value and calculating the distance between the feature points. The images used for comparison are the following two groups of images (Figs. 2 and 3).

Fig. 2. Group A: left image was photographed with a single-lens reflex camera, and the middle image is an original spherical image, and right image is corrected spherical image.

Fig. 3. Group B: left image was photographed with a single-lens reflex camera, and the middle image is an original spherical image, and right image is corrected spherical image.

Result and Discussion. The degree of coincidence between the matching degree of the histogram and the distance between the feature points in group A and group B is as follows (Tables 1 and 2). In the table, "Proposed" means Proposed method which adjust the brightness, and "Comparative" means Comparative method which does not correct the brightness.

Table 1. Histogram

	Proposed	Comparative
Group A	0.23	0.67
Group B	0.57	0.47

Table 2. Feature matching

	Proposed	Comparative
Group A	118.6	127.7
Group B	130.6	156.6

First, the value of the histogram in group A has decreased in the degree of coincidence in the proposed method, whereas in group B, the degree of coincidence of the histogram has increased. It is thought that the result at group A was originally due to the fact that the brightness of the bright part was further increased. On the other hand, in group B, the range the user was looking at is a dark part of the spherical image, and the part is brightened, so it is considered that the brightness has improved. If there is a difference in the brightness of the image in the entire circumference image, processing corresponding to that darkness is considered to be necessary. On the other hand, in both scenes A and B, it was found that the result of feature point matching improves in the proposed method. This is thought to be because the feature points became easier to understand by emphasizing the contrast even in the case of a bright image originally. From the above, it was found that there is a possibility of improving the appearance of the entire celestial image by emphasizing the contrast. On the other hand, it was suggested that adjustment of brightness requires processing to be divided between bright and dark parts of contents.

4.2   Spherical Image Synthesis

Detailed Procedures. In order to evaluate the validity of the method, we synthesize the spherical images taken at Ueno station and the spherical images taken in the train stopping at the station. Regarding the appearance at the time of viewing actually, evaluate the difference between the image created by the synthesis and the spherical image which is taken in real (ground truth image). As a comparison method, we compare it with a method of merely combining with spherical image selected by eye direction, and not transformed. For each composite image and ground truth image, feature points are extracted using A-KAZE feature detection in a range visible from the train window. Subsequently, the extracted feature points are visually associated with each other. This is because it is difficult to automatically match by the fact that the time of photographing was different or due to the influence of the color of the window. After that, in order to check the degree of coincidence between the composite image and the ground truth image, it is judged from relationships of the feature points matching. The amount of parallel movement and the amount of enlargement/reduction such that the group of feature points of one image coincide with the group of image feature points of the other are calculated. Also calculate the average pixel distance between feature points of each image. When the line of sight direction is perpendicular to the window, the angle is 0°, and the above indices are calculated when the gaze directions rotate 0°, 10°, 20°, 30°, respectively. The appearance of the entire sky image that was actually synthesized is as shown in the Fig. 4.

Result and Discussion. The amount of parallel movement, the amount of enlargement and reduction, and the average pixel distance between feature points between images are as shown in the following graph (Fig. 5). From this figure, it can be seen that composition shift is improved by using the image transformation method based on the Mobius transformation for each of the translation amount, the enlargement/reduction amount, and the average pixel distance between the

Fig. 4. Left image is photographed from the inside of the train, the middle is simply synthesized image, and the right image is synthesized image after conversion.

feature points between the images. As the center of sight axis rotated, every index was larger when synthesis is performed merely. In that method, this is because it takes feature points only in the part visible from the train window, so it seems to be because the range that takes the feature point is biased toward one side as the line of sight rotates. Also, we were seen to be out of the feature points on the pillar of the station's home because we specified the value of the Mobius transformation so that the wall of the station can be seen at the correct position. However, even in such a point, we see that the proposed method is superior to the image selection method alone.

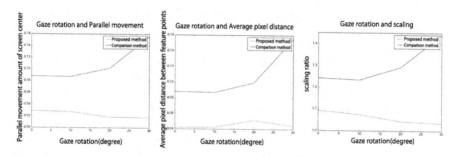

Fig. 5. The left figure is the scale, the middle image is the average moving distance of the image center, and the right figure is the average distance between the feature points.

5   Conclusion

In this paper, we propose a VR contents creation method to promote understanding of cultural properties based on spherical image materials taken under adverse conditions separated from the original context. We proposed a method of correcting the brightness of the spherical image taken and a method of synthesizing the spherical image for correction of the part different from the original appearance.

First, the brightness was adjusted. The luminance was adjusted by γ correction so that the overall contrast was smoothest. The degree of smoothness of the overall contrast is assumed to be the case where the cumulative frequency distribution of the histogram is the most linear. The degree of proximity to a straight line was determined by the least squares method. Subsequently, for a part different from the original appearance of the spherical image, it was synthesized with the spherical image taken at original place. First, the positional relationship between all the spherical image groups is obtained. hen, an image to be a composite material is dynamically selected according to the eye direction of the user. Based on the geometry of the real world, the selected image is transformed by Mobius transformation and synthesized.

We conducted experiments on the usefulness of each of the correction methods. First of all, regarding the experiment of the brightness correction method, We conducted experiments on the usefulness of each of the correction methods. In order to verify the brightness correction method, the appearance from the user of the spherical image is compared with the image photographed with the single lens reflex camera. We compared between the spherical image corrected by the brightness correction method and the spherical image not corrected. The similarity between the histogram and the similarity between the feature points is examined between the appearance of each spherical image and the image of the single lens reflex. As a result, in the histogram, it was found that when we were looking at the dark part in the spherical image, we improved it while when we were looking at the bright part, the similarity declined. On the other hand, in any case, it was found that the degree of coincidence of the feature points is increased, and it was found that the appearance is improved by smoothing the contrast. Regarding the synthesis method of spherical images, synthesis was performed on the actual spherical image and verified. The verification was based on comparing the error between the actually captured image (ground truth image) and the image simply synthesized without deformation and the error between the ground truth image and the image synthesized by the proposed method. Feature points of each image were determined, and the amount of translation and enlargement/reduction such that the feature points of one image coincided with the image feature points of the other image were calculated, respectively. We also calculated the distance between feature points. As a result, improvements were found in the proposed method for any indices.

In the limitation of the proposed method, the existing scenery can be synthesized in the reality image, but it can not be applied to things where the present figure can not be seen due to collection, damage or the like. In addition, it is difficult to deform if the distance between all the spherical images is too large, or when there are many substances that become disturbance. Also, when increasing the brightness, there are cases where the original image is too dark, it does not go well.

As futurework, I would like to apply this method to other cultural properties and carry out various archives. Also, I would like to confirm that the cultural properties actually archived in this way are exhibited and that the quality of the exhibition could be secured. Regarding the correction of brightness, because the correction parameters may be different depending on each cultural property, we want to verify the parameters.

Acknowledgement. This work was partially supported by the MEXT, Grant-in-Aid for Scientific Research(A), 25240057.

References

1. Miles, R.S., Alt, M.B.: The Design of Educational Exhibits. Psychology Press, Hove (1988)
2. Narumi, T., Hayashi, O., Kasada, K., Yamazaki, M., Tanikawa, T., Hirose, M.: Digital diorama: AR exhibition system to convey background information for museums. In: Shumaker, R. (ed.) VMR 2011. LNCS, vol. 6773, pp. 76–86. Springer, Heidelberg (2011). https://doi.org/10.1007/978-3-642-22021-0_10
3. Kondo, T., et al.: Practical uses of mixed reality exhibition at the national museum of nature and science in Tokyo. In: Joint Virtual Reality Conference of EGVE-ICAT-EuroVR (2009)
4. Nakano, J., et al.: On-site virtual time machine-navigation to past camera position and past picture superimpose on present landscape. Trans. Inf. Process. Soc. Jpn **52**(12), 3611–3624 (2011)
5. Tanaka, R., Narumi, T., Tanikawa, T., Hirose, M.: Motive compass: navigation interface for locomotion in virtual environments constructed with spherical images. In: 2016 IEEE Symposium on 3D User Interfaces (3DUI), pp. 59–62. IEEE (2016)
6. Tomasi, C., Kanade, T.: Shape and motion from image streams under orthography: a factorization method. Int. J. Comput. Vis. **9**(2), 137–154 (1992)
7. Furukawa, Y., Ponce, J.: Accurate, dense, and robust multiview stereopsis. IEEE Trans. Pattern Anal. Mach. Intell. **32**(8), 1362–1376 (2010)
8. Pan, Q., Reitmayr, G., Drummond, T.: ProFORMA: probabilistic feature-based on-line rapid model acquisition. In: BMVC, vol. 2, p. 6. Citeseer (2009)
9. Lee, J.-S.: Digital image enhancement and noise filtering by use of local statistics. IEEE Trans. Pattern Anal. Mach. Intell. **2**(2), 2165–2168 (1980)
10. Singh, K., Kapoor, R.: Image enhancement using exposure based sub image histogram equalization. Pattern Recogn. Lett. **36**, 10–14 (2014)
11. Boyack, J.R., Juenger, A.K.: Brightness adjustment of images using digital scene analysis. U.S. Patent no. 5,724,456, 3 March 1998
12. Glassner, A.S.: Principles of Digital Image Synthesis. Morgan Kaufmann, Burlington (2014)
13. Yasuda, K., Naemura, T., Harashima, H.: Thermo-key:human region segmentation from video. IEEE Comput. Graph. Appl. **24**(1), 26–30 (2004)
14. Kim, T., Hong, K.-S.: A practical single image based approach for estimating illumination distribution from shadows. In: Tenth IEEE International Conference on Computer Vision (ICCV 2005), vol. 1, pp. 266–271. IEEE (2005)
15. Schleimer, S., Segerman, H.: Squares that look round: transforming spherical images. CoRR, abs/1605.01396 (2016)

Information Design for Purposeless Information Searching Based on Optimum Stimulation Level Theory

Miwa Nakanishi[(⊠)] and Motoya Takahashi

Keio University, Hiyoshi, 3-14-1, Kohoku, Yokohama 223-8522, Japan
miwa.nakanishi@keio.jp

Abstract. There are increasing opportunities to engage in purposeless information searching, such as browsing purposeful information. A growing focus is also promoting interest in products and services by utilizing these opportunities. Thus, rather than the efficiency of information searching, the new concept of the "continuousness" of information searching is expected for information design. It has been suggested that searching for information without a defined purpose or use is done based on intrinsic motivation (IM). The theory of optimum stimulation level (SL) gives a U-shaped relationship between IM and SL. On the basis of this theory, we hypothesized that the continuousness of information is enhanced in an environment in which the user can enjoy searching for information at the optimum level of stimulation. In this study, we attempt to specify the requirements for a method of information design based on the aforementioned hypothesis. We assess the effectiveness of the method with an experiment in which participants engage in information browsing. The SLs of the referred pieces of information are quantified on a scale from zero to one. After analyzing the relationship between the SL and the physiological and psychological responses of the participants, we clarify (i) the optimum SL, (ii) the SL that most effectively enhances the IM of the participants to search for information, and (iii) the SL between one piece of information and the next at which the IM of the participants is increased.

Keywords: Purposeless information searching · Intrinsic motivation
Physio-psychological analysis

1 Introduction

Nowadays, with the spread of mobile terminals, the reasons why people search for information have diversified. In terms of the clarity of purpose, information searching can be divided roughly into two patterns. One pattern is typified by an information search (IS) that is performed with a clear purpose, such as a route search or an accommodation reservation. We deem this to be an "objective" IS in which the motivation of the user is directed to the desired information. Therefore, when designing information on such sites for purposeless IS, the main focus is usually on IS efficiency. By contrast, we deem an IS performed without a clear purpose, such as simple browsing, to be a "non-objective" IS in which the motivation of the user is directed to

© Springer International Publishing AG, part of Springer Nature 2018
S. Yamamoto and H. Mori (Eds.): HIMI 2018, LNCS 10905, pp. 132–143, 2018.
https://doi.org/10.1007/978-3-319-92046-7_12

the IS itself. In recent years, the expectation has been that a non-objective IS affords opportunities to promote the motivation to use products and services. Therefore, on such sites, rather than concentrating on IS efficiency, the importance from a new perspective of information design (ID) is to maintain IS sustainability, that is, to sustain the interests of users and to touch a lot of information.

As described above, in a non-objective IS, the motivation of the user is directed to the IS itself. In the field of psychology, such voluntary motivation in which the activity itself is the aim is called intrinsic motivation (IM) [1–7] to distinguish it from extrinsic motivation due to external factors such as compensation and punishment. In a non-objective IS, maintaining the user's IM at a high level is considered to be key in promoting IS sustainability.

One of the psychological theories related to the degree of IM is the theory of optimum stimulation level (OSL) [8–10] in which IM is viewed as having an inverted U-shaped relationship with stimulation level (SL). Replacing the novelty of information with its SL, if people touch too weak stimuli without novelty for themselves, they will be bored, but if they touch on themselves too novel stimuli, they become indifferent. However, there is an intermediate SL that induces IM, that is, the OSL. In a previous study [11], we considered familiar information as having too low an SL and unfamiliar information as having too high an SL. Furthermore, giving users information of differing SL revealed that IM increased the most in the group given information of intermediate SL.

In the present research, as a new ID targeting non-objective ISs, we applied OSL theory as a way to maintain user IM and promote IS sustainability. Specifically, in a non-objective IS, on the basis of the OSL theory, we hypothesized that it is highly likely that the user will stop the IS if she/he encounters information that is either too familiar or too unfamiliar. However, by continually encountering information whose SL for individual users is neither too low nor too high, high IM is maintained, and IS sustainability is promoted.

The purpose of this paper is to propose a method of ID aimed at maintaining user IM and promoting IS sustainability to give a new value to non-objective ISs. To do so, in the first experiment, we attempt to extract the ID requirements for promoting IS sustainability. Specifically, we used physiological psychology to investigate the effectiveness of giving information of how much range of SL to maintain user IM by analyzing the relationship between user IM and information SL in non-objective ISs. On the basis of this, we propose an ID that encourages IS sustainability, and in a second experiment, we verify the proposed ID.

2   Experiment

2.1   Method

In the first experiment, we observed and measured non-objective ISs by users. Our aim in doing so was to use physiological psychological indices to analyze the SL of the information being viewed and the extent of user IM at that time. Specifically, we had users engage in a non-objective IS, whereupon we analyzed the SL of the information that they viewed as time series and used physiological psychological indices to evaluate the degree of user IM.

Experimental Task. The task in this experiment was an IS using an experimental site that simulated a general site that users used for browsing. The presented information was taken from 24 categories covering hobbies and entertainment that are typical of people in their 20 s, these being the participants in this experiment. In addition, the experimental site consisted of four layers, and when a participant selected a category, it transited to the hierarchy of subcategory and reached the text information at the fourth hierarchy. Six types of text information were prepared for each meddle category. Each piece of text information consisted of around 1,000 characters.

Experimental Procedure. Our aim in the experiment was to observe the participants as they engaged in non-objective ISs, so we gave them the opportunity to browse while killing time without being conscious of the experiment. We first installed the measuring device and then asked the participants to wait in the resting state for 2 min. We then asked them to browse the site for 20 min while relaxing while we "prepared the experiment." We then asked them to wait again in the resting state for 2 min, whereupon we instructed them to search for particular keywords (e.g., "Ina Bauer") on the site. However, in this research, the latter IS was a dummy task and was not analyzed; it was the previous 20 min IS that we analyzed. After completing the experiment, we informed the participants of our actual intention with the experiment and asked them to complete a questionnaire.

Data, Participants, and Ethics. The first data set is the search log corresponding to the chronologically ordered information that the participants viewed. The second data set comprises subjective scores on the scale 0–100 for the degree of participant interest in each piece of information viewed. The third data set comprises the frequency of blinking, which is an index of interest and was measured using a digital video camera. The fourth data set comprises changes in cerebral blood volume (CBV), which is the Oxy-Hb concentration change in channel 16 of the prefrontal cortex (PFC) as measured by near-infrared spectroscopy. The data for subjective scores, blinking, and CBV changes were all standardized among the participants.

The participants were 15 students (11 men and 4 women) aged 20–25 who we instructed to refrain from caffeine and alcohol intake from the day before the experiments. Each participant gave written informed consent before the experiment.

Analysis Method. The SLs of the information that each participant viewed were quantified using the same method as in the aforementioned previous research. Specifically, prior to the experiment, we asked the participants to compare all pairs from the 24 categories in terms of "familiar" or "unfamiliar". On the basis of those data, each category was scaled by the modified Scheffe method and positioned on one dimension. Furthermore, to standardize the scale for all participants, the category with which a participant was most familiar was placed at zero, their least familiar category was placed at one, and each of the other categories was placed in the range 0–1 to quantify the SL of each category on a common scale for each participant. For example, the SLs of the information browsed by a particular (male) participant are shown in Fig. 1, and we draw the SL time series as the waveform shown therein.

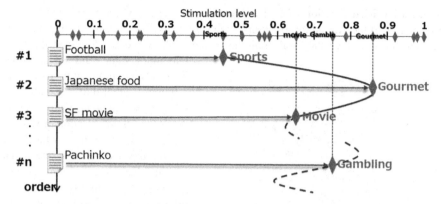

Fig. 1. Stimulus level (SL) of information browsed by a particular subject (time-series change)

2.2   Results

Characteristics of SL Time Series. As shown in Fig. 2, upon visualizing the SL time series of each participant (taking the SL on the horizontal axis and the order of the information viewed on the vertical axis), we see that each participant continued to search for information between a weak stimulus and a strong stimulus. Moreover, we found that the SL tended not to change suddenly between consecutive information browsings. From this, it seems that, in a non-objective IS, information is browsed while going back and forth with a certain SL as the boundary, and we reason that this SL is the OSL. Again, recalling the hypothesis of this research (namely that, in a non-objective IS, individual users are keen to access information whose SL to them is neither too low nor too high so that user IM is maintained and IS sustainability is promoted), we view this SL that is neither too low nor too high to be near the OSL.

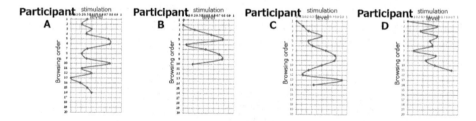

Fig. 2. Characteristics of changes in SL time series

Therefore, to embody the ID requirement of promoting IS sustainability, we first clarified quantitatively the OSL that was common to all participants and further examined the appropriate SL difference between consecutive information browsings.

Quantitative Identification of OSL. Therefore, taking the SL on the horizontal axis and the order of the viewed information on the vertical axis, we considered that the OSL is present in more than the local minimum value of the wave and less than the local maximum value, and we analyzed its existence probability using the following algorithm. Specifically, first, in the maximum value appearing at the i-th information browsing, we defined a section where the maximum value or less as one, and a section where the maximum value or more as zero. Similarly, in the minimum value appearing at the j-th time, we defined a section where the local minimum value or less as zero, and a section where the local minimum value or more as one. In this way, we evaluated as zero or one whether the OSL exists on every SL and accumulated it for all participants. Finally, by dividing the accumulated value by the total number of occurrences of the local maximum value and the local minimum value, we obtained the OSL existence probability (see Fig. 3).

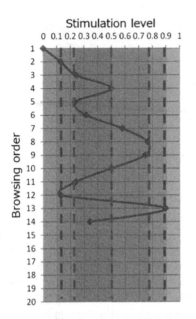

Fig. 3. Quantitative identification of optimum stimulation level (OSL)

From the analysis, we see that taking the SL on the horizontal axis and the OSL existence probability on the vertical axis presents the relationship between SL and OSL existence probability as an inverted "U". Therefore, for the sake of clarity, we grouped the SLs into five sections according to OSL existence probability and then examined the relationship between SL and IM from the physio-psychological reaction to viewing information in each section (see Fig. 4).

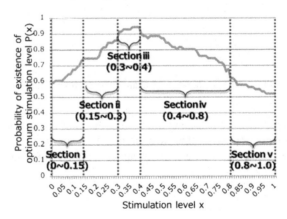

Fig. 4. OSL existence probability

Relationship Between SL and IM. Figure 5 shows the results of the subjective scores. Averaging the degree of interest in each section showed that the interest in Sect. 3 was the highest.

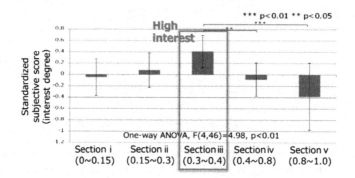

Fig. 5. Subjective scores

Figure 6 shows the blinking frequency. It is known that blinking frequency increases with SL. We investigated the increase (i.e., the average difference from the resting state) in blinking frequency while browsing information at each SL and found again that the interest in Sect. 3 was the highest.

Figure 7 shows the changes in CBV. The PFC is divided into several regions of which we focused on the dorsolateral PFC (DLPFC) that is involved in motivation and the medial PFC (mPFC) that is involved in spontaneity. By examining increases and decreases in Oxy-Hb concentration (average difference from resting state) that indicated activation of the DLPFC and mPFC while viewing information in each SL section, we found that spontaneity and motivation were highest in Sect. 3.

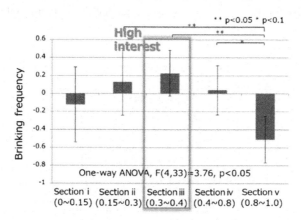

Fig. 6. Blinking frequency

Summarizing the results of the physio-psychological reactions suggests that IM would likely be highest when the information of Sect. 3 were viewed. From this, we

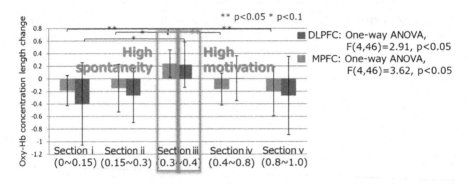

Fig. 7. Changes in cerebral blood volume (CBV)

consider the OSL common to all participants to be 0.3–0.4. Moreover, we found IM to be relatively high in the sections with SLs of 0.15–0.8. From the above, by designing the information such that the IS that goes back and forth within the SL range of 0.15–0.8 focusing on the OSL 0.3–0.4 is performed, it seems that user IM could be high and maintained at the level. This is the first requirement of ID for IS sustainability.

Quantitative Identification of Appropriate SL Difference. Another feature of the typical SL time series was that the SL tended not to change abruptly between consecutive information browsings. Therefore, we examined the appropriate SL difference between consecutive information browsings. Specifically, to identify the appropriate SL differences when the SL either increased or decreased, we analyzed the correlation between SL difference between information $n - 1$ and n and the IM intensity while browsing information n.

In Fig. 8, the SL difference between information $n - 1$ and n is recorded on the horizontal axis, and the subjective score at information browsing n is recoded on the vertical axis. Here for the sake of clarity, we divide the SL difference into three groups

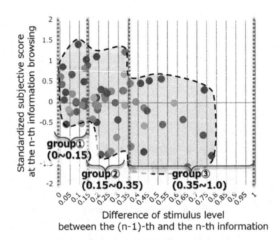

Fig. 8. Relationship between stimulation level (SL) difference from information $n - 1$ to n in the direction of increasing interest and the subjective score for browsing information n

and examine the relationship between SL difference and IM from that between the physio-psychological reaction to viewing information n and the SL difference in each group.

Figure 9(a) shows the subjective scores, Fig. 9(b) shows the blinking frequency as an indicator of interest, and Fig. 9(c) shows the CBV changes as an indicator of motivation and spontaneity. These results show that participant interest and spontaneity to information browsing were highest for category 1. This result suggests that IM is highest when the SL difference between consecutive browsing information is between zero and 0.15 in the increasing direction.

Similarly, we also examined the transition in the decreasing direction. From the distribution of subjective scores, we analyzed the SL differences by classifying them into four, as shown in Fig. 10, for the sake of clarity. The subjective scores and CBV changes suggest that IM is highest when the SL difference between consecutive browsing information is between -0.35 and -0.5 in the decreasing direction (see Fig. 11(a), (b) and (c).

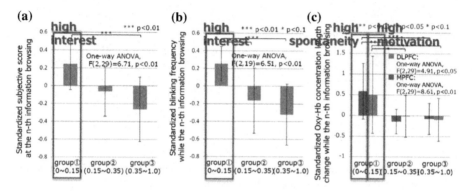

Fig. 9. (a) Subjective score for information browsing n. (b) Blinking frequency for information browsing n. (c) CBV change for information browsing n

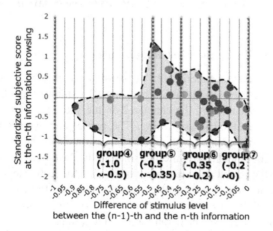

Fig. 10. Relationship between SL difference from information $n - 1$ to n in the direction of decreasing interest and the subjective score for browsing information n

Requirements of ID to Promote IS Sustainability. To summarize the ID requirements obtained from the experiment, with the most familiar category for each user
being zero and the least familiar category being one, we find an OSL of 0.3–0.4.
Furthermore, if the SL is 0.15–0.8 with a focus on that, the IM is maintained at a
relatively high level. This is the first requirement.

In addition, the desired SL difference is between zero and 0.15 in the increasing
direction (i.e., when transitioning to a less familiar category information) and between
−0.5 and −0.35 in the decreasing direction (i.e., when transitioning to a more familiar
category information). This is the second requirement.

From the above, it is conceivable that, by presenting the above requirements to a
site that is the target of non-objective ISs, the user IM could be maintained and the IS
persistence could be promoted.

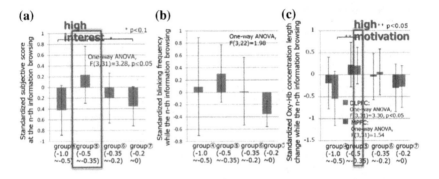

Fig. 11. (a) Subjective score for information browsing n. (b) Blinking frequency for information
browsing n. (c) CBV changes for information browsing n

3   Verification of Effectiveness of the Proposed Information Design

3.1   Experiment to Verify the First Requirement

Method. In the verification experiment, we verified whether IS persistence is promoted by maintaining user IM by embodying the ID requirements clarified from the
experiment that was discussed in the chapter 2 on the site.

The experimental method is as follows. The participants were asked to view as
much information as they wanted without restrictions on either the time or the number
of views. On the experimental site, the information displayed at time n was gathered
automatically from the information prepared in advance for each category based on the
SL data of each participant. For example, when transitioning from zero to 0.15 within
the SL range of 0.15–0.8 from information $n - 1$ to information n, the categories
included in the range were selected randomly, and furthermore, a random piece of
information included in the category was displayed.

The participants were asked to wait at the rest for 2 min after attaching the measuring device before carrying out the above tasks. During each task, blinks and CBV changes were measured. The participants were then asked to describe their introspection with a free sentence.

The subjects were 10 students (7 men and 3 women) aged 21–25. Two experimental conditions were prepared: one in which it was possible to view information in the whole SL range without applying the first requirement and another one in which it was possible to view information in the SL range of 0.15–0.8 by applying the first requirement. The SL difference between browsing information $n - 1$ and n was set randomly in both conditions. To eliminate the influence of ordering, the experimental order was controlled.

Results. We examined the following two indices of IS sustainability. Figure 12(a) shows the average number of pages that changed. It is understood that, when the first requirement was applied, the number of pages increased significantly compared to when it was not applied. Also, Fig. 12(b) shows the average number of free sentences. It can be seen that the number of free sentences is roughly proportional to the number of transitioned pages. This suggests that, in the case of an SL of 0.15–0.8 when the first requirement was applied, it was not that the number of pages was increased by successively shifting the page for reasons such as boredom but that the IS continued because of voluntary interest.

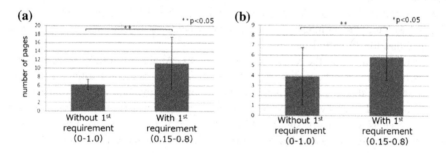

Fig. 12. (a) Average number of pages. (b) Average number of free sentences

Next, the results for the physiological index of endogenous motivation were analyzed. Analysis of blinks (Fig. 13(a)) as an index of interest showed that interest increased when applying the first requirement. Furthermore, analysis of CBV changes (Fig. 13(b)) as an indicator of motivation and spontaneity showed no statistical difference, but the average value was higher when the first requirement was applied. These results suggest that, when the first requirement is applied, the IM is higher than when it is not applied.

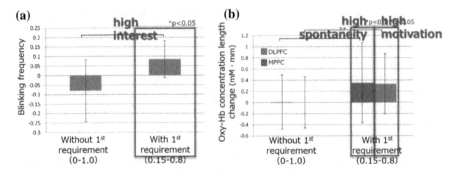

Fig. 13. (a) Blinking frequency. (b) CBV changes

3.2 Experiment to Verify the Second Requirement

Method. The participants in the experiment to verify the second requirement were the same as those in the experiment to verify the first requirement. There were two experimental conditions, namely, one applying only the first requirement and another one applying the optimum SL difference, that is, applying both the first and second requirements. All of the other experimental methods were the same as those in the experiment to verify the first requirement.

Results. We focused on the following two results as indices of IS sustainability. There were no statistically significant differences in the number of pages that transited (Fig. 14(a)) and the number of free sentences (Fig. 14(b)), but the average value suggests that the IS would last longer.

Next, we focused on the results for the IM index. Analysis of blinks (Fig. 15(a)) as indicators of interest showed no significant difference, but comparing the average values suggests that IM would increase when applying the first and second requirements compared to applying only the first requirement. Furthermore, analysis of CBV changes reveals statistically significant differences (Fig. 15(b)), suggesting the possibility of markedly increasing IM by applying the first and second requirements.

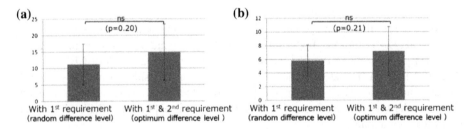

Fig. 14. (a) Average number of pages. (b) Average number of free sentences

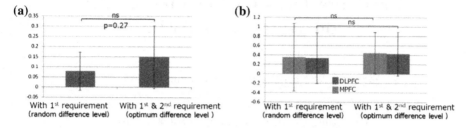

Fig. 15. (a) Blinking frequency. (b) CBV changes

4 Conclusion

In this research, we applied OSL theory as a way to maintain user IM and to promote IS sustainability in the ID for a non-objective IS. As a result, from the physio-psychological approach, it was found that, by designing the first and second requirements obtained from the experiment on the site, IM is maintained and the IS is sustained. As an application of this research to an actual situation, by acquiring personal characteristics such as the degree of user familiarity with various categories and providing a browsing environment in compliance with the requirements of the above ID, it seems possible to make an IS based on IM sustainably. This will lead to the creation of new opportunities to allow users to acquire information on products and services.

References

1. Deci, E.L., Ryan, R.M.: Self-determination theory and the facilitation of intrinsic motivation, social development, and well-being. Am. Psychol. **55**, 68–78 (2000)
2. Deci, E.L., Ryan, R.M.: Intrinsic Motivation and Self Determination in Human Behaviour. Plenum Press, New York (1985)
3. Harlow, H.F.: The formation of learning sets. Psychol. Rev. **56**, 51–65 (1949)
4. Harlow, H.F.: Learning and satiation of response in intrinsically motivated complex puzzle performance by monkeys. J. Comp. Physiol. Psychol. **43**, 493–508 (1950)
5. Harlow, H.F.: The nature of love. Am. Psychol. **13**, 673–685 (1958)
6. Harlow, H.F., Harlow, M.K., Meyer, D.R.: Learning motivated by a manipulation drive. J. Exp. Psychol. **40**, 228–234 (1950)
7. Heron, W.: The pathology of boredom. Sci. Am. **196**, 52–56 (1957)
8. Berlyne, D.E.: Conflict, Arousal, and Curiosity. McGraw-Hill, New York (1960)
9. Berlyne, D.E.: Structure and Direction in Thinking. Wiley, Oxford (1965)
10. Berlyne, D.E.: What next? Concluding summary. In: Intrinsic Motivation: A New Direction in Education, pp. 186–196 (1971)
11. Yasuma, Y., Nakanishi, M., Okada, Y.: Can "tactile kiosk" attract potential users in public? In: Proceedings of the 3rd International Conference on AHFE (Applied Human Factors and Ergonomics), on CD-ROM (2010)

User Interfaces for Personal Vehicle on Water: MINAMO

Shunnosuke Naruoka$^{(\boxtimes)}$ and Naoyuki Takesue$^{(\boxtimes)}$

Graduate School of Systems Design, Tokyo Metropolitan University,
6-6 Asahigaoka, Hino-shi, Tokyo 191-0065, Japan
ntakesue@tmu.ac.jp
http://www.sd.tmu.ac.jp/en/

Abstract. MINAMO (Multidirectional INtuitive Aquatic MObility) that we have developed is a personal vehicle on water. We implemented a maneuvering method by weight shift in order to allow the rider to move intuitively on water and to improve the efficiency of tasks on water. However, the maneuvering method by weight shift imposes a physical burden on the rider, so it may be unsuitable for long-term use. In addition, the rider needs some practice before the rider can maneuver it well. Therefore, in this study, we implement new user interfaces to maneuver the MINAMO and compared them experimentally to investigate appropriate UIs.

Keywords: Personal mobility · Vehicle on water · User interface
Maneuverability · Usability

1 Introduction

In recent years, many personal mobility operated by weight shift like Segway (Segway, Inc.) [1] and UNI-CUB (Honda Motor Co., Ltd.) [2] have been developed and attracted. Segway is introduced for security staff's transportation in Haneda airport, Japan. It is expected to expand the rider's view by raising his/her line of sights and to improve maneuverability. UNI-CUB appears in a music video of OK Go, a music artist from USA. They can use both hands freely during their ride on UNI-CUB, and they show fresh and good performance in the music video [3]. As mentioned above, a lot of such kinds of vehicles on land are developed [4] and commercialized. However, how about on water?

At the present time, the transportation on water is mainly ship. There are many works around water such as river cleaning and water quality survey by using ships. Ships are often propelled by the outboard motor at the rear of the ships and can move at steady speed while it's difficulty to turn on the spot and maneuver in a narrow space. Therefore, we think that the efficiency of water activities can be improved by implementing maneuvering method with weight shift that is introduced to the mobility on land. We have developed Multidirectional INtuitive Aquatic MObility (MINAMO, which means "water surface"

© Springer International Publishing AG, part of Springer Nature 2018
S. Yamamoto and H. Mori (Eds.): HIMI 2018, LNCS 10905, pp. 144–155, 2018.
https://doi.org/10.1007/978-3-319-92046-7_13

Fig. 1. The MINAMO

in Japanese) as shown in Fig. 1, which is maneuvered with weight shift [5]. The MINAMO enables the rider to freely move on water with hands-free maneuvering and it is expected that the efficiency of tasks on water is improved.

On the other hand, in Japan, Superhuman Sports Society chants slogan of 'unification of person and machine' to create the future where everyone can sport equally by the fusion of people and technology [6]. In Switzerland, a competition called "Cybathlon" was held in 2016 [7]. In this competition, disables who wear "motorized leg prosthesis" or "powered exoskeleton" perform tasks. Not only the performance of the competitor but also the technology contributes to the result. Thus, the relationship between people and machines is getting close, and various UIs are being developed.

In this study, we have implemented various new UIs to maneuver the MINAMO to investigate the appropriate UIs for water activities experimentally. We will summarize the evaluation of each UI on usability and develop MINAMO into more comfortable water vehicle.

2   Principle of Multidirectional Movement of MINAMO

Many mobile mechanisms that can move multidirectionally have been reported such as multidirectional personal mobility [2,4] and omnidirectional mobile robot [8]. In [4,8], the mechanism arrangements are explained. Propulsion arrangements in an underwater vehicle are introduced in [9].

The MINAMO realizes multidirectional movement on the water using four thrusters fixed as in Fig. 2. The x-y axes and the number of thrusters are defined as shown in the same figure. Each thruster can generate the propulsive force forward and backward. The multidirectional movement and turn are accomplished by adjusting the propulsion of these thrusters.

A method to calculate the output command to the thrusters (Fig. 3) is described below:

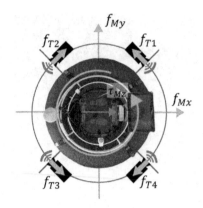
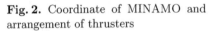
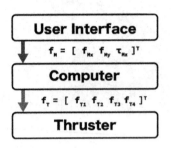

Fig. 2. Coordinate of MINAMO and arrangement of thrusters

Fig. 3. Derivation of propulsion

1. The rider commands the propulsive direction $\boldsymbol{f}_M = [f_{Mx}, f_{My}, \tau_{Mz}]^T$ to MINAMO through the UI.
2. The MINAMO's propulsive command $\boldsymbol{f}_M$ is converted to the thrusters' output command $\boldsymbol{f}_T = [f_{T1}, f_{T2}, f_{T3}, f_{T4}]^T$ by using the transformation matrix determined by the thruster arrangement as shown in Eq. (1).
3. The thrusters' output command $\boldsymbol{f}_T$ is given to the motor drivers.

$$\begin{bmatrix} f_{T1} \\ f_{T2} \\ f_{T3} \\ f_{T4} \end{bmatrix} = \frac{1}{4r} \begin{bmatrix} -\sqrt{2}r & \sqrt{2}r & 1 \\ \sqrt{2}r & \sqrt{2}r & -1 \\ \sqrt{2}r & -\sqrt{2}r & 1 \\ -\sqrt{2}r & -\sqrt{2}r & -1 \end{bmatrix} \begin{bmatrix} f_{Mx} \\ f_{My} \\ \tau_{Mz} \end{bmatrix} \tag{1}$$

where, r is the radius of the thruster arrangement.

3   User Interfaces of MINAMO

In this paper, five UIs of MINAMO, a force plate, a joystick, a gamepad, an inertial measurement unit (IMU) and Myo are evaluated. In this section, these UIs are described in detail.

3.1   Force Plate

Figure 4 shows a force plate used as an UI. It is made of the acrylic plate and four load cells that support the plate. The load cells are placed at the four corners of the force plate as shown in Fig. 5. The sensor values are expressed in Fig. 5 and as follows:

$$\boldsymbol{w}_{FP} = \begin{bmatrix} w_{LF}, & w_{LB}, & w_{RB}, & w_{RF} \end{bmatrix}^T \tag{2}$$

In addition, the total value $w_{sum} = w_{LF} + w_{LB} + w_{RB} + w_{RF}$ corresponds to the weight of the rider.

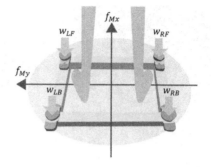

Fig. 4. Force plate   **Fig. 5.** Arrangement of load cells

Based on the above values, COG of the rider on the force plate is calculated and the propulsive direction vector $\boldsymbol{f}_{\mathrm{Md}} = \left[f_{\mathrm{Mxd}}, f_{\mathrm{Myd}}, \tau_{\mathrm{Mzd}} \right]^{\mathrm{T}}$ is set as follows:

$$f_{\mathrm{Mxd}} = k_{\mathrm{FPx}} \frac{(w_{\mathrm{LF}} + w_{\mathrm{RF}}) - (w_{\mathrm{LB}} + w_{\mathrm{RB}})}{w_{\mathrm{sum}}} \tag{3}$$

$$f_{\mathrm{Myd}} = k_{\mathrm{FPy}} \frac{(w_{\mathrm{LF}} + w_{\mathrm{LB}}) - (w_{\mathrm{RF}} + w_{\mathrm{RB}})}{w_{\mathrm{sum}}} \tag{4}$$

$$\tau_{\mathrm{Mzd}} = k_{\mathrm{FPz}} \frac{(w_{\mathrm{RF}} + w_{\mathrm{LB}}) - (w_{\mathrm{LF}} + w_{\mathrm{RB}})}{w_{\mathrm{sum}}} \tag{5}$$

They can be rewritten as a vector expression below:

$$\boldsymbol{f}_{\mathrm{Md}} = \frac{1}{w_{\mathrm{sum}}} \boldsymbol{K}_{\mathrm{FP}} \, \boldsymbol{S} \, \boldsymbol{w}_{\mathrm{FP}} \tag{6}$$

where,

$$\boldsymbol{K}_{\mathrm{FP}} = \begin{bmatrix} k_{\mathrm{FPx}} & 0 & 0 \\ 0 & k_{\mathrm{FPy}} & 0 \\ 0 & 0 & k_{\mathrm{FPz}} \end{bmatrix}, \qquad \boldsymbol{S} = \begin{bmatrix} 1 & -1 & -1 & 1 \\ 1 & 1 & -1 & -1 \\ -1 & 1 & -1 & 1 \end{bmatrix} \tag{7}$$

As a result, the rider can move in the direction with the COG and turn by providing weight on the diagonal line on the force plate such as "w_{RF} and w_{LB}" or "w_{LF} and w_{RB}".

The rider has to keep standing during the operation of this UI to move the COG of rider.

3.2 Joystick

A joystick (Fig. 6) that can detect the knob's inclination in the x-y axes and the rotation in the z axis is used as the second UI. Maneuvering the joystick corresponds with the propulsive direction of MINAMO as shown in Fig. 7. Therefore, multiplying the inclination ratio of the joystick knob by the gain is used for the MINAMO's propulsive command as follows:

$$\boldsymbol{f}_{\mathrm{Md}} = \boldsymbol{K}_{\mathrm{JS}} \, \boldsymbol{\theta}_{\mathrm{JS}} \tag{8}$$

JOYSTICK

MINAMO

Fig. 6. Joystick   **Fig. 7.** Joystick operation

where,

$$
\boldsymbol{K}_{\mathrm{JS}} = \begin{bmatrix} k_{\mathrm{JSx}} & 0 & 0 \\ 0 & k_{\mathrm{JSy}} & 0 \\ 0 & 0 & k_{\mathrm{JSz}} \end{bmatrix}, \quad \boldsymbol{\theta}_{\mathrm{JS}} = \begin{bmatrix} \dfrac{\theta_{\mathrm{JSx}}}{\theta_{\mathrm{JSx,\,max}}} & \dfrac{\theta_{\mathrm{JSy}}}{\theta_{\mathrm{JSy,\,max}}} & \dfrac{\theta_{\mathrm{JSz}}}{\theta_{\mathrm{JSz,\,max}}} \end{bmatrix}^{\mathrm{T}} \tag{9}
$$

The rider can operate this UI in a sitting state since the rider does not have to move the COG. On the other hand, the rider cannot operate the MINAMO without the hands.

3.3   Gamepad

We constructed a remote operation system using a gamepad shown in Fig. 8. The analog stick in the gamepad is used for the maneuvering. The left and right analog sticks are used for translation and turning, respectively. In addition, the arrow pad and LR buttons have the function to change the maneuvering mode and the gain for the propulsion. The gamepad makes it possible to maneuver the MINAMO without boarding it and to maneuver with a bird's-eye view as RC cars or ships.

The MINAMO's propulsive command by the gamepad is calculated as follows:

$$
\boldsymbol{f}_{\mathrm{Md}} = \boldsymbol{K}_{\mathrm{GP}}\,\boldsymbol{\theta}_{\mathrm{GP}} \tag{10}
$$

where,

$$
\boldsymbol{K}_{\mathrm{GP}} = \begin{bmatrix} k_{\mathrm{GPx}} & 0 & 0 \\ 0 & k_{\mathrm{GPy}} & 0 \\ 0 & 0 & k_{\mathrm{GPz}} \end{bmatrix}, \boldsymbol{\theta}_{\mathrm{GP}} = \begin{bmatrix} \dfrac{\theta_{\mathrm{GPx}}}{\theta_{\mathrm{GPx,\,max}}} & \dfrac{\theta_{\mathrm{GPy}}}{\theta_{\mathrm{GPy,\,max}}} & \dfrac{\theta_{\mathrm{GPz}}}{\theta_{\mathrm{GPz,\,max}}} \end{bmatrix}^{\mathrm{T}} \tag{11}
$$

θ_{GPx}, θ_{GPy}, θ_{GPz} are the angles of the analog sticks as shown in Fig. 8.

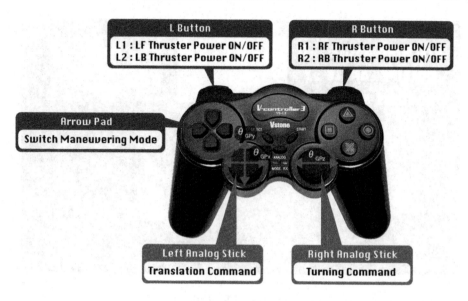

Fig. 8. Gamepad

3.4 Inertial Measurement Unit (IMU)

We implemented hands-free maneuvering using MINAMO's inclination on the water surface [10]. A high performance IMU is used for the inclination detection. This unit has a 3-axis gyroscope (angular velocity sensor), a 3-axis accelerometer, and a 32-bit microprocessor. The inertial information can be obtained by the serial communication of USB connection with PC, and the roll, pitch and yaw angles are calculated based on the inertial information received on the PC. In the touch display fixed to the front of MINAMO, the calculated roll and pitch angles and propulsive direction are displayed. At the same time, the rider can instantly recognize the inclination of 3D MINAMO model as illustrated in Fig. 9.

$$\boldsymbol{f}_{\mathrm{Md}} = \boldsymbol{K}_{\mathrm{IMU}}\,\boldsymbol{\theta}_{\mathrm{IMU}} \tag{12}$$

where,

$$\boldsymbol{K}_{\mathrm{IMU}} = \begin{bmatrix} k_{\mathrm{IMUx}} & 0 & 0 \\ 0 & k_{\mathrm{IMUy}} & 0 \\ 0 & 0 & k_{\mathrm{IMUz}} \end{bmatrix}, \qquad \boldsymbol{\theta}_{\mathrm{IMU}} = \begin{bmatrix} \theta_{\mathrm{pitch}}, \theta_{\mathrm{roll}}, \theta_{\mathrm{yaw}} \end{bmatrix}^{\mathrm{T}} \tag{13}$$

In this study, k_{IMUz} was set to zero since $\boldsymbol{\theta}_{\mathrm{IMU}}$ cannot be adjusted arbitrarily by the rider himself/herself.

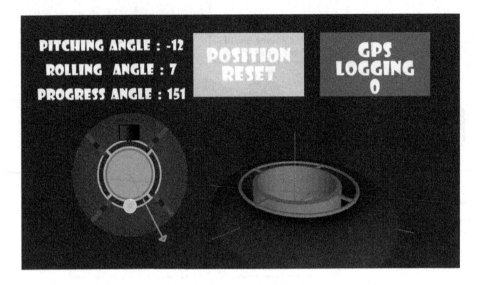

Fig. 9. GUI for IMU

3.5   Myo

Figure 10 shows Myo which is an armband type wearable device with myoelectric sensors, a gyroscope sensor, and an accelerometer [11]. Myo can easily detect gestures of arms, wrists and fingers. We built new hands-free maneuvering system using it.

In this study, we made gestures of 'spreading out hands', 'bending the wrist inward' and 'bending the wrist outward' correspond to 'moving forward', 'turning counterclockwise' and 'turning clockwise', respectively, as shown Fig. 11.

Since maneuvering by the hand gesture does not interfere with the posture of the rider, he or she can move with hands-free maneuvering without a physical burden.

Fig. 10. Myo

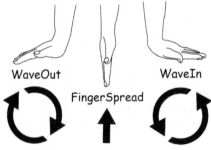

Fig. 11. Hand gesture on Myo

4   Experiments

In this section, the comparative experiments of UIs are described in detail.

4.1   Purpose of Experiments

Using the five types of UIs which were described in the previous section, the difference in the operation performance and the influence to the rider are evaluated experimentally. The purpose of the experiments is to examine the suitable UI for MINAMO.

──────────────────── UIs ────────────────────
(1) Force Plate   (2) Joy Stick   (3) GamePad
(4) IMU   (5) Myo

4.2   Expermental Procedure

The experimental environment is an outdoor swimming pool without flow as shown in Fig. 12. We set the following three procedures as evaluation criteria.

1. Go straight about 10 meters along to the poolside (Straight Running)
2. Turn it clockwise 90 degrees and go straight until it reaches the line of 3 lanes which is about the center of the pool (Cornering)
3. Scoop up the ball thrown from the poolside by using a spoon net (Task on the water)

Procedures 1 to 3 are performed using each UI and the period of time spent performing each procedure is recorded.

In the case of maneuvering by the inclination (IMU), since it does not have a turning function, the rider shall turn it by using the spoon net like an oar.

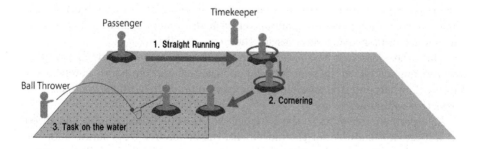

Fig. 12. Experiment environment

4.3   Experimental Results

The overlapped photos during the experiments are shown in Figs. 13, 14, 15, 16 and 17. The experimental results of time are summarized in Table 1.

As seen from Figs. 13, 14, 15, 16 and 17, in the cases of the force plate and Myo, the rider was standing on the MINAMO. On the other hand, in the cases of the joystick, the gamepad and IMU, the rider was sitting on the MINAMO.

Table 1. Experimental results of time

UI	Time[s]			
	Straight	Cornering	Task	Total
(1) Force plate	50.8	38.5	47.9	137.2
(2) Joystick	31.8	15.8	35.7	83.3
(3) Gamepad	29.3	18.9	32.3	80.5
(4) IMU	37.7	20.3	34.2	92.2
(5) Myo	62.2	30.7	26.4	119.3

4.4   Considerations

The differences in time for 'Tasks on the water' between UIs are smaller than those in time for 'Straight Running' and 'Cornering'. This may have been caused by the influences of the position where the ball was thrown in.

The periods of time for 'Straight Running' and 'Cornering', especially, in the cases of the joystick and the gamepad, were very short. The movements in the cases were also stable. Two reasons can be considered.

One of the reasons is that the maneuverability is not easily affected by the condition of the rider. In the cases of the force plate and Myo, the internal variation of the rider himself/herself is used as the command value. Therefore, it is considered that the maneuvering accuracy varies depending on the rider's physical condition. Meanwhile, the operation amounts in the joystick and the gamepad are generated by the inclination and the rotation of the potentiometer. They are easily adjusted in combination with the visual feedback. As a result, the riders can maneuver regardless of their own condition.

The other reason is the ease of turning. It can be said that the joystick and the gamepad have the translational and the turning commands independent from each other, and it is possible to command intendedly without confusing the respective commands.

As described above, while the UI which has the operation amount visualized like a joystick is suitable for the movement, the hands-free UI is suitable for the tasks on the water. Furthermore, it is expected that MINAMO will become a more comfortable water vehicle by constructing a system that the rider can more easily select the UI.

Fig. 13. Experiment in case of force plate

Fig. 14. Experiment in case of joystick

Fig. 15. Experiment in case of gamepad

Fig. 16. Experiment in case of IMU

Fig. 17. Experiment in case of Myo

5   Conclusions

Experimental results suggest that it is important to select the appropriate UI for the specified purpose in order to perform tasks on the water efficiently.

The maneuvering with the force plate or Myo often causes wrong instructions due to human errors. As a result, it took a relatively long time to accomplish the tasks. The reason may be that it is difficult for the rider to control the muscles of the arms and legs perfectly and the command includes noises sometimes. On the other hand, the UIs allow the rider hands-free operation.

As future works, it is necessary to construct a comfortable switching and combining system of UIs to improve the maneuverability.

References

1. Segway Japan. http://www.segway-japan.net/
2. Honda - UNI-CUB. http://www.honda.co.jp/UNI-CUB/specification/
3. OKGO Official Site. http://okgo.net/category/videos/
4. Miyakoshi, S.: Omnidirectional two-parallel-wheel-type inverted pendulum mobile platform using mecanum wheels. In: Proceedings of the 2017 IEEE International Conference on Advanced Intelligent Mechatronics (AIM 2017), pp. 1291–1297 (2017)

5. Kobayashi, D., Takesue, N.: MINAMO: Multidirectional INtuitive Aquatic MObility-improvement of stability and maneuverability. In: Proceedings of the 2014 IEEE International Conference on Robotics and Biomimetics (ROBIO 2014), pp. 741–746 (2014)
6. Superhuman Sports Society Official Site. http://superhuman-sports.org
7. Cybathlon Official Site. http://www.cybathlon.ethz.ch/en/
8. Zhang, Y., Huang, T.: Research on a tracked omnidirectional and cross-country vehicle. Mech. Mach. Theory **87**, 18–44 (2015)
9. Pugi, L., Allotta, B., Pagliai, M.: Redundant and reconfigurable propulsion systems to improve motion capability of underwater vehicles. Ocean Eng. **148**, 376–385 (2018)
10. Takesue, N., Imaeda, A., Fujimoto, H.: Development of omnidirectional vehicle on water (O-VOW) using information of inclination of the vehicle. Ind. Robot Int. J. **38**(3), 246–251 (2011)
11. Myo Official Site. http://www.myo.com/

Hearing Method Considering Cognitive Aspects on Evidence Based Design

Fuko Ohura[✉], Keiko Kasamatsu, and Takeo Ainoya

Tokyo Metropolitan University, 6-6 Asahigaoka, Hino, Tokyo, Japan
fu.a7te@gmail.com

Abstract. It is commonly believed that user understanding is important in the development of products and system, including the concept of human-centered design. However, in order to utilize these results as evidence, it is necessary to construct the evidence itself by a research method appropriately designed for the purpose and object. In this research, first of all, we summarize the type and features of the interview in the user research. In order to understand the user's perception, how the interview was used. The following uses examples to analyze of relationship between the purpose of the investigation and the characteristics of the investigation method.

Keywords: Interview method · Evidence-based design
Cognitive science approach

1 Introduce

1.1 Social Concern for User Research

In recent years, user understanding has been considered important in the design process of products and services, and appreciation of user research is increasing. In addition to the user survey methods such as the web questionnaire, the user interviews are now also highly valued by design companies and consulting companies. The major manufacturing industries have also begun to actively adopt. The following are the content and frequencies included in the design activities of IDEO, which is famous as a design consulting company. It is mainly divided into three parts: hear, create, and deliver [1]. 7 of 17 items included in hear are interviews of some form, and it is conceivable that the interview research is also regarded as an important position in the design process (Fig. 1).

User research is not proactively conducted only by organizations specializing in design approaches such as design consulting firms. Figure 2 shows the results of a survey on the implementation of user researches for 100 high-tech companies in the United States [2]. It is shows that usability testing, interviewing, and surveying appear to be the three most commonly used user research methods practiced today.

Today it is more common, and perhaps even necessary, for companies to incorporate user research into design and development processes.

© Springer International Publishing AG, part of Springer Nature 2018
S. Yamamoto and H. Mori (Eds.): HIMI 2018, LNCS 10905, pp. 156–166, 2018.
https://doi.org/10.1007/978-3-319-92046-7_14

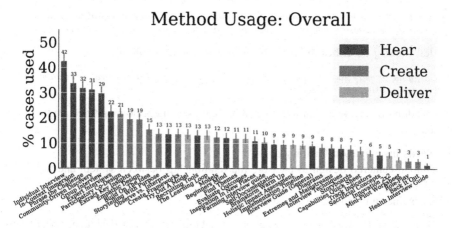

Fig. 1. Percent method usage by case in IDEO. Overall, users use methods from earlier design stages more frequently [1].

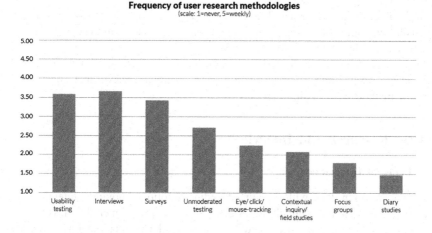

Fig. 2. Frequency user research methodologies in high tech companies. (Source: [2])

1.2 User Research

As mentioned in the previous clause, the interview for user's understanding is often used in companies, and it is necessary to organize the Utilization methods of the interview in order to more effectively utilize the method in various ways. However, interviews are also various, including interviews with reporters. The interview dealing with this research is an interview as user research. Therefore, it is necessary to confirm once again the purpose of user research and its significance.

Robert Schumacher defines user research in "The Handbook of Global User Research" as follows:

User research is the systematic study of the goals, needs, and capabilities of users so as to specify design, construction, or improvement of tools to benefit how users work and live [3].

Jeff Sauro j describes the purpose of user research as follows:

One of the primary goals of conducting user research is to establish some causal relationship between a design and a behavior [4].

In addition, as a field that user research often uses, the following three are the main trends.

- Needs investigation for innovation
- Human-centered design to measure the relationship between system and user
- Marketing
- Consumer understanding marketing perspective.

This paper focuses on user research on a human-centered design perspective to measure the relationship between the system and users. Surveys focusing on cognitive parts of people are sufficiently considered in the other two, but this time, by organizing the research method based on the human-centered design, it is possible to apply the application in other fields later Also possible.

In the field of human-centered design, in order to explore what ease of use, comprehensibility, comfort for users, it has often been researched from the cognitive aspect the interaction when a person touches an object.

In this research, we aim to systematize the interview method in such surveys and work on the disassembly of investigation cases using interviews that considers the interview method and cognitive aspects.

2   User Research and Interviews

2.1   Types of Interviews

Interviews are often done with the aim of finding emotions, thoughts, intentions, past behaviors, etc. that are difficult to directly observe [5]. Generally classified into Structured interviews, Semi-structured interviews, and Unstructured interviews from the degree of their structure [6]. A summary of each feature is shown in the Table 1.

Semi-structured interviews can pursue parts of interest to the investigators during the interview. This type of interview is widely used in qualitative research. Structured interviews are a convenient way to compare how different subjects answered the same question, or when multiple investigators conduct interview surveys on teams.

The unstructured interview is characterized in that the content of the question is not decided in particular and it is a free conversation, often used in combination with participation observation. Because we have to think and develop questions in response to the subject's remarks, the result of the investigator's skill and experience broadly varies.

Table 1. Characteristic of each interview type

	Characteristic	Data	Space
Structured interviews	An interview managed so that there is no difference for each survey for multiple survey subjects by setting question contents and answer format	Qualitative, Quantitative	Telephone, controlled space
Semi-structured interviews	A question is made according to the interview guide which decided rough direction, the question can be changed according to the flow of the dialogue, and it is possible to hear the opinion flexibly	Qualitative	Controlled space, field
Unstructured interview	Question contents are not decided, and you can explore problems from natural conversation	Qualitative	Controlled space, field, visiting place, everywhere

2.2 Previous Research

In user research, the research method is not limited to a single research method, and it is possible to try user understanding more effectively by combining a plurality of investigation methods according to the purpose of the investigation. For interviews, it is also important to conduct multiple interview methods and to consider combinations with other user research methods as shown in the Table 2. In the table, Nielsen Norman Group performing Evidence-Based User Experience Research, Training, and Consulting on his website, It summarizes the user's investigation method focusing on the product development phase [7]. In this way, you also need to consider what kind of investigation is effective in each phase in the design process.

Table 2. The table below summarizes these goals and lists typical research approaches and methods associated with each. (Source: [7])

	Product developement phase		
	Strategize	Execute	Assess
Goal	Inspire, explore and directions and opportunities	Inform and optimize designs in order to reduce risk and improve usability	Measure product performance against itself or its competition
Approach:	Qualitative and quantitative	Mainly qualitative (formative)	Mainly quantitative (summative)
Typical methods:	Field studies, diary studies, surveys, data mining, or analytics	Card sorting, filed studies, participatory design, paper prototype, and usability studies, desirability studies, customer emails	Usability benchmarking, online assessments, surveys, A/B testing

3 Hearing Method Considering Cognitive Aspects

From the characteristics of user research and user research in product development in the previous chapter, this chapter classifies interviews as patterns. In the next chapter, we will also try to model it in the actual research.

3.1 Purpose of Interview in Consideration of Cognitive Aspects

As an aim of the interview that considers cognitive aspects, focusing on how the human side receives in human-system interaction, to understand its structure and characteristics. Furthermore, it is supposed to be a hint to a design that is beneficial and effective for people.

3.2 Patternization of Interviews that Considers Cognitive Aspects

In an interview that takes cognitive aspects into consideration, contents related to human-system interaction are targeted. Therefore, the degree of reproducibility of the situation of interaction is a point to consider in investigation. We focused on "the temporal distance between interaction and interview" and "whether or not to consider factors involved in the environment at interaction" as factors influencing the reproducibility of the situation. Two axes of "immediate − retrospective − retrospective of self-experience" and "no presentation, system only, system + environment" were set, and each model corresponding to each of the biaxial matrix was taken as each model. We propose the five patterns shown in the Fig. 3. as a hypothetical model of interviews that considers cognitive aspects.

4 Case Analysis

In user research, the investigation method is rarely limited to one method, and it is possible to try user understanding more effectively by combining a plurality of investigation methods according to the purpose of the investigation.

In this chapter, we going to discuss two cases of the interview research actually conducted by the authors and the interview research used in the same laboratory project. Collect the findings in research design for user understanding by organizing and analyzing the investigation cases by combining the classified interview pattern with the user research method as shown in the Table 2 in the previous chapter. In order to understand the user's perception, how the interview was used. The following uses examples to introduce the purpose of the investigation and the characteristics of the investigation method.

4.1 Case Study Using Interview Method Considering Cognitive Aspects - 1

− Investigation of walking in an indoor space for low-vision people to investigate spatial recognition during walking

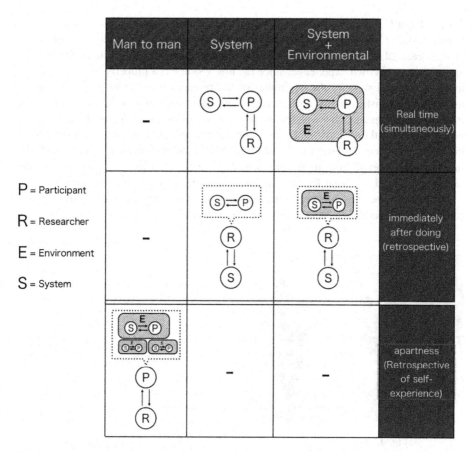

Fig. 3. Hypothetical models of interviews that considers cognitive aspects focused on "the temporal distance between interaction and interview" and "whether or not to consider factors involved in the environment at interaction"

4.2 Investigation Outline

In order to find out what requirements the low-vision people has in the space that easy to walk, we conducted an interview while walking and a retrospective interview after the behavior on the recognition of the space during indoor walking.

4.2.1 Investigation Procedure

First of all, we interviewed the viewing function status and the visual experience in everyday life in a given interview room, and then asked for a predetermined route in the facility to walk. Interview on how to see the space and behaviors that I was concerned while walking (Fig. 4). After walking, return to the predetermined interview room and interview again while looking back on the walking contents.

4.2.2   Result

Figure 5 shows the model of this investigation. The interview was divided into three phases, each of which was combined with a different model. In an interview on the appearance in his own experience, since the low vision has a property that each person looks different, I went to the beginning of the research in order to grasp the fundamental characteristics of its appearance. A major feature of this investigation is an interview accompanied by walking in real space that was conducted in the second phase. When walking, I encounter multiple pattern spaces. I asked in real time what kind of appearance and how I perceived it at that. Some participants, together with the investigators, re-exploring the unconscious cognitive part of the physical sense in the real space to find out what factors influence the formation of spatial cognition.

In the retrospective interview after walking, we did deep digging of parts that we could not hear during walking and looked back the whole.

Fig. 4. Research scene

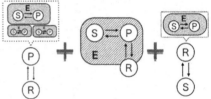

Fig. 5. Investigation model

4.2.3   Discussion

Although the participants of the research were spatial recognition of low vision people this time, it is difficult to grasp the characteristic behavior and cognition on how to recognize spaces themselves and speak themselves. Therefore, as a research method, we adopted a form of interviewing with walking.

The walking interview was done in the form of an almost unstructured interview. For that reason, the characteristics of investigators' skills, conversation on the spot, information obtained depending on the circumstances could change dramatically as seen in the characteristics of the unstructured interview given in Sect. 2. It is a way to gather information well by presetting some hypothesis and research theme and interviewing as much as possible on that axis.

The use of an interview accompanied by such behavior is thought to be effective when the influence of the environment is greatly influenced when using the system or the system can not be presented unless a person is enclosed in the environment (Such as space and traffic etc.).

4.3   Case Study Using Interview Method Considering Cognitive Aspects - 2

– Influence of visual and sound feedback of the feeling of sucking in the cleaner

4.3.1   Investigation Outline

It is an investigation of interfaces and sounds for the feeling that the cleaner is sucking. In this investigation, we aimed to explore changes in the cognition of functional functionality of products, such as how changes in visual stimuli and sound stimuli affect sucking sensation.

As an interview method, we conducted a retrospective interview after presenting the system.

4.4   Investigation Procedure

First of all, three stimuli stimulus patterns and three visual stimulus patterns are presented to the investigate participants and then they are completed on the two evaluation sheets shown in the Fig. 6. After that, the researcher inquires the participants about the reason for the evaluation value, how to receive the stimulus, etc. with reference to the evaluation sheet.

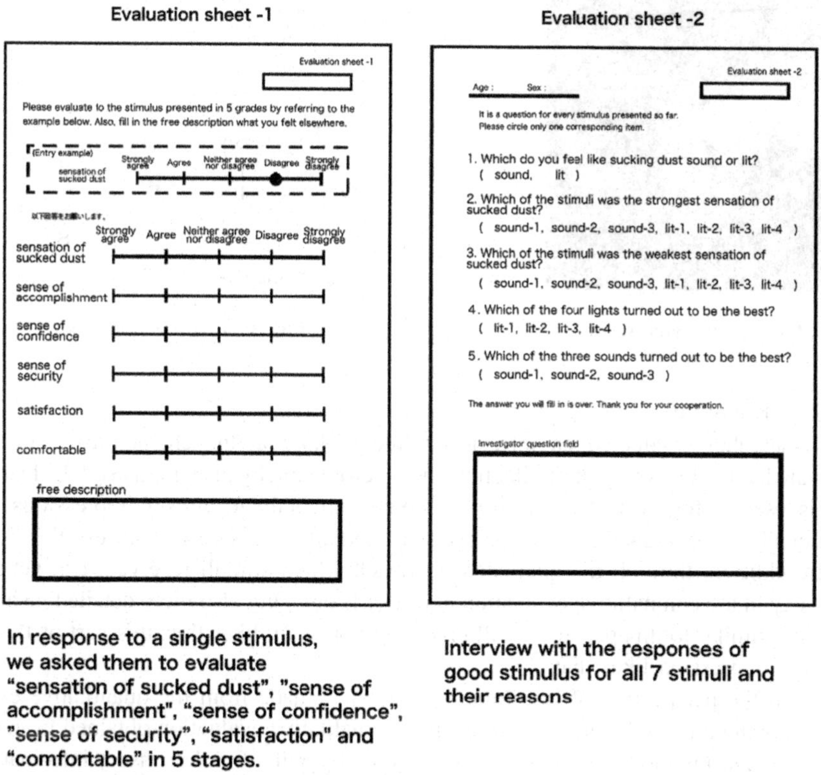

Fig. 6. Questionnaires

4.4.1 Result

The investigation model in this research is shown in the Fig. 8. The participant received stimulus of one element constituting the system in the laboratory (managed space) and asked the questionnaire to write the evaluation by SD method.

Therefore, it becomes a model as shown on the left side of the Fig. 7. Thereafter, the investigators interview the reasons for the evaluation and the contents of the interaction. At this time, the participants will reply while looking back at the time of interaction, so it becomes a model of "system × retrospective interview".

Since this time the questionnaire was filled out by the researcher, it is a questionnaire survey + "system × retrospective interview" The structure degree of the interview is based on the questionnaire, so it seems that it is close to semi-structured interview. As a result, in addition to the quantitative evaluation in the questionnaire, qualitative feedback of the number of investigators (10) × the number of items (5) regarding the reasons of each evaluation of the sheet 2 was obtained.

Fig. 7. Research scene

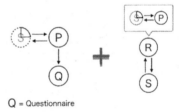

Q = Questionnaire

Fig. 8. Investigation model

4.4.2 Discussion

The researcher conducted an interview while looking at the characteristics of the evaluated value by asking the evaluation value beforehand by questionnaire. The feature of this research form is that we can quantitatively extract the features of impressions and then qualitatively draw out what has been decided after this assessment result.

In addition, through the qualitative feedback which stimuli received, can get not only stay in the comparative evaluation like A or B or C, but also more detailed insights which is similar to this part of A will give the impression likes this, this part of B will give the impression likes that.

From the perspective of cognition, the graph is made from the quantitative evaluation obtained from the questionnaire and the qualitative evaluation obtained from the interview. During model, I referenced the system diagram of multi sensory information processing and functional evaluation of Mr. Jingu [8] (Fig. 9).

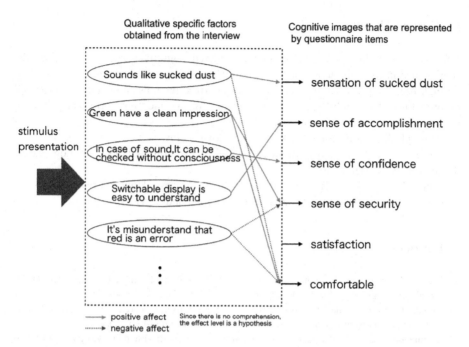

Fig. 9. Relationship diagram of interview and questionnaire considering cognitive aspect in this investigation

I use 6 kinds of words of impression to express the impressions which are prompted about the stimuli in this experiment, A square which made up of a dotted line is the processing activities performed by a unconscious person in the brain, the left arrow is stimulated, the investigators after consciously accept the stimulation, then consciously to evaluate 6 kinds of evaluation project.

In general, when a questionnaire is used to evaluate the impression, the result is usually not seen in the square of the dotted line. But when developing systems and products, it is very important to understand which elements are based on what kind of value judgments, or how emotionally affect the formation of impressions. This is not only an understanding of the user's cognitive benchmarks, but also helps to reduce the differences between the designers and the recipients.

The red and blue arrows in the figure from the qualitative data combine what kind of factors positively or negatively influence the evaluation as output.

For example, in this case, "sounds like inhaling garbage" play a role in the inhalation sense of vacuum, but on the other hand, it brings unpleasant feelings. Therefore, the design is carried out from the perspective of "avoiding the sound that can be directly associated with the garbage, while giving feedback to the inhalation".

5   Future Works

This paper focused on interviews that considered cognitive aspects, described the significance in user research, and modeled interviews and investigated case examples for their use. We picked up two cases and categorized them as an interview model and examined how the combinations and the features of the interview form worked in the research.

Through this study, future tasks will be shown below.

Issues for improving the accuracy of systematization: Analysis in more cases, Review of the model by it.

Other developments: Relationship with design process, How the research system changes due to differences such as objects, systems, and services.

References

1. Fuge, M., Agogino, A.: Pattern analysis of IDEO's human-centered design methods in developing regions. J. Mech. Des. **137**(7), 071405 (2015). research-article
2. Brent, T.: Product managers-user research on your user research (2017). https://www.linkedin.com/pulse/product-managers-user-research-your-brent-tworetzky
3. Schumacher, R.: The Handbook of Global User Research, 1st edn. Morgan Kaufmann Publishers, California (2009)
4. Sauro, J.: 4 Experiment types for user research (2013). https://measuringu.com/experimenting-ux/
5. Patton, M.: Qualitative Evaluation and Research Methods, 3rd edn, pp. 169–186. Sage Publications, California (1990)
6. Terashita, T.: Qualitative research methodology-for scientific analysis of qualitative data. Jpn. J. Radiol. Technol. **67**(4), 413–417 (2011)
7. Rohrer, C.: When to use which user-experience research methods. Nielsen Norman Group Homepage (2014). https://www.nngroup.com/articles/which-ux-research-methods/
8. Jingu, H.: Sensory Evaluation Learned by Practical Case, 1st edn. JUSE Press Ltd., Tokyo (2016). (in Japanese)
9. Ikeda, M.: Interview method in Cultural Anthropology. http://www.cscd.osaka-u.ac.jp/user/rosaldo/000602episte.html
10. Kuniavsky, M.: Observing the User Experience - A Practitioner's Guide to User Research, 22nd edn. Morgan Kaufmann Publishers, California (2003)
11. Sauro, J., Lewis, J.R.: Quantifying the User Experience: Practical Statistics for User Research, 2nd edn. Morgan Kaufmann Publishers, California (2016)
12. Mulder, S., Yaar, Z.: The User is Always Right: A Practical Guide to Creating and Using Personas for the Web, 1st edn. New Riders Press, Indiana (2006)

K-Culture Time Machine: A Mobile AR Experience Platform for Korean Cultural Heritage Sites

Hyerim Park[1], Eunseok Kim[2], Hayun Kim[1], Jae-eun Shin[1], Junki Kim[2], Kihong Kim[2], and Woontack Woo[1(✉)]

[1] KAIST UVR Lab, Daejeon 34051, Republic of Korea
{ilihot,hayunkim,jaeeunshin,wwoo}@kaist.ac.kr
[2] Augmented Reality Research Institute KAIST, Daejeon 34051, Republic of Korea
{scbgm,k.junki34,kihongkim}@kaist.ac.kr

Abstract. In applying MR (AR/VR) technologies to cultural heritage sites, the design and implementation of mobile MR applications have mostly lacked a holistic and systematic approach in the viewpoint of retrieving existing content and providing personalized experiences to visitors on the fly. To address this issue, we designed and implemented the KCTM application in which various types of content from different databases are brought together and reorganized by a newly proposed metadata schema that assigns spatio-temporal information to them for the purpose of visualizing them in MR environments. According to the user evaluation, most of the participants expressed strong interest in touring the heritage sites augmented with various related content.

Keywords: Cultural heritage · AR framework · Metadata schema

1 Introduction

Over the past two decades, various attempts have been made to actively incorporate AR and VR technologies to mobile applications for the purpose of digitally enhancing on-site experiences at cultural heritage sites. However, although these attempts considered how the technological elements for a mobile application should be placed and implemented for a better user experience, the type of content that should be given to the visitors and how such content should be organized atop the technology were not addressed. For example, the applications mostly focused on experimenting a visual overlay technology, such as marker detection, and feature tracking. Furthermore, a platform that can process retrieved data into a system, and effectively bind and link such data to suit the current context of the visitor is not available. To improve these issues, we designed and implemented KCTM application in which the retrieved and structured data would be shown considering users' various geo-context through augmented and virtual visualization system.

© Springer International Publishing AG, part of Springer Nature 2018
S. Yamamoto and H. Mori (Eds.): HIMI 2018, LNCS 10905, pp. 167–180, 2018.
https://doi.org/10.1007/978-3-319-92046-7_15

The 'K-Culture Time Machine' project is a research and development project for culture and technology hosted by the Ministry of Culture, Sports and Tourism in Korea from 2014 to 2016. It aims to develop technology for collecting cultural contents with spatial and temporal information, creating semantic correlation of them, and visualizing them on the AR and VR service platforms. In order to carry out the project, first, we developed data processing technologies that create, store, and retrieve cultural content based on the semantic relationship between them. Through these studies, various cultural contents of different organizations can be structured into new metadata-based cultural contents with temporal and spatial information, which can be provided to tourism or IT industries and to the public. This allows users to semantically search and utilize diverse cultural heritage contents.

Next, based on robust and context-aware visualization techniques, users can experience more immersive content through real or virtual worlds. In augmented reality visualization, the system enables real-time 6-DOF camera tracking and pose estimation by utilizing a 3-dimensional trackable map. This allows the user to track and identify the PoI (point of interest) and the OoI (object of interest) in real time in the field, and retrieve and view related information in AR. Virtual reality visualization is based on 360° panoramic image data obtained from the field. The system can render various contents, taking the camera posture of the user's wearable device into consideration such that the sense of presence can be improved even in the virtual environment. Developed technology was launched with the 'K-Culture Time Machine' application on the iOS App Store on May 23, 2017. The application runs on a trial basis in Changdeokgung Palace and is undergoing improvements in technology and services through user validation and feedback.

In the evaluation of the system, AR and VR visualization module processing speed was measured by FPS change. For each FPS change measurement, the FPS showed an average of more than 50. The augmented reality visualization module showed an average tracking speed of over 40 FPS. For the user evaluation, we conducted in-depth interviews during two days of open demonstration sessions from November 2 to 3, 2017 in the laboratory. A total of 43 participants experienced the application that was set in VR mode, and were asked to freely state their impressions on the experience they just had. Most of the participants expressed interest in viewing real-life locations and culturally meaningful monuments in a virtually and realistically recreated space. However, they stated that it was difficult to approach and acquire various hierarchical and multiple layers of information on PoIs and OoIs, and it would be more effective if such layers of information were organically linked and transmitted in a more interesting way.

The remaining sections are organized as follows. In Sect. 2, we will examine how augmented reality and virtual reality have been applied to cultural heritage sites and the problems encountered. Section 3 describes the entire system of the application and details the data processing and visualization modules, while Sect. 4 presents the system evaluation and user evaluation contents. Section 5 gives the conclusion and planning for future project.

2   Related Work

In the case of AR, the heritage industry has for some time considered the technology to be a key component in attracting and engaging visitors in unprecedented and innovative ways, thereby creating novel values [1]. Střelák et al. [2] showed through extensive user evaluations that AR technology is well received overall as a platform for digital guides at cultural heritage sites. However, the design and implementation of AR mobile applications for cultural heritage sites have mostly lacked a holistic perspective and a systematic approach in terms of how the effect of features specific to AR can be maximized while retrieving and providing existing content to visitors at timely moments during their individual, personalized experiences.

Casella et al. [3] conducted a state-of-the-art review of AR mobile applications for cultural heritage sites through observations held both online and offline. Before this work, available applications were mostly focused on introducing the technology itself and experimenting with basic components of AR for cultural heritage sites, such as visual overlay, marker detection, feature tracking, GPS, PoIs, location-based narrative, contextual data, and personalization. Starting with the Touring Machine made by Feiner, MacIntyre, and Höllerer in 1997, they noted the development of AR as a platform that could potentially reinvent methods of cultural mediation. The more recent but nonetheless outdated examples they reviewed, such as Archeoguide and iTacitus, laid the groundwork for the now recurring concepts of mobile AR applications that could only be realized with technology as yet unavailable. As mobile AR technology became more utilizable, museums began launching applications such as Streetmuseum, Londinium, and Phillyhistory with a common focus on visualizing related 2D and 3D contents of historic value over the physical realms of places they are related to.

More recent applications in the field have moved on from this exploratory stage to examine various design principles that can take full advantage of AR technologies at hand and thus provide a better sense of engagement to the visitors at heritage sites. de los Rios et al. [4] investigated how AR coupled with gamification, storytelling, and social media can engage visitors more actively by way of participation using a mobile device. In order to define appropriate functions that the mobile application should provide, it referred to actual use cases of TAG GLOUD. Rattanarungrot et al. [5] proposed a system of presenting cultural objects on an AR platform by accessing an aggregated RCH data repository. Although they considered how the technological elements for a mobile application should be placed and implemented for a better user experience, they did not address what type of content should be given to the visitors and how such content should be organized using the technology. Furthermore, when the systematic handling and delivering of data is a priority for mobile AR services in cultural heritage sites, the relation among the contents retrieved has not been sufficiently considered.

In the case of VR, various studies have been conducted to provide immersive experience without going directly to cultural heritage sites. For example, inside the virtual cave, users can experience various multimedia such as 2D and 3D animations, as well as 3D cinematography related to 'the Mogao Grottoes at Dunhuang' in an immersive and interactive way [6]. Students could explore a 360° video displaying the cultural heritage site, and solve problems designed by a gamification method [7]. Visitors could also experience real-scale graphics on the immersive stereoscopic display with hand gesture interaction [8]. However, these projects mostly focused on presenting an immersive display of cultural heritage sites. They did not utilize various multimedia from scattered databases related to the heritage site, and did not provide personalized experience in terms of considering various users' context.

3   KCTM Mobile Application

3.1   System Overview

The application provides mobile-based cultural heritage AR guides using vision-based tracking of cultural heritage sites. Through the embedded camera in a smartphone, the application can identify a cultural heritage site and provide related information and contents of the heritage site. At the same time, it can provide remote experience over time and space for cultural heritage or relics through a wearable 360° video-based VR. Users can remotely experience cultural heritage sites using a 360° video provided by installing an app in a smartphone HMD device, and can search information on historical figures, places, and events related to the cultural heritage site. Furthermore, 3D reconstruction of lost cultural heritage can be experienced.

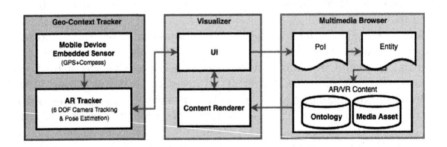

Fig. 1. KCTM application system diagram

We designed the system diagram of K-Culture Time Machine application as shown in Fig. 1. Originally, it comes from the modified AR reference model [9]. The modified AR reference is a supplementary model based on the AR reference model, which is an ISO standard, and provides a standardized workflow for interoperability with other augmented reality applications. In Fig. 1, the application obtains sensor information of the real or virtual world and the user through embedded sensors. Then, an AR tracker performs user localization and

PoI recognition through it. A user can select PoI by touching the UI button, and related PoI contents are retrieved through the data retrieval process. Finally, the content renderer generates the AR or VR scene to provide the content of the selected PoI to the user.

3.2   Geo-Context Tracker

AR Tracker. In order to visualize heterogeneous AR contents through the application, real-time camera tracking and pose estimation techniques are necessary. In our previous study [10], we proposed an all-in-one mobile outdoor AR framework and a real-time camera tracking prototype for cultural heritage sites. Through this AR framework, we addressed how to acquire a 3-dimensional trackable map and how to enable robust 6-DOF camera tracking and pose estimation. An ORB [11] keypoint-based standard SfM (structure-from-motion) pipeline was applied to reconstruct keypoints and camera pose of keyframes. In order to stabilize real-time camera tracking on a mobile device, the multi-threading technique was applied. The foreground thread tracks the inter-frame movement of ORB keypoints, and estimates the camera pose by solving 3d-2d correspondences. The background thread collects new candidate key-points, which are evaluated by a keypoint matching procedure. In our implementation, both threads work concurrently. Finally, this AR tracking module was stabilized and applied to the application. Figure 2 shows the overall procedure of our AR tracking module.

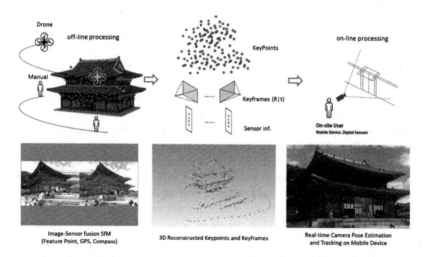

Fig. 2. System workflow diagram of real-time camera tracking module. (left) Image acquisition and SfM process, (middle) generated trackable map, and (right) real-time camera pose estimation and tracking demonstration on mobile device.

VR Tracker. The virtual reality visualization module was developed in order to solve the spatial constraint that the user has to be in the field and to provide an immersive experience in the wearable platform. Virtual reality visualization is based on 360–panoramic image data obtained from the field, allowing users to experience cultural heritage in a virtual environment. This module is implemented based on iOS, and the camera position estimation (6-DOF) technology is applied through 3D spatial feature map and image matching of the wearable device. This module is based on user's wide range of position information by tracking the direction and field of view (FoV) of the wearable camera. It applies real-time content rendering technology considering the camera posture of users' wearable devices for improving the sense of presence of the user.

3.3   Data Processing and Delivery

The process of contents retrieval for the selected PoI is divided into two processes. One process retrieves information about the cultural heritage itself and multimedia contents from the integrated ontology, which aggregates the existing cultural heritage databases. The other process retrieves multimedia contents created in modern times related to the cultural heritage concerned such as movie and drama. In the cultural heritage content part, the application uses Korean cultural heritage web-databases to deliver various cultural information and content related to the heritage site. First, we selected five cultural heritage databases that provide tangible and intangible cultural heritage information: 'Cultural Heritage Administration of Korea', 'Museum Portal of Korea', 'National Palace Museum of Korea', 'Encyclopedia of Korean Culture', and 'Culture Content'. Then, we employed the Korean Cultural Heritage Data Model (KCHDM) [12] to aggregate heterogeneous data from these five web-databases.

KCHDM comprises 5 super classes (Actor, Object, Place, Time, and Event) and 78 properties such that it represents the context of cultural heritage according to "who, when, where, what, and how"s. To construct the semantic information model of the targeted heritage site, we used the information modeling framework proposed by Kim et al. [13]. In the first phase of information modeling, we established a knowledge base comprising cultural heritage entities and their relationships, which show the cultural heritage features associated with PoIs by class and the inference rules of KCHDM. Then, we linked four types of web resources (i.e., text, audio, image, and video contents) to each cultural heritage entity. In the last phase, the information model was redesigned for mobile application users such that they can access it without unnecessary steps or duplicated contents.

The media asset database was built to contain multimedia content such as historical movie and drama related to the cultural heritage sites. In designing the metadata element and schema, we referred to three existing metadata sets: 'W3C core set of multimedia metadata [14]', 'Metadata Element and Format for Broadcast Content Distribution [15]', and 'Metadata Schema for Visualization and Sharing of the Augmented Reality Contents [9]'. Based on these references, we proposed the newly modified and extended metadata elements and schema [16].

It aims to systematically manage the spatial and temporal information of video in the AR system. The media asset database was written in MySQL and the video resource from media asset was incorporated into AR content when requested by the app. KCTM app contains information on a PoI, as well as information related to various cultural heritage sites including tangible and intangible heritage information linked to the PoI.

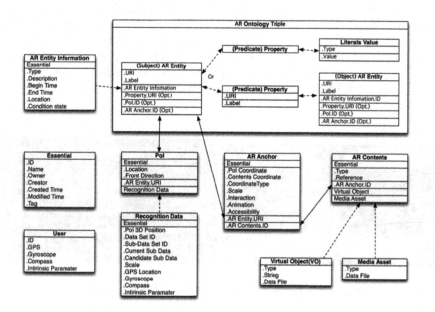

Fig. 3. Revised MAR contents metadata schema [17]

To retrieve cultural heritage information from the ontology and the relational database, we implemented the SPARQL query for KCTM application. In the case of offline retrieval, the application can retrieve related contents of PoI at the initialization process through the SPARQL query, and then it provides the contents for the selected PoI to the user without additional retrieval process. Otherwise, in the case of online retrieval, the application can provide only the contents related to the selected PoI when it is required by the user without the initialization process. Users can choose appropriate retrieval methods according to the network bandwidth and contents volume. The retrieved content is stored according to a standardized metadata structure (Fig. 3) [17]. According to previous studies on augmented reality metadata for cultural heritage [18], various types of AR contents can be integrated, and the reusability of other AR contents, which follow this standardized metadata structure, can be guaranteed.

3.4   User Interface and Content

The user interface has a wearable mode and a mobile mode, and mode switching is possible in each real and virtual space as shown Fig. 4. Although there is no difference between modes, it is implemented to allow users to select the optimal experience mode according to the situation. In order to move to the next place in each virtual space, users should touch the arrow in the direction of movement in a similar manner to a general 3D map service such as Kakao map load view and Naver map, and then the visualizer provides a 360° image display. When a user touches a PoI that exists in this environment, related information and multimedia contents can be checked. The wearable mode provides users with virtual reality experience by using smartphone HMD such as Samsung Gear VR. The user can select the target through eye gaze for a certain period of time in the user interface.

Fig. 4. Wearable mode (left) and mobile mode (right)

Fig. 5. Multimedia information of PoI in Changdeokgung Palace

The user can experience various multimedia information such as text, sound, photo, video, and 3D content in a specific object or place through interaction with the 2D UI of visualization module as shown Fig. 5. Especially, 3D content design and development was carried out to optimize 3D models of the existing cultural heritage database to the HMD mobile device visualization module. It was referenced from national cultural heritage portals such as 'E-museum' and

'Cultural contents'. Lightweight and resolution-conscious textures were produced such that 3D content could be operated in mobile devices. The optimized 3D model was stored and managed in the multimedia database. Users can experience lost cultural heritage by augmented reality and virtual reality through this content. For example, the 'Seungjungwon' building is a government office where all documents such as the letter of the king and the writings of the servants to the king should go through. It played a great role as the king's secretarial agency, sometimes exercising power over other institutions. However, it is lost and visitors cannot see the building in reality. Thus, based on the Dongwoldo, the picture of the palace of the late Joseon dynasty, it was restored through 3D contents and provided to users as shown Fig. 6.

Fig. 6. Virtual restoration of 'Seungjungwon' building

4   Evaluation

4.1   System Evaluation

The AR tracker processing speed was measured by FPS change according to real-time tracking. The VR tracker processing speed was measured by FPS change according to the speed of reading related data when moving in virtual space and the speed of reading information when manipulating the UI. For each FPS change measurement, the FPS showed an average of more than 50. The augmented reality visualization module showed an average tracking speed of over 40 FPS (Figs. 7 and 8).

4.2   User Evaluation

During two days of open demonstration sessions conducted in an indoor lab environment, we observed the behavior and response of various users as they experienced our mobile application. The demonstration sessions were held from November 2 to 3, 2017 at the KAIST UVR Lab. A total of 43 participants voluntarily used the app that was set in VR mode, wearing VR headsets that were adjusted according to their preferences. The age group of participants varied from elementary school students to teenagers and adults in their thirties and

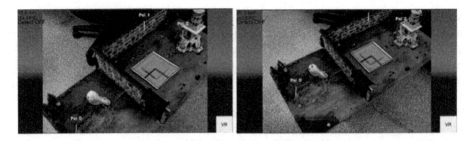

Fig. 7. FPS change in AR tracker

Fig. 8. FPS change in VR tracker

forties. The average duration of the trials was approximately 5 min. Overall, the participants showed much movement, turning their head and bodies around to fully grasp the 360° space of the reconstructed parts of the Sejongro and Changdeokgung Palace. They mostly tried to interact with all the points of interaction that were given as pointers or menu buttons within their field of view, which can be accessed by aiming and focusing their gaze on the red dot located at the center of their view. Through their means of interaction, they tried to proceed to all layers of information enabled by the ontology-based structure of the data system (Fig. 9).

After completing their turns at trying the VR part of the application, we carried out in-depth interviews with all the participants in which they were asked to freely state their impressions and opinions on the experience they just had. Most of them expressed interest in viewing real-life locations and culturally meaningful monuments in a virtually and realistically recreated space. They took note of the fact that an entirely fictional place was included within the real space. However, the overlying common factor in their experiences is the pronounced difficulty in navigating and following the content provided with each response they were required to make with their gazes. Owing to the fact that most of the participants had not used VR headsets prior to the demonstration session, controlling the hardware itself was the initial hurdle they encountered while trying to ease into the experience provided by the application. They also felt that accessing multiple layers of information on the PoIs within their field of view was not intuitive enough and required some strained effort on their part. Most importantly, they felt that text information and images that were augmented virtually in a consecutive order had no direct relation or meaning to the one that came before or after.

Fig. 9. Introduction of KCTM App for the user evaluation

5   Conclusion and Future Work

The 'K-Culture Time Machine' application uses real-time tracking of user's location and pose through AR and VR visualization technology, and provides various existing multimedia related to cultural heritage based on structured metadata schema and ontology. This is significant in that it experimented with how to utilize and visualize existing data on AR and VR platform, compared to existing apps that simply focused on creating and showing new virtual content related to cultural heritage sites. However, as mentioned in the user evaluation part, one of the critical weaknesses of our current system is that the ontology-based data organized through our own data structure cannot as yet be provided through an effective means to utilize the content in coherent, cohesive, and meaningful ways that also consider the context of the user and the paths they are given to navigate multiple PoIs. As a result, users felt that the information provided to them was often fragmented, and lacking a clear point of view to sufficiently guide them through an unfamiliar site, or reorient them to a site of historic value that they had already known.

In order to address this issue, we devised a framework that utilizes the merits of our ontology and metadata-based multimedia data retrieval system to remediate each content in a new context through storytelling in our follow-up study [19]. The framework focused on bringing contextual information and related multimedia content of differing PoIs within a single heritage site together by referring to principles of experience design and gamification in order to produce guide routes that connect multiple PoIs in a given order. After designing a themed narrative that embeds video data related to the PoIs, retrieved and filtered through the metadata system we created, we conducted an extensive between group user evaluation at the Changdeokgung Palace to compare the detection and search-based experience of the current KCTM application to an AR mobile application built around the narrative we proposed. The results showed that storytelling as a content remediating and route-creating method delivers a significant impact in enhancing the level of immersion and interest of users that consult additional digital material in an AR and/or VR environment to achieve the goal of enhancing their overall experience of a cultural heritage site.

In a follow-up project, we plan to include multiple narratives to upgrade our current version of applications in order to provide various route options that can at the same time structure and arrange the content of our data in meaningful ways and effectively link various PoIs within the site. The subsequent project began in June 2017. Its subject is 'Technology development for building cultural content creation and production infrastructure utilizing spatial data'. The objective of this research project is to reduce the cost of producing content using real world's spatial information. For this, we have started developing technology that converts mass spatial information produced by the country into a specialized spatial asset for the cultural content of the cultural industry. The motivation for these studies is as follows. There is a growing interest in the use of high-precision real world spatial data in games, movies, art, tourism, advertising, and entertainment, and the costs associated with data deployment are increasing. Thus, it is necessary to provide data acquisition infrastructure to reduce costs in such areas. The field of study consists mostly of four sections such as creating spatial contents, end-user service, standardization, and content demonstration services.

First, the creation of spatial culture content is related to the technology linking multi-layer space and diverse multimedia. It includes technologies to create user-generated spatial content in addition to conventional multimedia content. The second part allows users to semantically search multimedia based on temporal and spatial information and to visualize content on an end-user system. It also includes technological development that applies these content to commercial advertisement and personalized service. The spatial content standardization involves standardizing cultural content related to multi-layered space and guaranteeing copyright of them. Through these studies, affordable cultural content, which has very accurate spatial assets, can be served to the public. Finally, in the content demonstration services, we will implement a more user-centered service by adopting the multiple narratives mentioned above in order to provide more meaningful user experience in the various PoIs.

Acknowledgments. This research is supported by Ministry of Culture, Sports and Tourism (MCST) and Korea Creative contents Agency (KOCCA) in the Culture Technology (CT) Research & Development Program 2018.

References

1. Tscheu, F., Buhalis, D.: Augmented reality at cultural heritage sites. In: Inversini, A., Schegg, R. (eds.) Information and Communication Technologies in Tourism 2016, pp. 607–619. Springer, Cham (2016). https://doi.org/10.1007/978-3-319-28231-2_44
2. Střelák, D., Škola, F., Liarokapis, F.: Examining user experiences in a mobile augmented reality tourist guide. In: Proceedings of the 9th ACM International Conference on PErvasive Technologies Related to Assistive Environments, p. 19. ACM (2016)
3. Casella, G., Coelho, M.: Augmented heritage: situating augmented reality mobile apps in cultural heritage communication. In: Proceedings of the 2013 International Conference on Information Systems and Design of Communication, pp. 138–140. ACM (2013)
4. de los Ríos, S., et al.: Using augmented reality and social media in mobile applications to engage people on cultural sites. In: Stephanidis, C., Antona, M. (eds.) UAHCI 2014. LNCS, vol. 8514, pp. 662–672. Springer, Cham (2014). https://doi.org/10.1007/978-3-319-07440-5_60
5. Rattanarungrot, S., White, M., Patoli, Z., Pascu, T.: The application of augmented reality for reanimating cultural heritage. In: Shumaker, R., Lackey, S. (eds.) VAMR 2014. LNCS, vol. 8526, pp. 85–95. Springer, Cham (2014). https://doi.org/10.1007/978-3-319-07464-1_8
6. Kenderdine, S.: Pure land: inhabiting the mogao caves at dunhuang. Curator Mus. J. **56**(2), 199–218 (2013)
7. Argyriou, L., Economou, D., Bouki, V.: 360-degree interactive video application for cultural heritage education (2017)
8. Reunanen, M., Díaz, L., Horttana, T.: A holistic user-centered approach to immersive digital cultural heritage installations: case vrouw maria. J. Comput. Cult. Heritage (JOCCH) **7**(4), 24 (2015)
9. Kim, E., Woo, W.: Metadata schema for visualization and sharing of the augmented reality contents (2015)
10. Park, N.Y., Kim, E., Lee, J., Woo, W.: All-in-one mobile outdoor augmented reality framework for cultural heritage sites. In: 2016 12th International Conference on Signal-Image Technology & Internet-Based Systems (SITIS), pp. 484–489. IEEE (2016)
11. Rublee, E., Rabaud, V., Konolige, K., Bradski, G.: ORB: an efficient alternative to SIFT or SURF. In: 2011 IEEE International Conference on Computer Vision (ICCV), pp. 2564–2571. IEEE (2011)
12. Kim, S., Ahn, J., Suh, J., Kim, H., Kim, J.: Towards a semantic data infrastructure for heterogeneous cultural heritage data-challenges of korean cultural heritage data model (KCHDM). In: Digital Heritage 2015, vol. 2, pp. 275–282. IEEE (2015)
13. Kim, H., Matuszka, T., Kim, J.I., Kim, J., Woo, W.: Ontology-based mobile augmented reality in cultural heritage sites: information modeling and user study. Multimed. Tools Appl. **76**(24), 26001–26029 (2017)
14. Lee, W., Bailer, W., Bürger, T.: Ontology for media resources 1.0 (2012). https://www.w3.org/TR/2012/REC-mediaont-10-20120209/

15. Hyun, D.: Metadata element and format for broadcast content distribution (2014). http://mpeg.chiariglione.org/standards/mpeg-a/augmented-reality-application-format

16. Park, H., Woo, W.: Metadata design for ar spacetelling experience using movie clips. In: 2017 IEEE International Conference on Consumer Electronics (ICCE), pp. 388–391. IEEE (2017)

17. Kim, E., et al.: AR reference model for K-culture time machine. In: Yamamoto, S. (ed.) HIMI 2016. LNCS, vol. 9735, pp. 278–289. Springer, Cham (2016). https://doi.org/10.1007/978-3-319-40397-7_27

18. Kim, E., Kim, J., Woo, W.: Metadata schema for context-aware augmented reality applications in cultural heritage domain. In: Digital Heritage 2015, vol. 2, pp. 283–290. IEEE (2015)

19. Shin, J., Park, H., Woo, W.: Connecting the dots: Enhancing the usability of indexed multimedia data for AR cultural heritage applications through storytelling. In: Proceedings of the 15th International Workshop on Content-Based Multimedia Indexing, p. 11. ACM (2017)

Case Study on Motivation to Participate in Private Provision of Local Public Goods and Time Spent in the Region Measured Using GPS

Yurika Shiozu[1]([✉]), Koya Kimura[2], Katsunori Shimohara[2],
and Katsuhiko Yonezaki[3]

[1] Aichi University, Nagoya, Aichi 453-8777, Japan
yshiozu@vega.aichi-u.ac.jp
[2] Doshisha University, Kyotanabe, Kyoto 610-0394, Japan
[3] Yokohama City Univerity, Yokohama, Kanagawa 236-0027, Japan

Abstract. It is known that one or more dedicated people with strong leadership skills typically participate in successful community activities. It is called "ultra-altruistic motivation". These endeavors in various communities are often unsuccessful when the enthusiasm of the leaders and the residents toward the activities are not aligned. Some case study shows that the ultra-altruistic motivation works well and the private provision of local public goods has succeed. However, there are many kinds of local public goods, some cannot be provided by private sector, for example police, but some can be provided by private company like transportation services. Is it established even if the type of local public goods change?

This paper demonstrates that the members of a community non-profit organization (NPO) have a greater willingness to pay for local public goods than the residents themselves as determined through a contingent value method, but this is not necessarily driven by an "ultra-altruistic motivation." The preference regarding monetary cooperation toward public transportation services by NPO members was indicated to be irrelevant to the time spent in the region, as determined through GPS data.

Keywords: CVM · Ultra-altruistic motivation · GPS data

1 Introduction

It is known that one or more dedicated people with strong leadership skills typically participate in successful community activities. These endeavors in various communities are often unsuccessful when the enthusiasm of the leaders and the residents toward the activities are not aligned.

This paper examines two hypotheses: "community non-profit organization (NPO) members actively show willingness to contribute to the private provisioning of local public goods" and "there is a difference in time spent in the region between the group of community NPO members who chose to actively show their willingness to contribute to the private provisioning of local public goods, and the group of members who do not." With the cooperation of an NPO actually working on community

S. Yamamoto and H. Mori (Eds.): HIMI 2018, LNCS 10905, pp. 181–190, 2018.
https://doi.org/10.1007/978-3-319-92046-7_16

activities in a community, preferences regarding monetary cooperation toward public transportation services by the NPO members and the community residents, along with the GPS data of the NPO members, were obtained to examine these two hypotheses. From these data, willingness to pay for public transportation by the residents and the community NPO members were estimated using the contingent value method (CVM). Comparison of the estimation results revealed that the residents who are considering using the services themselves in the future showed more willingness to pay. No difference was observed in willingness to pay among the NPO members as driven by an "ultra-altruistic motivation." The GPS data of the individual NPO members were visualized on a map during the investigation period to derive the time spent in the region. Section 2 below reviews prior studies, and Sect. 3 describes the data used in this paper. Section 4 shows the method, and Sect. 5 provides the analysis results, and Sect. 6 gives a summary of the present study and future themes.

2 Prior Studies

A situation in which socially cooperative action is more desirable, but not offered, is called a social dilemma. In economics, such situations are defined such that the Pareto-inefficient solution is the dominant strategy and is analyzed using game theory. Using this theory, Axelrod explained that people choose a tit-for-tat strategy in social relationships [2].

In terms of community activities, Diekmann revealed that they will be successful if there are one or more selfless, willing leaders, but will fail without active participation [3]. Diekmann called this situation the volunteer's dilemma. Many studies have been conducted regarding this dilemma. Archetti and Sheuring pointed out that the larger the group is, the smaller the benefit per member [1]; therefore, the incentive for volunteerism will be reduced. A series of study results by Fujii et al. are also well known in Japan. They state that fostering the awareness that one's participation will lead to better problem-solving will make the volunteerism more active.

Hatori et al. explained that a person who selflessly promotes community projects, even singlehandedly, is driven by an "ultra-altruistic motivation," and these researchers quantitatively analyzed cases in which the existence of an ultra-altruistically motivated individual led to successful tourism projects [5].

Is "ultra-altruistic motivation" a subjective motivation? Shiozu et al. asked the name of the individual whom the community NPO members regarded as their leader [7]. With this result and through a social network analysis, based on the data collected through actual email correspondences and conversations, they derived the information sender (authority). Thus, the leader of the community activities can be determined through the recognition of others and actual communication.

A sociopsychological factor for individuals to participate in community activities is not necessarily altruistic motivation alone. Yamada and Hashimoto analyzed the subjective norm and self-efficacy by incorporating them into an agent-based model simulation [8]. The results of this analysis revealed that improved self-efficacy enhances the subjective norm and will lead to other people around the participants taking part in the community activities.

One community activity is the maintenance of public transportation. In traffic economics, Weisbrod defined a concept called an option value [9]. An option value describes one deeming the continuance of public transportation as being valuable, despite not using it currently, because one may use it in the future.

Participation in such community activities is said to be influenced by place attachment. Many studies have analyzed place attachment behaviors. In Scannell and Gifford, it is shown that plce attachment increases by environmental factors than the number of years of residence [6].

3 Overview of Area Investigation

An investigation was conducted in the Makishima area of Uji City, Kyoto Prefecture, which is located near one of three major metropolises in Japan. The population of the area is 15,223 as of January 2017, slightly greater than the previous year. Figure 1 shows a map of Uji City. The Makishima area is located on a plain in the northwest part of the city. Its geographical features make it easy to travel on foot or by bicycle. Two private railroads operate through the area. Buses serve in north–south directions, but there are no services available in the east–west directions.

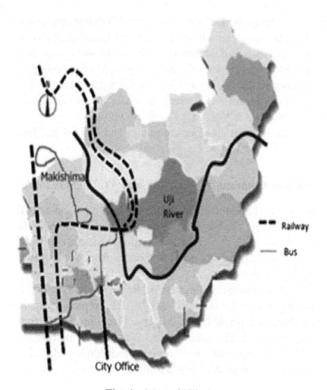

Fig. 1. Map of Uji city

Currently, many residents are able to travel by car, bicycle, or on foot, but some older residents would prefer a bus service. To that end, they are trying to revive the once discontinued bus routes, but this has not been realized.

3.1 Data Overview

With cooperation of the Makishima Kizuna Association of non-profit organizations (Community NPO), GPS data were collected from smart phones between January 1 to June 30, 2017 and used in this study (GPS investigation). Prior to its implementation, this investigation was ethically examined at Doshisha University, and the test subjects were informed that they could refuse to provide their data at any time.

A survey was conducted between October 15 and November 15, 2016 on 1,400 residents and 30 Community NPO members who consented to cooperate with the investigation. The survey asked the intention of the participants to pay for the cost of bus services and their reasons.

3.2 Activity Status Within the Region

The data obtained through the GPS investigation were sampled according to the following procedure to visualize the activity status within the region. First, individuals who have GPS data and responded to the survey were selected, and only acceleration data were extracted from the individual data. This can eliminate instances in which their smart phones were left at home or where the data were intentionally not provided. Next, to understand the activity status within the region, the data were further filtered by specifying the area based on the latitude and longitude corresponding to the applicable region. Of these narrowed data, those in which the GPS data were continuously obtained within a 15 min period were totaled by the individuals. These are visualized on the map as shown in Fig. 2. We use R Ver.3.4.2 for visualization.

This is defined as the time spent in the region and labeled as "Time" in Tables 1 and 2.

3.3 Participation in Community Activities and Willingness to Pay

The survey items inquired about the preference regarding paying annually if an annual payment was introduced for the bus services and the reason for their choice, as well as the preferred fare if bus services were offered without membership and the reason for their choice. There were 212 valid survey responses from the residents. Table 1 provides the descriptive statistics.

"PAYMENT" represents the willingness of the respondents to pay for the First Bid Price of the annual payment (1500 yen), and "FARE" represents their willingness to pay the First Bid Price of a single fare (200 yen). They chose 1 if they support the bid price and 2 if they opposed. The values 1 and 2 in "SEX" indicate male and female, respectively. "YOUNG" represents the number of children under 18, and "ELDER" indicates the number of seniors 65 years and older.

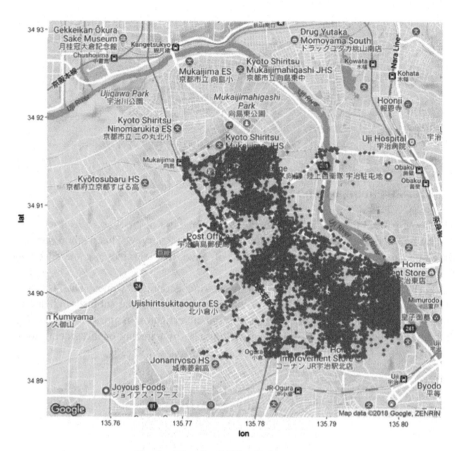

Fig. 2. Visualized "Time" data on map

Table 1. Descriptive statistics

Variable	Mean	Std. Dev.	Min	Max
PAYMENT	1.45	0.510418	1	2
FARE	1.15	0.366348	1	2
SEX	1.5	0.512989	1	2
AGE	59	9.67906	40	70
YOUNG	2.05	1.316894	0	4
ELDER	0.2	0.523148	0	2
TIME	6.786181	7.54375	0.027778	27.01319

There were 20 valid respondents in the survey from the Community NPO members. The descriptive statistics are as shown in Table 2. Time refers to the time spent in the region derived from the procedure described in 3.2.

Table 2. Descriptive statistics of NPO members

Variable	Mean	Std. Dev.	Min	Max
PAYMENT	1.608491	0.489243	1	2
FARE	1.283019	0.451532	1	2
SEX	1.570755	0.49614	1	2
AGE	58.34906	13.29971	20	80
YOUNG	0.400943	0.867856	0	4
ELDER	0.731132	0.740419	0	3

4 Measurement Method

The CVM is used to statistically estimate the willingness of the market, as a whole, to pay by hypothetically estimating the market based on a survey. This method has been developed to measure the economic value of goods and services not traded on the market, such as the envonment.

There are two main techniques for measuring economic values. One is for an investigator to present the test subjects with a scenario, and the subjects indicate the amount they would pay. The other is for the investigator to present the payment amount, and the test subjects show their willingness by agreeing or disagreeing to the

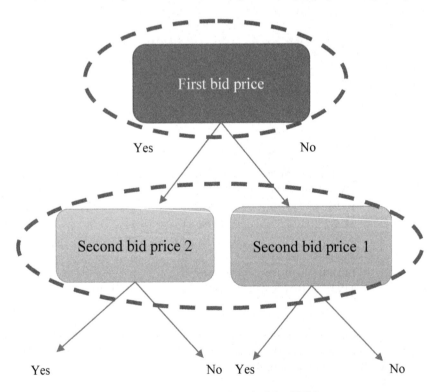

Fig. 3. Double bound method for CVM

amount. Because actual payment is not requested in either technique, the former technique in particular, is known to result in an over estimation, and thus, the latter technique has been more frequently used in recent studies.

In the latter technique, a method exists in which willingness is asked by presenting a second bid price higher (or lower) than the initial bid price, after the preference to the first bid price is made. This is called a multistep method. A type of multistep method in which the bid prices are presented twice is called a double bound method. This method allows a statistically valid estimation with fewer samples than that required in a single bound method (see Fig. 3).

A method that does not presume a normal distribution of the data obtained is called a nonparametric method, and a method that does make such a presumption is called a parametric method.

In transportation engineering and transport economics, the use of the CVM as a support to measure the option value is advocated.

This paper also estimates the willingness of the residents and community NPO members to pay for public transportation using a double bound method. Refer to [4] for details of these estimation methods.

5 Estimation Results

5.1 Interest in Community Activity and Willingness to Pay

The willingness of the residents and the Community NPO members to pay for public transportation is shown in Table 3. All results are statistically significant when the p value is 0.001 or 0.000. The mean values are the maximum bid prices, and all lower values have been excluded.

Table 3. Willingness to pay for community bus service between NPO and residents

	NPO	Residents
Annual payment (median)	1,637 yen (p = 0.002)	969 yen (p = 0.000)
Fare (median)	236 yen (p = 0.000)	226 yen (p = 0.000)
Annual payment (mean)	1,539 yen (p = 0.002)	1,089 yen (p = 0.000)
Fare (mean)	226 yen (p = 0.000)	209 yen (p = 0.000)

The NPO members are willing to pay the highest annual payment, approximately 1.8 times that of the residents' median value. It is also notable that the NPO members' willingness to pay is higher than the First Bid Price.

The NPO members also showed a higher willingness to pay the individual fare than the residents, but this amount was merely 1.04 times higher for the median value. The amount both groups were willing to pay exceeded the First Bid Price in both median and mean values.

Because the individual acknowledged as the leader by the NPO members and the Authority derived by Shiozu et al. were identical, the data on this individual were extracted, and the motivation to pay for the public transportation service was examined in the next step. As the result, the choices of the leader represented an altruistic motivation, and this individual is therefore deemed to have an "ultra-altruistic motivation."

The willingness to pay of the NPO members excluding this "ultra-altruistic motivated" individual was estimated as shown in Table 4.

Table 4. Willingness to pay for community bus services of NPO members excluding the leader

	Members excluding leader
Annual payment (median)	1,581 yen
Fare (median)	233 yen
Annual payment (mean)	1,514 yen
Fare (mean)	224 yen

Compared to Table 3, the values for annual payment and single fare payment are both higher than those by the residents. The willingness to pay for the single fare does not seem to be influenced by the existence of the leader. The willingness to pay the annual payment shows little difference, with a maximum difference of 56 yen.

5.2   Willingness to Pay and Place Attachment

Place attachment has been shown to be influenced by the environment in prior studies, rather than by the number of years of residency. Even is one's residence is in the region, if one's work or school is in another, the amount of time spent in the first region is smaller. In other words, if one spends less time in a region, it is possible to assume that one's interest in the region is less. On the other hand, traveling within the region may increase interest in regional issues, such as public transportation and road improvement. Thus, the time spent in the region is defined as an environmental factor in this paper.

The Kruskal-Wallis test was used to examine whether the time spent in the region differed by the groups that supported and opposed the First Bid Price of the annual payment. The results are shown in Table 5.

Table 5. Kruskal-Wallis test for payment

Payment		
Yes	11	99
No	9	11

Chi-squared = 1.571 with 1 d. f. probability = 0.2100

Considering the p-value of 0.2100, it can be said that there is no statistical difference between the preference of the First Bid Price and the time spent in the region.

Similarly, the group supporting the First Bid Price and the group opposed were examined using a Kruskal-Wallis test to determine the difference in their amount of time spent in the region. The results are shown in Table 6.

Table 6. Kruskal-Wallis test for fare

Fare		
Yes	17	164
No	3	46

Chi-squares = 2.356 with 1 d. f. probability = 0.1248

Considering the p-value of 0.1248, it can be said that there is no statistical difference between the preference of the First Bid Price and the time spent in the region.

6 Summary and Future Themes

This paper examined two hypotheses: "Community NPO members actively show willingness to contribute to the private provisioning of local public goods" and "there is a difference in the amount of time spent in the region between the group of community NPO members who chose to actively show willingness to contribute to the private provisioning of local public goods and the group of members who do not."

For the first hypothesis, it was verified that the NPO members try to actively contribute because the NPO members' willingness to pay is higher than that of the residents. However, this was not determined to be driven by an "ultra-altruistic motivation."

Testing of the second hypothesis revealed that the preference on monetary cooperation toward public transportation was not related to the time spent in the region. It was found that the preferences of the community NPO members were chosen regardless of the time spent in the region, whether monetary payment was required, and whether they have an interest in the region.

There have been cases in real community activities in which local public goods cannot be maintained without a monetary contribution by the residents. A mechanism that inspires more residents to become interested in community activities and promotes their participation, including monetary contributions, is needed.

The handling of GPS data was not examined closely in this analysis. A thorough investigation of the GPS data and a statistical examination are topics of future studies.

Acknowledgment. The authors acknowledge and thank the members of the Makishima Kizuna Association non-profit organization for their cooperation in preparation of this paper. This study was funded by JSPS Kakenhi (Grants-in-Aid for Scientific Research by Japan Society for the Promotion of Science) Nos. JP16K03718 and JP15H03364.

References

1. Archetti, M., Sheuring, I.: Review: game theory of public goods in one-shot social dilemmas without assortment. J. Theor. Biol. **299**, 9–20 (2012)
2. Axelrod, A.: The Complexity of Cooperation. Princeton University Press, Princeton (1997)
3. Diekmann, A.: Volunteers' dilemma. J. Conflict Resolut. **29**(4), 605–610 (1985)
4. Hanemann, M., Loomis, J., Kanninen, B.: Statistical efficiency of double-bounded dichotomous choice contingent valuation. Am. J. Agric. Econ. **73**(4), 1255–1263 (1991)
5. Hatori, T., Fujii, S., Suminaga, T.: An empirical study on determinants of pro-social behavior in a local community - individual and regional factors of ultra-altruistic motivation by "regional charismas". Kodo Keiryogaku **40**(1), 43–61 (2013). In Japanese
6. Scannell, L., Gifford, R.: Defining place attachment. J. Environ. Psycol. **30**, 1–10 (2010)
7. Shiozu, Y., Kimura, K., Yonezaki, K., Shimohara, K.: Does ICT promote the private provision of local public goods? In: Yamamoto, S. (ed.) HCI 2014. LNCS, vol. 8521, pp. 629–640. Springer, Cham (2014). https://doi.org/10.1007/978-3-319-07731-4_62
8. Yamada, H., Hashimoto, T.: Formation and expansion of community activity driven by subjective norm and self-efficacy. J. Jpn. Soc. Artif. Intell. **30**(2 SP-F), 491–497 (2015). In Japanese
9. Weisbrod, B.A.: Collective-consumption services of individual consumption goods. Q. J. Econ. **78**, 471–477 (1964)

Effects of Group Size on Performance and Member Satisfaction

Noriko Suzuki[1(✉)], Mayuka Imashiro[2], Haruka Shoda[3], Noriko Ito[2], Mamiko Sakata[2], and Michiya Yamamoto[4]

[1] Faculty of Business Administration, Tezukayama University,
7-7-1 Tezukayama, Nara City, Nara 631-8501, Japan
nsuzuki@tezukayama-u.ac.jp
[2] Department of Culture and Information Science, Doshisha University,
1-3 Tatara Miyakodani, Kyotanabe, Kyoto 610-0394, Japan
[3] Research Organization of Science and Technology, Ritsumeikan University,
1-1-1 Noji-higashi, Kusatsu, Shiga 525-8577, Japan
[4] School of Science and Technology, Kwansei Gakuin University,
2-1, Gakuen, Sanda, Hyogo 669-1337, Japan

Abstract. The effects of group size on performance and member satisfaction were assessed, with group size ranging from an individual to five members. Participants were 96 university students who engaged in a furniture-assembly task. Our results showed that group size had negligible effects on member satisfaction but strong effects on performance characteristics. As group size increased, performance characteristics, time-to-completion, and duration of interaction with materials decreased in an exponential manner, although member satisfaction tended to become saturated. The result for duration of interaction with materials suggested that the social loafing effect increased with the size of the group. We expect these results to be helpful in designing relationality for collaborative problem solving among people as well as between people and artifacts.

Keywords: Collaborative problem solving · Group size effect
Furniture-assembly task · Social loafing · Member satisfaction

1 Introduction

Both ATC21S [2] and OECD [14] have addressed collaborative problem solving as one of the 21st century skills for students [7]. In parallel with this research focus, several studies on science education have taken the learning-by-building approach in digital networks (e.g., [6]) as well as in the combination of the digital and physical worlds (e.g., [4,5]). Participation in collaborative problem-solving tasks is expected to enhance the skills of sharing common understanding, taking appropriate actions, and establishing group organization. On the other hand, group work in education has frequently incurred the social loafing problem as the number of participants increases [1].

© Springer International Publishing AG, part of Springer Nature 2018
S. Yamamoto and H. Mori (Eds.): HIMI 2018, LNCS 10905, pp. 191–199, 2018.
https://doi.org/10.1007/978-3-319-92046-7_17

Performance in group work sometimes becomes inferior to that in individual work as the participants increase. This is caused by participants making a smaller contribution in the presence of others, and it is called social loafing, also known as the Ringelmann effect. This effect has been verified by a rope-pulling experiment as well as a clapping and shouting experiment [9,11]. Similarly, conventional studies of the brainstorming task have pointed out that group performance sometimes produces a smaller number of ideas [3] or a lower portion of creative production [17]. On the other hand, it has also been reported that group performance is generally superior to that of an average individual, although it is often inferior to that of the best indivudual [8,10].

Does the social loafing problem occur in a collaborative problem-solving task involving building a physical structure? This paper focuses on how group size affects performance and member satisfaction while carrying out a collaborative physical task. In this paper's task, people were instructed to assemble a piece of furniture, a bed-side table, consisting of 6 wooden boards, 54 screws, and other hardware. We chose this task by referring to the TV-cart assembly task [13] and the large-structure assembly task [15]. Moreover, we analyzed performance in our assembly task using both behavioral and psychological indexes. In this setting, we investigated how different group sizes facilitated the individual's involvement in the furniture-assembly task.

In our previous study, participants carried out the furniture-assembly task as individuals or in two-person or five-person groups [16]. The results of that study suggest that the effects of social loafing emerge to a greater degree in five-person groups than in two-person groups. In this paper, we additionally had three- and four-person groups carry out this task. We compared three behavioral indexes (degree of completion, time-to-completion, and duration of interaction with materials) and three psychological indexes (degree of member satisfaction, degree of member contribution, and degree of familiarity with other members) using five group sizes, from individual to five-person group.

2   Method

2.1   Predictions

Physical performance. We predict that five-person groups will complete the task within a shorter time than groups of four or fewer persons because the members of five-person groups have many hands. Consequently, the degree of completion will show a higher ratio with five-person groups than with groups of four or fewer persons.

Psychological evaluation. We predict that the members of five-person groups will feel lower degrees of contribution, satisfaction, and familiarity than will the members of the smaller groups because they will engage in the task for a shorter duration of interaction with the materials.

2.2   Participants

A total of 96 graduate and undergraduate students (mean age: 20.625 years, SD: 1.481) participated in the experiment. They were randomly assigned to individuals, two-person, three-person, four-person or five-person groups. No participant was assigned a particular role in the task. Six individuals, nine two-person, six three-person, six four-person, and six five-person groups took part in the furniture-assembly task.

2.3   Procedure

Each group was instructed to assemble the furniture as soon as possible (Fig. 1 (right)). They had to build a bed-side table, OLTEDAL of IKEA International Group, using 6 boards and 8 kinds of screws and other hardware, 54 parts in total, with an electric screwdriver according to graphical instructions (Fig. 1 (middle)).

The characteristics of the task were (a) it took 30 min by a three-person group in a preliminary experiment, (b) it required the division of roles, e.g., carrying the boards, turning the screws with the screwdriver, and checking the instructions, and (c) it was difficult to explain the graphical instructions without any caption text.

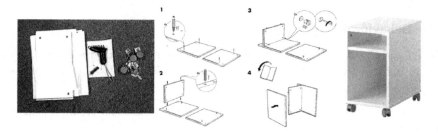

Fig. 1. Materials, a portion of the instructions and completed bed-side table for assembling this furniture (OLTEDAL of IKEA International Group)

2.4   Materials

We prepared a bed-side table consisting of flat wooden boards as the target object of this furniture-assembly task. Figure 1 (left) shows the experimental materials, consisting of 6 boards, 8 kinds of screws, fasteners, and castors (54 parts in total) as well as an electric screwdriver and an instruction sheet.

2.5   Parameters

Behavioral Indexes. The experimenter calculated the following individual or group performances as behavioral indexes.

1. **Degree of completion.** Whether participants persevered to succeed in building the bed-side table or gave up building it.
2. **Time-to-completion (TTC).** The amount of time required to complete the furniture-assembly task.
3. **Duration of interaction with materials.** The amount of time duration of interaction with the materials, i.e., the boards, the screws, the screwdriver, and the instruction, per minute for an individual, one two-person group and one five-person group. We extracted interaction behaviors of handling materials using the annotation software ELAN (EUDICO Linguistic Annotator [12]).

Psychological Indexes. The experimenter examined the results of the following questions, which were answered on a seven-point scale through a post-experiment questionnaire.

1. **Degree of contribution.** Four questions on the participant's degree of contribution to the task.
2. **Degree of satisfaction.** Two questions on the participant's degree of satisfaction with the task performance.
3. **Degree of familiarity.** Eight questions on the participant's degree of familiarity with the other participants.

3   Results

3.1   Behavioral Indexes

Table 1 and Fig. 3 show the results of a chi-square test of the degree of completion for six individuals, nine two-person, six three-person, six four-person, and six

Table 1. 2 × 5 contingency table for successful or unsuccessful groups among five group sizes

Group size		Successful groups	Unsuccessful groups
Individuals	Num. of groups	1	5
	Adjusted residuals	−3.4	3.4
Two-person	Num. of groups	6	3
	Adjusted residuals	−0.5	0.5
Three-person	Num. of groups	6	0
	Adjusted residuals	1.7	−1.7
Four-person	Num. of groups	5	1
	Adjusted residuals	0.6	−0.6
Five-person	Num. of groups	6	0
	Adjusted residuals	1.7	−1.7

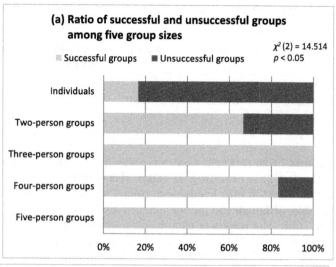

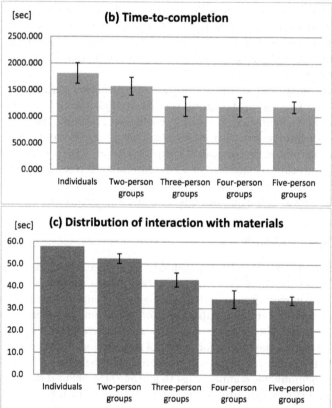

Fig. 2. Results of behavioral indexes: (a) ratio of successful and unsuccessful groups among five group sizes, (b) time-to-completion and (c) duration of interaction with materials

Table 2. Correlation coefficients between group size and duration of interaction with materials

	Group size	Duration of interaction with materials
Group size	1	–
Duration of interaction with materials	−.624*	1

* $p < .05$

five-person groups. From this test, there were significant differences among these five group sizes ($\chi^2(2) = 14.514, p < 0.05$). The results of residual analysis suggest that three-person and five-person groups always succeeded in building the bed-side table, although the individuals usually did not succeed.

Figure 2(b) shows the results of time-to-completion (TTC) for the five group sizes. As a result of ANOVA, there was a significant tendency in TTC among the three largest group sizes ($F(4) = 2.653, p = 0.054$). From multiple comparisons, there were significant differences between individuals and three-person groups ($p = 0.022$), between individuals and four-person groups ($p = 0.021$), and between individuals and five-person groups ($p = 0.021$). These results suggest that individuals had significantly longer time-to-completion than groups of three or more persons.

Figure 2(c) shows the results of duration of interaction with materials for the five group sizes. One successful group was selected as a sample from each group size. Table 2 shows Pearson's correlation coefficients between the group size and the duration of interaction with materials. The group size and the duration of interaction with materials correlated negatively ($r = -.67, p < .05$). The result showed that duration of interaction with materials decreased as the group size increased.

From these results, our predictions were partly supported.

3.2 Psychological Indexes

Figure 3 shows the results of the psychological indexes, i.e., the degree of satisfaction (upper) for six individuals, nine two-person groups, six three-person groups, six four-person groups and six five-person groups, as well as the degrees of familiarity (middle) and contribution (lower) for nine two-person groups, six three-person groups, six four-person groups and six five-person groups.

Figure 3(d) shows the results for degree of satisfaction. As a result of ANOVA, a significant tendency was found in the degree of satisfaction ($F(4,91) = 2.103, p = 0.087$). From multiple comparisons, there were significant differences between individuals and two-person groups ($p = 0.038$), three-person groups ($p = 0.005$), four-person groups ($p = 0.032$), and five-person groups ($p = 0.015$). These results suggest that the individuals felt less satisfaction through the furniture-assembly task than did the members of the groups.

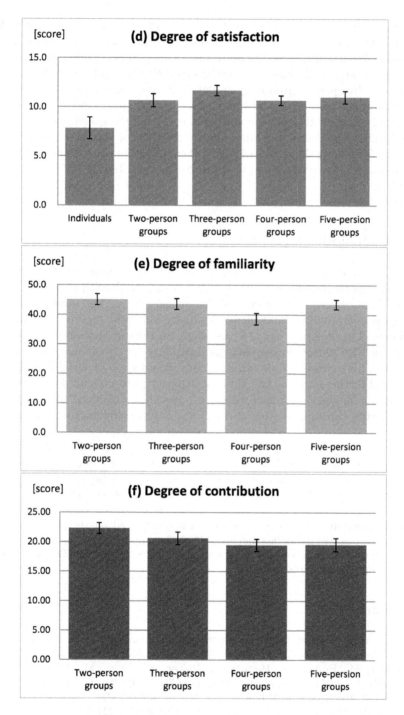

Fig. 3. Results of psychological indexes: (d) degree of satisfaction, (e) degree of familiarity, and (f) degree of contribution

Figure 3(e) shows the results for degree of familiarity. As a result of ANOVA, a significant tendency was found in the degree of familiarity ($F(3,86) = 2.466$, $p = 0.068$). From multiple comparisons, there were significant differences between two-person groups and four-person groups ($p = 0.016$) and between four-person groups and five-person groups ($p = 0.039$) as well as a significant tendency between three-person groups and five-person groups ($p = 0.065$). Furthermore, these results suggest that the members of four-person groups felt less familiarity in the furniture-assembly task than did the members of the other groups.

Figure 3(f) shows the results of degree of contribution. As a result of ANOVA, a significant tendency was found in the degree of contribution ($F(3,86) = 1.353$, $p = 0.263$).

From these results, our predictions were partly supported.

4   Conclusion

In this study, we examined how group size affects performance and member satisfaction in the furniture-assembly task through trials with individuals, two-person, three-person, four-person and five-person groups. The results show that both group performance and member satisfaction were superior to those for individuals. On the other hand, a case study on the distribution of duration of interaction with materials suggests the possibility that some members of four-person and five-person groups exhibited the phenomenon of social loafing.

We will further examine the kinds of behaviors participants produced when they did not interact with materials, e.g., providing directions with speech or hand gestures, cheering on other participants, or just watching the status of task completion.

As our future work, we will look for ways to increase the contribution of group members who engage in social loafing. For example, an observation robot or system might emit a signal, using sound or vibration, to provoke the awareness of members showing lower contribution as well as the other participants.

Acknowledgments. The findings of this study are based on the second author's graduation thesis. We thank 96 students of Doshisha University for their participation in the experiment. This work was supported by JSPS KAKENHI Grant Numbers JP16H03225 and JP15K00219.

References

1. Aggarwal, P., O'Brien, C.L.: Social loafing on group projects-structural antecedents and effect on student satisfaction. J. Mark. Educ. **30**(3), 255–264 (2008)
2. Binkley, M., Erstad, O., Herman, J., Raizen, S., Ripley, M., Miller-Ricci, M., Rumble, M.: Defining twenty-first century skills. In: Griffin, P., McGaw, B., Care, E. (eds.) Assessment and Teaching of 21st Century Skills, pp. 17–66. Springer, Dordrecht (2012). https://doi.org/10.1007/978-94-007-2324-5_2
3. Bouchard, T.J., Hare, M.: Size, performance, and potential in brainstorming groups. J. Appl. Psychol. **54**(1), 51–55 (1970)

4. Cira, N.J., Chung, A.M., Denisin, A.K., Rensi, S., Sanchez, G.N., Quake, S.R.: A biotic game design project for integrated life science and engineering education. PLoS Biol. **13**(3), e1002110 (2015)
5. Datteri, E., Zecca, L., Laudisa, F., Castiglioni, M.: Explaining robotic behaviors: a case study on science education. In: Proceedings of 3rd International Workshop on Teaching Robotics, Teaching with Robotics. Integrating Robotics in School Curriculum, pp. 134–143 (2012)
6. Ekaputra, G., Lim, C., and Kho, I.E.: Minecraft: a game as an education and scientific learning tool. In: The Information Systems International Conference (ISICO), pp. 237–242 (2013)
7. Fiore, S.M., Graesser, A., Greiff, S., Griffin, P., Gong, B., Kyllonen, P., Massey, C., O'Neil, H., Pellegrino, J., Rothman, R., Soule, H., and von Davier, A.: Collaborative Problem Solving: Considerations for the National Assessment of Educational Progress. National Center for Education Statistics (2017)
8. Hill, G.W.: Group versus individual performance: are N + 1 heads better than one? Psychol. Bull. **91**(3), 517–539 (1982)
9. Ingham, A.G., Levinger, G., Graves, J., Peckham, V.: The Ringelmann effect: studies of group size and group performance. J. Exp. Soc. Psychol. **10**, 371–384 (1974)
10. Laughlin, P.R., Hatch, E.C., Silver, J.S., Boh, L.: Groups perform better than the best individuals on letters-to-numbers problems: effects of group size. J. Pers. Soc. Psychol. **90**(4), 644–651 (2006)
11. Latane, B., Williams, K., Harkins, S.: Many hands make light the work: the causes and consequences of social loafing. J. Pers. Soc. Psychol. **37**, 822–832 (1979)
12. Lausberg, H., Sloetjes, H.: Coding gestural behavior with the NEUROGES-ELAN system. Behav. Res. Methods **41**(3), 841–849 (2009)
13. Lozano, S.C. and Tversky, B.: Communicative gestures benefit communicators. In: Proceedings of CogSci2004 (2004)
14. Organisation for Economic Co-operation and Development (OECD). PISA 2015 COLLABORATIVE PROBLEM SOLVING FRAMEWORK (2017). https://www.oecd.org/pisa/pisaproducts/Draft%20PISA%202015%20Collaborative%20Problem%20Solving%20Framework%20.pdf
15. Suzuki, N., Umata, I., Kamiya, T., Ito, S., Iwasawa, S., Inoue, N., Toriyama, T., Kogure, K.: Nonverbal behaviors in cooperative work: a case study of successful and unsuccessful team. In: Proceedings of CogSci2007, pp. 1527–1532 (2007)
16. Suzuki, N., Imashiro, M., Sakata, M., Yamamoto, M.: The effects of group size in the furniture assembly task. In: Proceedings of HCII2017 (2017)
17. Thornburg, T.H.: Group size & member diversity influence on creative performance. J. Creative Behav. **25**(4), 324–333 (1991)
18. Yamaguchi, S., Okamoto, K., Oka, T.: Effects of coactor's presence: social loafing and social facilitation. Jpn. Psychol. Res. **27**(4), 215–222 (1985)

Using Social Elements to Recommend Sessions in Academic Events

Aline de P. A. Tramontin[1], Isabela Gasparini[1(✉)],
and Roberto Pereira[2]

[1] Graduate Program in Applied Computing (PPGCA),
Department of Computer Science, Santa Catarina State University (UDESC),
Joinville, SC, Brazil
aline.tramontin@gmail.com, isabela.gasparini@udesc.br
[2] Department of Computer Science, Federal University of Paraná (UFPR),
Curitiba, PR, Brazil
rpereira@inf.ufpr.br

Abstract. Academic events bring together a large number of researchers and are composed of different types of sessions, which can cause overload of attention and difficulty deciding which sessions to participate. To deal with such problems, Recommender Systems can assist users by offering options that are appropriate for each user. This paper aims to present a recommender approach for sessions of academic events making use of social elements. We propose a recommendation using the academic event's co-authoring network to improve the quality of session recommendation based on the users' previous publications. For authors/participants who do not have publications in previous editions of the event, the recommendations will be generated through the Collaborative Filtering approach. In order to evaluate the viability of our approach, it was included in an Academic Event Application called AppIHC and participants were invited to answer a questionnaire about its use. The results indicate the approach is promising and other social elements could be included future versions.

Keywords: Recommender systems · Social elements · Co-authoring network
Academic events

1 Introduction

People keep connected in many ways, one of which is through social networks, which facilitate communication and interpersonal relationships. Social relations among individuals are often called social ties. Ties represent the existence of a relationship between two individuals [16].

An event has a characteristic to provide the encounter between people with specific purpose, constituting the main theme of the event and justifying its accomplishment. Attending an event is one of the essential components for the social network, and a person tends to attend events accompanied by his/her friends or peers [9, 15].

© Springer International Publishing AG, part of Springer Nature 2018
S. Yamamoto and H. Mori (Eds.): HIMI 2018, LNCS 10905, pp. 200–210, 2018.
https://doi.org/10.1007/978-3-319-92046-7_18

Participation in academic events also contributes to social relations. For Burt [19] the society can be viewed like a market in which people exchange every variety of assets and ideas looking for their interests. Social capital is the contextual complement to human capital. The social capital metaphor is that people who do better are somehow better connected. Increase social capital, that is, invest in social relationships waiting desired return is the main goal of events participation [18].

Academic (or Scientific) events bring together a large number of researchers and are composed of different types of sessions. Several topics are covered, being these subset of a big area of study, each session can contain related presentations with the big area. One of the goals in academic events participation is to increase academic collaboration networks. In these events, actors are researchers, friendships are collaborations, and organizers are members of a program committee [17]. In addition to the presentation of research papers, academic events are also aimed at connecting researchers and promoting potential collaborations [10]. Recommending events (or events sessions) becomes important because of the amount of options available and the frequency that a person has to make choices [13]. Usually, in academic events there are sessions occurring simultaneously, which makes the selection process a hard task.

This paper addresses recommending sessions of academic events to participants, using social elements. The paper is organized as follows: Sect. 2 presents the Theoretical Background. Section 3 shows the Related Work. Section 4 describes our Social Recommender System approach and a discussion about its use. Section 5 discusses an experimental study and its results. Section 6 presents the conclusion of the paper.

2 Background

A Recommender System (RS) is any system that provides personalized suggestions or to guide users in a personalized way to objects of their interest or that they may like [12, 26]. Traditional RS techniques (such as Content-Based Approach and Collaborative Filtering) consider aspects related to users and items to recommend and ignore contextual information [5, 27]. In this sense, Context-Aware RS (CARS) are RS that consider information about the context of users in order to improve the recommendation process. Context *"is any information that can be used to characterize the situation of an entity. An entity is a person, place, or object that is considered relevant to the interaction between a user and an application, including the user and applications themselves"* [2].

Context adds information to the representation of a user model with data relating to physical contexts (e.g. location, time), environmental contexts (climate, light and sound), information contexts (stock quotes, sports matches), personal contexts (social activity, who is in a room), contexts of applications (e-mails, websites visited) and system contexts (network traffic, status of printers) [11]. Jiang et al. [6] have identified that individual preferences and interpersonal influence are important contextual factors for social recommendations as they affect users' decisions about information retention. The social context can be directly incorporated into algorithms of Collaborative Filtering [1].

Due to the large volume of publications and user interactions on social media sites, there is a phenomenon called social overload. Social RSs deal with this overhead by presenting the most relevant data to users [4]. Social RSs are based on network structures, social cues and social tags, or a combination of these various aspects. In general, RSs based on social cues/tips and social tags are different from those based on structural aspects. Those based on structural aspects are used to suggest nodes and links within the network, those based on social cues/social tags, and social tags are used to recommend items, products, or social media content [1].

To improve the quality of the recommendation, Social RSs need information about the relationships of users of a social network, Social Network Analysis (SNA) is the study of social relations between a set of actors [12]. The main difference between network analysis and other approaches to the social sciences is the focus on the relationships between actors rather than the individual attributes of the actors [20]. The behaviors or opinions of individuals depend on the structures in which they are inserted; individual attributes are not analyzed, but the union of relations between individuals through their interactions with each other [21].

The concept of social bonds (or social ties) provides information about the structural properties of a user, as well as on the properties of isolated pairs [12]. Social ties can be categorized into strong ties (for example, trusted friends or family members) who share redundant, overlapping information. In contrast, weak ties (for example, known ones) share more diverse and new information. This information can be used in social recommendation systems to generate more specific recommendations, depending on the information a user wants [13].

According to Scott [22], the predominant approach in SNA is the mathematical approach, termed graph theory, where individuals and groups are represented by points (vertices) and their social relations are represented by lines, also called connections, links or ties [3].

Different studies [7, 10, 14, 15] show that using social information in recommender systems can improve the accuracy of a recommendation. Social online users prefer recommendations made by their friends than those provided by traditional RSs, who use anonymous people with similarities and preferences similar to them [4]. Thus, social recommendation systems present additional value and a new and individualized consumption of content is possible [8].

3   Related Work

This section presents papers that make use of social context in the recommendation process.

Pham et al. [10] proposes a modified version of Collaborative Filtering, which combines the social context extracted from social networks with the space-time context of conference participants, and delivers session and person recommendation services to the target user on mobile devices. The approach takes into account the mobility and sensitivity to location, time, user and social context. The social context refers to the community of on-the-spot researchers, co-authoring in publications and the citation network, research projects and links of mutual collaboration. The spatio-temporal context is considered crucial due to the movement of the participants in the place, that is, their preferences may change depending on the place and time. Pham *et al.* [10] use prediction of links for the formation of the neighborhood of users from the co-authorship and citation network, thus identifying researchers who have similar interest or who are working on similar topics with the target user.

Xia et al. [14] present a solution called Social Aware Recommendation of Venues and Environments (SARVE) to improve participation in conferences through recommendations on mobile devices. Using mobile device technology, SARVE recommends venues and sessions presentations to participants using socially aware and distributed community detection techniques. SARVE uses four types of context: location, time, user, and social relationships. The *location* involves the detection of the exact location of the presentation session. *Time context* uses a smart conference calendar with dates and times of available presentation sessions and allows users to enter their specific time data for the available presentation sessions. The *user context* (presenters and conference participants) captures their research interests. The *social context* is managed by its social ties and social popularity. The strength of the tie is measured based on the duration of the contact between the presenter and the user (participant). Using context data, social characteristics and research interests, Xia et al. [14] identify neighbors (participants who have similar interests) and use this information as a guide to detect relevant communities belonging to the presentation session venue at the conference.

Zhang *et al.* [15] investigate the problem of recommending events created in online social networks, and present three approaches to recommending events based on semantic similarities, relationships between users and the history of participation in events. In the first approach, they calculate the similarity between topic distributions in an event and a user profile, and the most similar events are recommended to the corresponding users. In the second approach, relationships with friends are considered for recommendation. The idea is that users with the same interests have a greater chance of attending the same events. In the third approach, event attendance history is used to construct a classifier for recommendation. Finally, they present a hybrid approach that combines the three approaches mentioned above. The hybrid approach uses weighted sum to calculate the similarity between an event and an user. The results show that the hybrid approach outperforms the other three methods. In addition, all four methods have greater accuracy of recommendation than the random method in both sets of data.

The recommendation of events/presentations that incorporate the social context is a subject under study still being explored. In this work we present a proposal to improve recommendations of presentations sessions in academic events by incorporating social elements relevant for this context.

4   Social Recommender System Approach

The studies presented in the previous section suggest that the context is used in event recommendations/presentation sessions, playing a crucial role with temporal data, location-based data and social data.

The main problems found in recommending events are the sparcity of evaluations and cold start. The sparcity of evaluations occurs when the amount of items evaluated is much smaller than the items available in the system, making it difficult to obtain similarities between people. The cold start problem occurs when a new user or item enter into the system, and there are no evaluations for an item or no items evaluated by the user, so it is not possible to recommend items or find similar users. In the case of RSs for events, the above problems are aggravated due to the short period of time that an event exists and the lack of history of participations and evaluations. Events created in online social networks maintain a history of participation, however, with a lack of user feedback.

Interpersonal influence is an important contextual factor, and it follows the idea that a person tends to attend events accompanied by his/her friends. Similarity between friends is also an aspect that contributes to the recommendation process, as well as the frequency of interactions and participation in events, important social characteristics that contribute substantially to the improvement of the recommendation process.

Different approaches can be used to recommend events, but most of the works analyzed use the Collaborative Filtering approach to recommend, and adds the dimensions of the context in the recommendation process.

Choosing the most relevant presentation/talk sessions and meeting potential collaborators with similar interests can be a tedious task at large events, especially because several parallel sessions usually occur. Academic conferences are dynamic, participants are moving, participating in different presentations in different locations and time. There is also the possibility of schedule changes. Presentations may be canceled due to the absence of the presenter [23], and so on.

Taking into account the mentioned aspects, this work presents a recommendation model that considers as social elements the relations of co-authorship. The social network is created from the co-authorship history of the researchers who have already published in the event, the first step is then, within the network, to calculate the tie strength between the authors, being this the differential in relation to the works presented in the Sect. 3. The calculation is made through the frequency of publications between two authors, inferring that, the greater the number of publications in co-authorship, the stronger the ties between the authors. Thus, the recommendation to the target user is generated based on the interests of the co-author with strong ties, and in publications, which the target user is not a co-author. The tiebreaker will be held by the most recent co-authorship. If the user does not have a strong tie in the network, the weak ties will also be analyzed.

When the co-authors identified in the co-authoring network have not published works in the current event edition, the recommendation will be made by calculating the centrality of the network, based on the premise that more central users influence the other members of the network. Figure 1 presents the basic recommendation procedure. The proposed model was partially added to a recommendation architecture developed in an event application in order to provide a better user experience through mobile technology.

In Fig. 1 the recommendation process starts from the user input data. In the sequence it is necessary to identify whether the user is an author in the event database, if so, it is verified through the Algorithm 1 the tie strength as mentioned previously.

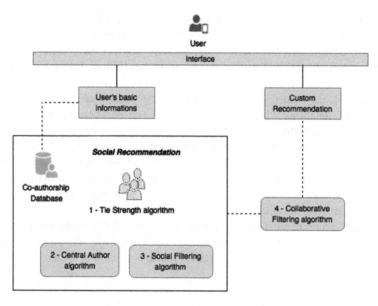

Fig. 1. Recommendation procedure

Having a strong link with a co-author, this author's works published in the current event edition will be recommended, but if the strong co-author does not have publications in the current event, it is checked if co-authors with weak ties have works in the current event to be recommended.

When the coauthors identified in the co-authoring network have not published works in the current event, the Algorithm 2 will be processed and the most central authors of the network will be selected. The minimum distance between the authors and the target user will be verified, thus selecting the closest central author. The works of this author in the current event will be recommended. If the central author closest to the target user has no publications in the current event, the recommendation will be generated by Social Filtering (Algorithm 3), that is, the items will be recommended based on similarity between previously identified co-authors. If the target user is not an author identified in the co-authorship database, the Collaborative Filtering (Algorithm 4) will be used through non-friends.

5   Experimental Study

In recent years, mobile applications (apps) have been used to assist attendees at events, and often offer functions such as calendar management, briefing on lecturers, and sessions. However few studies have been focused on personalization and recommendation of this type of apps [24].

The application, called AppIHC, is open source and was developed for Android Operational System using the guidelines of Material Design. The goal is to help users to get complete information about event programming.

The AppIHC has a content-based recommendation algorithm and a gamification module that aims to increase the use of the application and the participation in the event [25]. The model presented in Sect. 4 was partially implemented in the AppIHC and applied during IHC 2017 (Brazilian Symposium on Human Factors in Computing Systems).

Algorithm 1, mentioned in Sect. 4, verifies if the participants enrolled in the event are in the IHC database, which stores data of all full papers authors from previous editions of IHC (from 1998 to 2015), and if they have papers in the current event. Data are processed and a JavaScript Object Notation (JSON) file is generated with the list of papers to be recommended for each user. The file is imported into the application database, and the users, upon registering in the AppIHC, receives a notification indicating that they have been found in the database and prompting them to confirm their identity, so their data are filled in automatically.

The social recommendation by co-authoring was added implicitly so as not to conflict with the content-based recommendation. When using the AppIHC, the user receives recommendations through text and icon in the event schedule, presented per day according to Fig. 2.

(a)   **(b)**

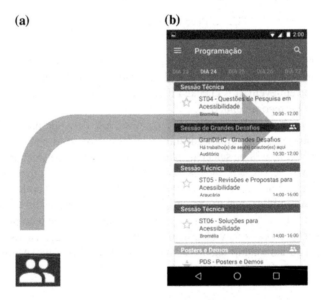

Fig. 2. (a) Social icon and (b) Social recommendation

The IHC 2017 had 250 registered participants, and 108 used the application. From the 250 participants, the algorithm identified 35 people in the event who are co-authors in the previous editions of IHC, but only 14 of them used AppIHC and then received co-authorship recommendation. We got indications that one of the reasons about the low application usage is because it was offered only for Android OS (and not for IOS). A questionnaire was sent to these 14 users, and 8 users answered the following questions (Table 1):

Table 1. Closed issues

	Yes	Not Rated	No
Q1- Did you notice the indication that in a given session there was your co-author (s)?	5	2	1
Q2- Did this information influence your decision-making process to which sessions to watch?	1	2	5
Q3- Do you find it useful to co-authoring to recommend sessions?	5	2	1

Analyzing the users' activity log, it was possible to identify that one of the five users who said they were not influenced by co-authoring recommendation selected the same session recommended as favorite. This could indicate that although the user does not consider s/he has been influenced by the recommender algorithm, s/he liked the recommended session.

The other questions were open and aimed to identify suggestions for improvements to the recommendation process, with and without the use of social elements, allowing the improvement of the recommendation model and the application.

In the context of social issues, users suggested to: 1. Allow the selection and visualization of co-authors' interests. 2. Create a network of friends by adding people from the academic community, not necessarily co-authors. 3. Share preferences, interests and works to be presented, with due agreement. 4. Recommend and receive recommendations from other users.

Thus, an improved version of the model has been developed, adding other social elements, such as adding friends from an online social network and sharing information among users of the application. Figure 3 exemplifies the improved model.

The improved Model, as shown in Fig. 3, allows users to share their preferences over the presentation sessions displayed in the AppIHC interface, and also allows the user to invite friends from an online social network, such as Facebook, to compose the network of friends. It also extracts relationship information to be used in the recommendation process: in this case, friends can make session recommendations for their App users (Algorithm 4) friends. Therefore, preferences of friends (Algorithm 6) and presentations of work from the same institution of the target user (Algorithm 5) will be considered in the recommendation process.

The co-authoring recommendation remains as in the model presented in Sect. 4, however, adding the concept of friendship. In this sense it is necessary to assign a weight to the recommendations, considering that a friendship relationship can be stronger than a coauthor relationship. These weights will be made to present the content considered most relevant.

The model presented in this section will be implemented and validated through experiments in further events.

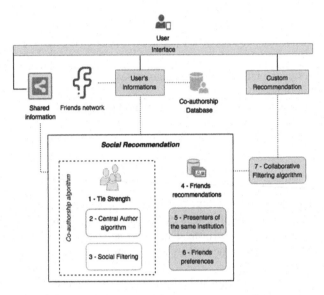

Fig. 3. Improved recommendation procedure

6 Conclusions and Future Work

This paper presents a proposal for a recommendation for sessions of academic events that considers co-authorship relations as social relations.

The literature indicates the need for deepen the studies in SRs of events/presentations sessions in the academic environment including social elements, such as the co-authorship network. Using the IHC co-authorship network, the recommendation model presented in this paper uses the strength of the link among co-authors based on the co-authorship frequency to generate the recommendations, the centrality of the authors of the network and the similarity between co-authors, and considers the possibility of the participant not being an author in publications of previous editions of the event, using the traditional collaborative approach.

This model was partially added in a mobile application (AppIHC) during the academic event IHC 2017. Recommendations were generated and added implicitly in the session schedule view. The users who actually received the recommendations by co-authoring answered a satisfaction questionnaire. Based on the analysis of responses and suggestions for improvements, the model was improved and other social elements were added, such as: the concept of social network and information sharing, allowing the addition of friends, recommending presenters from the same institution as the target user, allowing the user to recommend and receive recommendations from his/her friends. It is intended to apply the improved model in the IHC 2018 event to assess whether the use of these social elements actually improves the quality of the recommendation of sessions at scientific events.

References

1. Aggarwal, C.C.: Recommender Systems. Springer, Heidelberg (2016). https://doi.org/10. 1007/978-3-319-29659-3
2. Dey, A.K.: Understanding and using context. Pers. Ubiquit. Comput. **5**(1), 4–7 (2001)
3. Gabardo, A.C.: Análise De Redes Sociais: Uma Visão Computacional. Novatec Editora, São Paulo (2015)
4. Guy, I.: Social recommender systems. In: Ricci, F., Rokach, L., Shapira, B. (eds.) Recommender Systems Handbook, pp. 511–543. Springer, Boston (2015). https://doi.org/ 10.1007/978-1-4899-7637-6_15
5. Jannach, D., Zanker, M., Felfernig, A., Friedrich, G.: Recommender Systems: An Introduction. Cambridge University Press, Cambridge (2010)
6. Jiang, M., Cui, P., Liu, R., Yang, Q., Wang, F., Zhu, W., Yang, S.: Social contextual recommendation. In: Proceedings of the 21st ACM International Conference on Information and Knowledge Management, pp. 45–54. ACM (2012)
7. Macedo, A.Q., Marinho, L.B., Santos, R.L.T.: Context-aware event recommendation in event-based social networks. In: Proceedings of the 9th ACM Conference on Recommender Systems, pp. 123–130. ACM (2015)
8. Oechslein, O., Hess, T.: The value of a recommendation: the role of social ties in social recommender systems. In: 2014 47th Hawaii International Conference on System Sciences (HICSS), pp. 1864–1873. IEEE (2014)
9. Pascoal, L.M.L.: Um Método Social-Evolucionário Para Geração De Rankings Que Apoiem A Recomendação De Eventos (2014)
10. Pham, M.C., Kovachev, D., Cao, Y., Mbogos, G.M., Klamma, R.: Enhancing academic event participation with context-aware and social recommendations. In: IEEE/ACM International Conference on Advances in Social Networks Analysis And Mining (ASONAM), pp. 464–471. IEEE (2012)
11. Ranganathan, A., Campbell, R.H.: An infrastructure for context-awareness based on first order logic. Pers. Ubiquit. Comput. **7**(6), 353–364 (2003)
12. Ricci, F., Rokach, L., Shapira, B. (eds.): Recommender Systems Handbook. Springer, Berlin (2015). https://doi.org/10.1007/978-1-4899-7637-6
13. Seth, A., Zhang, J.: A social network based approach to personalized recommendation of participatory media content. In: ICWSM (2008)
14. Xia, F., Asabere, N.Y., Rodrigues, J.J., Basso, F., Deonauth, N., Wang, W.: Socially-aware venue recommendation for conference participants. In: Ubiquitous Intelligence and Computing, IEEE 10th International Conference on and 10th International Conference on Autonomic and Trusted Computing (UIC/ATC), pp. 134–141. IEEE (2013)
15. Zhang, Y., Wu, H., Sorathia, V.S., Prasanna, V.K.: Event recommendation in social networks with linked data enablement. In: ICEIS, vol. 2, pp. 371–379 (2013)
16. Vastardis, N., Yang, K.: Mobile social networks: architectures, social properties and key research challenges. IEEE Commun. Surv. Tutor. **99**, 1–17 (2012)
17. Licamele, L., Getoor, L.: Social capital in friendship-event networks. In: Sixth International Conference on Data Mining (ICDM 2006), pp. 959–964. IEEE (2006)
18. Lin, N., Cook, K.S., Burt, R.S. (eds.): Social Capital: Theory and Research. Transaction Publishers, Piscataway (2001)
19. Burt, R.S.: Closure as social capital. In: Social Capital: Theory and Research, pp. 31–56 (2001)
20. Mika, P.: Social networks and the semantic web. In: Proceedings of the 2004 IEEE/WIC/ACM International Conference on Web Intelligence, pp. 285–291. IEEE Computer Society (2004)

21. Marteleto, R.M.: Análise de redes sociais: aplicação nos estudos de transferência da informação. Ciênc. Informação **30**(1), 71–81 (2001)
22. Scott, J.: Social network analysis: developments, advances, and prospects. Soc. Netw. Anal. Min. **1**(1), 21–26 (2011)
23. Asabere, N.Y., Xia, F., Wang, W., Rodrigues, J.J., Basso, F., Ma, J.: Improving smart conference participation through socially aware recommendation. IEEE Trans. Hum.-Mach. Syst. **44**(5), 689–700 (2014)
24. Arens-Volland, A., Naudet, Y.: Personalized recommender system for event attendees. In: 11th International Workshop on Semantic and Social Media Adaptation and Personalization, pp. 65–70 (2016)
25. De Oliveira, B., Tramontin, A.P.A., Neves, E.S., Sohn, R., Ardjomand, L., Gasparini, I.: AppIHC: Uma proposta de aplicativo móvel para eventos científicos. In: Brazilian Symposium on Human Factors in Computational Systems (Simpósio Brasileiro sobre Fatores Humanos em Sistemas Computacionais), pp. 78–79 (2017)
26. Burke, R.D.: Hybrid recommender systems: survey and experiments. User Model UserAdapt. Interact. **12**(4), 331–370 (2002)
27. Verbert, K., Manouselis, N., Ochoa, X., Wolpers, M., Drachsler, H., Bosnic, I., Duval, E.: Context-aware recommender systems for learning: a survey and future challenges. IEEE Trans. Learn. Technol. **5**(4), 318–335 (2012)

Study of Experience Value Design Method by Movie Prototyping

Kazuki Tsumori[1(✉)], Takeo Ainoya[2], Ryuta Motegi[1],
and Keiko Kasamatsu[1]

[1] Department of Industrial Art, Graduate School of System Design,
Tokyo Metropolitan University, Tokyo, Japan
kzktmrl7@gmail.com, kasamatu@tmu.ac.jp
[2] VDS Co., Ltd, Tokyo, Japan

Abstract. In Human Centered Design (HCD), prototyping is very important. Prototyping is mainly used for verification of functions and design. In addition to this role, movie prototyping is also used for image sharing of products and services. Recently, movie prototyping is used in many companies and projects, but its production method has been hardly clarified. The purpose of this study is to analyze the existing movie prototype and to propose a method of making a movie prototype. In this study, we first analyzed three movies prototyping, Amazon Go, FUJITSU ROBOT FUTURE VISION, OTON GLASS, and extracted six points of descriptions. The extracted points are shape, function, how to use, persona, scene, and feeling. We analyze more movie prototypes based on those six depiction points and propose a method of movie prototyping with experience value design.

Keywords: Movie prototyping · Human Centered Design · UX design

1 Introduction

Prototyping in Human Centered Design (HCD) is very important. There are two main purposes.

1.1 Verification of Function and Design

In the early stages of product development, design can be verified by actually shaping its functions and ideas. This also means that feedback can be obtained early from the user. By repeating prototyping and user tests, it is possible to discover and correct design defects and problems of ideas at an early stage (Fig. 1).

1.2 Image Sharing Among Divisions

According to Tagawa [1], "There is no reference case in the project where innovation is required, there are many cases where you have no choice but to plan and design from the earliest stages." Innovative products and services that do not exist in the world are difficult to explain to others their attractiveness. Therefore, prototyping is also used as a

© Springer International Publishing AG, part of Springer Nature 2018
S. Yamamoto and H. Mori (Eds.): HIMI 2018, LNCS 10905, pp. 211–216, 2018.
https://doi.org/10.1007/978-3-319-92046-7_19

means for sharing intent within the project team and sharing images among internal divisions in the product development process (Fig. 2).

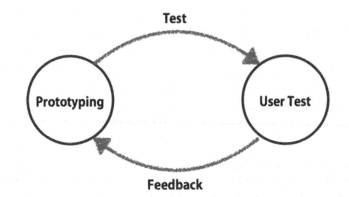

Fig. 1. Repeating prototyping and user test

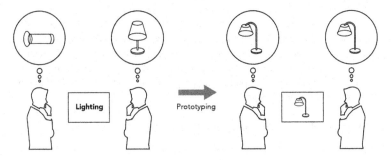

Fig. 2. Prototyping for image sharing among divisions

There are various types of prototypes, such as sketches, dirty prototypes, technical prototypes, styling prototypes, working prototypes, and movie prototypes. Movie prototypes have a role of verifying product usability and design. Also, the most characteristic aspect of the movie prototype is that it is suitable for depicting the user's behavior and feeling when using the product. It helps to image the experiential value obtained through the product by drawing personas, scenes of use, feeling. From this, it can be said that movie prototyping is good for smoothing the sharing of UX, which is easy to disperse images among internal divisions.

Meanwhile, the method of making the movie prototype has been hardly clarified. In the movie prototype for "COIN Chareeeen" made by the students of the Tokyo Metropolitan University, there were many explanations about the function of the product and it was effective to understand what kind of product it is. However, since the depiction of the scene where the user actually uses the product is insufficient, the experience value did not convey well to the viewers. In this way, as a result of producing the movie prototyping by looking for ways, it may happen that the depiction is different from the production intention, or that the thing to want to convey is not conveyed (Fig. 3).

Fig. 3. Movie prototype for COIN Chareeeen

2 Purpose

The purpose of this study is to propose a method of movie prototyping in experience value design. Currently, movie prototyping is used in many projects. However, its production method is hardly clarified. In this study, we analyze many movie prototyping and propose a method to create movie prototyping that matches the production purpose.

3 Case Study

3.1 Selected Movie Prototypes

In this study, we analyzed three movie prototypings as a case study.

1. Amazon Go
2. FUJITSU ROBOT FUTURE VISION
3. OTON GLASS.

These are movie prototypes depicting future vision before products and services are completed. It seems that these emphasis is on drawing a vision that they want to realize rather than as a promotion aspect.

3.2 Analysis

The movie prototype for Amazon Go, the word "JAST WALK OUT" frequently appears in video and narration. Although there is no detailed depiction on how to use the service in the video, it focuses on the point that it makes us experience "JUST WALK OUT" by interlocking with the application (Fig. 4).

The movie prototype for FUJITSU ROBOT FUTURE VISION does not explain detailed functions, but it depicts what kind of robot will do in the actual scene. Throughout the story is built in, and personas and their feeling changes are well drawn (Fig. 5).

Fig. 4. Movie prototype for Amazon Go (Source: [3]).

Fig. 5. Movie prototype for FUJITSU ROBOT FUTURE VISION (Source: [4]).

The movie prototype for OTON GLASS depicts a scene in which the problem is solved by the product after drawing the scene in which the current user is in trouble. This tells the value the product brings to the user. In addition, there are many depictions on the shape and structure of products (Fig. 6).

Fig. 6. Movie prototype for OTON GLASS (Source: [5]).

4 Discussion

From the analysis of the existing movie prototype, the following six points are considered as points for producing movie prototype.

1. What type of product (Shape)
2. What can be done (Function)
3. How to use (How to use)
4. Who uses it (Persona)
5. When to use (Scene)
6. How do you feel (Feeling).

Shape   Function   How to use   persona   Scene   Feeling

Fig. 7. Six points for producing movie prototyping

Table 1. Six points for producing movie prototyping in movie prototyping for FUJITSU ROBOT FUTURE VISION (Source: [4]).

	The woman trying to start a new business is shown.	The Woman are consulting with bankers about financing new business. There is a robot between them.	The robot is presenting finantial plans.	The robot installed at the entrance is watching a visitor.	The robot is analyzing visitor information.	A woman is communicating with a craftsman by the robot. The robot projects the hologram.
Shape		O		O		O
Function			O	O	O	O
How to use						O
persona	O					O
Scene		O				O
Feeling						

	Children are gathering in the library. Children are touching the hologram projected by the robot.	The robot is presenting infomations for children.	The robot safeguard a woman against traffic accidents. The woman is smiling with confidence	There is a robot at the entrance of the exhibition hall.	When a visitor touches the robot, the robot analyzes her information.	The organizer recommends products to visitors. The visitor is satisfied with the products.
Shape	O	O		O	O	
Function	O	O	O		O	
How to use	O				O	
persona	O					O
Scene	O		O	O	O	O
Feeling		O				O

"Form", "Function", and "How to use" are a depiction of the product itself. This depiction helps to understand the product. "Persona", "scene", and "Feeling" are depictions focusing on the experience of using the product. This depiction encourages empathy.

These will be applicable to movie prototypes in various fields such as product design, UI design, service design (Fig. 7).

Focusing on these six points, we reanalyzed movie prototype for FUJITSU ROBOT FUTURE VISION (Table 1). By extracting six points as in Table 1 also in other movie prototypes, it seems that it is possible to clarify the production purpose and expression points.

5   Future Work

We will analyze more movie prototypes and verify the six assumed points. In addition to extracting the scenes in which six points are drawn in the movie, if there are other points, we will analyze them.

We also analyze the method of presenting the movie prototype. By analyzing various movie prototypes such as live-action photography and animation, we will clarify the method of presenting the movie prototype suitable for the production purpose.

Based on the results obtained from those analyzes, we proposes a production method that everyone can make movie prototyping that matches the production purpose.

References

1. Tagawa, K.: Takram design engineering and its design process. Keio SFC J. **10**(1), 17–25 (2010)
2. Tanaka, K.: Practice of design engineering. In: Hitotsubashi University, Institute of Innovation Research (ed.), Hitotsubashi Business Review 2015 SPR, vol. 62, no. 4, pp. 36–51. Toyo Keizai Inc., Tokyo (2015)
3. Introducing Amazon Go and the world's most advanced shopping technology. https://www.youtube.com/watch?v=NrmMk1Myrxc
4. ROBOT FUTURE VISION. https://www.youtube.com/watch?time_continue=3&v=55iD9ItLM00
5. OTON GLASS. https://vimeo.com/175384517

Interaction Techniques and Pointing Task: A Preliminary Analysis to Understand How to Characterize the User Abilities in Virtual Environment

Eulalie Verhulst[1(✉)], Frédéric Banville[2], Paul Richard[1], and Philippe Allain[3]

[1] Laboratoire Angevin de Recherche en Ingénierie des Systèmes (LARIS), Universié d'Angers, Angers, France
eulalie.verhulst@gmail.com
[2] Département de Psychologie, Université du Québec Rimouski, Rimouski, QC, Canada
[3] Laboratoire de Psychologie des Pays de Loire (LPPL), Université d'Angers, Angers, France

Abstract. The study aims to detect how the skills of a participant using a specific interaction technique can be qualified with behavioural data as number of click or miss click. Three interaction techniques were used: the gamepad, the mouse and the Razer Hydra. Users were first trained then they had to complete a pointing task. We then created two subgroups of participants: one with good abilities to use the interaction technique and one with low skill with a clustering hierarchical analysis and then compared subgroups. The Fitts throughput score during a pointing task allow to differentiate users with good abilities from other for the gamepad and the Razer Hydra but not for the mouse. These results could help to understand how familiar is a user with an interaction technique.

Keywords: Interaction technique · Fitts throughput
Hierarchical clustering analysis

1 Introduction

Interaction techniques are the combination of software and hardware [1] and are used to interact with a Virtual Environment (VE). They are characterized by 3 concepts: navigation, selection and manipulation, and system control [2]. Navigation allows users to travel in the VE from place to place adjusting his/her point of view (i.e. steering) [3]. Selection is the action of picking a target in VE. System control refers to interaction between the user and software external functionalities such as interacting with a menu outside the VE.

The user performance when completing a virtual task depends on the interface. Indeed, the interaction techniques involve mental workload and modulate positively or negatively the user performance when the interaction technique

© Springer International Publishing AG, part of Springer Nature 2018
S. Yamamoto and H. Mori (Eds.): HIMI 2018, LNCS 10905, pp. 217–227, 2018.
https://doi.org/10.1007/978-3-319-92046-7_20

is not intuitive enough [4]. Workload is the mental resources needed at oncee for a task [4]. High workload affects performance because the human cognitive resources are limited in energy [5], if too much workload is requiring for the task realization, the user could miss information or commit errors. The elevation of the workload depends on the task complexity, environmental factors and the user abilities and knowledge [6]. Using the mouse involve only low workload during task realization because most people use it in their everyday life and perfectly know how to manipulate it. Other less known interaction technique could lead to more mental workload. There is a need to understand how different interaction techniques could impact the user's performance according his/her familiarity with it.

The use of an interaction technique is subtended by motors functions such as gesture control by hand [3] and cognitive functions [7,8] such as spatial abilities [9]. As the user uses an interaction technique, its use becomes automatic and no longer consumes as much attention-giving resources [10] and so, few workload. We can consider that individuals who used the computer mouse frequently for several years are experts. Thus, we can assume that an expert user with the mouse, will have better results during a task in VE because it will make fewer errors related to the usability of the interaction technique than a "novice" user. So, as we just discussed, the interface (or the interaction device) use a part of the cognitive functions to allow the interaction with the virtual world. On the other hand, to develop a useful virtual tool for neuropsychological assessment capable of giving the clinician an effective measurement of cognition, it is primary to understand the user abilities with the use of an interaction technique. It is important to be able to dissociate which part of cognition is used by the HCI and which part is really allowed to the cognitive task itself. For example, the principal variable used to qualify patient performance during virtual cognitive task is often time completion. So, a longer completion time was used to discriminate patients with mild cognitive impairments (MCI)from healthy elderly who completed the virtual task in less time [11]. However, completion time could be modulated by the user's abilities with an interaction technique. So, a user who never used a mouse before the virtual test could take more time than a familiar one even if their cognitive abilities are comparable.

The study aims to detect how the skill of a participant using an interaction technique can be qualified with behavioral and physiological data. To assess the user abilities with interaction technique, participants realized common activities in VEs (i.e., a training step and a pointing task) with several interation techniques, 2D interaction: gamepad (i.e., Xbox controller) and mouse. And 3D interaction: Razer Hydra. They also answered some questionnaires about their computer and video games usage.

2   Related Work

Navigation efficiency is mainly assessed by completion time and is linked to the user performance [12] where a long completion time is associated to a poor performance. Selection occurs in a 2D VE or in a 3D VE. During 2D selection, the user picks a target by moving the selection cursor in the x and y axes whereas during 3D selection he/she moves the cursor in x, y and z axes and must control the depth during selection. In 2D, the common selection techniques are pointing and drag-and-drop [13]. When pointing, the user put on the selection cursor on a target and then click on it whereas during drag and drop he/she selects an item and move it into the desired place before dropping it. Adults [13] and children [14] users are more efficient with pointing than with the drag and drop. Moreover, mental workload is higher during drag and drop tasks for the elderly than adults whereas there is no difference in mental workload for the 2 groups during pointing [15]. Workload can be assessed in a objective way by recording physiological data. Heart Rate (HR) and Heart Rate Variability (HRV) are sensitive to the different states of the autonomic nervous system and can be used to assess mental workload. HR is faster during complex tasks and high workload situation [16,17] whereas HRV is lower [18].

To assess the usability of a selection technique, ISO 9241-9 [19] proposes a standard pointing task where user must select as quickly as possible several targets with different positions and sizes (Fig. 1). The results can be analyzed with Fitts's Law [20], to predict the user performance with a selection technique according targets position and size (Eq. 1).

$$MT = a + b \cdot log_2 \left(\frac{D}{W} + 1 \right) \tag{1}$$

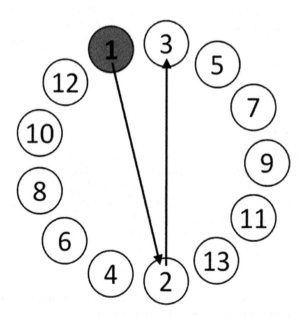

Fig. 1. Illustration of the pointing task in ISO 9241-9 standard

Where MT is completion time, D distance to the target and W its width. Log is the index of difficulty where a and b are determined by a linear regression. Fitts's Law was then adapted by Shannon Formulation and throughput (Eq. 2). Where De is movement amplitude in pixels and SD_x is the standard deviation of the distance between the selection and centre of the target in pixels. MT is the time to hit and select a target in second. Throughput (bits/second) assess the usability of a selection technique with a combination of velocity and accuracy.

$$TP = \frac{log_2\left(\frac{De}{4.133*SD_x}+1\right)}{MT} \qquad (2)$$

Few studies analyze the profile user during a virtual task and qualify them as novice or expert. Rosa et al. [21] isolated user profile associated with the most effective experience in VR using a correspondence analysis combined with cluster analysis. They showed that PC gamer profile had a better experience in VE than console gamer or non-gamer which is the profile with more cybersickness. Hourcade et al. [8] tested elderly subjects during a selection task with or without selection assistance software (i.e., PointAssist). They showed that an expert participant with the use of the mouse has better results without the PointAssist software. Assistance is not useful if users are expert with the technology. In addition, the authors showed positive correlations between computer and mouse use and target selection (click on the target). Individuals frequently using the mouse and computer are those who have done the selection task better and where assistance has been triggered least frequently.

Another study assessed participant during a daily activity of shopping in VE. They could interact in the virtual shop with gamepad. The novice subjects were isolated with t-modified test and took significantly more time to complete the task than expert participants [22].

3   Method

3.1   Participants

Thirty student volunteers were recruited from the local university. The participants were randomly divided in 3 groups. The first group, composed of 3 women and 10 men (age M: 23.7; SD: 3.5), used gamepad. The second, composed of 3 women and 6 men (age M: 23.9; SD: 5.3) used the mouse. The third, composed of 5 women and 3 men (age M: 23.6; SD: 3.9) used the Razer Hydra. No participants knew the Razer Hydra beforehand.

3.2   Tasks

Training. To understand how to use the interaction technique and be able to freely navigate and select items in VE, participants trained in a virtual apartement composed of 6 rooms. Time completion, distance travelled, number of clicks and miss clicks are recorded.

Pointing Task. We used a pointing task like the ISO 9241-9 standard. In this task 13 targets are positioned in circle and the participants clicked on each target. Targets are spheres of 16 cm width. Only the active target was displayed on the screen and participants received an audiofeedback when they missed the target (i.e., error). Time, errors and throughput are recorded. As we wanted to explore if a quick pointing task could be a useful task to discriminate participants, we only use one sequence of 13 targets during the task.

3.3 Apparatus

The experiment was conducted on a computer with Intel® Xeon® processor, a NVIDIA GeForce GTX 1080 and 32GO of RAM running Windows 10. The VE were displayed on a 50″ Samsung television with a 1630×768 resolution.

3.4 Procedure

After signing the protocol agreement, participants wore 3 sensors on the chest to collect HR and HRV data with the help of the investigator. ECG data were recorded through the (R)evolution BITalino board kit and the OpenSignals software. The ECG sensors were placed: under the right clavicula (+ electrode), under the left musculus pectoralis major (− electrode) and under the left clavicula (reference electrode) [23]. Then, participants were ready to begin the training step where they visit a virtual apartment. Guided by the researcher, participants visit the VE in the same order. When the visit is done, they can see three boxes in the kitchen. They have to select the three boxes, one by one, and drop them on a closed surface. Participants could spend more time acting in the VE and end the training when they feel comfortable with the use of the interaction technique. After the training step, they realized the pointing task and completed questionnaires about their use of PC and video games on a 5 points Likert scale.

4 Results

All data were analyzed with R software [24]. First, for each interaction technique, we conduct hierarchical clustering analysis (HCA) with he Agglomerative Nesting algorithm and average method of linkage. The HCA differentiated at least two subgroups. To describe the subgroups, we compared them using t-test if the data distribution was normal and Mann–Whitney U test if not. Moreover, confidence intervals were plotted as another result interpretation less dichotomous than the *p-value* [25]. To compare the 3 interaction techniques we used ANOVA or Kruskal-Wallis test depending on the distribution normality.

4.1 Profile Analysis

Gamepad Users. The cluster dendrogram could easily discriminate two groups (Fig. 2) called subgroup 1 ($n = 7$) and subgroup 2 ($n = 6$). Results are presented in the Table 1. Throughput for the subgroup 1 is higher than for the subgroup 2 ($t = 3.35; p = 0.008$). Subroup 1 realized the pointing task in less time than the subgroup 2 ($U = 31; p = 0.04$). Subgroup 1 spent less time completing the training step than the subgroup 2 ($U = 30; p = 0.04$). Subgroup 1 clicked more time in the training VE than the subgroup 2 ($t = 6.64; p = 0.001$). Subgroup 1 made more miss clicks during the training step than the subgroup 2 ($t = 3.41; p = 0.01$). Subgroup 1 played video game for more year than the subgroup 2 ($t = 4.25; p = 0.002$). Subgroup 1 played more often video game than the subgroup 2 ($U = 6; p = 0.01$).

Cluster Dendrogram

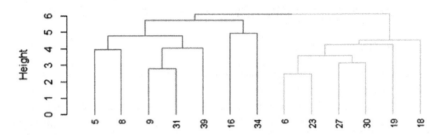

Fig. 2. Two subgroups of gamepad users can be discriminate viewing the cluster dendrogram. Participants number are displayed in the x axis.

Table 1. Mean (M), standard deviation (SD) and *p-value* for the gamepad users during pointing and training task.

	Subgroup 1		Subgroup 2		p
	M	SD	M	SD	
Pointing - Throughput	3.77	0.89	2.21	0.71	0.008*
Pointing - Time	29.68	5.71	42.83	4.72	0.04*
Pointing - Error	5.39	6.66	3.5	4.72	0.08
Pointing - HR	86	9.05	76.37	12.43	0.15
Pointing - HRV	792	76.38	690.33	116.87	0.1
Training - Time	4	0.45	5.26	1.22	0.04
Training - Distance	126.96	13	105.45	23.13	0.06
Training - Error	23.8	9.41	8.43	4.24	0.02*
Training - Click number	57.6	10.16	28.14	4.02	0.005*
Training - HR	85.89	9.42	78.45	11.30	0.15
Training - HRV	702.6	79.88	775	106.84	0.15

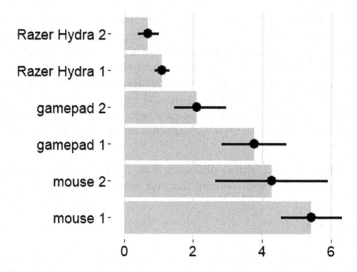

Fig. 3. Confidence intervals of throughput between group 1 and 2 with the 3 interaction techniques.

Mouse Users. The cluster dendrogram separated two groups of observations: subgroup 1 ($n = 6$) and subgroup 2 ($n = 3$). Throughput is almost higher for subgroup 1 ($M = 0.22$) than subgroup 2 ($M = 0.16$) but the difference is not statistically significant ($t = 2.89; p = 0.06$). Visualizing the confidence intervals between subgroup 1 and 2 of mouse users we can interpret that throughput for the subgroup 1 is not higher than the subgroup 2 (Fig. 3). Subgroup 1 took less time to complete the pointing task than the subgroup 2 ($U = 18; p = 0.02$). Subgroup 1 took less time during training VE than the subgroup 2 ($U = 0.5; p = 0.02$). Subgroup 1 use PC for longer than the subgroup 2 ($U = 3; p = 0.05$) and subgroup 1 played video game for longer than the subgroup 2 ($U = 0; p = 0.008$).

Razer Hydra Users. The cluster dendrogram separated 2 subgroups of observations: subgroup 1 ($n = 4$) and subgroup 2 ($n = 4$). Throughput is higher for the subgroup 1 than for the subgroup 2 ($t = 3.61; p = 0.01$). The subgroup 1 made less errors during the pointing task than the second ($U = 12; p = 0.05$) and realized the task is less time than the subgroup 2 ($U = 12; p = 0.05$).

4.2   Comparison of Interaction Techniques

Training. Razer Hydra users navigated during a longer time in the training VE than mouse users ($p = 0.02$) and gamepad users ($p = 0.004$). They also made more errors (i.e., miss click) than participants using the mouse ($p = 0.04$). Other variables were not significantly different.

Pointing Task. Throughput is higher for the mouse than the gamepad ($p = 0.001$) and the Razer Hydra ($p = 0.001$). Throughput gamepad is better than Razer Hydra ($p = 0.001$). Completion time of the pointing task is smaller for the mouse than the gamepad ($p = 0.001$) and the Razer Hydra ($p = 0.001$). Gamepad completion time is smaller than the Razer Hydra ($p < 0.001$). Mouse users made less errors (i.e., miss click) than the gamepad users ($p = 0.01$) and the Razer Hydra users ($p = 0.01$). Gamepad users made less errors than those of the Razer Hydra ($p = 0.01$). Other variables are not significantly different.

HR and HRV. No statistical difference was found in heart rate and HRV during the training step and the pointing task between interaction techniques.

5   Discussion

The study aims to understand how to characterize the user abilities with an interaction technique. To do it, we observed 3 groups of participants interacting with the gamepad, the mouse and the Razer Hydra. They realized a training and a pointing task in a VE.

Unsupervised clustering algorithms like HCA can organize observations at least in 2 subgroups according to several variables. Here, we took variables related to the use of the interaction (e.g., completion time, number of errors) and to workload (i.e., HR and HRV). The HCA discriminated two subgroups of users, within each group: a first subgroup with good skilled users and a second subgroup with less abilities with the use of the interaction technique.

Several variables are significantly different from subgroup 1 to subgroup 2, among these throughput is a recurrent one. Indeed, the subgroups 1 have a better throughput than the subgroups 2 of gamepad and Razer Hydra users. The subgroups 1 presented participants more familiar with video games or PC usage. For example, the subgroup 1 of gamepad users is characterized by video game players. These participants have a better throughput than subgroup 2 and explored more the VE of the training stage. Indeed, they clicked more on inactive and active items in the VE to see which reaction they could expect or not. They did it in less time than subgroup 2. In addition, the subgroup 1 of mouse users completed the pointing task in less time and they use PC for a longer time and play more video games than participants in subgroup 2. Throughput was not significant between subgroups of mouse users maybe because most of mouse users were already very familiar with the use of the mouse. Indeed, 8 of them use a computer every day and the last one use it several days in a week. The skill of a participant using an interaction technique could be calculated from several parameters like being accustomed with an interaction technique. For instance, console gamers are familiar with gamepad. Indeed, the more familiar is the user with an interaction technique, the more the control of the input device appears to be natural and easy [26]. A natural interaction technique may be not only a technology which is mapping real common gestures in the VE but is also a familiar one. Mouse and keyboard are perceived more natural than the

Razer Hydra [27] whereas mouse and keyboard are desktop-based and the Razer Hydra is a semi-natural interaction technique according Nabioyni and Bowman's taxonomy [28].

The results of the HCA show that the subgroup 1 in the several conditions is more skilled with the use of an interaction technique. Indeed, they have a better throughput or used common interaction techniques for longer than others. The calculation of throughput is a good way to discriminate skilled participants from less accustomed with the use of an interaction technique. Indeed, HCA mainly discriminates group from the throughput results and time completion results. We found no significant difference with ECG data between groups 1 and 2 or the 3 interaction techniques. The HCA didn't separate the groups from a workload measure and the mouse, the gamepad and the Razer Hydra seem to involved the same workload across users. As previous studies, the completion time is smaller with the mouse than with the gamepad and there are less errors with the mouse than with the gamepad [26,27]. The use of the mouse is associated with best results, in part because people use computer for several years and at least every day. The use of mouse cost few effort [28]. Razer Hydra users had lower performance at the pointing task than the mouse and the gamepad groups. These results are concordant with other studies where Razer Hydra is lower and made more errors during a pointing task than gamepad and mouse[29] or during a navigation task for elderly[30]. Razer Hydra is not a common interaction technique and users were not familiarized with it and need more time to be comfortable with it than with the gamepad or the mouse.

HCA is good statistic method to conduct profile analysis and then see by which variables there are the more characterized. To realize a short pointing task with only one sequence of trials, here 13 targets, may be an efficient way to understand the participant abilities with the use of an interaction technique. The users with higher abilities have a better throughput than the one with lower. Having a knowledge about the skill of the user with an interaction technique could help to have a better appreciation of the com- pletion time variable during cognitive test in VE. For instance, a gamepad user with good abilities (e.g., high throughput) who complete a virtual test with a long completion time may have more cognitive issues than a gamepad user with few skills with the use of this interaction technique. Indeed, the realization of complex tasks in a VE involve an amount of workload. A non-familiar user of an interaction technique would devote mental workload for both tasks realization and interaction technique use. So, his/her involved workload would be high and he/she may commit errors or take a long time to complete the complex virtual tasks even if the user has no cognitive issues.

Future studies should include more sequence trials during the pointing task to explore how many trials are necessary to distinguish the skills of mouse users because one sequence of 13 trials is not enough. There is also a need to compare the user performance of participants with and without good skills using an inter- action technique during cognitive virtual tasks. Indeed, the subgroup (i.e., good or bad abilities) may have higher, lower or not significant difference compared to the other.

References

1. Tucker, A.B.: Computer Science Handbook. CRC Press, Boca Raton (2004)
2. Jankowski, J., Hachet, M.: A survey of interaction techniques for interactive 3D environments. In: Eurographics 2013-STAR (2013)
3. Bowman, D.A., Kruijff, E., LaViola Jr., J.J., Poupyrev, I.: An introduction to 3-D user interface design. Presence: Teleoper. Virtual Environ. 10(1), 96–108 (2001)
4. Oviatt, S.: Human-centered design meets cognitive load theory: designing interfaces that help people think. In: Proceedings of the 14th ACM International Conference on Multimedia, pp. 871–880. ACM (2006)
5. Kahneman, D.: Attention and Effort, vol. 1063. Prentice-Hall, Englewood Cliffs (1973)
6. Chanquoy, L., Tricot, A., Sweller, J.: La charge cognitive: Théorie et applications. Armand Colin, Paris (2007)
7. Wong, A.W., Chan, C.C., Li-Tsang, C.W., Lam, C.S.: Competence of people with intellectual disabilities on using human-computer interface. Res. Dev. Disabil. 30(1), 107–123 (2009)
8. Hourcade, J.P., Nguyen, C.M., Perry, K.B., Denburg, N.L.: Pointassist for older adults: analyzing sub-movement characteristics to aid in pointing tasks. In: Proceedings of the SIGCHI Conference on Human Factors in Computing Systems, pp. 1115–1124. ACM (2010)
9. Santos, B.S., Dias, P., Pimentel, A., Baggerman, J.W., Ferreira, C., Silva, S., Madeira, J.: Head-mounted display versus desktop for 3D navigation in virtual reality: a user study. Multimed. Tools Appl. 41(1), 161 (2009)
10. Schneider, W., Shiffrin, R.M.: Controlled and automatic human information processing: I. Detection, search, and attention. Psychol. Rev. 84(1), 1 (1977)
11. Zygouris, S., Giakoumis, D., Votis, K., Doumpoulakis, S., Ntovas, K., Segkouli, S., Karagiannidis, C., Tzovaras, D., Tsolaki, M.: Can a virtual reality cognitive training application fulfill a dual role? Using the virtual supermarket cognitive training application as a screening tool for mild cognitive impairment. J. Alzheimers Dis. 44(4), 1333–1347 (2015)
12. Patel, K.K., Vij, S.K.: Spatial navigation in virtual world. In: Advanced Knowledge Based Systems: Model, Applications and Research. TMRF e-Book, pp. 101–125 (2010)
13. MacKenzie, I.S., Sellen, A., Buxton, W.A.: A comparison of input devices in element pointing and dragging tasks. In: Proceedings of the SIGCHI Conference on Human Factors in Computing Systems, pp. 161–166. ACM (1991)
14. Inkpen, K.M.: Drag-and-drop versus point-and-click mouse interaction styles for children. ACM Trans. Comput.-Hum. Interact. (TOCHI) 8(1), 1–33 (2001)
15. Gonçalves, A., Cameirão, M.: Evaluating body tracking interaction in floor projection displays with an elderly population. In: PhyCS, pp. 24–32 (2016)
16. Fallahi, M., Motamedzade, M., Heidarimoghadam, R., Soltanian, A.R., Farhadian, M., Miyake, S.: Analysis of the mental workload of city traffic control operators while monitoring traffic density: a field study. Int. J. Ind. Ergon. 54, 170–177 (2016)
17. Jimenez-Molina, A., Lira, H.: Towards a continuous assessment of cognitive workload for smartphone multitasking users. In: The First International Symposium on Human Mental Workload. Dublin Institute of Technology (2017)
18. Cinaz, B., Arnrich, B., La Marca, R., Tröster, G.: Monitoring of mental workload levels during an everyday life office-work scenario. Pers. Ubiquit. Comput. 17(2), 229–239 (2013)

19. ISO: DIS 9241-9: Ergonomic requirements for office work with visual display terminals, non-keyboard input device requirements. International Organization for Standardization (2000)
20. Fitts, P.M.: The information capacity of the human motor system in controlling the amplitude of movement. J. Exp. Psychol. **47**(6), 381 (1954)
21. Rosa, P.J., Morais, D., Gamito, P., Oliveira, J., Saraiva, T.: The immersive virtual reality experience: a typology of users revealed through multiple correspondence analysis combined with cluster analysis technique. Cyberpsychol. Behav. Soc. Netw. **19**(3), 209–216 (2016)
22. Verhulst, E., Richard, P., Richard, E., Allain, P., Nolin, P.: 3D interaction techniques for virtual shopping: design and preliminary study. In: Proceedings of the 11th Joint Conference on Computer Vision, Imaging and Computer Graphics Theory and Applications: GRAPP, vol. 1, pp. 271–279. SCITEPRESS-Science and Technology Publications, Lda (2016)
23. Němcová, A., Maršánová, L., Smíšek, R.: Recommendations for ECG acquisition using BITalino. In: EEICT Conference, vol. 1, pp. 543–547, April 2016
24. RStudio Team: RStudio: Integrated Development Environment for R. RStudio Inc., Boston (2016). http://www.rstudio.com/
25. Besançon, L., Dragicevic, P.: La différence significative entre valeurs p et intervalles de confiance. In: Alt. IHM, p. 10 (2017)
26. Vorderer, P.: Interactive entertainment and beyond (2000)
27. Seibert, J., Shafer, D.M.: Control mapping in virtual reality: effects on spatial presence and controller naturalness. Virtual Real. **22**, 1–10 (2017)
28. Nabioyuni, M., Bowman, D.A.: An evaluation of the effects of hyper-natural components of interaction fidelity on locomotion performance in virtual reality. In: Proceedings of the 25th International Conference on Artificial Reality and Telexistence and 20th Eurographics Symposium on Virtual Environments, pp. 167–174. Eurographics Association (2015)

Information and Learning

Development of a Blended Learning System for Engineering Students Studying Intellectual Property Law and Access Log Analysis of the System

Takako Akakura[1,4(✉)], Takahito Tomoto[2], and Koichiro Kato[3,4]

[1] Faculty of Engineering, Tokyo University of Science, 6-3-1 Niijuku,
Katsushika-ku, Tokyo 125-8585, Japan
akakura@rs.tus.ac.jp
[2] Faculty of Engineering, Tokyo Polytechnic University, 11583 Iiyama,
Atsugi-shi 243-0297, Japan
t.tomoto@cs.t-kougei.ac.jp
[3] Graduate School of Innovation Management,
Kanazawa Institute of Technology, 1-3-4 Atago, Minato-ku,
Tokyo 105-0002, Japan
[4] Graduate School of Engineering, Kanazawa Institute of Technology,
7-1 Ogigaoka, Nonoichi-shi, Ishikawa 921-8501, Japan
kkato@neptune.kanazawa-it.ac.jp

Abstract. In this paper, we analyzed access logs to this blended learning system. The blended learning system is divided into a login section, a menu section, a news section, a video viewing section, an exercise section, and a pdf output section. Therefore, we analyzed relationship between (a) access frequency, (b) video viewing time, (c) exercise system use time, and (d) pdf output count and students' achievement, respectively. The main results were as follows. (1) Of the 30 students who took the final exam, 22 students accessed the system one or more times. (2) The total access count was 81 times, the total video viewing time was 79 h 56 min 51 s, the exercise system use time was 22 h 20 min 58 s, the number of pdf output was 83 times (278 sheets). (3) Achievement superiors used all of the (a), (b), (c) and (d) more times or more hours than the achievement subordinate. (4) In the basic problems, there was a positive correlation between the students' achievement and video viewing, exercise system utilization time, respectively. (5) In the applied problems, there was a high positive correlation between the students' achievement and the use time of the exercise system. (6) In the problem with a low percentage of correct answers, a high positive correlation was found between the score and the use time of exercise system. From the above results, it seems that both video and exercise system are useful for acquiring basic knowledge, and exercise system is useful for developing applied abilities. In addition, it was considered that utilizing the exercise system also contributes to solving high difficulty problems.

Keywords: e-Learning system · Blended learning · Engineering education
Intellectual property law education

© Springer International Publishing AG, part of Springer Nature 2018
S. Yamamoto and H. Mori (Eds.): HIMI 2018, LNCS 10905, pp. 231–242, 2018.
https://doi.org/10.1007/978-3-319-92046-7_21

1 Introduction

Although the importance of intellectual property education within engineering departments is widely recognized [1, 2], engineering departments are unable to spend a sufficient amount of time on intellectual property education due to the fact that classes that are part of students' majors (and laboratory classes, in particular) take up large amounts of time [3]. The engineering department at university A offers a 30-h class on intellectual property law that is worth two units. However, due to time constraints, it is difficult to cover material related to applications of intellectual property law. Therefore, we developed and operated an e-learning system (referred to as simply the "system" in the following) based on a blended-learning approach that includes material on applications of intellectual property law intended to serve as a supplement to the material presented in classes. The system includes a recorded video of a face-to-face class that students can watch and also allows students to work on practice problems related to the video.

In this paper, we first describe the system that we developed and operated. Next, we analyze the results from the 2017 class to determine the approximate amount of time that students spent using the system, the ways that students used the system, and the relationship between system usage and knowledge acquisition among engineering students who were enrolled in the intellectual property law class. Next, we investigate which types of system features are useful for which kinds of knowledge acquisition. We also discuss the types of features that should be added to the system in the future and the types of usage methods that should be promoted.

2 Blended Learning System

2.1 System Overview

The class on intellectual property law offered by the engineering department at university A includes 6 h of lectures covering basic knowledge of the law and the Japanese legal system, 2 h of lectures on the Intellectual Property Act, 6 h of lectures on the Patent Act, 2 h of lectures on the Utility Model Act, 2 h of lectures on the Design Act, 4 h of lectures on the Trademark Act, 6 h of lectures on the Copyright Act, and a 2-h midterm exam. The lectures were all recorded on video, which was used as the core of the blended learning system. Recording the lectures on video allows students to watch the videos in the event that they miss class or want to review the material after attending class. Therefore, the main purpose of the system was to supplement learning. Furthermore, the system also provided a way for students who wished to go beyond the content delivered during the lectures to work on practice problems. Therefore, a secondary purpose of the system was to provide opportunities for practice based on the material in the lecture and opportunities for learning advanced material.

When students log in to the system through the login page, the system shows the home page (Fig. 1). When students select the learning content, the system shows the content list screen, after which students can access the lecture videos, pdf-format files of quizzes that were given during the lectures, and practice problem pages (Fig. 2). The system allows students to fast-forward and rewind the lecture videos (Fig. 3) as well as

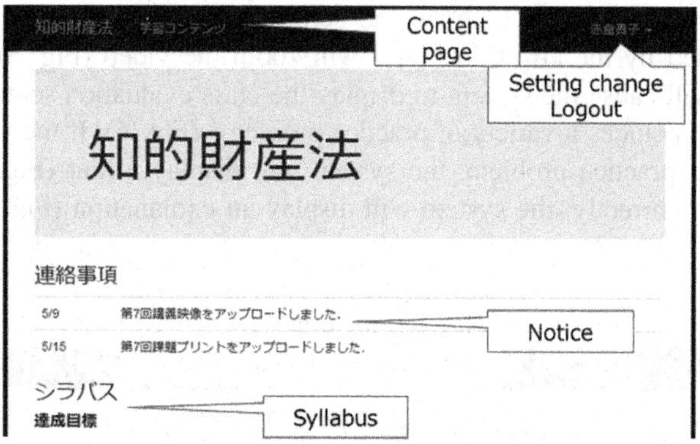

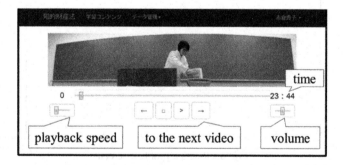

Fig. 1. Home page (publishes updates and syllabus, etc.)

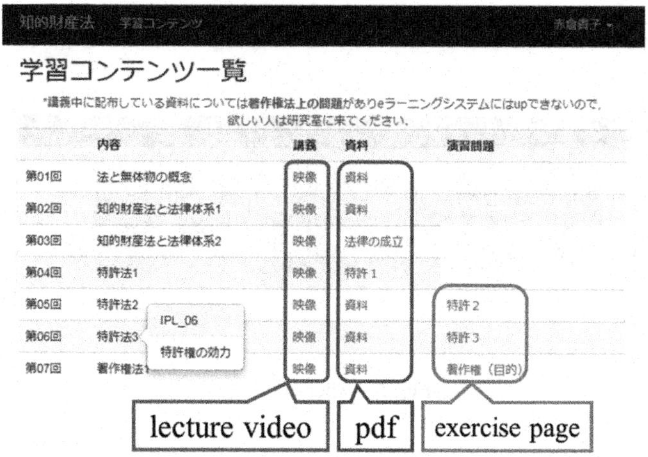

Fig. 2. Content selection screen

Fig. 3. Lecture video screen

zoom-in to the video by clicking on the screen. For example, left-clicking on the location indicated by the arrow in Fig. 4 will zoom the video (Fig. 5). Furthermore, right-clicking will cause the system to display the class evaluation screen. In addition, the system also includes a variety of practice problems (Fig. 6). If the student answers incorrectly on a practice problem, the system will display a hint (Fig. 7), and if the student answers correctly, the system will display an explanation (Fig. 8).

Fig. 4. Before zooming

Fig. 5. After zooming

Fig. 6. Practice problem screen

Fig. 7. Feedback screen (in the case of an incorrect answer)

Fig. 8. Feedback screen (in the case of a correct answer)

2.2   Usage of the System

System Access. Figure 9 shows a graph of the amount of traffic to the system on days for which at least one user accessed the system. In 2017, the number of students who took the final exam was 30 students. 22 students accessed the system at least one time, and 8 students did not access the system even once. The total number of times that the system was accessed was 93 times. The student who accessed the system the most accessed the system 12 times. Here, "access" refers to instances in which students used the system to watch videos or do practice problems, and does not include instances in which students only printed pdf files. Furthermore, situations in which the same student accessed the system multiple times in one day are counted as only one access in the results above. Since the midterm exam was held on May 29, traffic to the system increased around this time. In addition, the final exam was held on August 7. Accessing the system was completely voluntary for the students and was not compulsory. The results show that the students used the system relatively frequently.

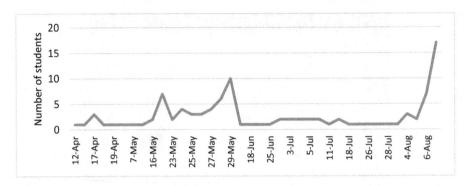

Fig. 9. History of number of students who accessed the system

System Use Time. Table 1 shows the total numbers of hours that students spent watching videos and the total number of hours that students spent using the practice problem system. The student who spent the largest amount of time watching videos spent 20 h, 37 min, and 24 s watching videos. Similarly, the student who spent the largest amount of time using the practice problem system spent 4 h, 55 min, and 25 s using the system.

Table 1. Total amount of system access time

Video system	Practice problem system
79:56:51	22:20:58

Table 2 shows the correlation between the number of hours students spent using the system and their final knowledge attainment (which was evaluated only in terms of their test scores, and did not include attendance or scores on practice problems given during lectures; referred to as "test scores" in the following).

Table 2. Correlation between the number of hours students spent using the system and their test scores

Number of accesses	Video system	Practice problem system
0.326	0.334	0.263

In all categories, there was a somewhat weak positive correlation. In addition, we also compared the amount of time students with high test scores and students with low test score spent using the system. Among the 30 students who enrolled in the class, 8 students who had the "best" scores and 8 students who had the "worst" or "bad" scores were identified. These two groups were referred to as the "top-ranking group" and "bottom-ranking group," respectively, and were compared. The results are shown in Table 3.

Table 3. Relationship between test scores and amount (standard deviation) of time students spent using the system

Test scores	Number of accesses	Video system	Practice problem system
Mean of students with high test scores (n = 8)	4.50 (3.251)	6:13:01 (7:09:27)	1:23:23 (1:48:1)
Mean of students with low test scores (n = 8)	1.75 (2.053)	0:36:47 (0:55:41)	0:12:14 (0:18:13)
p-value	0.063 (t-test)	0.063 (Welch's t-test)	0.107 (Welch's t-test)

The mean values show that the top-ranking group accessed the system more than the bottom-ranking group, spent a larger number of hours watching videos, and spent a larger number of hours using the practice problem system. Since the variances were large, it is not possible to state with certainty that the differences are statistically significant. However, the p-values in the tests for the differences of the mean values were approximately .06–.10, which indicates that the trend is significant. Although it is possible that the students who were more highly motivated or possessed stronger academic abilities used the system more frequently, the results show that there is a relationship between system usage and test scores.

Next, we investigate which types of knowledge acquisition were aided by the videos and practice problem system.

3   Relationship Between Knowledge Acquisition and System Use

3.1   Fundamental Skills and Application Skills

The Intellectual Property Management Skills Test (referred to as the Intellectual Property Test in the following) is one of the Japanese national examinations. For example, the test for level 3 is divided into the "academic subject" test and the "practical skills" test, which are both written tests. An example of a problem from the academic subject test is shown in Fig. 10.

Question
  Which of the following options (a, b or c) best fills in the blank to complete the following sentence about the objective of the Patent Act?

The purpose of this Act is, through promoting the protection and the utilization of inventions, to encourage inventions, and thereby to contribute to the development of <blank>.
  (a) society
  (b) culture
  (c) industry

(Test #16, Academic section) [4]

Fig. 10. Example of a problem from the "academic subject" test in the level 3 Intellectual Property Test [4]

The test consists of problems that should be answered based on actual experience with work related to intellectual property. Furthermore, the authors' practical skills test also required students to write the reason for their answers to the problems in the practical skills test, such as the example problem shown in Fig. 11.

Question

A home appliance manufacturer is considering filing a patent application for an invention relating to a particular type of towel rack. Which of the following behaviors (a, b or c) prior to filing for a patent is most likely to enable the company to obtain a patent ?

(a) Seven months have passed since the research division of the company finished the product design and created a prototype for the towel rack, but no announcements have been made and the product has not been released.
(b) Seven months ago the company demonstrated the towel rack at an industry exhibition, but since then no formal announcements have been made and the produc has not been released.
(c) Seven months ago the company started selling the towel rack to a limited number of suppliers, but no general announcements have been made and the product has not been released.

(Test #20, Skills section) [4]

Fig. 11. Example of a problem from the "practical skills" test in the level 3 Intellectual Property Test [4]

In this paper, we refer to knowledge that is based on the text of the law as "fundamental skills," and the ability to solve problems such as those in the practical skills test in the Intellectual Property Test as "application skills." Next, we investigated the relationship between test scores in the "fundamental skills" test and "application skills" test and the amount of time students spent using the system.

The correlation coefficients between the test scores earned by the students for different problem groups and the amount of time the students spent using the system is shown in Table 4. In addition, information regarding the problem groups, such as whether the problem group consisted of fundamental or application problems and the format of the problems, is shown in Table 5. Tables 4 and 5 show that the students who spent longer amounts of time watching the videos tended to have higher scores for the problems in the fundamental skills test, and the students who spent longer amounts of time using the practice problems system tended to have higher scores for the problems in the application skills test. Furthermore, the results demonstrated that the practice problem system was effective for problem groups 5 and 8, which were difficult groups (problems in which the percentage of students who answered correctly was low). This trend was observed even for problem group 3, which had the lowest percentage of students who answered correctly out of all of the problem groups in the fundamental skills test. In problem groups that consisted of a multiple-choice test, there was almost no correlation between the test scores and the amount of time that students spent watching videos or the amount of time that students spent using the practice problem system. In other words, we believe that the video watching system strengthens fundamental skills, and the practice problem system strengthens students' abilities to apply knowledge and answer difficult questions.

Table 4. Correlation between test scores and amount of time students spent using the system

	Number of accesses	Video system	Practice problem system
Problem group 1	0.120	0.149	0.070
Problem group 2	0.490	0.505	0.189
Problem group 3	0.149	−0.009	0.289
Problem group 4	0.173	0.436	0.108
Problem group 5	0.139	0.500	0.419
Problem group 6	0.267	0.031	0.303
Problem group 7	0.185	0.225	0.454
Problem group 8	0.311	0.146	0.425

Table 5. Problem groups for which test scores were analyzed

	Problem level	Type of problem	Patent Act	Utility Model Act	Design Act	Trademark Act	Copyright Act	System of Laws	Correct answer rate
Problem group 1	Basic	Multiple choice	O	O	O	O	O	O	0.614
Problem group 2	Basic	Simple descriptive	O			O			0.557
Problem group 3	Basic	Simple descriptive		O			O		0.534
Problem group 4	Basic	Simple descriptive			O			O	0.626
Problem group 5	Applied	Descriptive	O	O		O			0.282
Problem group 6	Applied	Descriptive			O	O		O	0.615
Problem group 7	Applied	Descriptive	O		O			O	0.489
Problem group 8	Applied	Descriptive	O			O	O		0.210

3.2 Relationship Between System Features and Problem-Solving Process Model

In a previous study, the authors proposed a "problem-solving process model" [5] for intellectual property law based on the problem-solving process model for physics [6] proposed by Hirashima et al. (Table 6). In this model, the problem solving process was broken down into the following steps: (1) Reading and understanding the problem text (surface-structure generation process), (2) applying the problem text to prior knowledge (formularization process), (3) organizing knowledge necessary for solving the problem (constraint structure in the solution-derivation process), (4) assembling prior knowledge and applied knowledge (solution structure in the solution-derivation process), and (5) answering the question (goal structure in the solution-derivation process). The model uses the fact that legal texts can be written as logical expressions [7]. Referring to this problem-solving process model reveals that the surface-structure generation process and the formularization process require the acquisition of knowledge related to basic legal terms. Listening to lectures and watching videos of the lectures helps students acquire this knowledge. The solution derivation process that follows involves organizing knowledge required for solving the problem, assembling prior knowledge and applied knowledge, and answering the question. The practice problem system helps students acquire these skills.

Table 6. Problem-solving process model

General solution method (problem-solving process)	Problem-solving process in physics (Hirashima et al.) [6]			Problem-solving process in intellectual property law [5]
Read and understand the question text	Process of generating a surface structure	Surface structure	Expresses relationships between the properties in the question text	Expresses relationships between the properties in the question text
Apply the surface structure to knowledge given in the question text	Formalization process	Formal structure	Expresses quantitative relationships in the surface structure	Expresses the surface structure using legal terminology
Organize the knowledge required to solve the problem	Solution-derivation process	Constraint structure	Quantitative relationships in the background to the question	Logical relationships between the properties in the question text
Assemble the existing knowledge and the given knowledge		Solution structure	Quantitative relationships for deriving a solution based on the properties	Logical structure for deriving a solution based on the properties
Answering the question		Target structure	Solution	Solution

3.3 Subjective Evaluation

Students were asked to write their opinions regarding the advantages and disadvantages of the system.

Several advantages of the system identified by the students include the fact that the system gives students the freedom to watch the videos any time, the fact that the system gives students the freedom to watch the videos anywhere, and the fact that the system allows students to repeat watching the videos at their own pace, including fast-forwarding and rewinding the video (each of these advantages were identified by multiple students). In addition, the following disadvantages were identified by the students, with each of the following six disadvantages identified by one student. One student said that "the ability to watch the lectures any time made me feel less motivated during actual lectures," which is an issue that must be investigated. In other words, there exist several students who feel that they can just watch the lectures afterwards at a leisurely pace since they are available on the system. However, if the system causes students to fail to pay attention during the lectures, this defeats the original purpose of the system. Therefore, it is necessary to think of measures for resolving this issue in the future, such as the possibility of including material that is only provided during the face-to-face lecture. Furthermore, it is also necessary to make technical improvements to the system in the future to address student complaints of problems such as slow downloads of video files and poor screen layout. Furthermore, usage of the system was not compulsory, and several students forgot about the existence of the system since the system was not promoted heavily. However, since there are no plans to make usage of the system compulsory in the future, promotion of the system will be limited to announcements during face-to-face lectures.

3.4 Discussion

The problems in the practice problem portion of the blended learning system were not reused in the final exam. Despite this, students who used the practice problem system for longer amounts of time had higher test scores in the problems in the application skills test. In other words, this implies that usage of the practice problem system may contribute to students' abilities to generalize learning activity, which consists of the process of using and applying acquired basic knowledge, to other problems. In the introduction of this paper, we mentioned that the engineering department cannot dedicate a large amount of time solely to intellectual property education since the curriculum includes a high number of laboratory classes. However, students do not have the chance to work on a satisfactory number of practice problems within the short amount of time available for the lectures. Therefore, the lectures are useful for helping students acquire basic knowledge, and the videos are useful for verifying and reviewing this knowledge. Students should use the practice problem system to strengthen their abilities to apply this knowledge.

In addition, although this paper does not cover methods for using the video system (including which sections of the lectures are viewed the most, and ways in which students use the zoom feature) or methods for using the practice problem system (including to what extent students read the explanations), we plan to analyze these topics and add the required features in the future.

Furthermore, the subjective evaluations provided by the students revealed that although the availability of the system provides several benefits, there is also a risk that students will neglect to pay attention during the face-to-face lectures. We plan to investigate methodologies for determining how to blend the system with face-to-face lectures in the future.

4 Conclusion and Future Work

In this paper, we reported the results of using an e-learning system based on a blended learning approach developed for students enrolled in a lecture regarding intellectual property law. We compared and analyzed the usage of the system and knowledge acquisition (test scores) among the students. The results show that usage of the practice problem system was effective for improving students' abilities to solve problems related to the application and practice of the knowledge once they have acquired knowledge in areas such as legal terms.

The authors are also developing a system intended for self-study among students who are not enrolled in face-to-face lectures [8]. We plan to study methods for combining these systems in the future.

Acknowledgments. This research was partially supported by a Grant-in-Aid for Scientific Research (B) (#16H03086; Principal Investigator: Takako Akakura) from Japan Society for the Promotion of Science (JSPS).

References

1. Furukawa, Y.: Intellectual property education in science departments. Patent **66**(1), 70–74 (2013). (in Japanese)
2. Kawakita, K.: Intellectual property training and creation. Patent **64**(14), 40–48 (2011). (in Japanese)
3. Iguchi, Y., Sera, K., Matsuoka, M., Muramatsu, H., Kagohara, H., et al.: Present status and future trends of intellectual property education. Patent **64**(14), 8–18 (2011). (in Japanese)
4. Upload Intellectual Property Training Research Center: Collection of Past Questions from the Intellectual Property Management Skills Test – Grade 3, 2017 edn. (2016). (in Japanese)
5. Akakura, T., Ishii, T.: Development and evaluation of a self-learning support system for Patent Act suited to the current state of intellectual property education in engineering departments. In: Proceedings of 2016 IEEE International Conference on Teaching, Assessment, and Learning for Engineering (TALE 2016) Bangkok, Thailand, December 2016, pp. 128–133 (2016)
6. Hirashima, T., Azuma, S., Kashihara, A., Toyoda, J.: A formulation of auxiliary problems. J. Jpn. Soc. Artif. Intell. **10**(3), 413–420 (1995). (in Japanese)
7. Tanaka, K., Kawazoe, I., Narita, H.: Standard structure of legal provisions. IPSJ SIG Tech. Rep. **93**(79), 79–86 (1993)
8. Akakura, T., Ishii, T., Kato, K.: Proposal of a problem-solving process model for learning intellectual property law using first-order predicate logic and development of a model-based learning support system. In: Proceedings of 11th annual International Technology, Education and Development Conference (INTED 2017), pp. 5145–5152, Valencia, Spain, March 2017

Development of an Asynchronous E-Learning System in Which Students Can Add and Share Comments on an Image of a Blackboard

Kazashi Fujita[1(✉)] and Takako Akakura[2]

[1] Graduate School of Engineering, Tokyo University of Science,
6-3-1 Shinjuku, Katsushika-ku, Tokyo 125-8585, Japan
5314063@ed.tus.ac.jp
[2] Faculty of Engineering, Tokyo University of Science,
6-3-1 Shinjuku, Katsushika-ku, Tokyo 125-8585, Japan
akakura@rs.tus.ac.jp

Abstract. The implementation rates of e-learning have increased in recent years, and most higher education facilities prioritize the development of e-learning content. Unlike face-to-face classes, asynchronous e-learning has the advantage that students can learn anytime and anywhere. However, it is difficult for students to share information about their degree of understanding and ask questions on an e-learning system. Previously, that issue was solved by implementing a bulletin board in the e-learning system. However, inputting text is a burden to students. Accordingly, we have developed a shared information system using images of a blackboard. This system allows the students to associate any area on the blackboard images with a comment while they watch a lecture video. Thus, the students can share information more smoothly than in a general bulletin board system. We evaluated the performance of the developed system in an experiment, and the results suggest the effectiveness of the proposal.

Keywords: E-learning · Bulletin board system · Blackboard image

1 Introduction

Japan is facing a declining birth rate and an aging population and, moreover, the population of Japan is decreasing at an accelerating pace. As a result, the number of private universities that have 18-year-old students has decreased. However, since an individualized lifestyle is now possible in various forms, the number of people seeking a qualification or personal improvement through university study is increasing. For that reason, universities should not only educate 18-year-old students but also educate and support other members of society. However, it is difficult for working people to devote time for study and to commute to a

© Springer International Publishing AG, part of Springer Nature 2018
S. Yamamoto and H. Mori (Eds.): HIMI 2018, LNCS 10905, pp. 243–252, 2018.
https://doi.org/10.1007/978-3-319-92046-7_22

university that is far away from the location of their employment. If universities mandate attendance in class, working adult students cannot take course credits and they may not be able to receive the education which they would like.

As a result of these demographic changes, the development of e-learning systems has been prioritized [1]. In 2013 the Ministry of Education, Culture, Sports, Science and Technology asked universities, "Do you think education utilizing information and communication technology is important at your university?" [2]. The answers "important" and "somewhat important" were given by 40.0% and 53.6% of universities, respectively. In addition, the implementation rates of e-learning increased from 14.0% in 2009 to 27.9% in 2015. These results show the importance of the development e-learning content.

Asynchronous e-learning systems are systems which allow users to select and learn from educational content such as lecture videos and complete exercises when they choose to, rather than having to learn at a specific time and location. So, e-learning has the advantage over face-to-face classes that students can learn anytime and anywhere. However, this approach does not allow students to share information about their degree of understanding or to ask questions. Thus, it is difficult for students to increase their motivation by comparing their own skills to those of others or by obtaining an appropriate answer to their questions from others. Consequently, the motivation of students tends to fall. To solve these problems, a bulletin board was implemented on an e-learning system in previous research. An electronic bulletin board is a kind of social networking service, which is one of the most important modes of communication in the present information society. In addition, this function is also useful for students who cannot interact with others in real time because this information is recorded into a database and the interaction log can be accessed asynchronously. These considerations show that installing a bulletin board in an asynchronous e-learning system is useful.

However, general bulletin board systems have a disadvantage in this context. For example, users may wish to comment on something written by the teacher on a blackboard. In this case, users would need to indicate what is being referred to using words like "the sentence to the right, third from the top." The resulting complexity of the comment cause it to be misunderstood by others. Therefore, we have designed a new interface for sharing comments among students in an e-learning system. In the study reported in this paper, we developed an e-learning system with the new interface and performed an experiment for its evaluation.

2   Previous Studies

Web-based bulletin boards for e-learning have been implemented previously. Umemura et al. [3] developed an asynchronous e-learning system to provide virtual group learning by displaying histories of communication during the viewing of a lecture video. Students can learn together as a group to share knowledge in this system.

Sawayama and Terasawa [4] developed a system that has functions utilizing a bulletin board system in question and answer style e-learning. The functions are "commenting", "asking a question" and "showing approval". Sawayama and Terasawa performed an experiment to evaluate this system, and found that it decreased the learners' academic workload. However general bulletin board system has the disadvantage that comments must be made using natural language only. Thus, if users would like to comment on something written by the teacher on the blackboard, they need to input a long sequence of words like "the sentence to the right, third from the top at time 15:20" to indicate what they are referring to. This complexity can cause others to misunderstand.

Ogawa et al. [5] developed a system that visualizes the correspondence between the scene on the lecture video and comments in the discussion, which supports collaborative lesson improvement in a dispersed environment. The evaluation experiment showed that the discussion was active and lecture improvement results were utilized in the problem sessions. By associating video times with comments, the content of comments can be better understood because a user can view appropriate comments while watching the video. It is also useful for students to be able to share knowledge about the video teaching material using the bulletin board. However, reproduction time alone does not provide sufficient information about what is being discussed unless you already have a deep understanding of the contents of the video. For that reason, it is necessary for readers to re-watch the video in order to understand the comments of others. However, this is problematic, in that students must self-identify commented points in the lecture video, and re-watching entire lecture videos can be an inefficient use of time.

3   Proposal

3.1   Comment Sharing Using Blackboard Images

In this study, we devised a new interface for sharing information by drawing rectangles on a blackboard image. Students can share comments by drawing rectangles on a blackboard image, which avoids the need to input lengthy text. A short comment and the contents enclosed in the rectangle on the blackboard also help the student to understand others' comments.

The comments about the blackboard image are associated with the timeline of the lecture video. So students can read comments, view the blackboard picture and watch the part of the lecture video associated with the comment.

3.2   Development of the System

The client side of the system was built using JavaScript, while the server side .was built using PHP and a MySQL database. Figure 1 shows the workflow of the developed system. A student logs into the system and is presented with a list of teaching materials, including a video of each lecture. After selecting a lecture, the learning screen (Fig. 2) is displayed.

At the top of the learning screen is the lecture video, and below it is the blackboard image. Students can comment while watching the lecture video. However,

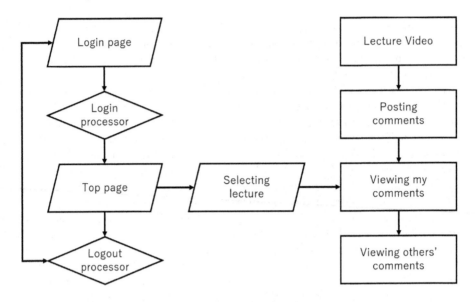

Fig. 1. The configuration of the system

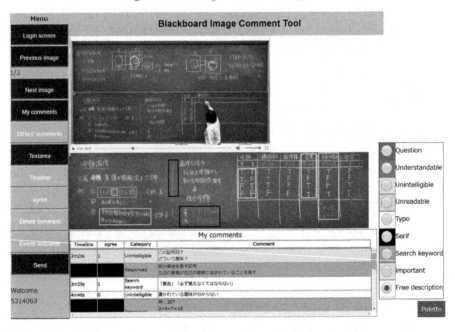

Fig. 2. Screenshot of the system

unlike face-to-face classes, this e-learning system allows the opinions of others to be "heard" during a lecture: below the blackboard image, it is possible to view the comments of others on the lecture. Timeline information from the video is automatically associated with a comment when the student posts it (Fig. 3).

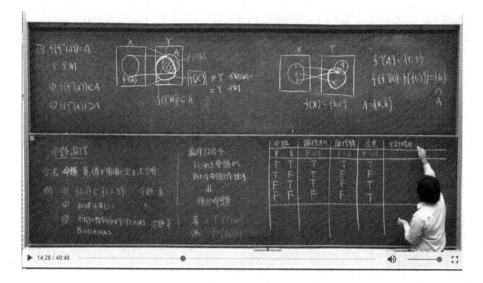

Fig. 3. Lecture video

3.3 Making Comments

The blackboard image shown in Fig. 4 has been the subject of several comments. Rectangles can be drawn by clicking and dragging the mouse over the image. This rectangle can then be associated with a category which indicates the gist of the comment. The color of the rectangle changes according to the comment's category. After drawing a rectangle, the student writes their comment and clicks a button to associate the rectangle with the comment. The rectangle is also associated with timeline of the video. This makes it easier for others to understand the content of the comment.

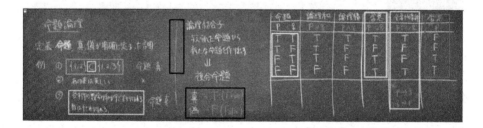

Fig. 4. Making comments on a blackboard image

3.4 Viewing Comments Associated with a Blackboard Image

The students can view comments posted by others with the comment viewer (Figs. 5 and 6). A list of comments exists for each lecture. When students select teaching material, a list of comments corresponding to that teaching material is displayed. If students click on one of their own comments, they can check, edit, or delete the information. In addition, if students click on a comment, the associated rectangle is displayed on the blackboard image and they can reply to comments left by other students. Furthermore, if students agree with another's comment, they can copy it to their own comment area by pressing the "agree" button. The number of times the "agree" button is clicked is recorded for each comment, and students can use this information to assess the tendency of others' thoughts. A video time is associated with all comments and, for that reason, it is possible to change that time simply by pressing a button.

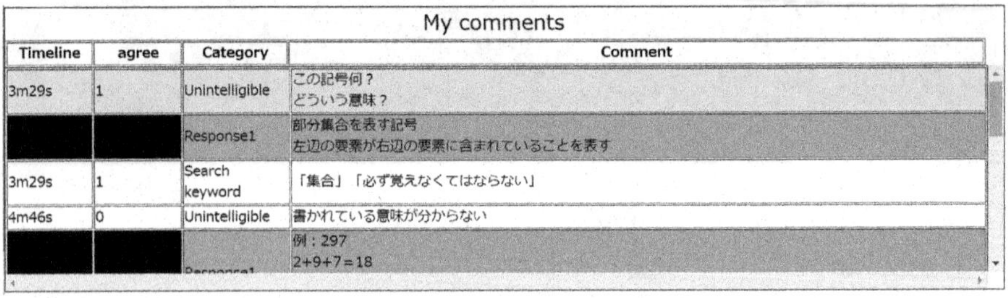

My comments			
Timeline	agree	Category	Comment
3m29s	1	Unintelligible	この記号何？ どういう意味？
		Response1	部分集合を表す記号 左辺の要素が右辺の要素に含まれていることを表す
3m29s	1	Search keyword	「集合」「必ず覚えなくてはならない」
4m46s	0	Unintelligible	書かれている意味が分からない
		Response1	例：297 2+9+7＝18

Fig. 5. Viewing my comments

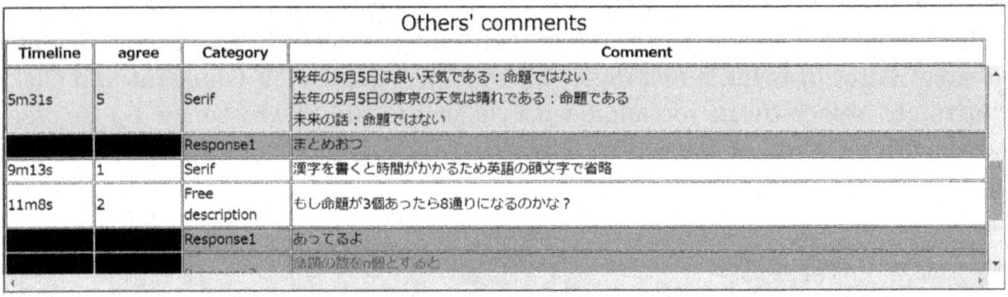

Others' comments			
Timeline	agree	Category	Comment
5m31s	5	Serif	来年の5月5日は良い天気である：命題ではない 去年の5月5日の東京の天気は晴れである：命題である 未来の話：命題ではない
		Response1	まとめおつ
9m13s	1	Serif	漢字を書くと時間がかかるため英語の頭文字で省略
11m8s	2	Free description	もし命題が3個あったら8通りになるのかな？
		Response1	あってるよ
			命題の数を n 個とすると

Fig. 6. Viewing others' comments

4   Evaluation of This System

4.1   Experiment Overview

We performed an experiment to evaluate the system. In the experiment, we used a video of a face-to-face mathematics class that we edited down to 40 min. Three blackboard images were used in this experiment. These were extracted from the lecture video using the rules "the period between the lecturer stopping writing on the blackboard to using the blackboard eraser" and "when there is no lecturer in front of the blackboard".

To evaluate the usefulness of the comment function for information sharing and the rectangle-drawing function for enhancing the understanding of comments, we offered students the use of the developed system and conducted a questionnaire survey. The subjects were eight college students of Tokyo University of Science. The questionnaire is a five-grade evaluation survey. Table 1 shows a part of the questionnaire. After subjects answered the pre-exposure questionnaire, we instructed them in how to use the system by demonstration. Subsequently, we gave orders that subjects make a comment two or more times and answers others' comments two or more times while watching the video. To ensure that all subjects replied to the same comments, we inputted some comments to the system before the experiment and displayed them during the experiment as "others' comments". In addition, we recorded an operation log while subjects used the system. After completing the work, we asked subjects to answer a post-exposure questionnaire.

Table 1. Part of the post-exposure questionnaire

Q1–3	Do you think it is necessary to give your opinion and feelings to others in e-learning?
Q1–5	Do you think that the presence of others is necessary for learning?
Q2–3	Do you think this system is easy to use?
Q2–6	Were you able to communicate your opinion by drawing a rectangle on the blackboard image and leaving your comment?
Q2–9	Do you feel the presence of others by reading their comments?
Q2–17	Do you think that the contents of your comment can be smoothly conveyed to others by drawing a rectangle on the image of a blackboard?
Q2–21	Were you able to understand the content of others' comments by looking at the rectangle on the image of a blackboard?
Q3–1	Was it a hindrance to your learning to draw a rectangle on the image of a blackboard?
Q3–2	Was it a burden on your learning to enter your comments?
Q3–3	Was it a burden on your learning to look at a rectangle in order to understand others' comments?
Q3–4	Was it a burden on your learning to answer others' comments?

4.2 Results and Remarks

In the five-grade evaluation 5 is the most positive and 1 is the most negative. Table 2 shows the responses to the questions shown in Table 1 along with their means and standard deviations.

4.3 Evaluation of This System as a Bulletin Board

The mean values for Q2–6 and Q2–9 were 4.25 and 4.00, respectively. These results show that the system has some of the functions of a bulletin board system and suggest that the system allows students to share comments. However, in this evaluation experiment, students were able to respond to others' comments, but

Table 2. Evaluations of subjects (A to H)

Question	A	B	C	D	E	F	G	H	Mean	S.D
Q1–3	4	3	4	5	1	2	2	2	2.88	1.27
Q1–5	4	5	3	4	1	3	2	2	3.00	1.22
Q2–3	4	4	2	4	2	4	5	3	3.50	1.00
Q2–6	5	5	4	5	4	4	3	4	4.25	0.66
Q2–9	5	4	5	5	2	4	2	5	4.00	1.22
Q2–17	5	4	4	4	3	2	2	3	3.38	0.99
Q2–21	5	5	5	4	1	2	2	2	3.25	1.56
Q3–1	2	1	2	1	4	2	1	4	2.13	1.17
Q3–2	3	1	2	1	1	2	1	3	1.75	0.83
Q3–3	1	1	5	1	1	2	3	2	2.00	1.32
Q3–4	2	1	2	1	1	2	3	2	1.75	0.66

they could not receive a response to their comments. Therefore, it was impossible to evaluate whether it is a system in which students can engage in a question and answer interaction. In order to evaluate this, it is necessary to conduct further evaluation experiments in a more realistic environment.

4.4 Evaluation on Facilitation of Information Sharing

The mean values for Q2–17 and Q2–21 were 3.38 and 3.25, respectively. This seems to indicate that this system does not enable the smooth sharing of information among students. However, as shown in Table 2, subjects A, B, C, and D, who tended to answer "5" or "4" to questions Q1–3 and Q1–5, answered "5" or "4" to questions Q2–17 and Q2–21. On the other hand, subjects E, F, G, and H, who tended to answer "1" or "2" to questions Q1–3 and Q1–5, answered "1" or "2" to questions Q2–17 and Q2–21. These results show that this system enables information to be shared smoothly among students who would like to share their opinions or feel the presence of others during learning.

4.5 Evaluation of the Impact on Learning

The mean values for Q3–2 and Q3–4 were both 1.75. In addition, the mean value for Q3–3 was 2.00 and the mean value for Q3–1 was 2.13. For these reasons using the system seems to add only a little to the burden of learning. However, the burden of learning is thought to fluctuate with difficulty of the video and the number of comments posted in advance. Therefore, when we conduct a long-term experiment using a variety of video materials, it will be necessary to investigate whether the system becomes a burden to students. In addition, in the free comments, the opinion that "it is an effective method to organize my ideas" was given. That opinion showed the possibility that the software could be helpful in independent learning. Since we didn't investigate any aspects of the acquisition of knowledge in the research, we need to investigate the influence of system use on learning.

4.6   Interface

The mean value for Q2–3 was 3.50. This evaluation is not bad, and by analyzing the log, we found that all subjects were able to post their comments. Thus, the system can be used without problems. However, many opinions on the interface were given in free comments. For example, one subject pointed out the troublesome operation of scrolling. From this, we conclude that it is necessary to improve the screen configuration. In addition, six subjects said that they would like to view others' comments without pausing the video and all eight subjects said that they would like to make their comments without pausing the video. It is, therefore, necessary to make an interface that allows this. We also plan to add a function that allows others' comments to be superimposed as text on the video.

5   Conclusion

We have developed a shared information system using blackboard images. This system allows students to associate any area on a blackboard image with a comment while they watch a lecture video. The intention is that students can share information more smoothly than in a general bulletin board system. In order to evaluate the effectiveness of the system in this respect, we conducted an experiment using eight subjects. The results show that students can share information smoothly using this system. It is especially effective for students who would like to share their opinions or feel the presence of others during learning. However, many comments about the interface indicated that this system could be substantially improved.

There are three challenges for the future. The first challenge is conducting the evaluation experiment with an increased number of subjects in order to investigate subjects for whom the system is effective. The second challenge is changing the interface to improve the usability of this system. The third challenge is adding functionality. At the development stage of the system, there were several functions that could not be added because we did not have the time or the technology. For example, we would like to automatically display others' comments using the timeline of a lecture video and provide graphical representations of category information and agreement data. The former functionality needs to be implemented, because many subjects said they wanted to view others' comments while watching the video. Although the latter functionality isn't something that the students wanted, it is necessary in order to investigate the effects of this data on the students. This functionality will be developed as a byproduct of a system of lesson improvement support to present the data created by students using this system to the lecturer in a graphical form.

Acknowledgements. This research was partially supported by a Grant-in-Aid for Scientific Research (B) (#16H03086; Principal Investigator: Takako Akakura) and a Grant-in-Aid for Scientific Research (A) (#15H01772; Principal Investigator: Maomi Ueno) from Japan Society for the Promotion of Science (JSPS).

References

1. The Ministry of Education, Culture, Sports, Science and Technology: Research about profit utilization of ICT which can be put higher educational facilities (in Japanese). http://www.mext.go.jp/component/a_menu/education/detail/__icsFiles/afieldfile/2014/05/19/1347641_01.pdf. Accessed 20 Jan 2017
2. The Ministry of Education, Culture, Sports, Science and Technology: Status on the reform of the contents of education at universities (in Japanese). http://www.mext.go.jp/a_menu/koutou/daigaku/04052801/005.htm. Accessed 20 Jan 2017
3. Umemura, T., Akahori, K., Akakura, T.: A development and evaluation of the e-learning system which gives learners an experience to feel attending to group learning. Jpn J. Educ. Technol. **29**(suppl.), 173–176 (2007). (in Japanese)
4. Sawayama, I., Terasawa, T.: The effect of connecting learners to each other on the transition of academic workload in e-learning for rote exercise. Jpn J. Educ. Technol. **38**(1), 1–18 (2014). (in Japanese)
5. Ogawa, H., Ogawa, H., Kakegawa, J., Ishida, T., Morihiro, K.: A study of trial use of a video sharing system for collaborative lesson improvement. Jpn J. Educ. Technol. **35**(4), 321–329 (2012). (in Japanese)

Proposal for Writing Authentication Method Using Tablet PC and Online Information in e-Testing

Daisuke Hayashi$^{(\boxtimes)}$ and Takako Akakura

Tokyo University of Science, 6-3-1 Niijuku, Katsushika, Tokyo 125-8585, Japan
hayashi@alumni.tus.ac.jp, akakura@rs.tus.ac.jp

Abstract. Existing research in examinee verification in e-Testing has been based on usage of a pen tablet, which has not become widespread in use. Therefore, in this paper, I have aim to introduce a tablet PC for examinee authentication in e-Testing. In an evaluation experiment for introducing the tablet PC, authentication accuracy exceeding that of the pen tablet on the false acceptance rate (FAR) associated with a false rejection rate (FRR) of 0% and also on the equal error rate (EER). In an evaluation experiment in an actual testing environment, FAR was 10% when FRR was 0%, and EER was 7.5%. Therefore, goals of 15% and no more than 10%, respectively, were achieved. From the above results, we believe that introduction of a tablet PC is feasible for examinee authentication in e-Testing.

Keywords: e-Testing · Tablet PC · Writing authentication
Online information · DTW · EER

1 Introduction

Currently, accompanying advances in information and communications technology, introduction of lectures using e-learning at higher education institutions is increasing [1]. However, not many universities award credit for simply taking lectures using e-learning. Instead, students are required to take an examination at a designated site to earn credit. Offering examinations at such designated times/places imposes an obstacle to students in society and students living far away from the examination site. Therefore, a method to alleviate temporal and spatial constraints in the above testing environments is necessary.

Accordingly, e-Testing, in which a test is administered on the Web, is effective for examinees living far away because there are no temporal/spatial restrictions. For e-Testing, it is possible to automate scoring and provide immediate feedback in adaptive tests and similar instruments while continuously estimating the examinee's ability [2]. This provides additional advantages, with adaptive testing able to optimally select initems specific to that ability level. In recent years, private qualifications for which an examination can be administered using

© Springer International Publishing AG, part of Springer Nature 2018
S. Yamamoto and H. Mori (Eds.): HIMI 2018, LNCS 10905, pp. 253–265, 2018.
https://doi.org/10.1007/978-3-319-92046-7_23

e-Testing at home are increasing; however, for university classes, introduction of e-Testing is not gaining acceptance. The reason is that it is considered to be easy to carry out fraudulent activities during testing.

Biometric authentication has been suggested as a method for identity authentication to prevent fraud [3]. This authentication can be divided into types using human physical features, such as the fingerprint or the face, and types using behavioral features, such as the voice or the signature. In e-Testing, it is necessary to acquire data for personal authentication frequently during a test. However, acquiring a fingerprint, which is a physical feature, for each question on an examination presents an obstacle to testing, making it impractical.

Meanwhile, examinee identity collation using signatures, which are behavioral features, has been a topic of serious study in Japan and abroad. Wavelet transformation, which is a method for frequency analysis, has been applied to signature authentication [4,5]. Many such studies use text-dependent signature verification (see [6], etc.), which applies DP matching to local maxima and minima for signatures. For text-based autonomous examinee verification, which also uses visual induction fields [7], shredded processing is applied to characters [8]. Although use of a probability distribution function for strokes in characters has been suggested (see [9], etc.), many characters must be registered for this purpose. Since this method is applied to long sentences, it cannot be applied as-is to personal authentication in e-Testing.

Therefore, Kikuchi et al. [10] have focused on handwriting authentication using handwritten answer letters for examinations in e-Testing. Specifically, this method has been described as "AIUEO" for a situation in which identity can be confirmed beforehand (for example, at the time of registration). In this method, five characters are registered (registration data), and comparison with the same characters (verification data) written at the time of the test is performed, thereby authenticating the examinee. Pre-registration of letters and answers for the tests are performed using a pen tablet. Kikuchi et al. [10] use writing pressure, which is online information, to authenticate the examinee. Since online information does not remain available as visible information, unlike information for character shapes, it is difficult for other people to imitate, and can be said to be suitable for individual authentication on the basis of offline writing information [11]. Research by Kikuchi et al. [10] has suggested that by performing individual authentication using answers on an examination, it is possible to perform sequential authentication without imposing a separate burden on examinees. However, since Kikuchi et al. [10] use a pen tablet, which has not become widely accepted, it is difficult to claim that e-Testing is gaining acceptance.

From the above, we propose a handwriting authentication method for online handwritten information using letters written on a tablet PC. Next, we confirm effectiveness of the proposed method in an evaluation experiment in which the tablet PC is introduced. Specifically, we confirm that the tablet PC is effective for writing authentication against single numeric characters (0 to 9). Finally, we aim to introduce a tablet PC for examinee certification in e-Testing, performing an evaluation experiment in an actual test environment. In this project, our writing certification method is called exam candidate authentication.

2   Writing Authentication Method

2.1   Candidate Authentication Using Characters in Answers

In this project, each time that a candidate fills in an answer, he/she performs authentication. The proposed examinee authentication method in e-Testing is shown in Fig. 1. A determination of whether the examinee is the expected examinee is made on the basis of data acquired from written numerals between 0 and 9 in answers entered onto a tablet PC during the test, in addition to identity verification in the form of an ID and a password at the start of the examination.

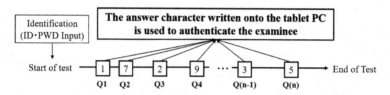

Fig. 1. Candidate certification in proposed e-Testing

2.2   Tablet PC

The tablet PC proposed in this project closely simulates the feeling of writing with a pen on paper since writing is possible while simultaneously watching the pen tip. Therefore, this aspect is believed to reduce the burden on examinees. In addition, market share for tablet PCs has increased fivefold between 2010 and 2015 [12]. Introduction of tablet PCs into public schools has increased 3.5-fold between 2014 and 2016 [13]. Because of this background, we believe that increased use of a tablet PC can lead to more widespread adoption of e-Testing, and, consequently, we have used a tablet PC.

2.3   Feature Value to Use

Figure 2 shows types of feature values obtained from a tablet PC and a pen tablet. Feature quantities used for authentication of answer characters are time T, x-coordinate, y-coordinate, writing pressure P, slope in the x-axis direction from the vertical direction of the writing surface (x-slope), and slope in the y-axis direction (y-slope). These types of online handwriting information can be acquired at intervals of 10 ms, and

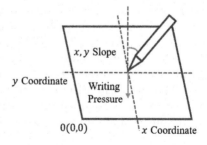

Fig. 2. Online information

can be stored as time-series information. Left-handed examinees respond similarly to right-handed candidates, and are treated by reversal of the sign for the x-slope.

2.4   Similarity Calculation Method

DTW [3] is used as the similarity calculation method. In DTW, the distance between two time series of different lengths. Since the distance becomes smaller as the values to be compared become smaller, it is possible to determine whether an examinee is the expected person ("principal") or an impersonator ("impersonator") on the basis of this distance. Although DTW is a typical technique for online signature verification [4,5,14], it has not yet been applied to examinee authentication using numerals between 0 and 9 in answers. We apply DTW to examinee authentication here.

2.5   Analysis Method

Analysis is performed according to the procedure described in Fig. 3. Personal distance is calculated by applying DTW to each set of verification data for each set of registration data, targeting the same numeral for each examinee. Inter-individual distance is calculated by applying DTW to each set of verification data for each set of registration data, targeting the same numeral and comparing each candidate against all other candidates. Personal distance and inter-individual distance are calculated for the six kinds of on-line writing information listed above.

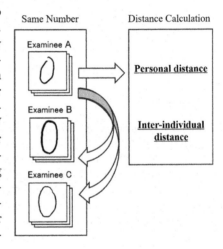

Fig. 3. Analysis procedure

2.6   Composite Distance Optimization

For time T, an x-coordinate, a y-coordinate, writing pressure P, x-slope, and y-slope, let the distance calculated by application of DTW be D_T, D_{Cx}, D_{Cy}, D_P, D_{Sx}, and D_{Sy}, respectively. To make the system robust against impersonation, it is important to combine multiple feature values [3], Therefore, a determination of personal identity will use a composite distance obtained by combining these distances. The composite distance expression is defined as (1). Arbitrary constants α, β, γ, δ, and ϵ are modified in increments of 0.1 to satisfy the expression (2). A determination of personal identity is based on composite distance.

$$D_{TCPS} = \alpha D_T + \beta D_{Cx} + \gamma D_{Cy} + \delta D_P + \epsilon D_{Sx} + (1 - \alpha - \beta - \gamma - \delta - \epsilon)D_{Sy} \quad (1)$$

$$0 \le \alpha \le 1, \quad 0 \le \beta \le 1 - \alpha, \quad 0 \le \gamma \le 1 - \alpha - \beta,$$
$$0 \le \delta \le 1 - \alpha - \beta - \gamma, \quad 0 \le \epsilon \le 1 - \alpha - \beta - \gamma - \delta \quad (2)$$

2.7   Evaluation Index

As an index for evaluation, EER [3] is widely used for authentication. EER is the value at which the FRR and the FAR agree. There is a trade-off relationship between FRR and FAR, with a decrease in one corresponding to an increase in the other. The closer the value of EER is to 0%, the higher the authentication accuracy.

In this project, subjects are certified or rejected according to the FAR obtained when FRR is set to 0% and the EER. The calculation procedure uses arbitrary constants α, β, γ, δ, and ϵ chosen such that FAR is minimized when FRR is initially set to 0%. FAR and EER are calculated with the obtained arbitrary constants. Setting FRR to 0% ensures that no non-impersonator will be rejected.

2.8   Candidate Authentication Flow

The flow of examinee authentication used in this project is shown in Fig. 4. Candidates are asked to write the numerals 0 through 9 beforehand in a face-to-face manner after identity verification by student ID card or similar, and the results are saved as registration data. Subsequently, e-Testing is performed, and the handwritten data values obtained from the answers are saved as verification data. Composite distance is then calculated by comparing the registration data with the verification data using DTW. If the composite distance is less than a threshold value, it is judged to be the expected examinee; if it is larger, impersonation is detected.

To evaluate the accuracy of the system, we set FRR to 0% for each examinee and calculate the values of α, β, γ, δ, and ϵ that collectively minimize FAR. The FAR and EER for each examinee are combined to calculate the overall FAR and EER. The calculated overall FAR and EER are used to evaluate the authentication accuracy as described in the next section.

3   Evaluation Experiment for Introducing Tablet PC

3.1   Outline of Experiment

In order to confirm that the tablet PC is effective for handwriting authentication using one numeral, an experiment comparing a tablet PC with a pen tablet was performed with 10 undergraduate and graduate students (including 1 left-handed person) as participants.

Written data were obtained by partitioning data from the same subject into two subsets with an interval of one month between collections. Four sets of numerals, totaling 40 characters, were used as the registration data. A further 40 characters (again, 4 each of the numerals) were acquired at a second time, one month after acquisition of the registration data. This second set was were treated as the verification data. The composite distance was calculated by comparing the original registration data with the verification data. The registration data

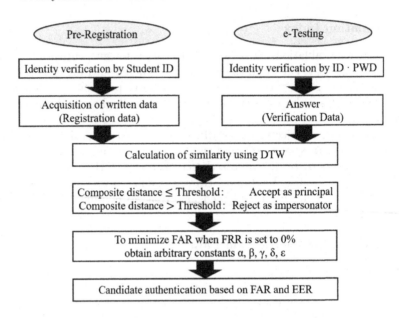

Fig. 4. Candidate authentication flow

were collected between August 23 and August 30, 2017, and the verification data were collected between September 25 and September 27 of the same year. At each session, each candidate wrote "A" ten times to become accustomed to the device before data were recorded.

The success criterion was for FAR (when FRR was set to 0%) and EER with the tablet PC to be lower than the same measures on the pen tablet.

3.2 Experimental Devices

In the evaluation experiment, a Surface Pro 4 and Surface pen were used for the tablet PC, and a Wacom Intuos 4, a Grip Pen, and an ASUS VivoBook X202E were used for the pen tablet. A comparison of the performances is shown in Table 1.

Table 1. Performance comparison

Device	Tablet PC	Pen tablet
Item name	Surface Pro 4	Wacom Intuos 4
	Surface pen	Grip pen
Writing pressure level	4096 level	2048 level
Slop detection level	±69 level	±60 level

3.3   Experimental Results and Analysis

The error rate curves for the tablet PC and the pen tablet are shown in Figs. 5 and 6, respectively. According to the experiment results in Table 2, when an arbitrary constant

Table 2. Experimental results

Experimental device	FAR (at FRR = 0%)	EER
Tablet PC	3.0%	2.5%
Pen tablet	3.6%	2.7%

was chosen so that FAR was minimized when FRR was set to 0% for the tablet PC, FAR was 3.0%, and EER was 2.5% with the same constants. In comparison, the pen tablet demonstrated an FAR of 3.6% and an EER of 2.7%.

The nominal advantage of the tablet PC over the pen tablet is 0.6% on FAR and 0.2% on EER, which shows that the tablet PC is at least as good as the pen tablet. These results confirmed that the tablet PC was effective for handwriting authentication, requiring analysis of only single numerals.

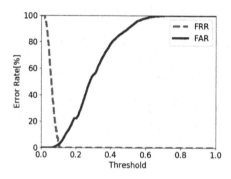

Fig. 5. Error rate curve: tablet PC

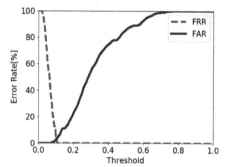

Fig. 6. Error rate curve: pen tablet

At the individual level, values for FAR and EER are shown in Table 3. In this table, there were three people with an EER for both the tablet PC and the pen tablet of 0%, demonstrating that personal authentication was performed perfectly for these subjects on both the tablet PC and the pen tablet. For the tablet PC, low values for EER and FAR were also demonstrated for other subjects, thereby showing that the effectiveness of the tablet PC can be confirmed even for each subject.

Table 3. FAR and EER for each subject

Evaluation index	Experimental device	No. 1	No. 2	No. 3	No. 4	No. 5	No. 6	No. 7	No. 8	No. 9	No. 10
FAR (at FRR = 0%)	Tablet PC	0.0	0.2	35.8	23.8	6.9	25.2	6.9	0.0	0.0	0.0
	Pen tablet	0.0	0.0	14.7	1.3	12.5	41.1	0.0	1.6	0.0	0.0
EER	Tablet PC	0.0	0.1	26.8	11.8	5.1	23.0	5.0	0.0	0.0	0.0
	Pen tablet	0.0	0.0	12.2	1.9	10.6	32.9	0.0	2.0	0.0	0.0

For each subject for which the tablet PC and the pen tablet were used, the value of an arbitrary constant was chosen such that the value of FAR was minimized when FRR was set to 0%, as shown in Tables 4 and 5, respectively. For both the tablet PC and the pen tablet, subjects Nos. 1 and 10 were able to perform individual authentication perfectly with 0% EER, as can be seen in Table 3 The coefficient α (time) is not used. The reason is that, for the complex distance optimization expression given by expression (1), numerical variation begins at ϵ due to the optimization of the composite distance being performed at an early stage.

Table 4. Arbitrary constant set for each subject on tablet PC

Any constant	No. 1	No. 2	No. 3	No. 4	No. 5	No. 6	No. 7	No. 8	No. 9	No. 10
α (time)	0.0	0.1	0.5	0.1	0.1	0.5	0.2	0.1	0.3	0.0
β (x-coordinate)	0.0	0.1	0.0	0.0	0.4	0.0	0.1	0.1	0.0	0.0
γ (y-coordinate)	0.0	0.2	0.0	0.2	0.0	0.1	0.1	0.0	0.0	0.0
δ (writing pressure)	0.2	0.2	0.0	0.2	0.0	0.0	0.1	0.1	0.1	0.0
ϵ (x-slope)	0.4	0.3	0.0	0.5	0.0	0.3	0.4	0.7	0.4	0.1
$1-\alpha-\beta-\gamma-\delta-\epsilon$ (y-slope)	0.4	0.1	0.5	0.0	0.5	0.1	0.1	0.0	0.2	0.9

Table 5. Arbitrary constant set for each subject on pen tablet

Constant	No. 1	No. 2	No. 3	No. 4	No. 5	No. 6	No. 7	No. 8	No. 9	No. 10
α (time)	0.0	0.6	0.0	0.5	0.2	0.5	0.1	0.0	0.1	0.0
β (x-coordinate)	0.1	0.0	0.1	0.4	0.0	0.1	0.1	0.0	0.0	0.0
γ (y-coordinate)	0.1	0.0	0.0	0.0	0.1	0.0	0.0	0.0	0.0	0.0
δ (writing pressure)	0.3	0.1	0.0	0.0	0.0	0.0	0.3	0.1	0.0	0.0
ϵ (x-slope)	0.1	0.2	0.7	0.1	0.5	0.0	0.1	0.2	0.1	0.2
$1-\alpha-\beta-\gamma-\delta-\epsilon$ (y-slope)	0.4	0.1	0.2	0.0	0.2	0.4	0.4	0.7	0.8	0.8

However, for the tablet PC, subjects Nos. 3 and 6 who both had high EER values, with high values for α (time). This reflects that it was difficult to perform personal authentication with handwritten data using the coordinates, writing pressure, and slope without time data, suggesting that the characteristics of the data changed between acquisition of the registration data and acquisition of the verification data. The same observation can also be made for the pen tablet for subject No. 6, who also had a high EER value for that device.

4   Evaluation Experiment in Actual Test Environment

4.1   e-Testing System

Figure 7 shows the e-Testing system evaluated in the experiment. The examinee fills in a solution number using the tablet PC within the frame at the center of the screen. Then, the examinee presses the send button, and the handwriting data for the entered number are transmitted to the server.

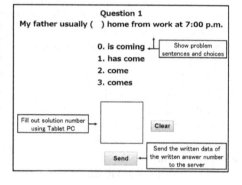

4.2   Outline of Experiment

In order to determine whether a tablet PC should be introduced for examinee authentication in e-Testing, we evalu-

Fig. 7. e-Testing system

ated accuracy of authentication using a tablet PC in an actual e-Testing environment. Forty English grammar questions (excerpted from [15]) that had been presented in the past in the multiple-choice (4 possible answers for each question) college entrance examinations were administered to the same 10 participants as in the experiment described in Sect. 3.

Questions 1 to 8 used answers in the range 0–3, questions 9 to 16 used 4–7, questions 17 to 24 used 8, 9, 0, 1, questions 25 to 32 used 2, 3, 4, 5, and questions 33 to 40 used 6, 7, 8, 9. This was done to disperse the answers equally among 0–9.

Written data were acquired from the same people using English tests that were given at intervals of one and a half months and two weeks apart from each other. Registration data used four sets of numerals between 0 and 9 acquired one and a half months and two weeks before the English test, each set comprising 40 characters. The verification data were acquired from an English test of 40 questions. Composite distance between the registration data and the verification data were calculated, and evaluation was performed. Verification data were collected between October 9 and October 11. In the same way as for the registration data, in order to ensure that the device was properly adjusted for the data, we acquired the verification data after "A" had been written 10 times.

The desired value was a FAR no more than 15% with FRR set at 0%, and an EER of 10% or lower. Guidelines for point-of-sale (POS) applications suggest an FRR of 0.5%when FAR is set to 20% or less [16], and an FRR of 0% when FAR is set to 25% or below [17]. Because POS applications allow face-to-face confirmation of identity, we have set values more strictly than suggested in POS guidelines. As the desired value for EER, we set a target of no more than 10%, yielding a discrimination rate of 90% or higher.

4.3   Experimental Result and Consideration

The error rate curve for the actual test environment experiment is shown in Fig. 8. According to the experimental results in Table 6, when constants were chosen such that FAR was minimized when FRR was set to 0%, FAR was 10.0%, and EER was 7.5%. This result satisfies the desired FAR of 15% or lower.

Table 6. Experimental results

FAR (at FRR = 0%)	EER
10.0%	7.5%

The EER value was lower than 10.0%, which was the desired ceiling, and a discrimination rate of 90% was achieved. Thus, it seems feasible to introduce a tablet PC for examinee authentication in e-Testing.

The values for FAR and EER for each examinee are shown in Table 7. Here, EER was 0% for two examinees, showing that these examinees were capable of performing individual authentication perfectly. Other examinees also demonstrated low values for EER, confirming the effectiveness of the tablet PC in e-Testing. In addition, FAR when FRR was set to 0% was also generally low.

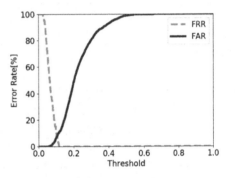

Fig. 8. Error rate curve: actual test environment experiment

However, there were candidates who demonstrated high values of FAR, such as candidates No. 3 and No. 5. This may have been due to the pressure of the test and the written data used for authentication being discrete data. The data may have resembled the data of other candidates.

Table 7. FAR and EER for each candidate

	No. 1	No. 2	No. 3	No. 4	No. 5	No. 6	No. 7	No. 8	No. 9	No. 10
FAR (at FRR = 0%)	0.0	14.1	34.1	13.0	44.4	13.6	28.8	3.6	12.7	0.0
EER	0.0	6.8	23.0	6.6	15.4	10.8	16.8	2.7	10.0	0.0

Also for each examinee, Table 8 shows the value of an arbitrary constant chosen so that FAR is minimized when FRR is 0%. Here, candidates No. 1 and

No. 10 have a perfect EER of 0% in Table 7. It can be seen that coefficients other than α (time) are also used. This indicates that authentication is possible from the coordinates, writing pressure, and slope without time data.

Table 8. Arbitrary constant set for each examinee

	No. 1	No. 2	No. 3	No. 4	No. 5	No. 6	No. 7	No. 8	No. 9	No. 10
α (time)	0.0	0.6	0.0	0.3	0.2	0.5	0.6	0.1	0.2	0.0
β (x-coordinate)	0.1	0.1	0.0	0.0	0.0	0.1	0.1	0.0	0.2	0.0
γ (y-coordinate)	0.1	0.1	0.1	0.2	0.1	0.0	0.0	0.2	0.3	0.0
δ (writing pressure)	0.3	0.1	0.0	0.1	0.5	0.1	0.1	0.2	0.0	0.0
ϵ (x-slope)	0.1	0.1	0.9	0.3	0.0	0.3	0.1	0.1	0.1	0.4
$1 - \alpha - \beta - \gamma - \delta - \epsilon$ (y-slope)	0.4	0.0	0.0	0.1	0.2	0.0	0.1	0.4	0.2	0.6

4.4   Test Analysis

The overall test results for 10 examinees are shown in Table 9 and test results for each candidate are shown in Table 10. To analyze how the test results affected authentication accuracy, we compare it with Table 7 in the previous section. The certification accuracy for candidate No. 3, who had the highest correct answer rate, is reflected in an EER of 23.0%, which was the highest among all examinees. In addition, the certification accuracy of candidate No. 10, who had the second-lowest answer rate, is reflected in an EER of 0.0%, which was the lowest among all examinees. These results suggest that test results and authentication accuracy are unrelated.

Table 9. Test results

Average score	Average accuracy rate	Standard deviation
26.7	66.7%	5.8

Table 10. Test results by examinee

Examinee	No. 1	No. 2	No. 3	No. 4	No. 5	No. 6	No. 7	No. 8	No. 9	No. 10
Score	27	29	38	33	25	30	21	26	17	21
Correct answer rate	67.5%	72.5%	95%	82.5%	62.5%	75%	52.5%	65%	42.5%	52.5%

5   Conclusion

Existing studies on examinee verification in current e-Testing have been based on a pen tablet, which has not gained widespread acceptance. We believe that because the pen tablet and the display are distinct, there may be an impact on the examinee's cognitive load. Therefore, in this paper, we have aimed to introduce a tablet PC system for examinee authentication in e-Testing. We have proposed authentication that uses a tablet PC and information stored online. To demonstrate its effectiveness, we verified the results of writing certification using a tablet PC in Sect. 3. In addition, Sect. 4 examined whether it was possible to introduce a tablet PC for examinee authentication in e-Testing. The following is a summary of each section.

In Sect. 3, a tablet PC and a pen tablet were compared. The objective was to verify that the tablet PC was effective for handwriting recognition using one numeral. Four sets of numerals, totaling 40 characters, were used as the registration data. A further 40 characters (again, 4 each of the numerals) were acquired at a second time, one month after acquisition of the registration data. This second set was treated as the verification data. The desired result was that the FAR and the EER of the tablet PC would be lower than the FAR and EER of the pen tablet when FRR was set to 0%. Experimental results showed that on a tablet PC (resp., pen tablet), FAR was 3.0% (3.6%) and EER was 2.5% (2.7%) when FRR was set to 0%. Thus, the FAR and EER for the tablet PC were lower than the same measures for the pen tablet, confirming that the tablet PC was effective for handwriting authentication.

In Sect. 4, we conducted 40-item English tests using a tablet PC, aiming to support tablet PCs for examinee authentication in e-Testing. The question format was multiple choice with four choices for each question. Registration data comprised 4 complete sets of numerals for each candidate, acquired one and a half months and two weeks before the English test, and the verification data from the 40-item English tests. The target was a FAR no higher than 15% with FRR set to 0%, and an EER of 10% or lower. The experiment results were a FAR of 10.0% and an EER of 7.5%, meeting the target for both FAR and EER. This is a discrimination rate of at least 90%, suggesting that it is feasible to introduce a tablet PC for examinee authentication in e-Testing. A future topic is to incorporate facial authentication in addition to writing certification to provide alternative verification for candidates with a low certification rate.

Acknowledgments. This research was partially supported by a Grant-in-Aid for Challenging Exploratory Research (#15K12427; Principal Investigator: Takako Akakura) and a Grant-in-Aid for Scientific Research (A) (#15H01772; Principal Investigator: Maomi Ueno) from Japan Society for the Promotion of Science (JSPS).

References

1. University ICT Promotion Council: Survey research survey report on ICT utilization in higher education institutions (2016, in Japanese)
2. Ueno, M., Nagaoka, K.: eTesting. Baifukan, Tokyo (2009). (in Japanese)
3. Hangai, S.: Biometric Textbook from Principle to Programming. The Institute of Image Information and Telecommunications, Corona Company, Tokyo (2012). (in Japanese)
4. Nanni, L., Lumini, A.: A novel local on-line signature verification system. Pattern Recogn. Lett. **9**, 559–568 (2008)
5. Nakanishi, I., Nishiguchi, N., Itoh, Y., Fukui, Y.: On-line signature verification based on subband decomposition by DWT and adaptive signal processing. Trans. Inst. Electron. Inf. Commun. Eng. **J87–A**(6), 805–815 (2004). (in Japanese)
6. Bovino, L., Impedovo, S., Pirlo, G., Sarcinella, L.: Multi-expert verification of hand-written signatures. In: Proceedings of the Seventh International Conference on Document Analysis and Recognition (ICDAR 2003), vol. 2, pp. 932–936 (2003)
7. Sawada, T., Ohashi, G., Shimodaira, Y.: Text-independent writer verification method by induction field in vision. J. Inst. Image Inf. Telev. **56**(7), 1124–1126 (2000). (in Japanese)
8. Ando, S., Nakajima, M.: A text-independent writer verification based on inclination of strokes. J. Inst. Electr. Eng. **120**(11), 1732–1737 (2000). (in Japanese)
9. Li, B., Sun, Z., Tan, T.: Online text-independent writer identification based on stroke's probability distribution function. In: Lee, S.-W., Li, S.Z. (eds.) ICB 2007. LNCS, vol. 4642, pp. 201–210. Springer, Heidelberg (2007). https://doi.org/10.1007/978-3-540-74549-5_22
10. Kikuchi, S., Furuta, T., Akakura, T.: Periodical examinees identification in e-test systems using the localized arc pattern method. Distance Learn. Internet Conf. **2008**, 213–220 (2008)
11. Nakamura, Y., Kidode, M.: Online writer verification using feature parameters based on the document examiners' knowledge. J. Inst. Syst. Control Inf. Eng. **22**(1), 37–47 (2009). (in Japanese)
12. Ministry of Public Management: Result of the survey on communication trend in Heisei 28 (2017, in Japanese)
13. Ministry of Education: Survey result on the actual situation of informatization of school in FY2007 school (summary) (2016, in Japanese)
14. Muramatsu, D., Hongo, Y., Matsumoto, T.: Online signature verification using user generic fusion model. Trans. Inst. Electron. Inf. Commun. Eng. **J90–D**(2), 450–459 (2007). (in Japanese)
15. Shinoda, S., Yoneyama, T.: English Grammer. Dictionary Vintage, Iizuna Shoten, Tokyo (2010). (in Japanese)
16. Nelson, W., Turin, W., Hastie, T.: Stastical methods for on-line signature verification. Int. J. Pattern Recogn. Artif. Intell. **8**(3), 749–770 (1994)
17. Lee, L.L., Berger, T., Aviczer, E.: Reliable on-line human signature verification systems. IEEE Trans. Pattern Anal. Mach. Intell. **18**(6), 643–647 (1996)

Proposal of a Framework for a Stepwise Task Sequence in Programming

Kento Koike[1](✉), Takahito Tomoto[1], Tomoya Horiguchi[2],
and Tsukasa Hirashima[3]

[1] Faculty of Engineering, Tokyo Polytechnic University, Kanagawa, Japan
c1418030@cs.t-kougei.ac.jp
[2] Graduate School of Maritime Sciences, Kobe University, Kobe, Japan
[3] Graduate School of Engineering, Hiroshima University, Hiroshima, Japan

Abstract. The execution process for solutions in problem-solving usually involves multiple operations; however, in many cases, the learner does not recognize why individual operations are performed. Therefore, the importance of understanding the grounds for performing individual operations has been pointed out. To fully understand individual operations in task presentation for learning support, it is necessary to discuss not only the properties of a single task, but also the kind of task sequence that should be designed within a continuous task. In general, the order and grain size of tasks presented in programming learning are determined by the teacher. However, if tasks are not presented in a stepwise manner, the learner may have difficulty or may have an incomplete understanding. In continuous tasks, such as programming learning, we assert that a framework based on the qualitative consideration of various properties, such as the order and granularity of tasks presented to learners, is necessary. Therefore, in this study, we examined the properties of sequential tasks to provide a framework for designing stepwise task sequences for programming learning support.

Keywords: Programming learning · Gradual learning
Gradual knowledge acquisition

1 Introduction

In the execution process for solutions in problem-solving, multiple operations is typically performed; however, in many cases, the learner does not recognize why individual operations are performed in the solution. Therefore, the importance of understanding the basis for why individual operations are performed has been pointed out [1]. To fully clarify individual operations in task presentation for learning support, it is necessary to discuss not only the properties of a single task, but also the kind of task sequence that should be designed within a continuous task. In general, the order and grain size of tasks presented to learners in programming learning are determined by the teacher. However, if tasks are not presented in a stepwise manner, the learner may have difficulty or may have an incomplete understanding. Therefore, for continuous tasks, such as programming learning, we assert that a framework based on the

S. Yamamoto and H. Mori (Eds.): HIMI 2018, LNCS 10905, pp. 266–277, 2018.
https://doi.org/10.1007/978-3-319-92046-7_24

qualitative consideration of various properties, such as the order and granularity of tasks presented to learners, is necessary. In addition, for each task, it is beneficial for the learner to understand the previously learned tasks and to gradually progress to the developed state in order to understand the influence of individual operations in combination. Such an approach is known as "gradual learning."

In the study of progressive learning in physics exercises, a framework called Graph of Microworlds (GMW) has been proposed to design these task sequences [2]. GMW has been proposed based on ICM (Increasingly Complex Microworlds). ICM is a learning support method that not only focuses on experiencing and understanding a phenomenon in a learning environment based on simulations, but also gradually introduces complexity to lead to an understanding of complicated phenomena. In addition to gradually building complex phenomena, GMW supports differences between phenomena gradually experienced by learners.

Consideration of stepwise task sequences (e.g., physical exercise), like GMW, is equally important for gradual programming learning. Therefore, in this research, we examined the nature of tasks to provide a framework for designing stepwise task sequences in programming learning support.

2 Previous Research

Watanabe et al. [3] have conducted learning support using continuous tasks. They proposed a method for stepwise reading by giving the learner the entire code using a technique called stepwise abstraction. For example, each line of the following three-line program is simple assignment: (1) store the value of A in C, (2) store the value of B in A, (3) store the value of C in B. However, by recognizing three rows at once, the learner can identify a specific part, such as "swapping two variables." When the learner notices this part, if a program with the if-statement (4) if A > B at the beginning of the three lines is presented, the learner can detect a program containing parts called "swapping two variables." Therefore, the learner can understand the function of the expanded part by considering only the function of the if statement extended to parts described by "swapping two variables." In this way, this method supports the gradual acquisition of parts by providing the whole program to the learner, summarizing the parts that the learner considers a series of operations from the whole, and considering its meaning. Watanabe et al. insist on the importance of continuous tasks in this stepwise abstraction. By using continuous tasks, learners who solved a certain task within a complex task included a solution for solving the task stepwise. In addition, presenting a simpler task with a narrow solution as an auxiliary problem to learners who are unable to solve a certain task encourages students to understand the solution. Therefore, Watanabe et al. built a task sequence in programming and supported the migration of issues, like in GMW.

In this paper, we discuss task sequences in programming with reference to the task sequence described by Watanabe et al.

3   Task Structure

To construct task sequences, the task itself needs to be structured. In this chapter, we discuss the structuring of problems in programming. First, we point out the limitations of previous research. Next, we focus on the difference between the problem-solving process in general and in programming and discuss the elements to be addressed in the problem-solving process in programming. Finally, we describe the structure and definition of the task based on the elements to be solved.

3.1   Definitions and Problems in Previous Research

To structure the task, Watanabe et al. [3] defined three tasks: (i) the purpose of the code, (ii) the actual process (code), and (iii) verification of the understanding of the task. In addition, the tasks are related to is-a relations or part-of relations and support the transition of learners.

Watanabe et al. discussed the association between a certain process and a specific purpose. However, several processes can achieve the same purpose, and accordingly it is insufficient to discuss one particular process. To discuss multiple processes associated with a purpose, it is necessary to discuss the behavior of the process by which a certain purpose is achieved. Additionally, to position the purpose of a certain task, the purpose as a whole is required. In such a case, a task cannot be reused for other purposes and does not have the property of reusability. In order for a task to be reusable, it is necessary to discuss the functions of the subject, rather than the purpose. Therefore, in this study, we discuss the characteristics of functions, behaviors, and processes separately for the task, and we aimed to make tasks reusable.

3.2   Problem-Solving Process

In general, in the problem-solving process, it is necessary to achieve the target end state by applying the operation sequence to the initial state (Fig. 1). Similarly, in the problem-solving process in physics, it is necessary to achieve the end state by applying the operation sequence for the phenomenon to the initial state. At this point, the difference between the initial state and the end state is observed as a behavior of a certain operation sequence. For example, in Fig. 2, the object M is moving to the right from the initial state to the end state. The movement of object M can be considered a change caused by a series of operations, that is, a behavior. In physics, such a behavior can be interpreted as a function, such as "uniform motion." Furthermore, it can be interpreted not only as "uniform motion," but also as "rightward motion" and "frictionless movement." In other words, the functions that can be inferred from behaviors are not uniquely limited, and multiple interpretations are possible depending on the conditions at each point.

The properties of a task in problem-solving in physics are understood in terms of a superficial difference, such as "an object is hanging on a thread/hanging on a spring." Despite this superficial similarity, the solution to be applied differs substantially. However, the learner assesses the properties of the task from these surface features and applies a similar or a different solution to the task to which the same solution can be applied. In these cases, Horiguchi et al. [4] argue that it is important not only to

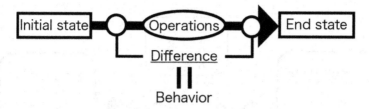

Fig. 1. General overview of the problem-solving process.

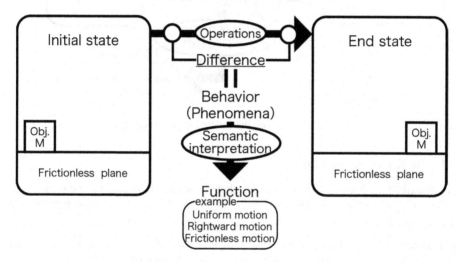

Fig. 2. Problem-solving process in physics.

determine the surface features of the task (i.e., the surface layer structure), but also to understand the structural features (i.e., the deep layer structure), to which appropriate principles and laws can be applied.

As with physics, even in the target programming in this research, the target end state is achieved by giving the source code as the operation sequence from the initial state of the variable. The source code in this initial state and the end state, that is, the difference caused by processing, can be observed as a behavior (Fig. 3). It is possible to interpret meaningful behaviors as functions. However, it is difficult to interpret behaviors as a function in the form of "C = A" in Fig. 3 without an excess or deficiency. That is, functions and behaviors do not necessarily correspond. If functions and behaviors do not correspond, learners can only obtain a superficial understanding by grasping functions alone. Therefore, they do not develop an understanding of the operation series that should be specifically applied. Horiguchi et al. emphasized that an understanding of functions is an understanding of the surface structure, and an understanding of behavior and processing is an understanding of the deep structure. Based on the points made by Horiguchi et al., we considered that an understanding of functions is an understanding of the surface structure and an understanding of behavior and processing is understanding of the deep structure.

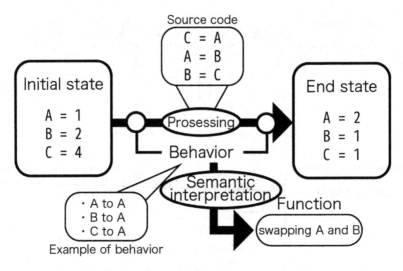

Fig. 3. Problem-solving process in programming.

3.3 Basic Concept in Tasks

As described in Sect. 2, a task includes three properties: processing, behavior, and function. In this section, in order to handle tasks, we discuss the relationships among processing, behavior, and function after positioning each property.

Processing. The definition of processing in this research is an operation series that causes a difference between the initial state and the end state, that is, the source code. In addition, primitive operations are handled as statements, including assignment statements, if statements, and for statements. In addition, it handles its own abstract language as a method for expressing processing in tasks owing to the focus of this study on a general understanding that does not depend on a specific language in programming learning.

Behavior. The definition of behavior in this research is the difference occurring between the initial state and the end state of the variable. Variables include iterators, such as for statements. That is, we do not distinguish between specific scopes, such as global or local variable states. The difference between the initial state and the end state of a variable is considered to follow a certain rule that determines the end state via a process, regardless of the value of the variable. Therefore, the behavior itself is described as a set of constraints that can be read from changes in variables between input and output. For example, in Fig. 3, three constraints, "A is B," "B is A," and "C is A," are necessarily satisfied between the initial state and the end state. However, it is difficult to observe such constraints by relying on the input state of variables in recursive processing, such as for statements. Therefore, in this paper, the discussion of behavior is not detailed in the stepwise task sequences.

Function. The aim of the function in this research is to interpret a certain behavior after estimating the function series based on the goal. Sasajima et al. [5] also stated that a function does not interpret a behavior without excess or deficiency, but implies a certain behavior. On the other hand, Sasajima et al. point out that a valid function is one that interprets a behavior after understanding the ultimate goal. In other words, if a function is regarded as a part, when a structure one level above exists as a whole, for the first time, that part has meaning in the context of a structure, e.g., when the whole function is swap, "temp" meaning, but swap itself has no meaning. The handling of such functions has the advantage that the position of the function is set in the series, so that erroneous function positioning is not possible. However, if a target sequence and its function are positioned, it becomes difficult to reuse the function by directly applying it to other target sequences. Therefore, in this research, we assumed that functions do not necessarily connect with a single goal on a one-to-one basis, and estimate the possible overall goals for each function and position the function.

Relationships Among Processing, Behavior, and Function. Let us describe the relationships between processing, behavior, and function. For example, the behaviors of two different processes, simple sort and selection sort, are considered to be similar in terms of rearranging the array. Additionally, considering each function, the same function is obtained in the interpretation of "rearranging the arrays." However, the learner sees these two as the same function, and sometimes judges that it is equivalent in behavior and processing. This is an understanding of a surface structure, as pointed out by Horiguchi et al. [4]. To promote an understanding of a deep structure, it is necessary to understand behavior and processing. Therefore, the task sequences in this research are based on the development of processing.

4   Construction of Task Sequences

In this chapter, as described earlier, we will examine the stepwise task sequence for gradual learning based on the development of processing.

An increase or decrease in the processing of the problem-solving process is equivalent to an increase or decrease in the required solution. By learning stepwise from a solution with a smaller grain size, it is possible to increase the transparency of the basis for executing individual processing [1]. In the task sequences in this chapter, it is assumed that the function is given to the learner, and the processing is handled as a correct answer. Therefore, in order to construct stepwise task sequences based on processing, it is necessary to design functions to be given as tasks in order to create relationships that gradually expand individual processing. Therefore, in this research, with reference to GMW [2], we consider stepwise task sequences based on the development of progressive processing. In GMW, a Microworld is a simulation of a specific phenomenon that limits the scope of handling so that the learner can understand the underlying laws and principles, mainly in physics. In this research, the task is shifted stepwise. By this transition, the learner understands the difference between the tasks and learns gradually. A task migration method includes (1) migration by modifying a part of the process in the same sequence and (2) migration by developing the process to a different sequence. In these transitions, learners can learn to acquire other functions by reworking the processing or to reuse them from the already learned sequence. In addition, the relationship between tasks in task series is classified into three types: (1) part-of relations, (2) is-a relations, and (3) formal part-of relations uniquely defined in this research.

The migration of tasks (1) to (3) includes (a) migration in the same sequence by extending part of the process and (b) migration to another sequence by modifying part of the process (learning-transfer). However, in this paper, we cannot conduct a detailed discussion of the kind of transition relation that should be defined for the link in (b), and it is defined as a migratable link. Furthermore, in this paper, we made prototypes of task sequences for simple sorting, selection sorting, bubble sorting, and insertion sorting.

First, the relationships (1)–(3) will be explained based on the task sequence of simple sorting (Fig. 4). In simple sorting, looking first at the relationship between A1 and A2, since A2 is included in A1, it is understood that it is a part-of relation. Next, looking at the relationship between A2 and A2-a, since A2-a is defined by excluding a part of the behavior of the outer for-loop in A2, A2-a is a specialization of A2. Therefore, this relation is an is-a relation. Then, looking at the relationship between A2-a and A3, A3 performs the same behavior as A2-a, but exhibits a substantial semantical difference depending on the presence or absence of an outer for-loop. Although such task sequences are effective for learning differences from the presence or absence of concepts, they cannot be established in the absence of the relationship between A2 and A3. Therefore, in this research, this relation was regarded as a formal part-of relation.

Using the above relationships, assignment sort, bubble sort, and insert sort task sequences were constructed (Figs. 5, 6 and 7). First, learning-transfer will be discussed using task sequences of simple sorting and selection sorting. In the figures, A5 is a common problem between a task sequence and selection sort. Additionally, A4 and B4 are in the migratable link, whose behavior is the same. The same is true for task sequences of the simple sort and bubble sort. On the other hand, the migration between simple sort and insertion sort task sequences is described in Figs. 4 and 7. Between A4 and D3, D3-a, A4 and D3, and A4 and D3-a, a migratable link is established between D3 and D3-a, and a migratable link is established between D3 and D3-a. The relationship between D3 and D3-a is defined by the same learning goal; accordingly, the relationship is unidirectional. However, since A4, D3, and D3-a are equivalent in behavior, it can be said that mutually migratable links are established.

The selection sort and bubble sort task sequences can be constructed using the description of the simple sort task sequence. As a result, this construction method is effective for constructing other task sequences.

However, the insert sorting task sequence cannot be described conventionally in terms of negating processing in if conditional statements and requiring controlled exit from for-loops. For this reason, we constructed an insertion sort by adding else and break statements. By adding a new notation, we were able to increase the number of tasks that can be dealt with by the construction method of this task sequence. Additionally, recursive processing and other functions can be described in the future. However, it is difficult to understand the essential structure of the process because the notation becomes cumbersome, and accordingly caution is needed.

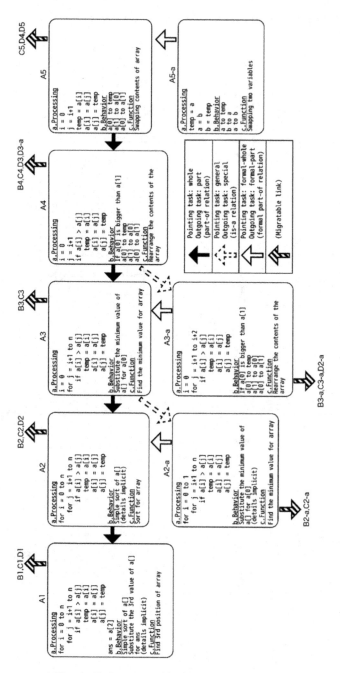

Fig. 4. Task sequence for simple sort algorithms.

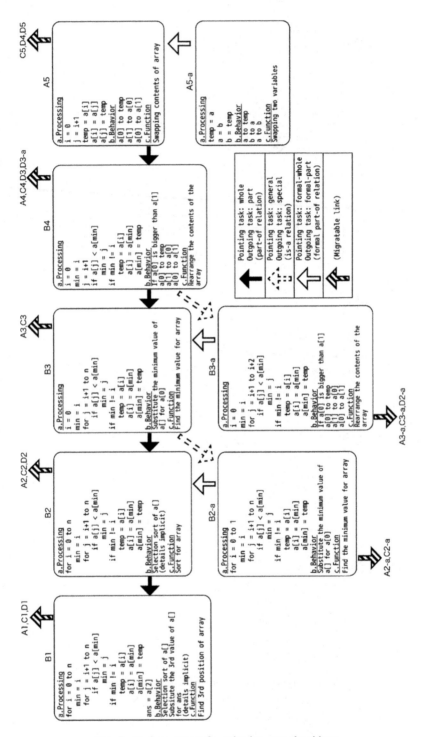

Fig. 5. Task sequence for selection sort algorithms.

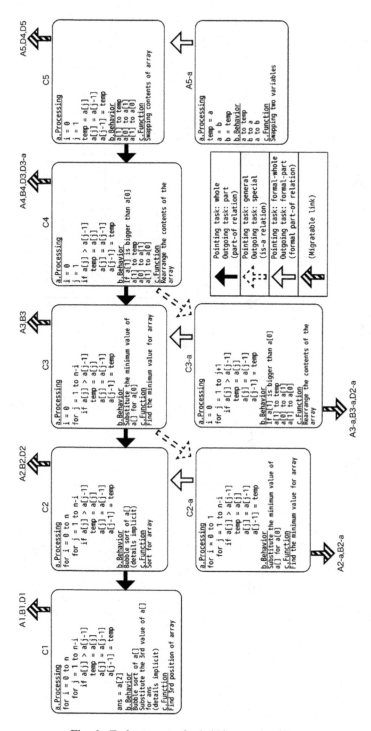

Fig. 6. Task sequence for bubble sort algorithms.

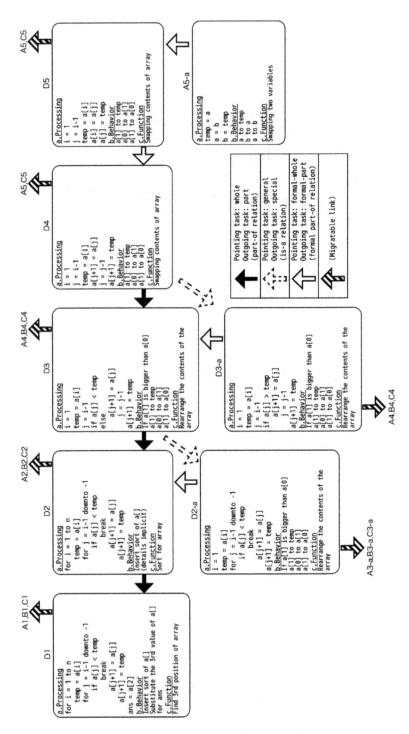

Fig. 7. Task sequence for insert sort algorithms.

5   Discussion

In this research, we aimed to build stepwise task sequences in programming. We point out that processing, behavior, and function and the relationships among these features describe the nature of problems in programming, and we examined the trial production of problem series using these properties. In addition, we constructed multiple task sequences and verified the construction method. Moreover, we showed that it is possible to migrate tasks in multiple task sequences.

To enable the migration of the task to the learner, an understanding task is required to grasp the problem. In addition, the authors think that the skills necessary for learners in the migration of tasks are different depending on links between tasks. Therefore, future studies, with reference to the approach of Watanabe et al. [3], should focus on the addition of understanding tasks to each task and examination of necessary understanding skills in links between tasks. In addition, it is possible that tasks with similar behaviors, like those of A5 and C5 in Figs. 4 and 6, can be regarded as common semantic structures from a certain perspective, and this should be evaluated in future studies. Therefore, we believe that a sequence in the semantic structure will be established, and we consider that each task is established as an instance.

References

1. Hirashima, T., Nakamura, Y., Ikeda, M., Mizoguchi, R., Jun'ich, T., Tsukasa, H., Yuichi, N., Mitsuru, I., Riichiro, M., Jun'ichi, T.: MIPS: A Problem Solving Model for ITS. J. Japanese Soc Artif. Intell. **7**, 475–486 (1992). (in Japanese)
2. Horiguchi, T., Hirashima, T.: Graph of microworlds: a framework for assisting progressive knowledge acquisition in simulation-based learning environments. In: AIED2005, pp. 670–677 (2005)
3. Watanabe, K., Tomoto, T., Fujimori, S., Akakura, T.: Design and development of the function to set auxiliary problems in the process of program meaning deduction. IEICE Technical report no. 116, pp. 85–88 (2017). (in Japanese)
4. Horiguchi, T., Tomoto, T., Hirashima, T.: The effect of problem sequence on learners' conceptual understanding in mechanics. SIG Adv. Learn. Sci. Technol. **78**, 1–5 (2016). (in Japanese)
5. Sasajima, M., Kitamura, Y., Ikeda, M., Mizoguchi, R., Munehiko, S., Yoshinobu, K., Mitsuru, I., Riichiro, M.: Design of a functional representation language FBRL based on an ontology of function and behavior. J. Jpn. Soc. Artif. Intell. **11**, 420–431 (1996). (in Japanese)

Analysis of Student Activity in a Virtual Seminar Using a Seminar Management System

Yusuke Kometani[1(✉)], Masanori Yatagai[2], and Keizo Nagaoka[3]

[1] Faculty of Engineering/Information Technology Center, Kagawa University,
2217-20 Hayashicho, Takamatsu City, Kagawa 761-0396, Japan
kometani@eng.kagawa-u.ac.jp
[2] School of Literature, Kyoritsu Women's University, 2-2-1 Hitotsubashi,
Chiyoda-ku, Tokyo 101-8437, Japan
myatagai@kyoritsu-wu.ac.jp
[3] School of Human Sciences, Waseda University, 2-579-15 Mikajima,
Tokorozawa, Saitama 359-1192, Japan
k.nagaoka@waseda.jp

Abstract. In recent years, there has been growing demand for collaboration among experts from different fields to address social problems. In this research, we aim to build a "virtual seminar", which is a learning form through remote communication in which education and research communities among multiple universities share mutual management resources with the aim of fostering collaborative skills among university students. As an information infrastructure system, we developed and operated a seminar management system (SMS) equivalent to a lecture learning management system. Using the SMS, we conducted a virtual seminar between two universities and clarified the actual situation of learning in the virtual seminar and problems in the learning environment based on comparative analysis of self/peer assessment data of students.

Keywords: University education · Seminar activity · Virtual seminar
Seminar management system · Communication skills · Cooperation skills

1 Introduction

In recent years, cooperative work skills among experts in different fields such as science, technology, and sociology have become important. As an example of this, corporate society requests that universities cultivate fundamental competencies for working persons, which consist of three competencies (action, thinking, teamwork) and 12 capacity elements [1]. To develop these skills in a university setting, it is necessary to provide an environment for students to work together outside the frameworks of their individual faculties and departments. For this reason, we are aiming to build a virtual learning environment in which students who have different academic backgrounds and cultures can communicate without limitations due to spatial constraints.

Seminars are held by the education and research communities of universities [2]. Each seminar in Japan generally consists of a few teachers and 10 to 20 students. Seminars are intended to be opportunities for acquiring high level expertise through

© Springer International Publishing AG, part of Springer Nature 2018
S. Yamamoto and H. Mori (Eds.): HIMI 2018, LNCS 10905, pp. 278–287, 2018.
https://doi.org/10.1007/978-3-319-92046-7_25

peer teaching and learning. Collaborative learning methods conducted by students under the support of faculty in such seminars are evaluated as effective human resource development methods beyond mere expert knowledge learning [3]. In recent years, research empirically investigating the learning effect of seminars has shown that activities that are suited to student situations, such as support for group activities and job hunting activities, are useful for improving student generic skills [4].

In this research, we conceive, propose, and develop a system that supports seminar activities by a single community in order to expand the educational function of seminar activities [5–8]. In addition to this, we consider that promoting exchanges among communities, which was not so active so far, will lead to the development of new educational functions.

Each seminar is focused on its own specific knowledge. Therefore, we believe that communication between a wide variety of seminars would be effective for the development of cooperation skills. In this study, we call the generic activities performed through cooperation among multiple seminars a "virtual seminar."

We define **"Virtual Seminar" as a collaborative learning method implemented by multiple seminars having diverse academic cultures**. Joint seminars that connect technically multiple seminars with a TV conference system are becoming familiar. However, many of the implementations are extension of normal seminar function among seminars sharing same expertise. The learning goals of Virtual Seminar are totally different from those of existent joint seminars. The goals are to know future customers and colleagues at university education stage by exchanging completely different expertise associating with diversity of cultures and values on equal terms, and to have a panoramic perspective.

The purpose of this study is to develop a system for supporting the construction and operation of virtual seminars. The design requirements include the combination of seminars, learning tasks, exchange environment, and group composition for discussion. In the operation of a virtual seminar, a mechanism is needed to appropriately assess the collaboration skills training. However, because seminars are diverse, it is difficult to define a unified design model and evaluation model that takes into account the diversity of expertise. Thus, by utilizing behavioral data in practice, it is necessary to consider a descriptive approach for expanding building support and operational support on the basis of data analysis.

To realize the construction and operation support of virtual seminar based on the data, a framework for storing data about the seminar activities is needed. We define the system as a seminar management system (SMS) that has data generation and data analysis functions for seminar activities. This study is thus an attempt to develop an SMS. We have already run a virtual seminar, and the aim of this paper is to elucidate the design guidelines for virtual seminars through the analysis of the data obtained from the SMS.

2   Concepts and Implementation of SMS

Figure 1 shows an SMS compared with other management systems. Learning management systems (LMSs) are responsible for the operation and management of lessons and have been introduced in many universities. Meanwhile, social networking services

(SNSs) are systems for supporting the formation of virtual learning communities. This study is intended for data management of a community formed from multiple seminars (community), and it is necessary to employ different management schemes from the past.

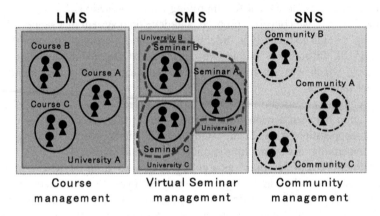

Fig. 1. Support by SMS

Figure 2 shows an implementation of the virtual seminar in this study. The SMS is responsible for storing the role and action history data that link the different seminars. In this study, this includes the comments in the seminar activities, scale rating, and the ability to accumulate action history data through video recording and other means. To enable communication with learners in remote locations, video conferencing and text chat functions are added to the SMS.

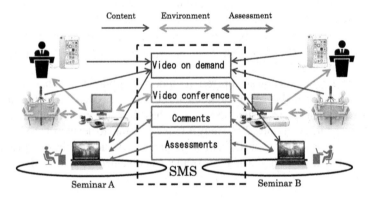

Fig. 2. Model of a virtual seminar mediated by SMS

3   Practice

3.1   Overview

Data stored in the SMS obtained by running a virtual seminar are analyzed to reveal the design guidelines for virtual seminars. The virtual seminar was conducted with the cooperation of seminars at Waseda University (seminar A) and Kyoritsu Women's University (seminar B) from April 2017 to July 2017. Table 1 shows a schedule of the seminars. Adlib speech is an activity in which students give a speech on a specified theme for a specific length of time. To train the ability to summarize a story in a short space of time, the speech time is set to 1 min. Virtual group discussion (VGD) is a discussion conducted in groups between students who belong to different seminars. To communicate with other learners in remote locations, students use text chat.

Table 1. Schedule of virtual seminars

#	Date	Time (min)	Content
1	4/27	45	Adlib speech
2	5/11	45	Adlib speech
3	5/18	45	Adlib speech
4	6/1	90	Adlib speech + VGD
5	6/8	45	VGD
6	6/22	90	Guest speaker
7	6/29	90	Research report
8	7/20	90	VGD
9	7/27	90	VGD

Figure 3 shows a scene of the virtual seminar. Seminars at Waseda University, Kyoritsu Women's University, and the University of Kitakyushu are connected via

Fig. 3. Scene of a virtual seminar between multiple universities

Skype, an Internet communication tool. Faculty belonging to the University of Kita-kyushu serve as facilitators, and the seminars of the other two universities are interacting. At this time, students at Waseda University are giving a speech, whereas students at Kyoritsu Women's University are inputting assessments/comments using smartphones.

3.2   Adlib Speech

Figure 4 shows the random selection function of the speech theme. This function randomly presents a pre-accumulated speech theme. Students do not know the theme in advance. Therefore, students are required to structure their speech instantaneously.

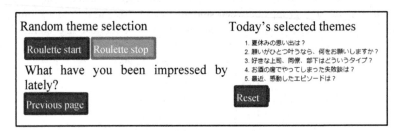

Fig. 4. Random theme selection function of adlib speech activity

Figure 5 shows an input screen for speech evaluation. Assuming various users in the virtual seminar, the user interface of the system in previous research [7, 8] was updated to support input by smartphones. Assessment from each viewpoint, free description of good points and points for improvement, and comprehensive assessment for how good the speech is can be input. Assessments/comments collected by this function are aggregated and are fed back to the speaker instantaneously.

Figure 6 shows the feedback screen. Participants can use smartphones to check assessment results. Assessment can be visualized using a radar chart. Self assessment,

Fig. 5. Input screen for assessment   **Fig. 6.** Feedback screen

peer assessment, and teacher assessment overlap in one chart. Free description data such as good points and points for improvement are also summarized.

3.3   Virtual Group Discussion

VGD is a method of discussing by groups composed of mixtures of seminars from multiple universities (virtual group composition). Figure 7 shows a schematic diagram of VGD. VGD basically uses text chat to allow students in remote areas to discuss at the same time in multiple groups. The video conference system is used for announcing each discussion result. This method enables discussion among students with different academic cultures, group reconstruction in limited class hours, observation of discussion by other groups asynchronously, and so on.

Figure 8 shows the screen of text chat by SMS. Individual rooms are created so that groups can be changed dynamically and each group can share one room.

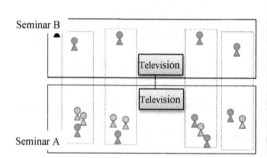

Fig. 7. Concept of virtual grouping   **Fig. 8.** Text chat screen on SMS for VGD

3.4   Assessment for Each Activity

Users can set evaluation items for each activity using SMS. Table 2 shows the evaluation items used for the speech activity. It consists of nine items on the behavior of the speaker and the impression given to the audience.

Table 2. Assessment Items for adlib speech

#	Query
Qp1	Was the voice volume appropriate?
Qp2	Did the speaker seem confident?
Qp3	Did the speaker smile enough?
Qp4	Did you feel the speaker gave consideration to the listeners?
Qp5	Did the speaker use eye contact with the audience?
Qp6	Was the structure of speech easy to understand?
Qp7	Was speech entertaining?
Qp8	Could you imagine the situation?
Qp9	Overall rating of the speech

Table 3 shows the evaluation items for discussion activities. This is a self-evaluation item. Learners evaluate what kinds of difficulties arise during discussion activities. Using these data, it is possible to develop better activities.

Table 3. Assessment items for virtual group discussion

#	Query
Qd1	It was difficult to listen to the opinions of others faithfully and to tell their opinions faithfully
Qd2	It was difficult for everyone to participate in discussions equally
Qd3	It was difficult to make the argument exciting
Qd4	It was difficult to make various opinions
Qd5	It was difficult to sufficiently compare and consider each claim
Qd6	It was difficult to firmly control the flow of discussion
Qd7	It was difficult to build on opinions constructively

4 Analysis of Seminar Activity Using SMS Data

To clarify the design guidelines for virtual seminars, we investigated the following research questions.

1. What points are important in the virtual seminar in a remote environment compared with a face-to-face (FTF) environment?
2. What points are important when making groups in VGD?

The data collected through running the seminars mentioned in Sect. 3 were analyzed in order to clarify these two research questions.

4.1 Comparison Between FTF Environment and Remote Environment

To address research question (1), FTF environment and remote environment were compared based on data about the adlib speech activity in Table 1. Since historical data in an FTF environment were accumulated by the SMS in previous research [7, 8], it is possible to compare the FTF and remote environments.

(i) Comparison of average values of assessment items by t-test

Average values in the FTF environment (FTF group) and those in the remote environment (virtual group) were compared by t-test. Table 4 shows the result. We found that the evaluation of non-verbal communication tended to be lower in the virtual group than in the FTF group (Qp1–Qp7). Thus, one point that needs improvement in virtual seminars is to improve non-verbal communication to make the remote learning environment more similar to the FTF learning environment.

However, the difference in assessment regarding the impression felt by the audience was not significant (Qp8–Qp9). We assume that these impressions strongly depend on the content of the speech. This is therefore not a matter of differences in learning environment.

Table 4. Comparative analysis by t-test between the FTF and virtual groups

#	Average FTF	Average virtual	$d.f.$	t	p	Significance
Qp1	4.68	4.47	239	2.13	0.034	*
Qp2	4.56	4.36	223	2.01	0.046	*
Qp3	4.55	4.25	235	2.85	0.005	**
Qp4	4.42	4.03	217	3.65	0.000	**
Qp5	4.58	4.17	230	4.41	0.000	**
Qp6	4.55	4.12	222	4.15	0.000	**
Qp7	4.35	4.13	282	2.19	0.030	*
Qp8	4.44	4.30	274	1.38	0.169	n.s.
Qp9	3.43	3.36	214	1.09	0.278	n.s.

*, 5% significance; **, 1% significance; n.s., not significant
Qp1–Qp8, seven-point Likert scale; Qp9, five-point Likert scale

(ii) Comparison of determinants based on multiple regression analysis.

To apply multiple regression analysis, the overall rating of speeches (Qp9) is treated as an objective variable, and assessment values of items related to verbal and non-verbal speech techniques are treated as explanatory variables. By analyzing and comparing the determinants between learning environments, we seek to clarify the important factors for improving the remote learning environment.

As a result of the analysis (Table 5), we found commonalities and differences between the two environments. Scores of determinants "speaker confidence (Qp2)" and "consideration of listeners (Qp4)" are commonly high, while scores of the determinant "voice volume (Qp1)" were low. For differences between the learning environments, "eye contact (Qp3)" and "smiling (Qp5)" were significant as determinants in the FTF group but not significant in the virtual group. Since it is difficult to recognize some expressions through a TV conference system, expressions might not be important for learners. Based on these results, we believe that it is necessary to support the audience's recognition of facial expressions and eyes in a remote environment (Table 6).

Table 5. Comparison of determinants by learning environment based on non-standardized partial regression coefficients

	FTF group			Virtual group		
	Partial regression coefficient	p	Significance	Partial regression coefficient	p	Significance
Intercept	1.34	0.000	**	1.32	0.000	**
Qp1	0.01	0.470	n.s.	0.06	0.364	n.s.
Qp2	0.11	0.000	**	0.16	0.012	*
Qp3	0.15	0.000	**	0.01	0.798	n.s.
Qp4	0.14	0.000	**	0.15	0.009	**
Qp5	0.05	0.023	*	0.08	0.203	n.s.
	Adjusted coefficient of determination, 0.441			Adjusted coefficient of determination, 0.385		

Table 6. Inter-group comparison of difficulty of discussion activities using multiple comparison (Tukey honestly significant difference)

Group (I)	Group (J)	Difference of average score (I-J)	p	95% confidence intervals	
				Lower	Upper
B	A	0.083	0.996	−0.896	1.063
A2B2	A	0.542	0.714	−0.799	1.883
A2B1	A	1.0578 *	0.046	0.012	2.104
A2B2	B	0.458	0.721	−0.690	1.607
A2B1	B	0.975 *	0.009	0.190	1.759
A2B1	A2B2	0.516	0.675	−0.690	1.722

*, 5% significance; five-point Likert scale

4.2 Impact of Group Composition on Discussion Activities

To clarify research question (2), the relationship between the group composition and the evaluation of VGD was analyzed. Seven items (shown in Table 3) were designed to ask about the difficulty of learning activities. As a result of running the seminars, four types of group could be compared. Group A was composed of only seminar A members, group B was composed of only seminar B members, group A2B2 was a mixed group consisting of two members from seminar A and two from seminar B, and group A2B1 is a mixed group consisting of two members from seminar A and one from seminar B.

The results of the multiple comparison test (Tukey honest significant difference) found that differences between group A2B1 and group A and between group A2B1 and group B were significant. Thus, differences in the number of participants between seminars in one group led to difficulties in group discussion in a virtual seminar. To correctly assess discussion abilities or ideas through a discussion, group composition should be unified for all groups in VGD.

5 Conclusion

Data collected from running a virtual seminar were analyzed for the purpose of developing an SMS to support the construction and operation of virtual seminars. We found that it is possible to analyze factors that contribute to improving a virtual seminar. To improve the virtual seminar, we found that it is important to consider the following:

(1) Support functionality for non-verbal communication is needed
(2) In VGD, consideration is necessary to ensure there is no bias in the number ratio among different seminars.

In the future, we intend to carry out functional improvements in the SMS to increase the variety of data acquired in the SMS for a deeper understanding of virtual seminars. For example, it is important to develop functions for seminar rooms suitable for virtual seminars and to configure the automatic grouping function of VGD. In addition, it will be necessary to enhance contents of seminar activities and enhance assessment items by seminar activities in future.

Acknowledgements. This research was partially supported by Grants-in-Aid for Scientific Research (No. 26350288 and No. 16K01126) from the Japanese Ministry of Education, Culture, Sports, Science and Technology.

References

1. Ministry of Economy, Trade and Industry: Fundamental Competencies for Working Persons (2006). http://www.meti.go.jp/policy/kisoryoku/FundamentalCompetenciesforWorkingPersons. ppt. Accessed 6 Feb 2017
2. Mouri, M.: Possibility and necessity of FD concerning class of seminar style. Kagawa Univ. Jpn. **15**, 1–6 (2007). (in Japanese)
3. McGuire, J.R.: Engineering education in Japan: my experience. In: Proceedings of Frontiers in Education Conference, pp. 368–371 (1996)
4. Fushikida, W., Kitamura, S., Yamauchi, Y.: The effects of lesson structures in undergraduate seminars on growth in generic skills. J. Jpn. Soc. Educ. Technol. **37**(4), 419–433 (2014). (in Japanese)
5. Nagaoka, K., Kometani, Y.: Seminar activity as center of university education - SMS: seminar management system, proposal and state of development. Research report, Japan Society for Educational Technology, JSET16–1, pp. 307–314 (2016). (in Japanese)
6. Kometani, Y., Nagaoka, K.: Development of a seminar management system. In: Yamamoto, S. (ed.) HCI 2015. LNCS, vol. 9173, pp. 350–361. Springer, Cham (2015). https://doi.org/10. 1007/978-3-319-20618-9_35
7. Kometani, Y., Nagaoka, K.: Construction of a literature review support system using latent Dirichlet allocation. In: Yamamoto, S. (ed.) HIMI 2016. LNCS, vol. 9735, pp. 159–167. Springer, Cham (2016). https://doi.org/10.1007/978-3-319-40397-7_16
8. Kometani, Y., Nagaoka, K.: Development of a seminar management system: evaluation of support functions for improvement of presentation skills. In: Yamamoto, S. (ed.) HIMI 2017. LNCS, vol. 10274, pp. 50–61. Springer, Cham (2017). https://doi.org/10.1007/978-3-319-58524-6_5

Development of a Mathematical Solution Environment to Understand Symbolic Expressions in Mathematics

Kai Kurokawa[1(✉)], Takahito Tomoto[2], Tomoya Horiguchi[3],
and Tsukasa Hirashima[4]

[1] Graduate School of Engineering, Tokyo Polytechnic University,
Kanagawa, Japan
m1765003@st.t-kougei.ac.jp
[2] Department of Applied Computer Science, Faculty of Engineering,
Tokyo Polytechnic University, Kanagawa, Japan
[3] Graduate School of Maritime Sciences, Kobe University, Hyogo, Japan
[4] Graduate School of Engineering, Hiroshima University, Hiroshima, Japan

Abstract. This paper describes problems in mathematics learning, the development of a system to address them, and its effectiveness and considerations. We realize a function that converts symbolic expressions entered by learners to graphical representations and a function for manipulating the converted graphic. The conversion function enables the visualization of an input symbolic expression as a graphic. If the symbolic expression contains an error, the function visualizes the error so that learners can become aware of it. Through their operations, learners deepen their understanding of how the symbolic sentence influences the graphic and clarify their understanding of the relation between the symbolic expression and its graphical representation.

Keywords: Mathematics education · Error visualization · Learning by error
Learning support system

1 Introduction

This paper describes the development of a learning support system for understanding mathematics by expression transformation and active manipulations.

In an ordinary teaching form, since the ratio of learners to teachers is N to 1, the teacher cannot individually diagnose each learner's perception and provide appropriate feedback. Therefore, the learner cannot sufficiently learn whether an idea is correct or erroneous, or the underlying explanation. Even if there is an error in the learner's response, it is not noticed, and trial and error cannot be performed for correction. In order to solve this problem, it is necessary to construct a learning environment that can diagnose and provide feedback appropriate to each learner. However, in mathematical learning, since learners often have a superficial understanding, simple diagnosis and feedback may result in an insufficient understanding of errors. In problem-solving in mathematics, learners cannot interpret a sentence based on the unique symbolic description; accordingly, those who do not aim for a deep understanding are limited by

© Springer International Publishing AG, part of Springer Nature 2018
S. Yamamoto and H. Mori (Eds.): HIMI 2018, LNCS 10905, pp. 288–299, 2018.
https://doi.org/10.1007/978-3-319-92046-7_26

a superficial understanding. In normal learning, it is possible to memorize an example and the way in which it is solved; if a similar problem is presented, it can be solved by applying the procedure corresponding to the past problem. Even a learner who gains such a superficial understanding may not fully understand the meaning of his own solution or may not be able to explain it [1]. In order to solve this problem, it is important to understand what the problem sentence or solution means in graphical form and to understand how to express a graphical representation as a sentence or a mathematical expression. In addition, it has been pointed out that thinking about a situation represented by one sentence described by a learner, discovering an error, and obtaining an endogenous awareness are important for correcting and understanding knowledge [2–5].

Therefore, in this research, a concept expressed by symbols, such as letters, mathematical expressions, and sentences, is represented by symbolic expression, and a concept handled by mathematics is represented by a graphical depiction composed of points and lines in order to promote an understanding of the relationship between this symbolic expression and graphical expression. Therefore, the purpose of this work was to define a concept expressed in mathematics by symbols, such as letters, mathematical expressions, or sentences, as a symbolic expression, and a concept expressed as a graphic composed of points, lines, and the like, which is handled in mathematics as a graphical expression, in order to promote an understanding of the relationship between this symbolic expression and the graphical expression.

To deepen the understanding between symbolic representation and graphical representation, we used a system in which a function described one expression by a learner and visualized it by converting it to the other expression type. In this system, the learner describes the solution method using a symbolic expression, and a feedback function is developed such that the system visualizes the learner's solution as a graphical expression. In this paper, we deal with a planar coordinate geometry problem with a locus as a theme. Recent work has suggested sentences handled in mathematics aimed at promoting the understanding of symbolic expressions and graphic expressions (point P (x, y), $AP = BP$, etc.). We developed a system that generates graphics that can be operated. We found that converting a symbolic expression in mathematics to a graphical representation results in a voluntarily awareness of the relationship between the two expressions. Based on this, learners can think about their own descriptions and consider ways to detect errors.

In this paper, we adopt a function in which the system presents a graphical representation of a correct answer corresponding to each sentence of the answer and an answer template in a form in which a part of the symbolic text is blank, and receives inputs from the learner. A system with these functions enables learners to create an "endogenous awareness" of errors and to understand the relationship between a symbolic representation and graphical representation. As a result, an answer can be considered graphically for a symbolic sentence, and these graphical contents can be understood symbolically, consequently leading to a mathematical understanding.

2 Conventional Work

Nakahara [6] proposes five categories of mathematical expressions: realistic, operation, graphical, linguistic, and symbolic. Using these expressions, it is difficult for learners to grasp the quantitative relationships included in mathematical sentences. The roles and effects of graphical representation have been clarified in conventional work [9]. Recent studies of graphical representations have indicated that they have multiple roles and effects. For example, they can lessen the role of working memory in children learning mathematics, produce concrete models, make it easier to find related information, and make features of a problem clearer [7]. Furthermore, graphical representations more clearly express problem structures, provide a basis for correctly solving problems, allow tracing learners' knowledge, and clearly show implicit information [8].

However, despite research themes aimed at promoting the understanding of symbolic expressions by exploiting graphical representation, there has been no change in the status quo, and graphical expressions are not commonly used. Additionally, many studies have focused on graphical expressions [6], but the understanding of their relationship with symbolic expressions is unclear.

Nakahara [6] argues that it is important for learners to capture problems and solutions using various expression methods. Even if an answer can be described by a symbolic expression, a learner who cannot imagine a graphic represented by the form of the answer cannot be said to have a sufficient understanding. Accordingly, it is claimed that it is important to develop various expression methods for learners to deepen their mathematical understanding in mathematics learning. There is a lack of training approaches in which symbolic expression errors are visualized in graphical form and students are prompted to think about their relationships in a trial-and-error manner. Additionally, the visualization of errors in previous studies has only been performed by the simulation (execution) of results based on the learner's thinking.

3 System Design

In this paper, we develop a system that realizes a function for converting symbolic expressions described by learners into graphical representations as an initial stage of environmental development that enables the mutual conversion between symbolic representation and graphic expression. The internal design of the system is summarized in Fig. 1. As mentioned in Sect. 2, in mathematics learning, it is important to understand each component of the answer sentence. The system diagnoses the correctness or incorrectness of each component, rather than the whole sentence, for the learner's mathematical solution, and returns a graphic that conveys the content to the learner. Like source code compilation in programming learning, we diagnose the correctness of each sentence, and graphically display the contents to the learner.

3.1 Learning Task

Although the scope of this work covers high school mathematics, college students do not fully understand the relationship between symbols and graphics, even for

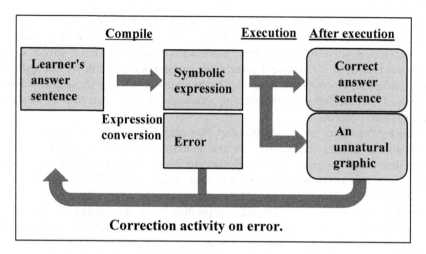

Fig. 1. Internal design of the system.

mathematics in the high school range. A sentence expressing the concept/quantity relation handled in this range (a sentence such as "point P is (x, y)" used in mathematics textbooks "from the condition of problem") is obtained with a numerical/restricted part by preparing an answer template in which a part is blank, supplementing words and numerals suitable for the blank, and combining the template sentences to create an answer sentence. In addition, in this paper, we shall make a simple sentence containing one or more graphically meaningful elements and limit it to a single element to the extent possible. An example would be a sentence containing "a statement defining a point" and "a relation of that point." In this case, the point definition becomes one element and is made visible in the graphic. Here, the point definition is given as point coordinates (x, y), and the point relationship is a sentence, such as "AP = BP," which stipulates that the line segments AP and BP are of equal length. We refer to this type of statement as a short sentence. We refer to such a point P as (x, y), AP = BP, with the restricted part of the sentence as a blank "answer template" that is used when the learner designs the answer. When describing an answer by a learner, an answer template is selected, the learner's words, numerical values, variables, and the like are inserted into the blank space of the template, a single sentence for one line of the whole answer is completed, and answers can be described. Completion of the whole answer can be realized by combining this simple sentence and then expressing the answer.

3.2 Feedback Through Error Visualization

In an ordinary classroom lecture where the ratio of learners to teachers is N to 1, it is extremely important to individually diagnose the cause of wrong answers by learners and to provide appropriate feedback for each learner having difficulty. Often, only the learner's answer is deemed correct or incorrect, and the correct answer is often presented. However, this level of feedback is insufficient for a trial-and-error approach, and the learner cannot fully reflect on the reason for an incorrect answer. For the learner to look back on the error itself rather than negative feedback indicating that an answer was incorrect, positive feedback (error visualization) indicating what is wrong, even if an answer is correct, is effective. Learning from mistakes is possible by this approach. The visualization of errors can be applied to the situation in which the learner's answer is affirmed, and the learner is prompted to spontaneously detect an error by observing the visualized contents.

Many works have suggested that this "learning from errors" approach plays an important role in knowledge correction and understanding, particularly when learners notice their errors themselves [2–5]. Recent studies have suggested that error visualization is a way of grasping action and reaction dynamics [10–12], inferring error visualization in geometric proofs [13], visualizing errors as three-dimensional models [14], and visualizing errors in English composition using animation [15]. To produce an "intrinsic awareness" of errors, it is effective to indicate the kind of conclusion that results from answers and to show the learner contradictions that arise.

In this work, according to the answer sentence constructed by the learner using the answer template in the mathematical problem, a corresponding graphical representation is generated one sentence at a time. As a result, when the wrong answer template is used or when inappropriate content is entered in the answer template, a graphic with strange elements (constraints) is generated so that the learner introspectively notices the error.

3.3   Manipulation Function of Visualized Graphics

In studies of the visualization of errors in physics (dynamics) learning and related research described in the previous section, in addition to the solution expressed by the learner, a method for simulating errors using ordinary physical laws or the like is used. Therefore, even if the learner's answer is insufficient, the behavior can be generated using the physical law. On the other hand, in mathematics, when the learner's answer is insufficient, there is a problem in which an appropriate graphic cannot be generated. For example, for the single sentence "Position P is (x, y)," it is impossible to uniquely determine the original location of point P (point P is correct if it is arranged on the xy-plane). Therefore, the system will be placed on random coordinates, such as $(3, 2)$, as an example. However, in this case, we cannot visualize the difference between the original sentence and the simple sentence "point P is $(3, 2)$." Therefore, simply visualizing the content of the sentence is not enough to determine whether the learner happened to obtain such a picture by chance or whether the graphic was definitively written.

Therefore, in this research, the visual output is designed so that the learner can operate it within the range of sentence constraints for the sentence in the current situation (a simple sentence). As a result, the learner can confirm the range of the drawing obtained by the current answer sentence and deepen their understanding. Additionally, despite being a trajectory problem, it is expected that point P exposes the error in the manipulation by not including the trajectory or fixing it.

4 System Implementation

4.1 Learning Range

In this system, the subject range is mathematics II "graphics and equations;" that is, the problem ranges up to the ideas of a locus and trajectory in planar coordinate geometry (e.g., calculations of the distance between points). In the system, problems in the above-mentioned range of topics in actual mathematics textbooks, following the guidelines in Sect. 3.1, "point P (x, y)," "AP = BP," and so on were analyzed. We prepared an answer template sentence containing graphical elements of about one or two clauses. Each problem (three questions) prepared by the system can be answered by combining six simple sentences. A simple sentence prepared in the system consists of a "prerequisite" combining the definition of points and question by points, "conditional expression" summarizing conditions on points, "calculation formula" consisting of the precondition and conditional expression, "result" of the "calculation formula," and "conclusion" derived from the "result".

4.2 Learner's Activities in the System

First, the system presents a problem list that the learner can tackle. The learner then selects the problem to be addressed and shifts to the problem solution of mathematical learning (Fig. 2). The problem that can be addressed in this system is one question (problem in Fig. 2), namely, calculating unknown coordinates from the distance between a point and another point, and two questions (three total questions) for finding the trajectory of a point. The learner's activities within the system are as follows.

1. Check mathematical problem sentences given to the system.
2. Press the "answer start" button to operate functions, such as pull-down and answer construction.
3. Since the model answer graphic for the sentence contents of process 1 (the first line of the overall answer) is displayed on the right graphic, the observation and manipulation contents of that graphic are confirmed (right graphic of Fig. 3).
4. After confirmation, a sentence matching the graphic is selected from the template list on the lower left portion of the screen and the contents of the selected template are displayed below the list.
5. According to the contents of the selected template, the system creates a text box for blank input, so that each word or numerical value corresponding thereto is inserted in the blank. Then, the "build answer" button is pressed.
6. When the "build answer" button is pressed, the system generates the selected template and the drawing with the contents in the blanks on the left side of the screen (Fig. 3, Left); check the observation and the manipulation content of that graphic.
7. Steps 4 to 6 are repeated when there is an uncomfortable feeling (error) regarding the behavior of the graphic based on the symbolic text, with reference to the model solution drawing.

8. When answer 6 is correct, it becomes impossible to manipulate the template, and the "answer confirmation" button at the center of the screen (between drawing drawings) becomes operable. After sufficiently confirming that the relationship between the answer graphic and the model answer graphic is consistent, pressing this button shifts the solution to the next process.

By repeating the above flow, the learner constructs an answer.

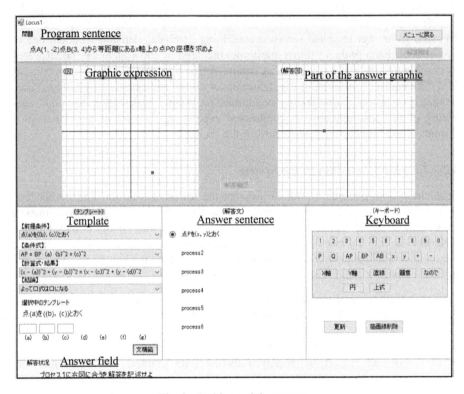

Fig. 2. Problem-solving screen.

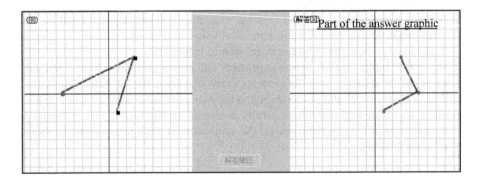

Fig. 3. Conversion graphic (right) and model answer graphic.

4.3   Interface

The user interface of the system is shown in Fig. 2. The learner's problem answer is obtained by providing an input to the template using the keyboard at the lower right portion of the screen. This is necessary because when allowing the learner's free input, it is possible to enter sentence content that the system cannot interpret. In this research, focusing on the understanding of the relationship between symbolic expression and graphical expression, which is not a logical consistency of the solution, it is used not as a free construct by the learner, but as a meaningful mistake choice. The interface was made to enable recognition of the difference between the graphical meaning of the group mean and the graphical meaning of meaningful choices of correct answers. A sentence informing the learner's current answer situation is displayed at the bottom of the screen, and a "return to menu" button for returning to problem selection, an "update" button for initializing the state of the problem being worked on, and a "delete line drawing" button for erasing only the line are prepared.

4.4   Realization of Graphical Generation and Manipulation Based on Sentences

In this system, graphical elements are stored in the database in correspondence with prepared template sentences. For example, for the simple sentence "the point P is ((a), (b))," it is described in the database that "it is a point moving on the xy-plane" and "AP = BP." The distance moved by point P, equivalent to the distance between point A and point B, is described for the simple sentence. The system refers to this database for the sentence used by the learner and calculates and generates a graphic according to the constraint. When the graphic cannot be uniquely determined due to the constraint, a graphic is generated as an example within the constraint, and the graphic can be manipulated within the constraint. (The elements of the graphic described in each template are as follows.)

In "prerequisites," points can be generated from the information given from the problem and the definitions of points suitable for the problem by the learner. It judges whether the input value is a numeric value or a variable and generates a graphic that has a constraint matching the content. It describes what point, which points, and what constraints are added regarding the distance between points, such as "AB = BP" and "AP:BP." Assuming that the constraint is (x, y) under the precondition, the point can be moved along a straight line according to the conditional expression "AB = BP," and is determined as one point if it is (x, 0) or (0, y). A graphic is drawn. In "calculation formula," mathematical expressions consisting of "conditional expressions" or mathematical expressions of "results" derived from the "calculation formula" are prepared. For example, when a numerical value is incorrect when assigned to a "conditional expression" in the "calculation formula," a graphic is generated referring to the wrong numerical value. Figure 4 shows an example of errors when inputting incorrect coordinates for given coordinates by formulas. In "conclusion," since the coordinates of the point and the expression of the straight line are determined based on the "result," if a wrong input is made, a different point is generated.

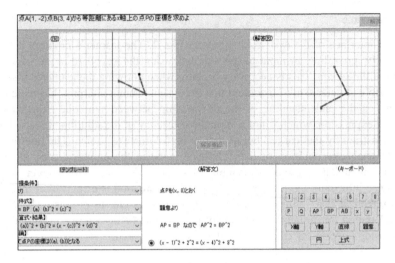

Fig. 4. Visualization of errors in answer sentences.

5 Evaluation Experiment

5.1 Purpose

We will conduct an experiment to verify the validity and effectiveness of the feedback function of the learning system in mathematics developed in this paper. In addition, since this experiment is scheduled to be performed ex post, this report will describe its progress.

5.2 Method

To confirm the effectiveness of system functions, we also used a system without an expression conversion function. Subjects were 18 university students who had taken at least one mathematics course intended for those in math-heavy fields of study. These subjects were divided into an experimental group (9 subjects) and a control group (9 subjects). Before applying the system, the experimental procedure was explained to the subjects. The system was operated by an author of the study, and the method and manipulation of the system were described, including a tutorial on how problems are handled in the case of an incorrect answer.

The experimental procedure was as follows: preliminary system testing before learning (15 min), the first half of system learning (30 min), and post-test 1 (15 min). The contents were as follows: Question 1 is a contradiction evaluation problem including four questions, namely, a question to show the symbolic statement given question 2, question 3 is a blank column assist question, and question 4 is a description problem. It is a quiz questionnaire with 21 evaluation points. A delayed posttest is expected to be implemented 4 weeks after the first day. In the delayed follow-up test, after a 15-min test (first day and common content), the questionnaire is administered.

5.3   Provisional

The expected effects of this experiment on learners are as follows:

- Expression conversion from symbols to graphics is effective for mathematical understanding.
- Visualization of errors leads to mathematical understanding.
- Manipulating graphics leads to an understanding of the source symbolic sentence.

5.4   Results

In this paper, we do not conduct the delayed follow-up test, and so we report the results of the pre-ex post-test on the first day.

Table 1 shows the scores and pre- and post-test results for the experimental and control groups obtained in the experiment described in Sect. 5. All tests were evaluated with a significance level of $p < 0.01$.

Table 1. Average number of correct answers for the control group and experimental group.

	Experiment group		Control group	
	Pre-test	Post-test	Pre-test	Post-test
Q1 (max: 4)	2.78	3.00	2.67	2.78
Q2 (max: 4)	2.22	2.89	2.11	2.56
Q3 (max: 7)	1.33	2.89	1.56	1.89
Q4 (max: 6)	1.22	3.56	1.56	2.11
Total	7.56	13.33	7.89	9.33

Table 1 shows the average number of correct answers in the pre/post-tests for subjects of the experimental and control groups. From Table 1, the average number of correct answers immediately after system learning in the experimental group (from pre to post-testing) improved by 5.78. In particular, in the description problem of Q4, the learning effect by this system is evident. In the control group, the average number of correct answers immediately after system learning (from pre- to post-testing) improved by 1.44. This shows that system learning outperforms the average number of correct answers as compared to the system without the function for converting symbolic representations to graphic representations.

Furthermore, a variance analysis confirmed significant differences among individuals in both groups. In an analysis of variance, significant differences were confirmed within the individual. Since an interaction effect was confirmed, a simple main effect analysis was performed to determine the source of the interaction. In the simple main effect analysis, significant differences were detected in both the experimental group ($p < 0.001$) and the control group ($p < 0.05$) from pre- to post-testing. In other words, significant differences were obtained in the post-test after learning in this system compared with the control group, indicating that the proposed system was effective.

6   Discussion

In this work, we proposed a method to improve understanding of the relationship between symbolic and graphical expressions. In particular, learners describe answers using symbolic expressions, and the system develops a function to visualize the answers by graphical expressions. We demonstrated that the proposed system is suitable for mathematical learning. Pre- and post-testing showed that subjects could produce answers from problem solving, along with graphic answers. We also confirmed that points of confusion at the time of testing were corrected after learning in the system. These results suggest that learners themselves experienced expression transformation to diagrams, leading to the manipulation of figures.

7   Summary

We proposed the transformation of mathematical expressions and graphic manipulations as a method for improving learner understanding of mathematics. The selected target expressions were symbolic and graphic, two forms of expression that are important for understanding mathematics. Using the proposed system, we conducted experiments to verify its effectiveness at improving understanding in mathematics.

We evaluated the effectiveness for mathematical learning by providing graphical feedback to learners using the proposed system. The proposed method of converting mathematical expressions promoted learner understanding, as evidenced by the experimental results. Furthermore, by manipulating graphics, we were able to support understanding of motion constraints and quantity relations in graphics included along with symbolic sentences.

Our findings are summarized as follows:

1. At the time of learning, there were significant differences in test results between the system with functions for converting sentences into graphics and the control group. This demonstrates that the method is suitable for mathematical learning.
2. Learner efforts to correct errors were indicated by graphical manipulations. In immediate post-tests, learners presented their own ideas by drawing graphics.
3. In the evaluation experiment, we will plan a delayed posteriori test as an extension of this experiment; we need to consider the whole experiment further based on a comparison with the experiment results obtained on the first day.

Future work will focus on further improvements in awareness of errors due to observation and manipulation by considering conversion from graphics to symbols.

Acknowledgements. This work was partially funded by Grants-in-Aid for Scientific Research (C) (15K00492), (B) (K15H02931), and (B) (K26280127) in Japan.

References

1. Fujimura, N.: Acquisition and utilization of knowledge and meta - cognition, Makiko Sannomiya Metacognition - Higher order cognitive function supporting learning ability, Kitaoji Shobo, Kyoto (2008). (in Japanese)
2. Perkinson, H.J.: Learning from Our Mistakes: A Reinterpretation of Twentieth-Century Educational Theory, p. 305. Keiso shobo, Tokyo (2000). (in Japanese)
3. Hirashima, T., Horiguchi, T.: Attempt to visualize errors from orienting learning from error. Trans. Jpn. Soc. Inf. Syst. Educ. 21(3), 178–186 (2004). (in Japanese)
4. Hirashima, T.: Aiming at interaction giving awareness of error. Hum. Interface Trans. Hum. Interface Soc. 6(2), 99–102 (2004). (in Japanese)
5. Tomoto, T., Imai, I., Horiguchi, T., Hirashima, T.: A support environment for learning of class structure by concept mapping using error-visualization. Trans. Jpn. Soc. Inf. Syst. Educ. 30(1), 42–53 (2013)
6. Nakahara, T.: Study of Constitutive Approach in Mathematics and Mathematics, p. 389. Seibunsha, Kasugai (1995)
7. Van Essen, G., Hamaker, C.: Using self-generated drawings to solve arithmetic word problems. J. Educ. Res. 83(6), 301–312 (1990)
8. Diezmann, C.M., English, L.D.: Promoting the use of graphicals as tools for thinking. In: NCTM 2001 Yearbook: The Roles of Representation in School Mathematics, pp. 77–89. NCTM (2001)
9. Doishita, A., Shimizu, H., Ueoka, T., Ichisaki, M.: Teaching strategy in the problem solving: conducting a survey of the pupils on pictures and graphicals and putting it into practice. Jpn. Soc. Math. Educ. 68(4), 18–22 (1986). (in Japanese)
10. Horiguchi, T., Hirashima, T.: Simulation-based learning environment for assisting error-awareness - management of error-based simulation considering the expressiveness and effectiveness. Trans. Jpn. Soc. Inf. Syst. Educ. 18(3), 364–376 (2001). (in Japanese)
11. Horiguchi, T., Hirashima, T.: Simulation-based learning environment for assisting error-correction. Jpn. Soc. Artif. Intell. 17(4), 462–472 (2002). (in Japanese)
12. Imai, I., Tomoto, T., Horiguchi, T., Hirashima, T.: A classroom practice of error-based simulation to improve pulils' understanding of mechanics: the "challenge to Newton!" project. Trans. Jpn. Soc. Inf. Syst. Educ. 25(2), 194–203 (2008). (in Japanese)
13. Funaoi, H., Kameda, T., Hirashima, T.: Visualization of an error in solution of geometry proof problems. Jpn. J. Educ. Technol. 32(4), 425–433 (2009). (in Japanese)
14. Matsuda, N., Takagi, S., Soga, M., Horiguchi, T., Hirashima, T., Taki, H., Yoshimoto, H.: Error visualization for pencil drawing with three-dimensional model. IEICE Trans. Inf. Syst. 91(2), 324–332 (2008). (in Japanese)
15. Kunichika, H., Koga, T., Deyama, T., Murakami, T., Hirashima, T., Takeuchi, A.: Learning support for English composition with error visualization. IEICE Trans. Inf. Syst. 91(2), 210–219 (2008). (in Japanese)

Adaptive Interface that Provides Modeling, Coaching and Fading to Improve Revision Skill in Academic Writing

Harriet Nyanchama Ocharo[1] and Shinobu Hasegawa[2(✉)]

[1] School of Information Science, Japan Advanced Institute of Science
and Technology, Nomi, Japan
harriet.ocharo@jaist.ac.jp
[2] Research Center for Advanced Computing Infrastructure,
Japan Advanced Institute of Science and Technology, Nomi, Japan
hasegawa@jaist.ac.jp

Abstract. Revision is an important aspect of academic writing. Students may learn to revise their articles by observing already published articles in their research field. However, these articles are in their final form, so the students have no way of learning from the revision process that led to the final articles. In our previous research, we constructed a corpus of articles written by former students in our laboratory. The articles included the raw drafts and the feedback from the supervisor in the form of comments. Students can then learn from this corpus. However, it is important to recognize that the needs of novice students are different from those of more experienced students. For improving their revision skill, our main idea is to apply the cognitive apprenticeship theory which proposes several methods to teach cognitive skills to novices, such as modeling, coaching and scaffolding/fading. In modeling, the novice student is presented with a conceptual model of the processes required to accomplish a task, and the revision of an article is an example of such a task. In coaching, a student is provided with hints to help them accomplish a task. In our case, the student uploads his/her own article and is provided with hints for improving that article. In fading, the hints provided become fewer and are presented in less detail as the student's revision skill improves. In this paper, we present the design for an interface that shows the revision process in academic writing from the initial drafts to the final drafts so that the students can learn revision skill from practical observation. In addition, the design we present is adaptive to the cognitive needs of the students.

Keywords: Cognitive apprenticeship theory · Revision skill
Academic writing · Adaptation · Writing tools

1 Introduction

Academic writing skill is an important skill to have especially for students in higher education. Students need to communicate their research clearly and concisely when writing school reports, conference papers, journal articles, grant applications, theses or

© Springer International Publishing AG, part of Springer Nature 2018
S. Yamamoto and H. Mori (Eds.): HIMI 2018, LNCS 10905, pp. 300–312, 2018.
https://doi.org/10.1007/978-3-319-92046-7_27

dissertations. For this reason, computer tools to assist students in the learning of academic skill have been a subject of research for decades.

Writing is a cognitive activity consisting of three stages: planning, translating and revising (Flower and Hayes 1981). Much of the research in the area of academic writing tools is focused on equipping students with grammatical skills (especially in the case of English as a second language) and technical or scientific writing skills. Some of the tools also assist students with the planning stage. There is not much research on the use of software tools to assist students in the revision process of academic articles. This is the motivation for us to focus on how to improve the revision skill of research students in higher education using an adaptive interface.

Revision skill is important in helping improve the quality of an academic article. Revision is more than just editing or proofreading a document to fix spelling and punctuation. It might involve restructuring the arguments, reviewing of evidence, refining or reorganization of the entire article. However, it is a difficult skill to learn because like any cognitive skill, learning it is an implicit process. It may be learned directly from language teachers or by co-authoring papers with supervisors and other students in the same learning environment such as a common laboratory (Hyland 2000). However, research students are often pressed for time and may end up copying the writing style from bibliography. The problem is that these published articles they learn from are in their final form, so the students have no way of learning from the revision process that led to the final articles.

In addition to providing a way for students to learn from the revision process itself, there is need to consider the differences in the skill levels of the students. Research has shown that in the writing and subsequent revision process, there are important differences between novices and experts (Hayes et al. 1987). Novices may be those whose writing is judged to be of poorer quality, such as first year college students. Experts may be those whose writing is judged to be of higher quality, such as professors or advanced college students (Kozma 1991).

In this paper, we present the design for an interface that utilizes a corpus of academic article drafts from the initial drafts to the final drafts so that the students can learn from it. One of the original points we present in the interface is adaptation to the cognitive needs of the novice, intermediate and advanced students.

2   Research Background

2.1   Computer Based Writing Tools and Interfaces

Writing is centered around the cognitive processes of planning, translating and reviewing (Flower and Hayes 1981). Planning involves generating ideas and information that might be included in the composition, setting goals and organizing the retrieval of information from memory. Translating is the process of converting ideas into a textual output, and reviewing involves evaluating and revising the text.

There is a growing number of tools and interfaces to support students with these cognitive processes in the writing process. Examples of computer tools to support planning include idea prompters and organizers, structure or outline organizers etc.

Examples of tools to support the translation from ideas into text include English-as-a-Second-Language tools, abstract summarizers, etc. To support the review process, there are several tools that provide spelling and grammar checking functions, text analysis etc. This paper is a contribution in the tools that support the overall revision of ideas, not just proof reading. A general analysis reveals that some of the tools do not take into account the underlying cognitive processes and the needs of novice, intermediate or advanced students.

2.2   The Cognitive Apprenticeship Theory

In the revision process, there are important differences between novices and experts. During revision, novices tend to focus on superficial aspects of revision such as grammar, spelling, and punctuation. Experienced writers, on the other hand revise more on a global scale - their primary goal is to shape the argument (Sommers 1980). Another difference is that novices fail to detect problems in the text that need revision while experts easily detect both global and local problems in the text (Hayes et al. 1987).

Not only should the needs of novices and experts be considered, but ways to help novice students *improve* their revision skill should be considered as well. Since revision is an implicit cognitive process, there is need to make the process explicit so novices can learn from it. The Cognitive Apprenticeship Theory (CAT) (Collins et al. 1989) proposes a way to "make thinking visible." Novices can be taught cognitive skills through the methods of modeling, coaching, scaffolding, articulation, reflection, and exploration. An important aspect of the cognitive apprenticeship theory is *fading*, where the support given to the novice student fades as their skill level rises.

The following items are core teaching methods in CAT:

- **Modeling** - In modeling, an expert performs a task so that students can observe his actions and build a conceptual model of the processes required for task accomplishment.
- **Coaching** - students are engaged in problem-solving activities that require them to appropriately apply and actively integrate subskills and conceptual knowledge. The expert coaches students by providing hints, feedback, and reminders to assist students to perform closer to his level of accomplishment. The content of coaching interaction is related to specific problems that students face while carrying out a task.
- **Scaffolding** - Scaffolding is coupled with fading, the gradual removal of the expert's support as students learn to manage more of the task on their own.
- **Articulation** - an expert encourages students to explicate their knowledge, reasoning, and problem solving strategies. Such activities provide the impetus for students to engage in the refinement and reorganization of knowledge.
- **Reflection** - the expert provokes students to compare their problem solving processes with his own, with that of other students, and with an internal cognitive model of the relevant expertise.

- **Exploration** - the expert pushes students to be independent learners. Students are only set general goals. At the same time, they are encouraged to identify personal interests and pursue personal goals.

The CAT has been implemented in various online tools and interfaces to adapt to the cognitive skill level of students. CAT methods were implemented the CAT methods in the domain of learning Smalltalk, a programming language (Chee 1995). Another researcher (Kashihara et al. 2008) applied CAT methods in web-based navigational learning. They showed a promising method where learners can adjust the scaffold level in accordance with their metacognitive skill. The general conclusion is that if such systems are well designed and used judiciously, they can make a positive contribution towards achieving learning goals.

In the case of revision skill, it is important to support students of various levels. Novice students would benefit most from a revision process **model** - a conceptual model of the processes required to accomplish the revision of an article. Intermediate students require **coaching**. The expert (in our case the system) coaches the students by providing hints, feedback, and reminders to assist students to perform the revision process. As the students improve their revision skill, they will benefit from the **fading** aspect of scaffolding. Fading is the gradual removal of the expert's support as students learn to manage more of the task on their own.

3 Design of Adaptive Interface Based on the Cognitive Apprenticeship Theory

3.1 Learning Content

Providing an adaptive interface for students to learn revision skill requires the provision of learning content as well. What constitutes learning content in the case of revision skill? Related research suggests that a corpus is one way for novice students to learn from experienced researchers (Narita 2012). In our previous work (Ocharo et al. 2017), we presented the case for using a corpus to support learning writing skill for novice students.

In the previous research, a corpus of articles written by former and current students in our laboratory was built. The corpus contains not only the final copies of the articles but also the initial drafts and the corresponding feedback in the form of comments by the supervisor. The comments by the supervisor prompt and suggest the revisions to be undertaken by the students. Some of the comments such as those on the grammar or formatting of the document may distract students, especially novices, from focusing on the overall revision of the content (Sommers 1980). Aside from grammatical comments, the drafts also contain content-related comments. These are comments that require time and effort to revise as they involve clarifying, explaining or refining the idea or topic that the article is based on. We successfully applied a machine learning approach to automatically classify the comments in our corpus as *content-related* or not. The content-related comments and the corpus of articles will be used as hints and feedback to help current students during their own revision process.

The corpus is part of a tool called TRONA (Topic-based **R**evisi**ON** **A**ssistant). The overview of the architecture of the TRONA is illustrated in Fig. 1. The corpus contains 42 articles with a combined total of 245 drafts, all from students in our laboratory. The corpus also contains 7,809 comments extracted from the articles, of which 36% are content-related. The articles are either in Word or PDF format.

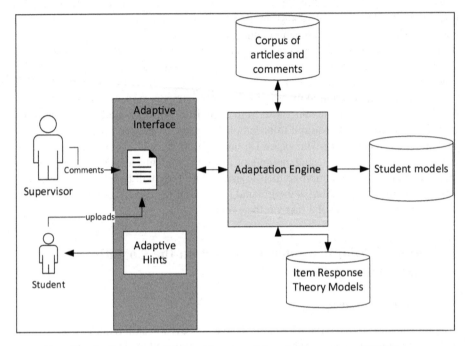

Fig. 1. Overview of the architecture of the adaptive system, TRONA

To initiate the revision process, a student uploads their draft into the system. The supervisor gives feedback in the form of comments to improve the draft. The adaptation engine estimates the student level based on the number and type of comments in the draft (the adaptation engine is discussed in detail in Sect. 3.3). The newly uploaded draft is added to the corpus of articles and comments. The student model is updated with the estimated user level. Depending on the level of the student, appropriate hints are provided to help the student resolve the comments and thus improve their draft.

3.2   The Design of the Adaptive Interface

There are several approaches to realize adaptation. Adaptation can be interface-based, learning-flow based or content-based (Burgos et al. 2007). The design we propose applies content-based adaptation, where the content presented to users differs depending on their cognitive level. As illustrated in Fig. 2, skill level estimation is carried out by the adaptation engine. The student is then classified as novice, inter-mediate or advanced. For novice students, the interface provides modeling, for intermediate, coaching and for the advanced, fading. Mapping of the CAT methods to the actual content presented by the interface is implemented as below:

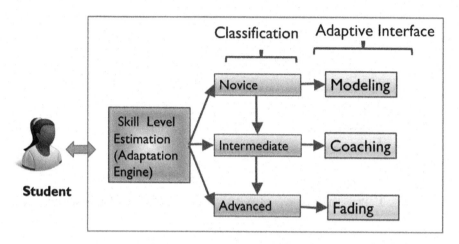

Fig. 2. Design of the adaptive interface

Modeling for novices:

- Revision process overview
- Checklist of common comments by novices.

Coaching for the intermediate students:

- Hints: similar comments and the corresponding comment ranges are shown from the first draft to the final draft.
- Checklist of common comments by intermediate students.

Fading as intermediate skill level rises towards advanced:

- Hints: Examples of similar comments but without the corresponding comment ranges
- Checklist of common comments by intermediate students.

First time users of the system are assumed to be novices until they upload their article. For novice students, **a conceptual model** of the processes required for to accomplish the revision of an article is presented. In terms of the actual content, the student chooses the type of article they need to work on, such as a conference paper, journal article, research proposal etc. If the student chooses a conference paper, then previous conference papers and an overview of those papers' revision processes are presented to the student. Information such as the number of drafts, number of comments and duration of revision are displayed (please refer to Fig. 3). The student can then can click on a specific conference paper to see it in detail, and can read each draft version from the first one to the last. The system also presents a checklist of the most common comments found in novice drafts so that the novice students can keep in mind when preparing their own articles (please refer to Fig. 4).

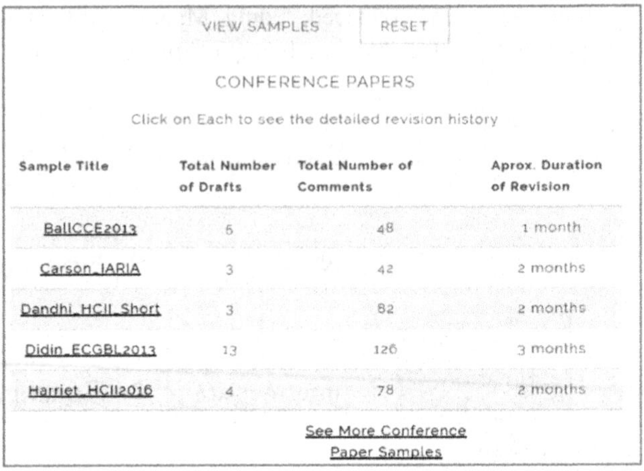

Sample Title	Total Number of Drafts	Total Number of Comments	Aprox. Duration of Revision
BallCCE2013	6	48	1 month
Carson_IARIA	3	42	2 months
Dandhi_HCII_Short	3	82	2 months
Didin_ECGBL2013	13	126	3 months
Harriet_HCII2016	4	78	2 months

VIEW SAMPLES   RESET

CONFERENCE PAPERS

Click on Each to see the detailed revision history

See More Conference Paper Samples

Fig. 3. Providing a conceptual model of the revision process for the novice

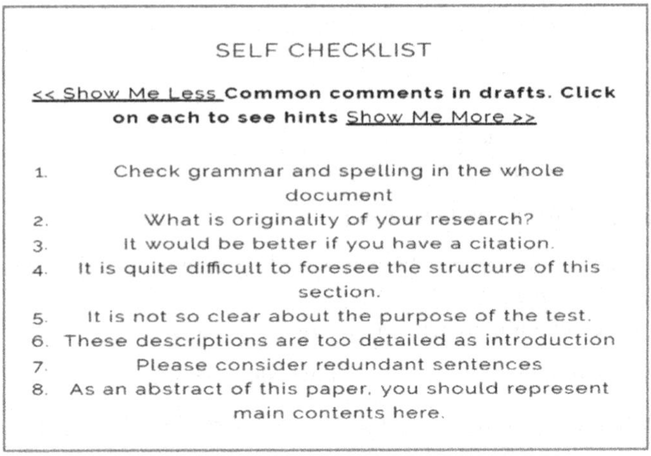

SELF CHECKLIST

<< Show Me Less **Common comments in drafts. Click on each to see hints** Show Me More >>

1. Check grammar and spelling in the whole document
2. What is originality of your research?
3. It would be better if you have a citation.
4. It is quite difficult to foresee the structure of this section.
5. It is not so clear about the purpose of the test.
6. These descriptions are too detailed as introduction
7. Please consider redundant sentences
8. As an abstract of this paper, you should represent main contents here.

Fig. 4. Showing a checklist of the most common comments for novices

Once the student has written their own article, they upload it to the system where the supervisor gives feedback in the form of comments. Based on the number and type of comments, the student is evaluated as either remaining a **novice**, or moved to **intermediate** or **advanced** categories.

The interface then displays hints on how to solve them (**coaching**) by searching the corpus for similar comments in drafts by the previous students. For each comment, the corresponding texts (comment range) in the drafts are displayed that shows the changes made to the comment range from the initial draft to the final article. For the novice student, they need more specific examples to follow to resolve their comments. This includes several examples of closely matching comments with links to the full drafts so that novices can examine them in detail.

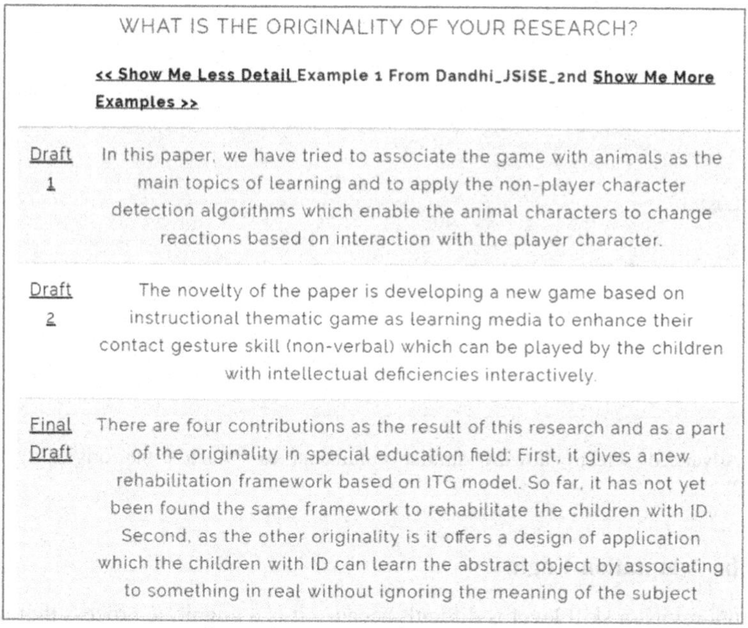

WHAT IS THE ORIGINALITY OF YOUR RESEARCH?

<< Show Me Less Detail Example 1 From Dandhi_JSiSE_2nd **Show Me More Examples >>**

Draft
1

In this paper, we have tried to associate the game with animals as the main topics of learning and to apply the non-player character detection algorithms which enable the animal characters to change reactions based on interaction with the player character.

Draft
2

The novelty of the paper is developing a new game based on instructional thematic game as learning media to enhance their contact gesture skill (non-verbal) which can be played by the children with intellectual deficiencies interactively.

Final
Draft

There are four contributions as the result of this research and as a part of the originality in special education field: First, it gives a new rehabilitation framework based on ITG model. So far, it has not yet been found the same framework to rehabilitate the children with ID. Second, as the other originality is it offers a design of application which the children with ID can learn the abstract object by associating to something in real without ignoring the meaning of the subject

Fig. 5. Intermediate screen example showing hints to resolve "What's the originality of your research."

However, for the intermediate students, the hints provided to help the student resolve a particular comment differ in the level of detail depending on the revision skill of the user. If a student has lower skill, then they are given more hints. The level of detail increases, e.g. they are shown more examples. The student can choose to see more or less detail, and this **interaction data** is saved and used to update the student model. Figure 5 is an example of a typical screen presented to an intermediate student to help resolve the comment, *"What's the originality of your research?"*.

The more advanced students with the highest skills are presented with the least details. As the student's revision skill level increases, the amount of detail presented to the student decreases so as to avoid over-reliance on the tool (**fading**). In the case of the same comment *"What's the originality of your research?"*, the advanced student is presented with just examples of similar comments but without the corresponding comment ranges as shown in Fig. 6.

Students can choose to request more or less hints at any point. They can also choose to see "models" of previous examples, as well as self-check tips of the most common comments. This information is used update the user model in order to provide a better estimation of the student's skill level. As the student's skill level changes, the type of hints they get also changes.

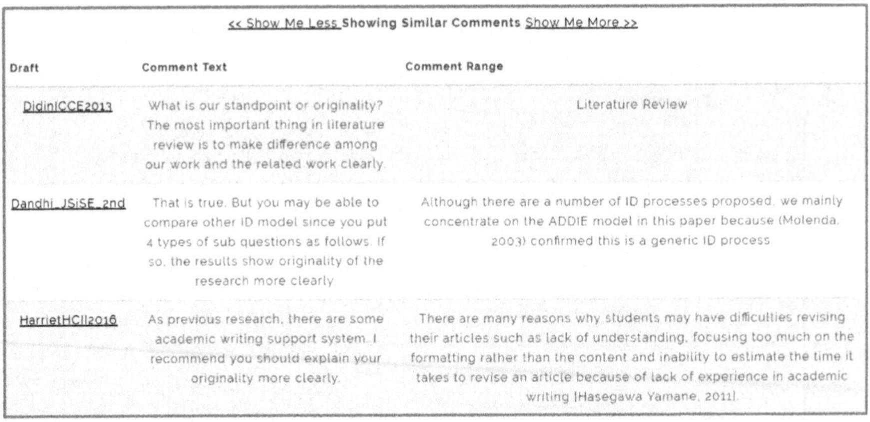

<< Show Me Less Showing Similar Comments Show Me More >>

Draft	Comment Text	Comment Range
DidinICCE2013	What is our standpoint or originality? The most important thing in literature review is to make difference among our work and the related work clearly.	Literature Review
Dandhi_JSiSE_2nd	That is true. But you may be able to compare other ID model since you put 4 types of sub questions as follows. If so, the results show originality of the research more clearly.	Although there are a number of ID processes proposed, we mainly concentrate on the ADDIE model in this paper because (Molenda, 2003) confirmed this is a generic ID process.
HarrietHCII2016	As previous research, there are some academic writing support system. I recommend you should explain your originality more clearly.	There are many reasons why students may have difficulties revising their articles such as lack of understanding, focusing too much on the formatting rather than the content and inability to estimate the time it takes to revise an article because of lack of experience in academic writing [Hasegawa Yamane, 2011].

Fig. 6. Advanced screen showing similar comments to "What's the originality of your research."

3.3 The Adaptation Engine

Estimating revision skill level is difficult because it is a cognitive process that involves many complex processes. In addition, it takes many years to acquire. The outcome of revision can be measured, but there are many different types of academic articles with different styles in different fields. Therefore, it is hard to come up with objective standards of measurement. The Item Response Theory (IRT) offers a way to evaluate the cognitive ability of a student in a certain subject by taking into account the student's ability and the difficulty of the questions in the test (Baker 2001).

This research will apply the IRT to evaluate the revision skill of the student. The IRT is a psychometric instrument for measuring abilities, attitudes, or other variables. There are several advantages of using IRT, such as it allows people to be compared to one another, even though they may have completed different items, allowing for computer-adapted testing such as in the case for online tests - (Merrouch et al. 2014). It is also simple enough that it can be used by many people without formal training in psychometrics. This makes it especially suitable for use in the adaptation engine to evaluate revision skill based on the number of content-related comments (items) that a student has in their draft.

The comments to be used in the IRT Model are the most common or most critical content-related questions. A number of the most common comments will be selected as the items. If a student has such a comment in their document, they deemed to have scored "incorrectly" according to the IRT model. If such a comment is absent in their document, then they have done "well" as it indicates their higher level of revision skill. Therefore, it is deemed a "correct score."

The IRT incorporates not only student model but also comment model. This makes IRT suitable as the type of support given to the student depends not only on their skill level but also on the type of comment as well. To realize adaptation we need to estimate the skill level of the student while considering the difficulty of the comment.

Suppose we obtain comments from students' drafts that are coded 1 for the absence of a particular comment and 0 for its presence. In Table 1, adapted from (De Boeck and Wilson 2004) we list results from five students. C denotes comment, S denotes student.

Table 1. An example of students' score result

	c1	c2	c3	c4	c5	c6	c7	c8	c9
s1	1	1	1	0	0	0	0	1	0
s2	0	0	1	0	0	0	0	1	1
s3	0	0	0	1	0	0	1	0	0
s4	0	0	1	0	0	0	0	0	1
s5	0	1	1	0	0	0	0	1	0

The goal is to test the students' revision skill and classify the students into novice, intermediate or advanced groups. We can test the students' skill by looking at the total score, but the problem is that it depends on the comments considered. If the comments are *easy to revise*, many students will appear advanced, and if all the comments are *hard to revise*, even advanced students will be considered novices.

The relationship between the probability of correctly answering an item $P(\theta)$, and the estimated ability of the student (θ), is expressed by a function called the Item Information Function (refer to Eq. 1) and plotted as the item characteristic curve (ICC). The probability of a student answering an item correctly $P(\theta)$, is expressed by Eq. 1:

$$P(\theta) = \frac{1}{1 - e^{-a(\theta - b)}} \tag{1}$$

Where:

- e: 2.718
- a: the discrimination parameter, an index of how well the item differentiates low from top ability students; typically ranges from 0 to 2, where higher is better
- b: the difficulty parameter, an index of what level of examinees for which the item is appropriate; typically ranges from -3 to $+3$, with 0 being an average examinee level
- θ is an ability level.

By utilizing available IRT evaluation software, we can use a model to estimate the parameters a and b for each comment. The ICC of the comments in Table 1 can be visualized as in Fig. 7, which also shows the estimated difficulty of the comments. A student with a higher ability (θ) has a higher probability of answering a question correctly.

The probabilities in Fig. 7 represent the expected scores for each item along the ability continuum. For this particular model, the midpoint probability (0.5) for each item corresponds to the estimated difficulty parameter.

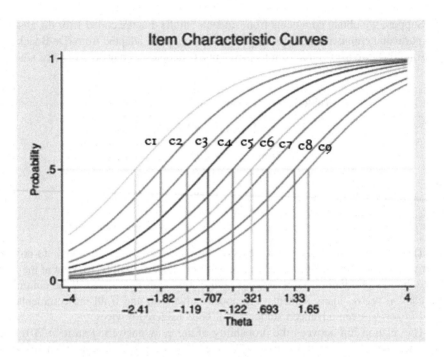

Fig. 7. Item characteristic curves showing the difficulties of the comments (1) to (9): adapted from (De Boeck and Wilson 2004)

Once the ICCs of the items have been calculated, the students' ability can then be estimated. The following equation (Eq. 2) is used to calculate the student's ability, θ (Baker 2001).

$$\theta_{s+1} = \theta_s + \frac{\sum_{i=1}^{N} -a_i[u_i - P(\theta_s)]}{\sum_{i=1}^{N} a_i^2 P_i(\theta_s) * Q_i(\theta_s)} \tag{2}$$

Where:

- θs: Learner ability within iteration S, the value of θ is theoretically between $-\infty$ and $+\infty$ but in reality is limited between -3 and 3.
- i: Current asked item
- N: Number of items
- u_i: The learner response to item i
  - $u_i = 1$ for a correct answer
  - $u_i = 0$ for a wrong one
- $P_i(\theta s)$: The probability to give a correct answer to question i in iteration s
- $Q_i(\theta s)$: The probability to give an incorrect answer to question i in iteration s

Initially, the θs on the right side of the equal sign is set to some arbitrary value, such as 1. After each iteration, the estimation gets more precise until a stable competence value, and a low error value, are obtained. Using available statistical software, we can estimate the ability of a student (θ) by supplying pre-calculated parameters (a & b) and the student's comments.

4 Conclusion and Future Work

In this paper, we presented the design for an adaptive interface that implements the cognitive skill learning techniques of modeling, coaching and fading to improve the revision skill of students writing academic articles. The current goal of our research is to provide a better interface to provide support to students during their revision process, although the final goal is to improve students' revision skill.

Future work will involve evaluating the effectiveness of the design presented in this paper. We can obtain subjective feedback by asking students whether the hints and feedback they receive are useful or not. In addition, we will present the students with a series of hints, some adaptive to their skill level while some will be non-adaptive, and then ask them which feedback they prefer. They have the option of requesting more or less feedback, and of answering whether the feedback is useful or not. We will analyze the responses obtained and other interaction data for insights on the effectiveness of the adaptive interface. For objective evaluation, we will provide each subject in the experiment with a number of comments to resolve. Some of the hints will be provided through a random approach and some through the adaptive interface. For example, a student can revise five comments through the random approach, and five comments by the interface. Based on the result of the revision, the quality of the revised text can be given a score. Thereafter, we will compare the average score of the random approach vs the interface approach. The results of the evaluation will give us insights into the effectiveness of the adaptive design presented in this paper.

References

Baker, F.: The Basics of Item Response Theory (2001). http://echo.edres.org:8080/irt/baker/. Accessed 19 Jan 2018

Burgos, D., Tattersall, C., Koper, R.: How to represent adaptation in e-learning with IMS learning design. Interact. Learn. Environ. **15**(2), 161–170 (2007)

Chee, S.Y.: Cognitive apprenticeship and its application to the teaching of smalltalk in a multimedia interactive learning environment. Instr. Sci. **23**, 133–161 (1995)

Collins, A., Brown, J.S., Newman, S.E.: Cognitive apprenticeship: teaching the crafts of reading, writing, and mathematics. Knowing Learn. Instr.: Essays Honor Robert Glaser **18**, 32–42 (1989)

De Boeck, P., Wilson, M.: A framework for item response models. In: De Boeck, P., Wilson, M. (eds.) Explanatory Item Response Models: A Generalized Linear and Nonlinear Approach, pp. 3–41. Springer, New York (2004). https://doi.org/10.1007/978-1-4757-3990-9_1

Flower, L., Hayes, J.R.: A cognitive process theory of writing. Coll. Compos. Commun. **32**(4), 365–387 (1981)

Hayes, J.R., Linda, F., Schriver, K.A., Stratman, J., Carey, L.: Cognitive processes in revision. Adv. Appl. Psycholinguistics **2**, 176–240 (1987)

Hyland, K.: Disciplinary Discourses: Social Interactions in Academic Writing. Person Education, Harlow (2000)

Kashihara, A., Shinya, M., Sawazaki, K., Taira, K.: Cognitive apprenticeship approach to developing meta-cognitive skill with cognitive tool for web-based navigational learning. In: Seventh IASTED International Conference on Web-Based Education, pp. 351–356. Innsbruck, Austria (2008)

Kozma, R.B.: Computer-based writing tools and the cognitive needs of novice writers. Comput. Compos. **8**(2), 31–45 (1991)

Merrouch, F., Hnida, M., Idrissi, M.K., Bennani, S.: Online placement test based on item response theory and IMS Global standards. Int. J. Comput. Sci. Issues **11**(5), no. 2, 1–10 (2014)

Narita, M.: Developing a corpus-based online grammar tutorial prototype. Lang. Teacher **36**, 23–29 (2012)

Ocharo, H.N., Hasegawa, S., Shirai, K.: Topic-based revision tool to support academic writing skill for research students. In: Proceedings of the Tenth International Conference, pp. 102–107. ThinkMind, Nice (2017)

Sommers, N.: Revision strategies of student writers and experienced adult writers. Coll. Compos. Commun. **31**(4), 378–388 (1980)

Generating Learning Environments Derived from Found Solutions by Adding Sub-goals Toward the Creative Learning Support

Takato Okudo[1]([✉]), Tomohiro Yamaguchi[1], and Keiki Takadama[2]

[1] National Institute of Technology, Nara College, Nara, Japan
okudo@nii.ac.jp, yamaguch@info.nara-k.ac.jp
[2] The University of Electro-Communications, Tokyo, Japan
keiki@inf.uec.ac.jp

Abstract. This paper proposes a learning goal space that visualizes the distribution of the obtained solutions to support the exploration of the learning goals for a learner. Subsequently, we examine the method for assisting a learner to present the novelty of the obtained solution. We conduct a learning experiment using the creative learning task to identify various solutions. To allow analyzing how to create the subject's own learning goals for subjects, several measurement items related to the success of the task which is not instructed to the subjects are setup. In the comparative experiment, three types of learning feedbacks provided to the subjects are compared. The first type is presenting the learning goal space with obtained solutions mapped on it. The second type is directly presenting the novelty of the obtained solutions mapped on it. The third type is presenting some value that is slightly related to the obtained solution. In the experiments, each subject continues to learn the way to find solutions creatively. Therefore, in a creative learning task, these types of learning feedbacks support the learner continuously to set the learning goals.

Keywords: Learning support system · Sub-reward · Continuous learning
Learning goal space · Creative learning

1 Introduction

The progress of information technology will lead to the replacement of approximately half of the jobs performed by humans with computers in the near future. The remaining jobs that are of technical difficulty for both artificial intelligence and computers require high creativity or social skills. This paper describes the creativity of both humans and the computer system. According to Boden [2], "creativity is the ability to come up with ideas or artifacts that are novel and valuable." Previous research on human creativity suggests that "one process of creating ideas involves making unfamiliar combinations of familiar ideas, requiring a rich store of knowledge" (Frey, p. 26) [3]. However, it is extremely difficult for a computer to acquire the creativity of a human. This is so because it is unclear how to combine familiar concepts via unfamiliar approaches. To solve this problem, we focus on a method to utilize the higher creativity of humans compared with that of computers. We propose a mechanism based on a framework of a

© Springer International Publishing AG, part of Springer Nature 2018
S. Yamamoto and H. Mori (Eds.): HIMI 2018, LNCS 10905, pp. 313–330, 2018.
https://doi.org/10.1007/978-3-319-92046-7_28

continuous learning support system provided for a human learner to perceive creativity based on his/her own learning. In the proposed mechanism, the support system generates an already derived learning achievement by combining the original achievement with the solution determined by the learner. Consequently, the learner can reflect his/her own learning trace in the learning goal space. Based on the learning trace, the support system makes the learner aware of the unclear learning results and sense of values. We propose three types of support methods. The first type is the visualization of the learning traces to support the discernment of creativity on learning. We design a learning goal space to visualize a learning trace [9]. It is the distribution of the learning goals attained by a learner who learns an original achievement and of the derived goals generated by the support system. This makes it easier to reflect the learning orientation by showing the position of the learning goal relative to the learning trace. The second type is a discovery support for obtaining unknown solutions by generating a derived achievement based on the negation of the shortest solution of the learner. This encourages the learner to identify his/her unclear solutions. The third type is the generation of a derived achievement by justification of the determined redundant solution. This encourages the learner to notice his/her unclear sense of values.

2 Background

This section describes the theoretical background of this research. After the research on the creativity is described, we summarize an overview of continuous learning because it is the basic framework of the creative learning process, and then we describe the creative learning skill.

2.1 The Definition of Continuous Learning

Continuous Learning is not defined clearly. Some Industrial and organizational psychologists try to conceptualize this term. Smita and Trey [8] reviews research on continuous learning. There are three levels, individual, group and organizational level. One of conceptual definitions of continuous learning is follows; "Continuous learning at the individual level is regularly changing behavior based on a deepening and broadening of one's skills, knowledge, and worldview". Our previous research on Continuous learning is described in detail in [11, 12].

2.2 The Research on Creativity

We consider creativity at the base of J.P. Guilford's approach. Guilford [4] says creativity has primary characteristics, sensitivity to problems, fluency in generating ideas, flexibility and novelty of ideas, and the ability to synthesize and reorganize information [7]. Sensibility to problems is the skill to find the problem. We consider that it is the skill to comprehend the learning task. Fluency in generating ideas shows how many ideas a human create. Flexibility is the skill to create various ideas. Novelty of ideas is the skill to create unusual ideas. The ability to synthesize and reorganize information is the skill to utilize a thing for the divergent purposes. We consider it needs to focus on the interpretation of the meta-learning process. Then we describe several characteristics which are concerned in their research. As sensibility to problems in the task, the learner can comprehend the structure of the learning environment through the trial and errors. As fluency in generating ideas in the task, the learner can find many solutions since the task gives the achievement to him/her. As flexibility in the task, the learner can find the various solutions by seeing his/her learning trace in the learning goal space.

2.3   Intellectual Stimulation

Intellectual stimulation is defined as the leader's behavior for stimulating the follower's creativity in the field of business psychology [6]. The intellectual simulation involves such leader behaviors as questioning old assumptions, traditions, and beliefs, stimulating new perspectives and ways of doing things, and encouraging the expression of ideas and reasons [10]. These leader's behaviors is classified as simulating for the follower's awareness and encouraging the follower's new endeavor [6]. We aim to implement the system with the role of intellectual stimulation for encouraging the learner to perform creatively. This paper proposes the component technology to implement two leader behaviors, simulating for the follower's awareness and encouraging the follower's new endeavor. The next section explains how to incorporate the component technology with the role of the intellectual stimulation into the continuous learning process for extension to the creative learning process.

2.4   The Creative Learning

The creative learning is defined as the continuous learning with discoveries of unusual solutions from achievements in the learner's own. In their previous research [11], the human designed the achievements for a learner as the sequence of mazes. However, creative learner needs a new achievement continuously. In other words, it is necessary for the creative learner to discover the new achievements by himself/herself, but it is not easy. So we propose the interactive mechanism between a human learner and the learning support system in which the system derives the achievements from his/her found solutions with two kinds of heuristics. Once the learner found an unusual solution, the system can derive the new achievements from the unusual one.

The Creative Learning Process. Figure 1 shows the flow of the creative learning process based on the continuous learning process.

This process consists of triple cycle. Innermost cycle is called a trial. A trial is defined as a transition sequence from start state to encountering either a goal state. In this cycle, a learner repeats an action and one's mental process including awareness until the learner results in either success or fail of the task. Second cycle is called an achievement. An achievement is defined as a unit of the main task which is the learning of a maze with the start and the invisible goal. In this cycle, when the trial ends by the encounter with a goal, the learner finds the solution of the achievement. Then, the learner reflects the trial by the reflection of viewing one's learning traces on the learning goal space. The learning goal space has the role of simulating for the follower's awareness as intellectual stimulation. This process is described at Section Designing the Learning Goal Space.

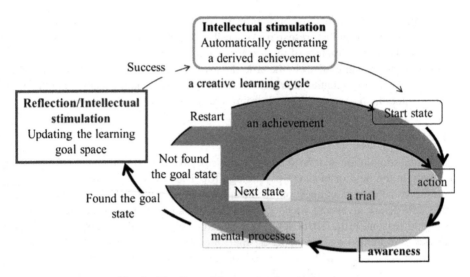

Fig. 1. The flow of the creative learning process

If current trial is not accomplished, the learner restarts the trial from the start state. Outmost cycle is the creative learning cycle. When the learner accomplished current achievement, the system generates a derived achievement according to the learner's solution, and then, the learner can challenge next new achievement. Automatically generating a derived achievement process has a role of encouraging the follower's new endeavor as intellectual stimulation. Section Designing Automatically Generating the Derived Achievement describes this process.

The Creative Learning Skill. The creative learning skill is defined as the learning skill to try to find more creative solutions on the given tasks or problems having optimal or entrenched solution. We propose the interactive mechanism consisting of two parts. The human part is to find a new solution from the achievement. The support system's part is to generate a new achievement derived from the human learner's solution by adding the sub-reward on it randomly to support the learner to find more creative solutions.

3   Designing the Creative Learning Support System

3.1   The Learning Environment by an Maze Model

As the learning environment for a human learner, we adopt a grid maze from start to goal since it is a familiar example to find the path through a trial and error process. First, we define a maze, a path in the maze, and a solution in the maze. A maze is the shape of two-dimensional maze defined by three kinds of states (start, goal and normal state) and the walls surrounding the states. In detail, it is described later as the maze model. A path consists of states and action transition sequence from the start state to the goal state. A solution is a path of the achievement of the maze.

A maze model for creative learning consists of five elements, state set, transitions and walls, action set, and rewards. Figure 2 shows the structure of a two-dimensional grid maze.

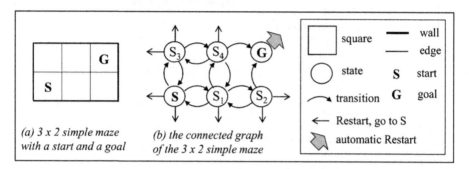

Fig. 2. The structure of a 3 × 2 simple maze

The n x m grid maze with four neighbors consists of the n x m n number of 1 × 1 squares. It is called a simple maze surrounded by walls in a rectangle shape. Figure 2(a) shows a 3 × 2 simple maze with a start and a goal. In a grid maze, every square touches one of their edges except for a wall. Each square in a maze model is called a state. A state can be visited at once. Transitions between states in a maze model is defined whether corresponding square with four neighbors, {up, down, left, right} is connected or not connected by a wall. They are represented as the labeled directed graph as shown in Fig. 2(b). Action set is defined as a set of labels to distinguish the possible transitions of a state. In a grid maze, the learner can take four kinds of actions: up, right, down, left. Note that a trial is a transition sequence from the start state to encountering either a goal state or a wall, and the action toward a wall results in the transition to the start state to restart the trial. Transition to the goal state results in the success of the achievement, then the learner finds a solution and obtains a main reward (+1).

3.2   Designing the Creative Learning Support System

This section describes the way to automatically generate a new achievement as shown in Fig. 1. First we describe a stage and an achievement in the creative learning task. A stage is a set of achievements of the same maze shape. An achievement of the creative learning task is defined as the learning of a maze to find a path from the start state to the goal state. An achievement consists of a maze shape and generated sub-rewards if any. It is a unit of the learning which is either an original achievement which consists of only maze shape or a derived achievement which contains generated sub-rewards. Figure 3 shows the flow of generating a new achievement by the system.

The inner loop in Fig. 3 shows the interactive process of generating a new achievement by the system from the solution the learner searched. The inner loop is corresponding to the flow of a creative learning cycle shown in Fig. 1. The learner follows the achievement cycle including one's mental process until finding a solution.

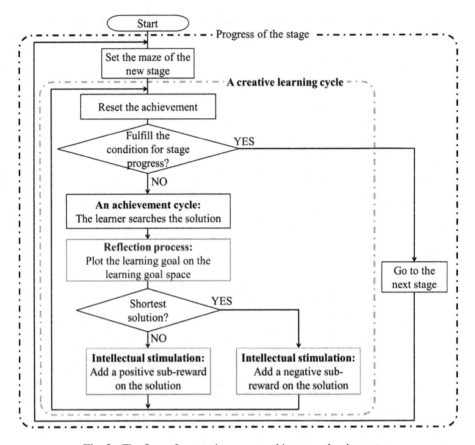

Fig. 3. The flow of generating a new achievement by the system

After a solution is found by the learner, if it is a new one, it is displayed on the learning goal space as the found learning goal for the reflection of the learner, and then the system derives the achievement by adding a sub-reward according to the type of the solution. The system resets the achievement and the learner tries it. The outer loop in Fig. 3 shows the progress of the learning stage. The outer loop is not shown in Fig. 3. The loop provides opportunity to perform an achievement without sub-rewards to the learner. It is two ways. One is when the learner decides to leave a current stage. Another is the decision of the system when the condition for the stage progress is complete. In this paper, the sequence of the rectangle-shaped maze shape of the stage is predefined.

3.3   Designing Automatically Generating the Derived Achievement

Classification of the Solutions for Creative Learning. This section describes the classification of the solutions for creative learning. First, we mention the size of solution. Table 1 shows the classifying solutions based on the length of them.

Table 1. Classifying solutions based on the length of them

The types of a solution	Description
The shortest solutions	The shortest paths from a start state to an encountering goal state
The redundant solutions	The paths besides shortest solutions and longest solutions

We classify them whether it is shortest or not. Note that to make a redundant solution into the learning goal, it is necessary to introduce some optimality. Second, we introduce the optimality of a solution to define the quality of solution. This paper adopts average reward reinforcement learning framework. In it, optimal solution is defined as a solution with the maximum average reward. Note that average reward of it is the sum of the acquired rewards divided by the solution length. Therefore, the shortest solution with acquiring rewards has a tendency to be optimal.

In the field of reinforcement learning, the way to find an optimal solution has been investigated in recent years. However, it is not the end of learning on creative learning. So we focus on the learning after an optimal solution found, and also focus on redundant, i.e. non-optimal solutions to utilize them since they have not been drawn attention as learning goals. Next subsection describes how to derive the learning goal from a shortest solution and from a redundant solution.

Deriving a New Achievement by Negating a Shortest Solution. This subsection describes the method to generate a derived achievement. When a learner identifies the shortest solution, the system adds a negative sub-reward on one of the transitions in the identified path to negate it. This roughly negation of the optimal solution derives a new achievement creatively as the remaining redundant solutions encourage the learner to be creative by avoiding the negative sub-reward to obtain new solutions. Note that the negative sub-reward is randomly placed in the path, and its value is -1.

Deriving a New Achievement by Justifying a Redundant Solution. This subsection describes the method to generate a derived achievement based on the justification of a non-optimal solution. When a learner obtains a redundant solution, the system adds a positive sub-reward on one of the transitions in the identified path to justify it. This crude justification of the redundant solution yields a new achievement creatively because there may be better solutions with a positive sub-reward than this redundant solution. Note that the positive sub-reward is placed randomly in the path, and its value is $+1$.

3.4   Designing the Learning Goal Space

The learning goal space is the space in which found solutions are positioned to display. The learning goal space has the vertical axis as the solution quality and the horizontal axis as the solution cost. The solution cost is the cost for implementing the solution. The solution quality expresses how many sub-goals the leaner is able to achieve in the process of finding solution. These two axes are generalized to the learning cost and the learning quality. The learning cost partially depends on each learner. In this paper, we simplify the learning cost, and define the solution cost as independent axis instead of it. A point on the learning goal space is a set of solutions which the solution quality equals the solution cost. In Sect. 4.3, we discuss about the application of this proposed system to the other tasks. Figure 4 shows the illustrated example of the learning goal space.

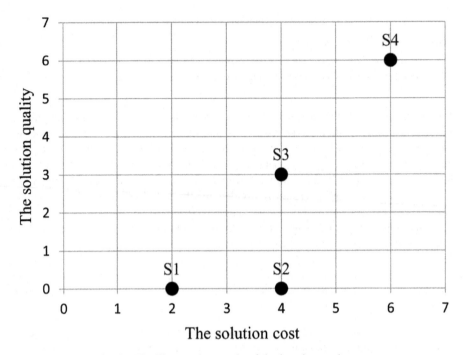

Fig. 4. The illustrated example of the learning goal space

Si is the ith solution found by the learner. The number of S expresses the order in which the learner finds them. The transition from S1 to S2 shows that the direction of learning is right, and it means the learning only increases the solution cost towards the horizontal learning goal. S3 is transited from S2 towards the vertical direction in which the solution quality only grows. S5 is transited from S3 towards the direction simultaneously to increase both the solution cost and the solution quality.

4   Experiment

The aim of the experiment is to evaluate effectiveness of three ways to show the learning record. The comparative experiment is conducted for four subjects. All subjects all university students in their twenties.

4.1   Experimental Setup

Experimental Task. We explain the experimental task on two points of view, the subject point of view and the experimenter point of view.

First, we explain the experimental task on the subject point of view. This experimental task is a maze task with multiple paths from start to goal. The instructions about the score are three points: if the subject finds the goal, the subject obtains score +1; if the subject goes through the green rectangle, the subject obtains score +1; if the subject goes through the purple rectangle. The termination condition of the experimental task is to complete all the stages. The completing condition of the stage is not shown to the subject. During the experiment, a countdown relating to the completing condition is shown to the subject. The countdown is finished, the stage progresses to the next stage. When the subject finds the goal, the learning record by the method described later is shown.

Next, we explain the experimental task on the experimenter point of view. The score is added on the path of the solution. If the subject finds the solution having the unknown path length, the score as the positive sub-reward +1 is added. If the subject finds the solution having the known path length, the score as the negative sub-reward −1 is added. The termination condition of the experimental task is to complete five stages. The completing condition of the stage is to find the designated number of the path-length kinds. Table 2 shows the simple maze size and the initial value to complete the stage on each stage.

Table 2. The simple maze size and the initial value to complete the stage

Stage number	1	2	3	4	5
Simple maze size	4 × 4	4 × 5	5 × 5	5 × 6	6 × 6
The initial value to complete the stage	5	7	9	11	13

The Flow of the Experiment. The flow of the experiment is as follow.

Step 1. The subject reads through the instruction for the experiment task.
Step 2. The subject starts the experiment task when he likes to start it. He is able to read the instruction for the experimental task during the experimental task.
Step 3. The subject takes the experimental task.
Step 4. The subject answers the questionnaire after the experimental task.
Step 5. The subject answers the hearing investigation.

Next, the flow of the experimental task is as follow.

Step 3-1. The subject takes either four actions, up, down, left, right, with four way controller in the stage.
Step 3-2. When the subject finds the path from start to goal, the learning record is shown.
Step 3-3. The subject makes the learning record closed.
Step 3-4. If the subject meets the completing condition or if the limited time is over, next is Step 3-6.
Step 3-5. Return Step 3-1.
Step 3-6. The stage progresses to the next stage. The learning record is reset. If the stage is final, the experimental task is ended.

The stage progresses to the next stage. The learning record is reset. If the stage is the final stage, the experimental task is finished. If not, return Step 3-1.

The GUI of Creative Learning Support System. Creative learning support system consists of two windows, Maze window and Learning goal space window. Figure 5 shows Maze window before reaching the goal state, and Fig. 6 shows Maze window after reaching the goal state once.

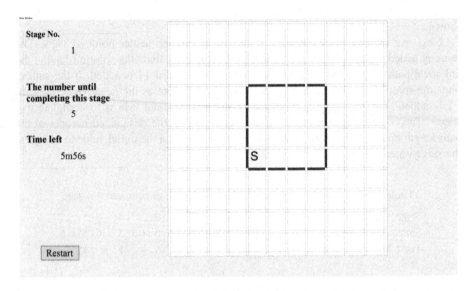

Fig. 5. Maze window before reaching the goal state

As shown in Fig. 5, Maze window says Stage number, The number until completing this stage, and Time left, and has the grid area and Restart button. The grid area in Maze window shows the maze surrounded by black rectangles. Start state labeled as S in the square is visible in the maze, while goal state labeled as G is invisible. As shown in Fig. 6, goal state becomes visible after reaching the goal state once. The green rectangle expresses a positive sub-reward, and the purple rectangle expresses a negative sub-reward. The gray square is the visited state. Figure 7 shows Learning goal space window.

As shown in Fig. 7, Learning goal space window shows the found solution in the grid area, the learning goal space, The number until completing this stage, Previous stage button and Current stage button. The learning goal space has two kinds of points, red and blue. A red point is the learner's recent found solution. Blue points are the learner's found solutions except for the learner's recent found solution. The learner can see the previous the learning goal space with pushing Previous stage button above OK button and below the learning goal space. The learner pushes Current stage button right next to Previous stage button to come back the learning goal space of the current stage. This experiment compares the learning goal space with two methods described in next subsection.

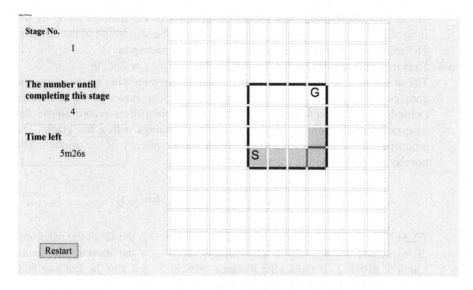

Fig. 6. Maze window after reaching the goal state once (Color figure online)

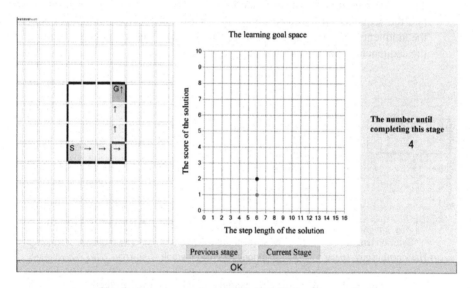

Fig. 7. Learning goal space window (Color figure online)

The Compared Methods. This experiment is conducted by three compared method. Three compared methods are as follow. The sequence of findings is a natural number showing what number the path from start to goal is found.

(i) Showing the scatter chart in the learning goal space
   The learning goal space is a two-dimensional space consisting of two axes, the score and step length of a found solution. Figure 8(a) shows the representation

example of the learning goal space. The position of the solution found by the subject is decided by the score of the solution and the step length of the solution. The solution is plotted as a point in the learning goal space.

(ii) Showing the uniqueness record graph in the sequence of findings
The uniqueness record is a two-dimensional space consisting of two axes, the uniqueness and the sequence of findings. Figure 8(b) shows the representation example of the uniqueness record graph. The uniqueness record shows the uniqueness of the solution in the sequence of findings with a bar graph. The uniqueness is the quantitative value expressing the novelty of the solution. The novelty of the solution is calculated as follow [5].

$$\text{Novelty}(s, \mathbb{S}) = \frac{1}{|\mathbb{S}| - 1} \sum_{s_i \in \mathbb{S}}^{N-1} dist(s_i, s) \tag{1}$$

Where Novelty $(s, \mathbb{S})$ is the novelty of the solution, $\mathbb{S}$ is the set of the solutions, $|\mathbb{S}|$ expresses the number of the solutions in $\mathbb{S}$, s is the most recent found solution, $dist(s_i, s)$ is the Euclid distance between s_i and s in the learning goal space. This is why the learning goal space is not a grid space.

(iii) Showing the score record graph in the sequence findings
The score record is a two-dimensional space consisting of two axes, the score and the sequence of findings. Figure 8(c) shows the representation example of the uniqueness record graph. The score record shows the scores of the solution in the sequence of findings with a bar graph.

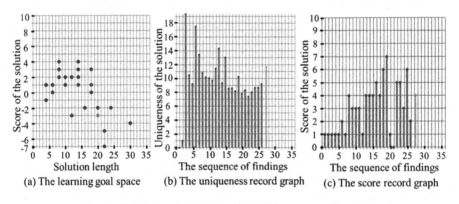

Fig. 8. Each representation example of the compared methods

The difference between the learning goal space and the score record graph is the number of the learning goals shown to the subjects. The axes of the learning goal space are the score of the solution and the step length of the solution. The both axes are the learning goals. On the other hand, the axes of the score record graph are the score and the sequence of findings. While the score is the learning goal, the sequence of findings is not it.

The difference between the learning goal space and the uniqueness record graph is the way to show the novelty of the solutions. In the learning goal space, the positions of the found solutions express the novelty. Showing the novelty in the learning goal space is implicit. On the other hand, in the uniqueness record graph, the quantitative values express the novelty of the solutions. Showing the novelty in the uniqueness record graph is explicit.

The difference between the score record graph and the uniqueness record graph is whether the learning goal is simple or complex. In the score record graph, the score of the solution is accumulative score acquired by the finding the sub-reward and the goal. The score is simply calculated. On the other hand, in the uniqueness record graph, the uniqueness is calculated by the Eq. (1) using the score and the step length. The uniqueness is complexly calculated.

The Instruction for the Subjects. A brief summary of the instruction for subjects includes the following points:

- The objective of the task is to complete all stages.
- If you reach the goal, you obtain +1 score.
- If you go through the green rectangle grid, you obtain +1 score.
- If you go through the purple rectangle grid, you obtain −1 score.
- A time limit is six minutes in each stage.
- You must look at {the learning goal, the score record, the uniqueness}* window. *One of three compared methods is shown.
- Step length is the number of actions from start to goal. (Only the learning goal space and the uniqueness score)

  Uniqueness expresses how unique the path is.

The Measurement Items. The major measurement items of the experiment are as follows:

(a) The number of the trials
(b) The number of the found solution
(c) The number of the path
(d) The number of the learning goal
(e) Novelty on the learning goal space
(f) The number of the path length kinds

Note that the number of the path length kinds (f) shows the accumulative number of the all completing conditions. In the case that the subject completes all stages, the number of the path length kinds (f) is forty-five. Novelty on the learning goal space (e) is the average novelty on the learning goal space of all stages.

4.2   Experimental Results

We analyze the quantitative effectiveness of compared methods. Table 3 shows the experimental results. Note that measurement items (a)–(e), and (f) are described in Sect. 4.1.5.

In Table 3, subject 1 performs the experimental task with looking the learning record in the learning goal space. Subject 2 performs the experimental task with looking the learning record in the uniqueness record graph. Subject 3 and subject 4 perform the experimental task with showing the learning record in the score record graph. Subject 1, subject 3 and subject 4 complete all stages without within the limited time. Subject 2 does not complete stage 5 because he is over the time limit. As shown in Table 3, subject 1 of the learning goal space obtains the most number in four measurement items, the number of the trials (a), the number of the found solutions (b), the number of the path (c) and the number of the learning goals (d) out of six. Subject 2 of the uniqueness record graph obtains the second most number in the same four measurement items out of six. Both subject 3 and subject 4 of the score record graph obtains the fewest number in the same measurement items out of six. Figure 9 shows the line chart of the experimental results in all stages. The linear approximation in the graphs shows the tendency for the collection of data that is obtained from four subjects.

Table 3. The experimental results

Subjects (Condition)	Measurement items					
	(a)	(b)	(c)	(d)	(e)	(f)
Subject 1 (The learning goal space)	148	141	119	98	7.21	45
Subject 2 (The uniqueness record graph)	132	129	112	90	7.11	44
Subject 3 (The score record graph)	92	88	84	73	7.01	45
Subject 4 (The score record graph)	100	94	91	81	7.22	45
The average of subject 3 and subject 4	96	91	88	77	7.11	45

Figure 9 shows the line charts of the experimental results in all stages. The overview of following results, Fig. 9(a), (c) and (d) show that the slope of linear approximation is positive, and Fig. 9(b) shows the slope of linear approximation is positive. These suggest that each subject continues creative learning with progress of the stage.

Figure 9(a) shows the line chart of the rate of obtaining the learning goals. The rate of the learning goals shows the ratio of the number of the learning goals to the number of the found solutions. The lower limit is 0. The upper limit is 1. The slope of the linear approximation in Fig. 9(a) is 0.035. As shown in Fig. 9(a), the rate of obtaining the learning goals of subject 1 prominently increases. The slope of that of subject 1 is 0.11 that is bigger than that of the linear approximation. The slope of that of subject 2 is −0.03 and only negative. This suggests that the subjects other than subject 2 learn to fill the learning goal space.

Figure 9(b) shows the line chart of the rate of finding the same path in all stages. The rate of finding the same path is the ratio of the number of the same path to the number of the found solutions. The slope of the linear approximation in Fig. 9(b) is −0.022. As shown in Fig. 9(b), the rate of finding the same path of subject 1 and subject 2 decrease with progress of stage. The slope of that of subject 1 is −0.049. The slope of that of subject 2 is −0.026. They are smaller than the slope of linear approximation. The slope of that of subject 3 is 0.004 and only positive. This suggests that the subjects other than subject 2 learn the way to find solutions not to find the same path.

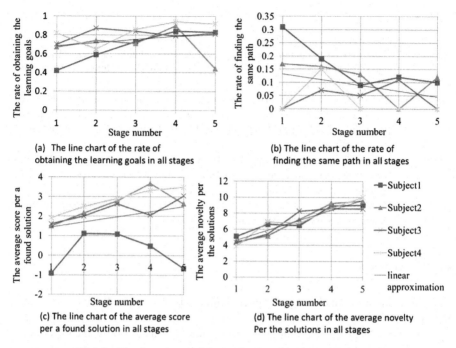

Fig. 9. The line charts of the experimental results in all stages

Figure 9(c) shows the line chart of the average score per a found solution in all stages. The slope of the linear approximation in Fig. 9(c) is 0.27. As shown in Fig. 9(c), the line chart of the average score of subject 1 is the different from the forms of the other subjects. The slope of that of subject 1 is −0.011 and only negative. This suggests that the subjects other than subject 1 learn the way to find solutions to maximize the score of a solution.

Figure 9(d) shows the average novelty per of the solutions found in each stage. The slope of linear approximation in Fig. 9(d) is 1.3. As shown in Fig. 9(d), the form of line chart of all subjects is similar. This suggests that all subjects learn the way to find solutions to maximize the novelty of a solution.

We describe the result of the questionnaire and the hearing investment after the experimental task. Table 4 shows the overview of the answers in the questionnaire and the hearing investment.

As shown in Table 4, both subject 1 of the learning goal space and subject 2 of the uniqueness record graph consciously decide the way to search solutions, and search the solutions. The both subjects keep concern about the meaning of the compared methods shown to them in the latter stages. On the other hand, two subjects of the score record graph do not have concern about the compared method in the latter stages. From these results, it is supposed that showing the learning goals and feedback of the learning results by the form having some complexity are effective to continuation of the interests of the learner. The detail about it is discussed in the next section.

Table 4. The overview of the answers in both the questionnaire and the hearing investigation

Subjects	The overview of the answer
Subject 1	Subject 1 takes random actions and searches the solutions to make clear where the score adds in the former four stages. He searches the solutions to fill a blank space in the learning goal space
Subject 2	Subject 2 tries to make clear where the score adds to obtain more visible green rectangle in the former three stages. In both stage 4 and stage 5, he tries to find the solutions having different path length with the found solutions to obtain higher uniqueness
Subject 3	Subject 3 searches the solutions to obtain higher score first. After she realizes that obtaining high score is not the completing condition of the stage, she searches the solutions without thinking
Subject 4	Subject 4 think that she is able to complete the stage if she finds many solutions. She searches the solutions without thinking

4.3   Discussions

The Effect of the Learning Goal Space Which Visualizes the Novelty of a Found Solution. To evaluate the effect of the learning goal space, we discuss whether each subject sets the heuristic learning goals or not, and the way performs the experimental task continuously toward the task accomplishment. As shown in Table 4, subject 1 to whom the novelty of the found solutions by the learning goal space was indirectly presented sets the heuristic learning goals through all stages. Subject 1's heuristic learning goals are to make clear the position of added score and to fill the blank space in the learning goal space. First we focus on the former heuristic learning goal. As shown in Fig. 9(b), it seems that the subject 1 observes how to add the score on the found solution in both stage 1 and stage 2 by finding the same path. It seems that subject 1 takes random actions with finding the different path with the found paths in both stage 3 and stage 4. Next, we focus on the latter heuristic learning goal. As shown in Fig. 9(a), it seems that the subject 1 tries to fill the blank space in the learning goal space by taking advantage of the learning goal space. In addition, as shown in Fig. 9(c), it seems that the subject 1 is not conscious of obtaining the score. Like subject 1, subject 2 to whom the novelty of the found solutions was presented sets the heuristic learning goals through all stages. As shown in Table 4, the heuristic learning goals are to maximize the score and to maximize the uniqueness. As shown in Fig. 9(c), it seems that subject 2 try to maximize the score in former three stages. As shown in Fig. 9(d), it seems that subject 2 searches the solutions to try to maximize the uniqueness in both stage 4 and stage 5. In contrast to them, other subjects to whom the novelty of the found solutions were not presented set their learning goals differently. Subject 3 sets the heuristic learning goal in former three stages. As shown in Table 4, the heuristic learning goal is to maximize the score. As shown in Fig. 9(c), it seems that subject 3 try to maximize the score. Subject 4 does not set the heuristic learning goal. As shown in Table 4, subject 4 does not set the learning goal. Therefore, it is suggested that the subject who are presented the learning goal space (subject 1) or presented the novelty of the found solution (subject 2) keep finding the solutions according to their learning goals until the final stage in the experiment.

The Choice of the Distance Function. We discuss whether the Euclidean distance is appropriate to the distance function between the most recent found solutions and a set of the found solutions or not. Lehman speaks of the distance function as domain-dependent measure [5]. In the fields of clustering, the appropriate definition of the distance is generally dependent on presence or absence of correlation between axes of space [1]. If the correlation is large positive or negative, the Mahalanobis distance is appropriate to the distance function. In the case of no correlation, the Euclidean distance is appropriate. In this experiment, the correlation coefficients between two axes of the learning goal space of four subjects are 0.5, 0.5, 0.3, and −0.3 (Overall correlation coefficient is 0.2). The two correlation coefficients out of four have moderate positive correlation, and the other two correlation coefficients have poor correlation. Therefore, the appropriate distance function is the Mahalanobis distance. However, the appropriate distance function is the Euclid distance in this creative learning task because it is necessary to measure the novelty of the found solution in the process incrementally generating the known solution.

Application to the Other Tasks. In this sub-section, we discuss about the application of the proposed system to the other tasks. We show the necessary conditions for implementing the creative learning support system in the task. The necessary conditions for implementing the creative learning support system in the task are as follows.

(i)   quantifiable solution cost
(ii) quantifiable solution quality
(iii) enough solutions
(iv) enough learning sub-goals

(i–ii) is about the axes of the learning goal space. They show that the axes are quantifiable. (iii) shows that it is necessary to plot enough solutions on the learning goal space for the purpose of displaying the learning trace of the learner. (iv) is for the system to generate the task by adding the sub-reward as the sub-goal.

Future Works. There are several future works. First one is to discuss effects of each support methods of the learner's reflection for the measurement items. Second, we will define the value of sub-reward. The value of sub-reward V at sub-reward r_i will be given by

$$V(r_i) = \frac{N_u(r_i)}{N_s(r_i)} \tag{2}$$

where N_u is the number of the unknown solutions containing the sub-reward r_i and N_s is the number of the solutions containing the sub-reward r_i. The reason why the equation shows the value of sub-reward is that the value of sub-reward becomes larger as increasing the number of the findable unknown solutions with the sub-goal as the cue. Third one is to modify the way to add the sub-goal. The current way for setting a sub-goal is on a random transition in the found solution. However it does not intentionally lead to the unknown solution. To realize it, we will select two kinds of transitions, one is with the maximum number of passing in the found solution, the other is the least visited transition nearby the found solution.

5   Conclusions

This paper presented the interactive method for human to creatively learn under the learning support system. We described the way to design the learning support system towards acquiring the creative skill on learning. We proposed the learning goal space which visualizes the distribution of the found solutions on it, then examined the method for supporting the learner's exploration of the learning goals to present the novelty of a found solution. As the experimental results, each subject continues to learn the way to find solutions creatively. From discussion, on the creative learning task, it is suggested that our supporting method by presenting the novelty of a found solution directly or that of indirectly by the learning goal space support continuously to set the learner's own learning goals. Future work is to make clear the definition of creative learning, especially the definition of "unusual solutions".

Acknowledgements. This work was supported by JSPS KAKENHI (Grant-in-Aid for Scientific Research ©) Grant Number 16K00317.

References

1. Jain, A.K., Murty, M.N., Flynn, P.J.: Data clustering: a review. ACM Comput. Surv. **31**(3), 264–323 (1999). https://doi.org/10.1145/331499.331504
2. Boden, M.A.: The Creative Mind: Myths and Mechanisms. Routledge, Abingdon (2003)
3. Frey, C.B., Osborne, M.A.: The future of employment: how susceptible are jobs to computerization? Oxford Martin School Working Paper, no. 7, pp. 1–72 (2013)
4. Guilford, J.P.: Creativity research: past, present and future. In: Frontiers of Creativity Research: Beyond the Basics, pp. 34–64. Bearly Ltd., Buffalo (1987). http://www.cpsb.com/research/articles/creativity-research/Creativity-Research-Guilford.pdf
5. Lehman, J.: Evolution through the search for novelty. University of Central Florida, Ph.D. Computer Science (2012)
6. Leung, K., Chen, T., Chen, G.: Learning goal orientation and creative performance: the differential mediating roles of challenge and enjoyment intrinsic motivations. Asia Pac. J. Manag. **31**, 811–834 (2013)
7. Finke, R.A., Ward, T.B., Smith, S.M.: Creative Cognition: Theory, Research, and Applications. The MIT Press, Cambridge (1992)
8. Smita, J., Trey, M.: Facilitating continuous learning: review of research on individual learning capabilities and organizational learning environments. In: The Annual Meeting of the AECT International Convention, Louisville (2012). http://www.memphis.edu/icl/idt/clrc/clrc-smita-research.pdf
9. Okudo, T., Takadama, K., Yamaguchi, T.: Designing the learning goal space for human toward acquiring a creative learning skill. In: Yamamoto, S. (ed.) HIMI 2017. LNCS, vol. 10274, pp. 62–73. Springer, Cham (2017). https://doi.org/10.1007/978-3-319-58524-6_6
10. Ono, Y.: The study of charisma and transformational leadership in terms of follower's view. Bus. Rev. Kansai Univ. **58**(4), 53–87 (2014). (in Japanese)
11. Yamaguchi, T., Takemori, K., Tamai, Y., Takadama, K.: Analyzing human's continuous learning processes with the reflection sub task. J. Commun. Comput. **12**(1), 20–27 (2015)
12. Yamaguchi, T., Tamai, Y., Takadama, K.: Analyzing human's continuous learning ability with the reflection cost. In: Proceedings of 41st Annual Conference of the IEEE Industrial Electronics Society (IECON 2015), pp. 2920–2925 (2015)

Investigation of Learning Process with TUI and GUI Based on COCOM

Natsumi Sei[1(⊠)], Makoto Oka[2], and Hirohiko Mori[1]

[1] Tokyo City University Graduate Division, Tokyo, Japan
{g1791801,hmori}@tcu.ac.jp
[2] Tokyo City University, Tokyo, Japan
moka@tcu.ac.jp

Abstract. In this paper, opportunity to use computer for learning has been on the rise. Therefore, we investigate the effect of the types of user interfaces how learning process is affected by using computer. An analysis was conducted for the verification using a cognitive model COCOM as a reference. Similar means for solving problems were taken for both interfaces when participants stay initial stage. Adopting other measures when getting familiar with such problems, however, it has been proved that participants find out a correct answer by accumulating sub-goal achievement in case of GUI and they figure out a correct answer while taking various measures such as searching of clues and accumulation of sub-goal achievement in case of TUI.

Keywords: Tangible User Interface (TUI) · COCOM · Cognitive process
Interface

1 Background

Along with spread of PCs in common, number of interface types for operating computer has increased. Among, Tangible User Interface (TUI) has been studied which is able to operate PCs by directly touching physical objects. TUI has made intuitive operation possible. Therefore, it is believed that users are possible to operate computer without learning its method of operation and focus on original learning purpose. However, it is believed that a learning system using TUI has not been spread commonly because it has not been recognized how affect people's learning process.

2 Purpose

In this paper, When TUI and Graphical User Interface (GUI) were used, how an adopted interface may affect learning process was focused. It is believed learning and its method suitable for a specific interface is clarified by clarifying the effects making it possible to support achieving learning purpose. In the process, actions are classified based on Contextual Control Model and difference in the actions by interface is verified by transition of the classified actions.

© Springer International Publishing AG, part of Springer Nature 2018
S. Yamamoto and H. Mori (Eds.): HIMI 2018, LNCS 10905, pp. 331–339, 2018.
https://doi.org/10.1007/978-3-319-92046-7_29

3   Related Works

Learning using PCs includes GUI's Light-bot [1]. It is operated with a mouse while watching PC screen. Light-bot is a product aiming at learning a concept of programming. There is also Osmo coding [2] based on TUI. Osmo coding is capable of moving characters on a screen by arranging panels on which illustrations of actions are depicted. It is also possible to learn a concept of programming similarly as Light-bot. Thus, GUI and TUI are different interface products but working on the same issue. However, they are not such products that have been developed by focusing on how a specific interface affects specific learning. It is believed that interface is able to support achievement of learning purpose by focusing on a relationship between interfaces and learning contents and figuring out advantages. Akahori [3] compared and verified tendency of each learning tendency when using PC, tablet terminal, and paper, in order to examine tendency of learning with PCs. It has been proved that paper is excellent in recognition of letters and text data and that PC and tablet terminal is excellent in input of letters and recognition of photos and the like, respectively. It has been also revealed that paper provides sense of learning completed and tablet terminal is suitable for motivation of further learning. However, it has not been focused on what kinds of difference in concept may appear at the time of learning.

4   Experiment

4.1   Objective of Experiment

We conducted the experiment to clarify the effect on learning process when a specific interface was used, difference in learning process is to be reviewed by using TUI and GUI.

4.2   Experimental Policy

As a learning task, logical circuit learning was selected in which GUI and TUI showed the same appearance. Subjects were asked to prepare a logic circuit based on a truth table. The three logical symbol used in this experiment was the "AND", the "OR" and the "NOT". In addition, the lead wire objects and the lamp objects, to connect among each gate and to confirm the output of the circuit respectively, were also prepared.

GUI is operated with a mouse using "click" and "drag" while watching on a PC screen. TUI is operated using cube- shaped blocks to create a logic circuit by applying them in 2-dimensional place. Asking subjects to voice what the subjects asked to "think aloud", and, their behavior and verbal protocol data were recorded by VCR.

4.3   Experimental Methodology

Subjects were asked to solve problems using a system to learn logic circuit. Operation manual of the interface was written out on a paper for subjects to be able to read again during experiment and instruction of the logic circuit was also written on a paper likewise.

Experiment was performed for 6 weeks (6 times) by increasing the difficulty every week. Subjects were asked to solve problems taking for around an hour in each experiment. A personality assessment test was conducted in the first week in order to recognize their original strategies. They were asked to solve problems regarding a logic circuit in the second and later experiments. In the second week, such problem was set for them to fill in an output part in a truth table in reference to a logic circuit in order for them to get familiar with a way to read truth tables. From the third and later week, they were asked to solve problems to create a logic circuit from a truth table. The problems of 3th are make logic circuit from truth table. Logical symbol of 3th is 2 pieces, 4th is 3pieces, 5th and 6th are no specified pieces for increase the difficulty level.

In the 6th, number of available logic symbols was not designated for the problem with a setting of two lamps for output. Because of the increased two output lamp, subjects were required for applied skill different from the case of problems provided in the 5th or before.

5 Analytical Methods

We focus Contextual Control Model which makes analysis based on subjects' behavior. It is one of cognitive models.

5.1 Contextual Control Model

In order to analyze what kinds of subject behavior will take, an analysis is made in reference to Contextual Control Model (COCOM) [4] by Hollnagel. There are five modes for COCOM.

Scrambled control mode: A control to select of random or panic.

Opportunistic control mode: A control to select next action just based on the current situation.

Explorative control mode: A control to seek for new ways at a venture without any other option available.

Tactical control mode: A control to select next action according to regulations provided in advance.

Strategic control mode: A control at high level in consideration of overall situation.

An analysis was conducted using an applied model which was adapted for problem solutions in reference to five modes advocated by COCOM.

Scramble control: The state that the user has no idea of what to do to solve the problem.

Explorative control: Their sub-goal is not clear and not to go toward the goal directly but to just try to find something tentatively.

Opportunistic control: Though subjects think looking at only one or a few sub-goals toward the goal, they do not know what kind of action should be done to achieve them.

Tactical control: Subjects think several sub-goals sweeping some part of path to the goal and knows what kind of action is needed to achieve sub-goals though they have not found the whole Tactical control: Subjects think several sub-goals sweeping some part of path to the goal and knows what kind of action is needed to achieve sub-goals though they have not found the whole path for the goal yet.

Strategic control: The path toward the goal is established more detail than tactical control, and subjects know what action are required to achieve it.

Then, we focus how the five modes of COCOM may transit in problem solving process. It was frequently observed that subjects tried ideas different from thoughts they had in the process of problem solving. With a concept to regard these thoughts as a cluster, how they bring about a transition in the cluster is reviewed.

Without any time limit for the experiments required, such a scrambled control mode to work on problems in panic did not occur. In addition, because the strategic control mode occurred when they were working on an easy problem and did not transit to other mode, the mode was excluded from the current analysis.

6   Results of the Analysis

6.1   Probability of Occurrence of Mode by COCOM

We analyzed the subjects' mode transition based on COCOM by dividing the whole period of the experiment into 3 stages to focus on their learning process as follows:

Initial stage: The third week
The subjects learn the basis of the logic circuits.

Latter stage: The 4th and 5thweeks
The subject becomes accustomed to solve the problems of the logic circuit.

Practical stage: The 6th (using 2 lams)
The subjects are asked to solve some advanced problems.

We focused what kinds of transition occurred in each stage. It was also observed that many participants reset their thoughts. It was an action observed when they tried to switch their thought of the time to another. We regarded their thoughts between resets as a cluster of thoughts. We examine what kinds of transition were brought about in the cluster of thoughts. In the course of examination, such cases were observed that two to up to five modes were transited in a single cluster of thoughts. Probability of transition occurrence is shown in Tables 1 and 2. Transition is shown from the first line to lower lines in sequence of occurrence.

Table 1. Mode transition with GUI

1	Opp	Opp	Opp	Exp	Exp	Exp	Tac	Tac	Tac
2	Opp	Exp	Tac	Opp	Exp	Tac	Opp	Exp	Tac
Initial	8.33	0.00	50.00	0.00	0.00	0.00	16.67	0.00	25.00
Later	24.19	0.00	27.42	1.61	0.00	3.23	9.68	1.61	14.52
practical	25.81	0.00	19.35	3.23	0.00	0.00	6.45	0.00	9.68

1	Opp	Opp	Opp	Exp	Exp	Exp	Tac
2	Exp	Exp	Tac	Opp	Opp	Tac	Opp
3	Opp	Tac	Opp	Exp	Tac	Opp	Tac
Initial	0.00	0.00	0.00	0.00	0.00	0.00	0.00
Later	0.00	1.61	4.84	0.00	0.00	0.00	3.23
practical	0.00	0.00	19.35	0.00	0.00	0.00	3.23

1	Opp	Tac	Exp	Tac	Exp	Exp	Opp
2	Tac	Opp	Tac	Exp	Opp	Opp	Tac
3	Opp	Tac	Opp	Opp	Exp	Tac	Opp
4	Tac	Opp	Tac	Exp	Tac	Opp	Tac
5							Opp
Initial	0.00	0.00	0.00	0.00	0.00	0.00	0.00
Later	1.61	3.23	1.61	0.00	0.00	0.00	1.61
practical	12.90	0.00	0.00	0.00	0.00	0.00	0.00

6.2 Initial Stage

In GUI, it was the most frequent cases that transited from Opportunistic to Tactical control mode followed by cases that transited from Tactical to Tactical control mode. From a perspective of mode transition from Opportunistic to Tactical control mode, it is understood that problems are solved by setting sub-goals until a correct answer is obtained while considering how a logic circuit is created in order to achieve the sub-goals. We are also understand that steady measures for new sub-goal were not figured out immediately after resetting in many cases. From the fact that it is the most common case that transits from Tactical to Tactical control mode, however, we are also understand that procedures until sub-goals are achieved have been figured out in many cases even immediately after resetting. In consideration of the fact that this type of transition is frequent, it is believed that they try to progress to a stage that a correct

Table 2. Mode transition with TUI

1	Opp	Opp	Opp	Exp	Exp	Exp	Tac	Tac	Tac
2	Opp	Exp	Tac	Opp	Exp	Tac	Opp	Exp	Tac
Initial	5.56	0.00	16.67	5.56	11.11	11.11	0.00	0.00	33.33
Later	11.90	0.00	16.67	2.38	26.19	11.90	7.14	2.38	7.14
practical	19.23	7.69	11.54	7.69	7.69	15.38	7.69	0.00	7.69

1	Opp	Opp	Opp	Exp	Exp	Exp	Tac
2	Exp	Exp	Tac	Opp	Opp	Tac	Opp
3	Opp	Tac	Opp	Exp	Tac	Opp	Tac
Initial	0.00	11.11	0.00	0.00	5.56	0.00	0.00
Later	0.00	0.00	0.00	4.76	4.76	2.38	0.00
practical	0.00	0.00	3.85	0.00	3.85	0.00	0.00

1	Opp	Tac	Exp	Tac	Exp	Exp	Opp
2	Tac	Opp	Tac	Exp	Opp	Opp	Tac
3	Opp	Tac	Opp	Opp	Exp	Tac	Opp
4	Tac	Opp	Tac	Exp	Tac	Opp	Tac
5							Opp
Initial	0.00	0.00	0.00	0.00	0.00	0.00	0.00
Later	0.00	0.00	0.00	2.38	0.00	0.00	0.00
practical	0.00	0.00	0.00	0.00	3.85	3.85	0.00

answer is obtained by accumulating achievement of sub-goals. However, procedures to achieve sub-goals in an accumulative manner have not been clarified in some cases. From this fact, it is also understood that GUI has got to a correct answer by accumulating achievement of sub-goals at the initial stage. It is characteristic that Explorative control mode did not occur and that transition of three or more modes was not observed in any case. For the reason why Explorative control mode did not occur, it is believed that any thought with a challenging spirit to try everything possible in an explorative manner was not induced because such thought of foreseeing was dominant due to accumulated achievement of sub-goals observed in many cases. In addition, there was no case that transited three or more modes because there was not motivation to transit through various modes because of the foreseeing thought.

In TUI, transition from Tactical to Tactical control mode was most frequently observed. Even though transition from Opportunistic to Tactical control mode as frequently occurred in GUI was also observed frequently, the percentage was at a level less than 20%. From the fact that transition from Tactical to Tactical control mode was frequently observed, we are also understand that they tried to get to a correct answer by repeated accumulation of sub-goals at the initial stage like the GUI. Different from GUI, however, transitions from Explorative to Explorative control mode as well as from Explorative to Tactical control mode were also observed frequently. Therefore, we are also understand that they have tried to find out clues by creating a logic circuit as a trial without setting a sub-goal. In case of a transition from Explorative to Tactical control mode, in particular, it is believed that they transited to Tactical control mode by a clarified sub-goal because clues were found in Explorative control mode. In addition, there were cases that transited through three or more modes from the initial stage in TUI, and transition of Opportunistic, Explorative, and Tactical control modes were often observed. While they often got to a correct answer by accumulating achievement of set sub-goals similarly to GUI, such actions were also observed that they got to a correct answer by taking different measures such as creation of a logic circuit as a trial.

From the above, we are understand that subjects tried to get to a correct answer by accumulating achievement of sub-goals in many cases of GUI and TUI, but such different approach to create a logic circuit as a trial was observed in TUI alone.

6.3   Later Stage

In GUI, transition from Opportunistic to Opportunistic as well as from Opportunistic to Tactical control mode became to be observed frequently. In this case, it is believed that Opportunistic control mode was repeated being unable to clarify measures for achieving sub-goals even though subjects tried to solve the problem by setting a sub-goal similarly to the procedures at the initial stage. It is believed as one of the causes that whether measures in hand is appropriate or not has become to be checked by GUI frequently before judging it in mind since they have got familiar with use of GUI recognizing that it has become easier to check it in that way than solving problems by foreseeing situations. Therefore, it is believed that sub-goals were achieved at the later stage by checking whether measures in hand was appropriate or not for the sub-goals by GUI rather than considering it in mind. Even though it has increased to transit to Opportunistic control mode due to increased use of GUI, it is believed that the fundamental concept to get to a correct answer by accumulated sub-goal achievements has not changed. In consideration the fact that transition to Explorative control mode has been seldom observed in spite of increased transition by three or more modes, it is believed that accumulation of sub-goal achievements is repeated with almost no challenge as a trial similarly to that at the initial stage.

Transition from Explorative to Explorative control mode was observed most frequently in TUI. We are understand that subjects create various circuit diagrams trying to find out clues without setting a goal. It is believed that subjects have become to use a method for seeking for clues such as consideration based on a circuit diagram created as a trial rather than adopting a thought to set up sub-goals for problems with increased difficulty. It is also understood that creation of a circuit diagram as a trial is not a simulation of a result but an action to create a circuit diagram dependent on TUI just like in GUI.

From the above, it is proved that results of the circuit diagram are checked promptly by using interface at later stage in cases of both GUI and TUI. Since different approaches were observed in TUI at later stage while similar actions were often observed at early stage, it has been proved that subjects became easy to be affected by use of each interface as they got used to the interface. Therefore, it has been revealed that GUI tries to get to a correct answer being dependent on interface while accumulating sub-goal achievements and that TUI tries to get to a correct answer being dependent on interface while seeking for clues to get to a correct answer.

6.4   Practical Stage

In GUI, transition from Opportunistic to Opportunistic and from Opportunistic to Tactical control mode, as well as transition through Opportunistic, Tactical to Opportunistic control mode has increased. In the transition through Opportunistic, Tactical and back to Opportunistic control mode, even though measures for sub-goals had been clarified, there were wrong and unable to be clarified eventually, and therefore it is believed that it became difficult to have explicit measures for problems required for application. Further, when mode transition ended up in Opportunistic control mode, the next mode transition started with Opportunistic control mode in many cases since later stage. It is believed that other procedures were performed by setting a new sub-goal because any clear procedure for the previous sub-goal was found but it is understood that measures for the new sub-goal are also not clear. In addition, as mode transition to occur Explorative control mode was seldom observed, it is understood that sub-goal had been set up continuously in mind without seeking for clues.

In TUI, we are understand that patterns to transit two modes appeared evenly except for a pattern of transition from Tactical to Explorative control mode. In particular, patterns of transition to start with Opportunistic and Explorative control mode were observed frequently. In other words, it is believed that various solutions such as a method to seek for clues and to accumulate sub-goals had been performed.

From the above, it has been proved that GUI tries to get to a correct answer also in application problems by accumulating sub-goal achievements and that TUI tries to get to a correct answer by combining various methods such as to seek for clues and to accumulate sub-goals.

7   Conclusion

In order to verify a learning process at the time when TUI and GUI were used, comparison experiments were performed for the interfaces. An analysis was performed on actions of the participants during the experiments in combination with an analysis based on COCOM.

In GUI, the participants got to a correct answer at initial stage by achieving a sub-goal with transition from Tactical to Tactical control mode which was frequently observed. At later stage for application, transition from Opportunistic to Tactical as well as from Opportunistic to Opportunistic control mode increased. Even though mode transition changed, such tendency did not change that tries to get to a correct answer by setting a sub-goal and accumulating the achievements. In other words, subjects tried to get to a correct answer by the same methods in case of GUI both before and after they got familiar with the interface.

In TUI at initial stage, the participants got to a correct answer by also achieving a sub-goal with transition from Tactical to Tactical control mode which was frequently observed like a case of GUI. At later stage when a pattern to start with explorative control mode increased, however, they became in frequently take a method to seek for clues in addition to a method to accumulate sub-goal achievements. At a stage of practical stage, it was proved that they tried to get to a correct answer by taking various measures with various transition occurred evenly. It was proved that they took different methods between before and after they got used to the interface.

8   Application

In consideration of the fact that an intention to get to a correct answer is induced by using GUI to set a subgoal until getting to a correct answer, it is believed to be suitable for such problems that the final completion is achieved by accumulating subgoals just like programming.

In consideration of the fact that an intention to get to a correct answer is induced by using TUI to take various methods, it is believed to be suitable for problems aiming at thinking from various directions like active learning.

References

1. Lightbot (2016). http://lightbot.com/
2. Osmo coding (2016). https://www.playosmo.com/
3. Akahori, K.: Effectiveness in cognition through interface differences of papers, PCs and tablet terminals as learning devices 7(2), 261–279 (2013)
4. Hollnagel, E.: Human Reliability Analysis: Context and Control. Academic Press, Cambridge (1994)

Information in Aviation and Transport

Measuring the Effects of a Cognitive Aid in Deep Space Network Operations

Edward Barraza[1,2(✉)], Alexandra Holloway[1], Krys Blackwood[1], Michael J. Gutensohn[3], and Kim-Phuong L. Vu[2]

[1] The Jet Propulsion Laboratory, Pasadena, CA, USA
edward.barraza@jpl.nasa.gov
[2] California State University, Long Beach, Long Beach, CA, USA
edward.barraza@student.csulb.edu
[3] Rollins College, Winter Park, FL, USA

Abstract. Cognitive aids have long been used by industries such as aviation, nuclear, and healthcare to support operator performance during nominal and off-nominal events. The aim of these aids is to support decision-making by providing users with critical information and procedures in complex environments. The Jet Propulsion Lab (JPL) is exploring the concept of a cognitive aid for future Deep Space Network (DSN) operations to help manage operator workload and increase efficiency. The current study examines the effects of a cognitive aid on expert and novice operators in a simulated DSN environment. We found that task completion times were significantly lower when cognitive aid assistance was available compared to when it was not. Furthermore, results indicate numerical trends that distinguish experts from novices in their system interactions and efficiency. Compared to expert participants, novice operators, on average, had higher acceptance ratings for a DSN cognitive aid, and showed greater agreement in ratings as a group. Lastly, participant feedback identified the need for the development of a reliable, robust, and transparent system.

Keywords: Cognitive aid · Decision support · Technology acceptance
Expertise · Deep Space Network

1 Introduction

Cognitive aids have long been used by industries such as aviation, nuclear, and healthcare to support operator performance during nominal and off-nominal events. The concept of cognitive aids includes checklists, flowcharts, sensory cues, safety systems, decision support systems, and alerts (Levine et al. 2013; Singh 1998). Some examples of cognitive aids are the Quick Reference Handbook for pilots, the National Playbook for air traffic controllers (ATCos), and the Surgical Safety Checklist for surgeons (Playbook 2017; Catalano 2009). They are all designed to guide decision-making by providing users with important information in complex environments. The Jet Propulsion Lab (JPL) is exploring the concept of a cognitive aid for Deep Space Network (DSN) operations.

© Springer International Publishing AG, part of Springer Nature 2018
S. Yamamoto and H. Mori (Eds.): HIMI 2018, LNCS 10905, pp. 343–358, 2018.
https://doi.org/10.1007/978-3-319-92046-7_30

1.1 The Deep Space Network

The DSN is a global network of telecommunications equipment that provide support for interplanetary missions of the National Aeronautics and Space Administration (NASA) and other space agencies around the world (DSN Functions n.d. 2017). Over the next few decades, the network will experience a significant surge in activity due to an increase in the total number of antennas, an increase in missions, higher data rates, and more complex procedures (Choi et al. 2016). To meet the projected demand, JPL is managing a project called Follow-the-Sun Operations (FtSO). A number of automation improvements designed to help manage the workload and efficiency of operators are in development. One of these improvements is the application of complex event processing (Johnston et al. 2015).

Broadly, complex event processing is a machine learning method of combining data streams from different sources in order to identify meaningful events and patterns (Choi et al. 2016). One complex event processing application is the detection of operational deviations from the norm. The system matches ongoing situations in real-time to previous incidents and then provides procedural resolution advisories. The information output of CEP is intended to guide operator decision making. In doing so, the information aids operator cognition. CEP will be just one of an array of tools that are available for LCOs to perform their job.

1.2 Cognitive Aids

A cognitive aid is a presentation of prompts aimed to encourage recall of information in order to increase the likelihood of desired behaviors, decisions, and outcomes (Fletcher and Bedwell 2014). Cognitive aids include, but are not limited to, checklists, flow-charts, posters, sensory cues, safety systems, alerts, and decision support tools (Levine at al. 2013; Singh 1998). A large body of research has demonstrated the concept is applicable anywhere users operate in stressful working conditions with high information flow and density.

Arriaga et al. (2013) evaluated an aid that assisted operating-room teams during surgical crisis scenarios in a simulated operating room. They were interested in seeing if a crisis checklist intervention would improve adherence to industry best practices. A total of 17 operating-room teams were randomly assigned to manage half the scenarios with a crisis checklist and the other half from memory alone. They found that the use of crisis checklists was associated with a significant improvement in adherence to recommended procedures for the most common intraoperative emergencies, such that 6% of steps were missed when checklists were available as opposed to 23% when they were unavailable. Thus, crisis checklists have the potential to improve surgical care.

There is evidence that decision support tools can reduce workload in addition to improving performance (Van de Merwe et al. 2012). Van de Merwe et al. (2012) evaluated the influence of a tool named Speed and Route Advisor (SARA) on ATCo performance and workload in an air traffic delivery task. In a simulation, the experimenters captured accuracy by measuring the adherence of aircraft to the expected approach time, the controllers' subjective workload through a self-assessment measure, and controllers' objective workload through the total number of radio calls and device inputs. SARA provided speed and route advisories for every inbound flight, thereby providing controllers with the information to issue a single clearance to each aircraft to manage traffic delivery.

1.3   Current Study

We are not aware of any research that investigates operator performance in DSN operations. Further, there is no research that explores how a cognitive aid affects the workload of LCOs or provides a measure of their acceptance of an aid. The current study examines the impact of a cognitive aid on the performance and workload of expert and novice LCOs in a DSN task. We had the following research questions:

1. What effect will the cognitive aid have on operator performance in the DSN framework?
2. Will experts and novices perform differently using the cognitive aid?
3. What effect will the cognitive aid have on operator subjective workload in the DSN framework?
4. Will novice operators have higher acceptance ratings for the cognitive aid compared to experts?

Participants were asked to monitor, detect, and resolve any issues they encountered during simulated tracks. On some tracks, a cognitive aid detected issues and provided resolution advisories. We measured their performance, acceptance using the perceived usefulness scale of the TAM, and workload using a composite NASA Task Load Index (TLX) score. While the cognitive aid in DSN operations is still a concept, it is important to understand its utility and effects early in its development.

2   Method

2.1   Participants

Eight employees (3 females, 5 males) of the Deep Space Network were recruited for this experiment. To examine different levels of expertise, four participants were expert LCOs from the Goldstone Deep Space Communications Complex; the other four participants were DSN Track Support Specialists from the Space Flight Operations Facility at JPL. The Track Support Specialists were selected as proxy for novice LCOs because they are familiar with the terminology, displays, and data constituents. However, they do not perform the same tasks as LCOs. Participants had 9 months – 40 + years of experience working in the DSN. The expert group had an average of 30.5 years of DSN experience; novice group had an average of 2.75 years of experience. Participants were between 35–70 years of age (M = 53.71). The average age for the expert group was 58.75 years; average age for novice group was 46 years (one participant did not disclose). Participants were volunteers and were not compensated for their time.

2.2 Design

The simulation employed a 2 (Group: Experts/Novice) × 2 (Assistance: Cognitive aid/No Cognitive aid) mixed factorial design. Group was a between-subjects variable. Assistance was a within-subjects variable. The dependent measures include performance, subjective workload, and operator acceptance. Performance was assessed with the following variables: system interactions (clicks on the interface display) and efficiency (seconds). Subjective workload was assessed with the NASA TLX. The TAM Perceived Usefulness scale was distributed to assess participant acceptance of the cognitive aid.

2.3 Materials

Participants were tested using the Deep Space Network Track Simulator. The Deep Space Network Track Simulator is a medium fidelity simulator that reproduces the cruising phase of a track. This simulator ingests archival data and presents this information on up to three current DSN displays (see Fig. 1). Three displays were determined by a subject matter expert to be the minimum displays operators required to diagnose and resolve the four simulated issues. The displays are the Signal Flow Performance screen, Current Tracking Performance screen, and the Radio Frequency Signal Path screen. All screens were presented in a single display using a 27-inch Apple Thunderbolt monitor (see Fig. 2). The screen was recorded to capture the interactions enacted by participants. The simulator captured the total task times and tracked the displays that were clicked.

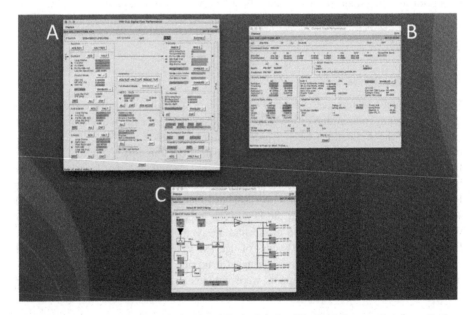

Fig. 1. Deep space network track simulator displaying the (A) signal flow performance display, (B) Current tracking performance display, and the (C) radio frequency signal path display.

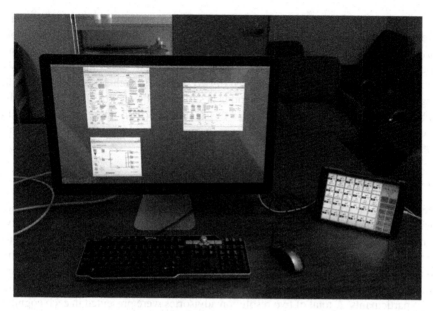

Fig. 2. The link control station setup.

There were four scenarios, each duplicating an issue that LCOs currently manage during operations. Each scenario was approximately 10 min long and with fault injection occurring approximately 3 min in to the track. Archival data was used to simulate a track. Scenarios were counterbalanced between participants.

Other materials used in the JPL study include an informed consent form, a demographics questionnaire, and the NASA TLX (Hart and Staveland 1988) and the Perceived Usefulness Scale (PU scale; Davis 1989). The NASA TLX is a subjective measure of workload that uses a 1 to 100 scale. The NASA TLX assesses workload across six different dimensions: mental demand, physical demand, temporal demand, effort, performance, and frustration. The PU scale uses a 1 to 7 scale that assess various dimensions of perceived usefulness of technology. Participants also filled out a post-questionnaire at the conclusion of experimental trials.

3   Procedure

One participant was run at a time. Upon arrival, participants were instructed to read and sign the informed consent form and fill out the demographics questionnaire. Participants were then briefed for 10 min on the study. They were briefed on the task and the displays that would be available to them.

After the briefing session, participants completed a training scenario which exposed them to the simulation environment through a training track. During this scenario, the participants were able to ask the simulation manager questions about the functionality of the simulator. After the conclusion of the training scenario, participants were asked if they had questions about the procedures.

The participants had two experimental blocks. The two blocks varied in assistance level. The two assistance levels were cognitive aid and no cognitive aid conditions. Participants first ran in one experimental block and then the other. The experimental blocks were counterbalanced across participants. Each experimental block was approximately 30 min long and consisted of two scenarios, each a maximum of 10 min long. A detailed breakdown of a scenario is described in the next section. At the end of each scenario, participants were required to fill out the NASA TLX. At the end of the second experimental block, participants were debriefed and thanked for their participation.

3.1 Cognitive Aid

The variable of Assistance was counterbalanced by either including or omitting a cognitive aid in the second experimental block. For half of the participants, the first two scenarios included no cognitive aid (i.e., participants had to diagnose and resolve issues themselves), and the second experimental block provided participants with cognitive aid assistance, where the tool diagnosed issues and provided resolution advisories three minutes in to the scenario (See Fig. 3). The opposite order was given to the other half of the participants. A total of two resolution advisories were presented in each cognitive aid recommendation, but only one advisory resolved the simulated issue. Clicking the "+" revealed step-by step procedures for the advisory. Whether the correct advisory was listed first or second was also counterbalanced between scenarios.

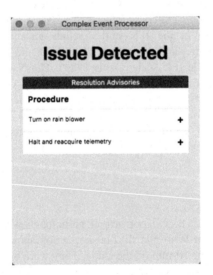

Fig. 3. The cognitive aid (CEP) prompt.

4 Results

Descriptive statistics are reported as well as the effects of separate 2 (Group: Expert/Novice) × 2 (Assistance: Cognitive Aid/No Cognitive Aid) mixed ANOVAs, with group as the between-subjects variable. Although the sample size was small, due to the specialized qualifications of the participants, ANOVAs were conducted to help identify trends in the data. The dependent variables are system interactions, as indicated by the number of clicks participants performed during each trial, efficiency, as indicated by the total task time for each trial, and workload, as indicated by subjective NASA TLX ratings for each trial. Significance was set at $p < .05$, but we also report trends if $p < .10$.

4.1 System Interactions

System Interactions were captured by the number of clicks participants enacted to troubleshoot scenarios, after controlling for ambiguous clicks (i.e. clicks that had no strategic intent), and for cognitive aid clicks (i.e. clicks that were enacted on the cognitive aid display) (See Table 1). The 2 (Group: Expert/Novice) × 2 (Assistance: Cognitive aid/No cognitive aid) mixed ANOVA yielded a non-significant main effect of group, $F(1, 6) = 5.56$, $MSE = 2.31$, $p = .06$, $\eta p^2 = .48$. Experts ($M = 3.94$, $SE = .57$) tended to click on their displays more than novice participants ($M = 2.15$, $SE = .54$). The effect of assistance was not significant, $F(1, 6) = .37$, $MSE = 4.77$, $p = .56$, $\eta p^2 = .06$. Participants clicked on their displays about three times, both in conditions with no cognitive aid, ($M = 3.38$, $SE = .75$), and in conditions with a cognitive aid, ($M = 2.71$, $SE = .57$) (See Table 1). No significant interaction was found, $F(1, 6) = .25$, $MSE = 4.77$, $p = .64$, $\eta p^2 = .04$ (See Fig. 4).

Table 1. Descriptive statistics from the 2 (Group: Expert/Novice) × 2 (Assistance: Cognitive Aid/No Cognitive Aid) mixed ANOVA for system interactions

Assistance	Group	System interactions (clicks)
Cognitive Aid	Expert	M = 3.88
		SD = 1.79
	Novice	M = 1.54
		SD = 1.41
	Total	M = 1.54
No Cognitive Aid	Expert	M = 4.00
		SD = 1.78
	Novice	M = 2.75
		SD = 2.39
	Total	M = 3.38
Total	Expert	M = 3.94
	Novice	M = 2.15

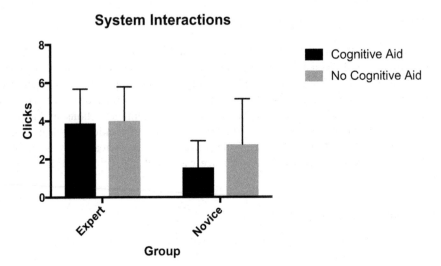

Fig. 4. Mean number of clicks by group and assistance for system interactions.

During cognitive aid trials, participants had the option of adhering to or ignoring the tool. We examined the number of times participants chose the correct resolution as their initial resolution strategy in cognitive aid conditions. Participants chose the correct resolution first in 9 (56%) of 16 cognitive aid trials. Of these 9, experts were responsible for 5 of these trails and novices for 4. All but one trial (performed by a novice) were eventually resolved.

4.2 Efficiency

Efficiency was captured as the total task time, in seconds, for each trial. Each scenario was approximately 600 s in length (10 min). Participants were required to resolve issues within the 600 s scenario, otherwise the simulator would timeout. Timeouts were examined for frequency of occurrence by group. Out of a total of 32 trials, eight (25%) timed out. Of these eight, three timeouts occurred in the expert group and five occurred in the novice group. One novice participant was responsible for three of five timeouts in the novice group. More specifically, seven timeouts occurred in trials where participants received no cognitive aid assistance and one timeout occurred in the novice group trial with cognitive aid assistance.

We performed separate 2 (Group: Expert/Novice) × 2 (Assistance: Cognitive Aid/No Cognitive Aid) mixed ANOVAs to examine task time measures with and without timeouts, see Table 2 for means. For the analysis with timeout trials, the time on task was set as the limit of 600 s. There was a significant main effect of assistance, $F(1, 6) = 26.97$, $MSE = 4261.12$, $p < .01$, $\eta p^2 = .82$. Participants were faster at resolving issues when they had the assistance of a cognitive aid ($M = 281.06$, $SE = 32.9$) than when they did not ($M = 450.56$, $SE = 35.41$). The effect of group was not significant, $F(1, 6) = .71$, $MSE = 14425.49$, $p = .43$, $\eta p^2 = .12$. Experts were not significantly faster at resolving issues ($M = 340.44$, $SE = 42.46$) compared to novices

(M = 391.19, SE = 32.9) (See Table 2). No significant interaction was found, $F(1, 6)$ = .03, MSE = 4261.12, p = .88, ηp^2 = .00 (See Fig. 5).

Table 2. Descriptive statistics from the 2 (Group: Expert/Novice) × 2 (Assistance: Cognitive Aid/No Cognitive Aid) mixed ANOVA for efficiency

Assistance	Group	Efficiency with timeouts (seconds)	Efficiency without timeouts (seconds)
Cognitive Aid	Expert	M = 253.13	M = 253.13
		SD = 86.23	SD = 86.23
	Novice	M = 309.00	M = 263.83
		SD = 99.39	SD = 50.76
	Total	M = 281.06	M = 257.71
No Cognitive Aid	Expert	M = 427.75	M = 323.38
		SD = 82.65	SD = 99.13
	Novice	M = 473.38	M = 340.33
		SD = 115.01	SD = 156.09
	Total	M = 450.56	M = 330.64
Total	Expert	M = 340.44	M = 288.25
	Novice	M = 391.19	M = 302.08

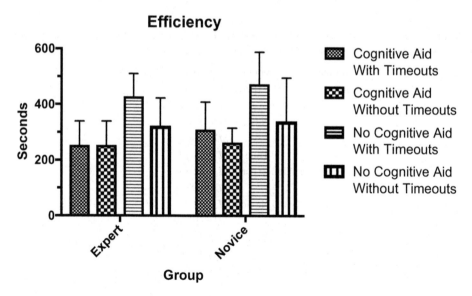

Fig. 5. Mean task completion time by group and assistance for efficiency, with and without timeouts.

A second 2 (Group: Expert/Novice) × 2 (Assistance: Cognitive Aid/No Cognitive Aid) mixed ANOVA was performed with the timeout data coded as missing values. For this analysis, there was no main effect of group, $F(1, 5)$ = .04, MSE = 15225.67,

$p = .84$, $\eta p^2 = 01$. However, the numerical pattern remained the same as the previous analysis: experts were numerically faster ($M = 288.25$, $SE = 43.63$) at resolving issues compared to novices ($M = 302.09$, $SE = 50.38$). The effect of assistance was also not significant, $F(1, 5) = 3.13$, $MSE = 5907.78$, $p = .14$, $\eta p^2 = .39$. Participants were not significantly faster at resolving issues when they had the assistance of a cognitive aid ($M = 258.48$, $SE = 28.30$) than when they did not ($M = 331.85$, $SE = 47.76$), see Table 2. No significant interaction was found, $F(1, 5) = .006$, $MSE = 5907.78$, $p = .94$, $\eta p^2 = .00$ (See Fig. 5).

4.3   Subjective Workload

Subjective workload was captured by the NASA Task Load Index at the end of each trial. All the ratings for the seven dimensions were combined to provide one composite score for each participant. The higher the TLX rating, the higher the workload. A 2 (Group: Expert/Novice) × 2 (Assistance: Cognitive aid/No Cognitive Aid) mixed ANOVA did not yield a significant main effect of group, $F(1, 6) = .01$, $MSE = 766.6$, $p = .93$, $\eta p^2 = .00$, nor assistance, $F(1, 6) = .27$, $MSE = 198.51$, $p = .62$, $\eta p^2 = .04$. Experts reported a mean workload rating of 41.26 and novices 39.93. Conditions with a cognitive aid were reported to have a mean workload of 38.77. Mean workload in conditions with no cognitive aid was 42.42 (See Table 3). No significant interaction was found, $F(1, 6) = .03$, $MSE = 198.51$, $p = .87$, $\eta p^2 = .01$ (See Fig. 6).

Table 3. Descriptive statistics from the 2 (Group: Expert/Novice) × 2 (Assistance: Cognitive Aid/No Cognitive Aid) mixed ANOVA for subjective workload

Assistance	Group	Workload (NASA TLX)
Cognitive Aid	Expert	$M = 40.05$
		$SD = 29.53$
	Novice	$M = 37.50$
		$SD = 18.74$
	Total	$M = 38.77$
No Cognitive Aid	Expert	$M = 42.48$
		$SD = 18.91$
	Novice	$M = 42.36$
		$SD = 18.69$
	Total	$M = 42.42$
Total	Expert	$M = 41.26$
	Novice	$M = 39.93$

4.4   Technology Acceptance

Acceptance was measured by the Perceived Usefulness (PU) scale taken from the Technology Acceptance Model (Davis 1989). The acceptability items were all rated on a scale from 1 "Extremely Likely" to 7 "Extremely Unlikely" and a 4 as "Neither". Lower numbers indicate a greater acceptance of technology. One-way ANOVAs with

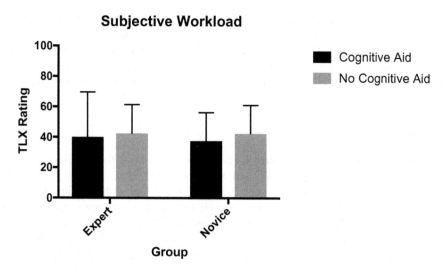

Fig. 6. Mean workload ratings by group and assistance for subjective workload.

(Group: Expert/Novice) as a between subject variable did not yield any significant effects of group, see Table 4 for means, F-ratios, and p-values. Overall, expert rated questions one through five slightly above a 4, indicating indifference to Complex Event Processing (CEP or cognitive aid). Question six was the only question that experts rated the CEP has more acceptable (M = 2.75) and with the lowest variability (SD = .98). Although novices rated all questions below a 4, indicating greater acceptance of a cognitive aid when compared to expert ratings, these differences were not significant.

Table 4. Descriptive statistics from the perceived usefulness scale (lower numbers indicate greater acceptance to technology). CEP refers to complex event processing (cognitive aid).

Technology acceptance scale: perceived usefulness	Expert	Novice	F-ratio; P-value
1. Using CEP in operations would enable me to accomplish tasks more quickly	M = 4.25 SD = 2.75	M = 2 SD = 0	$F(1, 6) = 2.67$ $p = .15$
2. Using CEP would improve my operational performance	M = 4.25 SD = 2.75	M = 2.25 SD = .5	$F(1, 6) = 2.04$ $p = .20$
3. Using CEP in operations would increase my productivity	M = 4.5 SD = 2.38	M = 3 SD = 1.41	$F(1, 6) = 1.17$ $p = .32$
4. Using CEP would enhance my effectiveness in operations.	M = 4.25 SD = 2.75	M = 2.5 SD = 1	$F(1, 6) = 1.42$ $p = .28$
5. Using CEP would make it easier to conduct operations	M = 4.5 SD = 2.38	M = 2.5 SD = .58	$F(1, 6) = 2.67$ $p = .15$
6. I would find CEP as a useful tool in operations	M = 2.75 SD = .98	M = 2.75 SD = .98	$F(1, 6) = .65$ $p = .45$

A frequency distribution of participant ratings for each question of the Perceived Usefulness scale revealed two patterns. First, expert participants tended to be split in their ratings such that half of expert participants clustered on one end of the scale and the other end of the scale. This pattern is consistent for all questions. On the other hand, novice participants tended to cluster together in their responses in agreement to accept CEP.

4.5   Post Questionnaire

Participants were asked to complete a six-question survey at the conclusion of the simulation. The first five questions used a five-point Likert scale from 1 "Strongly Disagree" to 5 "Strongly Agree" and a 3 as "Neither". The sixth question was open ended. One-way ANOVAs with (Group: Expert/Novice) as a between subject variable did not yield a significant main effect for any questions, see Table 5 for means, F-ratios, and p-values. Both groups gave mostly neutral responses to four of five questions. The fifth question, which asked if a tool like CEP would be useful in Follow-the-Sun operations, had the highest agreement among the two groups (M = 3.88).

Table 5. Descriptive statistics from the post questionnaire (higher numbers indicate greater agreement). CEP refers to complex event processing (cognitive aid).

Post questionnaire	Expert	Novice	F-ratio; P-value
1. I trust the CEP system	$M = 2.5$	$M = 3.5$	$F(1, 6) = 1$
	$SD = 1.91$	$SD = .58$	$p = .36$
2. The CEP system lowered my workload	$M = 3$	$M = 3.25$	$F(1, 6) = .06$
	$SD = 1.83$	$SD = .96$	$p = .82$
3. The CEP system increased my situation awareness	$M = 2.5$	$M = 3.25$	$F(1, 6) = .87$
	$SD = 1.29$	$SD = .96$	$p = .39$
4. A tool like CEP will be useful in Follow-the-Sun operations	$M = 3.5$	$M = 4.25$	$F(1, 6) = .7$
	$SD = 1.73$	$SD = .5$	$p = .44$
5. The simulated environment was believable	$M = 3$	$M = 3.75$	$F(1, 6) = 1.42$
	$SD = 1.15$	$SD = .5$	$p = .28$

A frequency distribution of participants' ratings for the first three questions and question 5, in the post questionnaire revealed that experts tended to have greater variability in their ratings, with two experts giving ratings of 2 or lower, and 2 experts giving ratings of 3–5. Novice participants' ratings clustered together closer to the middle of the scale. For question 4, seven of eight participants gave ratings that agreed or strongly agreed when asked if they though CEP would be a useful tool in Follow-the-Sun Operations.

Question six asked participants to share any other thoughts on complex event processing. Of eight participants, seven submitted responses. Overall, there were three major themes. The first theme encapsulates the responses that were entirely positive in regard to CEP. The second theme encapsulates positive responses but also captures some dependencies. Lastly, the third theme captures the desire for validation of the concept.

5 Discussion

The Deep Space Network is the system that provides support for all interplanetary missions of space agencies around the world. As DSN demands continue to grow, there will be an increased need to understand how the technologies designed to address those demands affect operations. The Follow-the-Sun paradigm shift will surely see a spike in complexity and in the number of tracks that are attended to by a human operator (Choi et al. 2016). To help operators, automated systems will likely be employed. Therefore, there should be research aimed at understanding how human operators interact with those technologies and what factors contribute to overall system success.

The purpose of this study was to better understand the impact of a cognitive aid, operationally known as Complex Event Processing (CEP), on LCO workload, performance, and acceptance. We focused on two groups of LCOs. The first group were the expert LCOs who have many years of experience with configuring, monitoring, operating, and troubleshooting DSN tracks. The second group were novice proxies, composed of DSN TSS who specialize in monitoring, but do not configure, operate, nor troubleshoot tracks. Each participant was asked to interact with a DSN simulator where we presented them with four scenarios that they had to resolve. Half of the conditions provided them with the assistance of a cognitive aid and half of them did not. The sample size was small, so there were few significant effects. The results show numerical trends that distinguish experts from novices in their system interactions and efficiency. However, most of these differences were not statistically significant. This study was limited in that the population of DSN operators in the US are small, and this in turn limited the sample size of participants in the study. Therefore, trends evident from this study need to be investigated further in future studies.

The first and second research questions were: What effect will the cognitive aid have on operator performance in the DSN framework? Will experts and novices perform differently using the cognitive aid? Results indicate that neither the effects of group nor assistance were statistically significant for the total number of clicks. Experts tended to perform more clicks compared to novices, but they tended to engage in the same number of clicks for cognitive aid than no cognitive aid conditions. Novices had slightly less clicks in conditions with a cognitive aid than in conditions without. Rather than limiting their strategies to cognitive aid advisories, experts tended to rely more on their internal troubleshooting schemas. The observed pattern is consistent with the idea of expert-based intuition (Salas et al. 2010). This theory proposes that in the later stages of experience, the decision-maker draws on a deep and rich knowledge base from extensive experience within a domain such that decisions become intuitive. Intuition, by definition, occurs without outside assistance. It is the product of "affectively charged judgments that arise through rapid, non-conscious, and holistic associations" (Dane and Pratt 2007). Thus, experts may rely on intuition more than cognitive aids.

There was a significant effect of assistance for total time on task, when timeouts were taken into consideration. Task completion times were significantly lower when cognitive aid assistance was available when compared to when it was not. This suggests that regardless of group, participants tended to resolve issues faster when a cognitive aid was available. The cognitive aid therefore facilitated detection and resolution of issues by narrowing down the causes and resolutions to those issues. This finding is consistent with the idea that a cognitive aid is associated with significantly improved operational performance (Arriaga et al. 2013). The effect of group was not found to be significant. Although experts were numerically faster at resolving issues than novices, they also tended to engage in more clicks, indicative of trying to solve the problem on their own. It could be the case that if the expert group trusted the cognitive aid more, they would also see a larger benefit in the time to detect and resolve the problem.

The second analysis did not include the 600 s ceiling value for timeouts and did not yield any significant effects. This suggests that the effect of assistance found in the previous analysis was driven by observed timeout times. However, the numerical trend in the data remained the same: experts were more efficient than novices at resolving issues. A descriptive analysis of timeouts revealed that most of them occurred in the novice group during conditions where cognitive aid assistance was not available. This is not surprising, as the novice group are not as familiar with the procedural nature of troubleshooting tracks. Recall that the novice group was composed of DSN Track Support Specialists, a role that is largely responsible for supervising configurations and not to directly manipulate the displays to solve issues. It is no surprise that 62.5% of timeouts (5 of 8) occurred with this group as their schemata for LCO tasks is limited. The expert group was responsible for 37.5% of timeouts (3 of 8). With more time, the experts would have likely been able to solve the problem. Thus, even for the expert group, a cognitive aid has the potential of improving the efficiency of a solution, if it is used.

The third research question was: What effect will the cognitive aid have on operator subjective workload in the DSN framework? No significant effects of group nor assistance were found for workload. Participants in both groups rated their workload about the same, regardless of assistance condition. The mean workload ratings of expert and novice participants across both assistance conditions was 40.6. In Grier's (2015) cumulative frequency distribution of global TLX scores, a rating of 40.6 is greater than 30% of all scores. Furthermore, if only scores for monitoring tasks are considered, as were implemented in this study, a rating of 40.6 is only above 25% of observed scores. According to this analysis, both groups were experiencing low workload levels. The biggest numerical difference was observed with the novice group: participants experienced lower levels of workload in conditions where they had cognitive aid assistance. This suggests that a cognitive aid may potentially benefit novice workload levels to a greater degree than experts. However, this conclusion cannot be made based solely on the results of the present study.

The final research question was: Will novice operators have higher acceptance ratings for the cognitive aid compared to experts? Responses to post questionnaires showed that expert participants had mixed feelings about the usefulness of a cognitive aid in operations. A frequency distribution showed that expert ratings tended to be split, such that half felt they were likely to accept CEP, and the other half felt unlikely to accept CEP. One expert participant consistently rated all questions a six and another expert participant rated all questions a seven. Experts' mean ratings resulted in neutral responses to five of six questions pertaining to the Perceived Usefulness scale. This makes sense – experts take pride in their ability to perform their jobs. A tool that diagnoses and provides resolutions advisories on their behalf, in essence, performs an aspect of their job for them. However, experts generally agreed that a cognitive aid could be a useful tool in Follow-the-Sun operations. This indicates that experts acknowledge that CEP can be a useful tool in future operations.

Novices had higher mean acceptance ratings for complex event processing, and their responses were more clustered together. All novices had the greatest agreement when asked if they believed using CEP in operations would enable them to accomplish tasks more quickly. In the post questionnaire, both groups utilized the halfway point more than they did on the Perceived Usefulness scale. This may be due to the fact that the PU scale is a seven-point scale while the post questionnaires used a five-point scale.

Participant open-ended feedback provided insight to opinions about complex event processing. The majority of responses captured positive dispositions toward CEP. Overall, participants indicated there was utility for CEP in Follow-the-Sun operations, training, and learning. However, areas for improvement were identified. CEP should be robust in its diagnosis and recommendations to facilitate operator trust. Additionally, the CEP needs to provide system transparency to keep LCOs in the loop. One expert participant, who rated all PU scale questions a 7, expressed that CEP will be useful as it matures over time. A second expert participant, who rated all PU scale questions a 6, expressed the need for real world validation of CEP. These opinions on CEP highlight important system attributes for trust in human-machine interactions (Sheridan 1988). Operators communicated the need for a reliable, robust, and transparent system, which coincide with Sheridan's (1988) list of attributes that facilitate trust in the human-machine environment.

References

Levine, A.I., DeMaria, J.S., Schwartz, A.D., Sim, A.J.: The Comprehensive Textbook of Healthcare Simulation. Springer, New York (2013). https://doi.org/10.1007/978-1-4614-5993-4

Singh, T.D.: Incorporating cognitive aids into decision support systems: the case of the strategy execution process. Decis. Support Syst. **24**(2), 145–163 (1998)

Playbook: Federal Aviation Administration (2017). https://www.fly.faa.gov/Operations/playbook/current/current.pdf. Accessed 2 June 2017

Catalano, K.: The world health organization's surgical safety checklist. Plast. Surg. Nurs. **29**(2), 124 (2009)

DSN Functions: Jet Propulsion Laboratory (n.d.). https://deepspace.jpl.nasa.gov/about/DSNFunctions/#. Accessed 17 Apr 2017

Choi, J.S., Verma, R., Malhotra, S.: Achieving fast operational intelligence in NASA's deep space network through complex event processing. In: SpaceOps 2016 Conference, Daejon, Korea (2016)

Johnston, M.D., Levesque, M., Malhotra, S., Tran, D., Verma, R., Zendejas, S.: NASA deep space network: automation improvements in the follow-the-Sun Era. In: 24th International Joint Conference on Artificial Intelligence, Buenos Aires, Argentina (2015)

Fletcher, K.A., Bedwell, W.B.: Cognitive Aids: design suggestions for the medical field. In: Proceedings of the International Symposium on Human Factors and Ergonomics in Health Care, vol. 3, no. 1, pp. 148–152 (2014)

Arriaga, A., Bader, A., Wong, J., Lipsitz, S., Berry, W., et al.: Simulation-based trial of surgical-crisis checklists. New Engl. J. Med. 368(3), 246–253 (2013)

Van de Merwe, K., Oprins, E., Eriksson, F., Van der Plaat, A.: The influence of automation support on performance, workload, and situation awareness of air traffic controllers. Int. J. Aviat. Psychol. 22(2), 120–143 (2012)

Hart, S.G., Staveland, L.E.: Development of NASA-TLX (task load index): results of empirical and theoretical research. Adv. Psychol. 52, 139–183 (1988)

Davis, F.: Perceived usefulness, perceived ease of use, and user acceptance of information technology. MIS Q. 13(3), 319–340 (1989)

Salas, E., Diazgranados, D., Rosen, M.A.: Expertise-based intuition and decision making in organizations. J. Manag. JOM 36(4), 941–973 (2010)

Dane, E., Pratt, M.G.: Exploring intuition and its role in managerial decision making. Acad. Manag. Rev. 32, 33–64 (2007)

Grier, R.: How High is High? A Meta-Analysis of NASA-TLX Global Workload Scores. Proc. Hum. Factors Ergon. Soc. Ann. Meet. 59(1), 1727–1731 (2015)

Sheridan, T.: Trustworthiness of command and control systems. IFAC Proc. Vol. 21(5), 427–431 (1988)

Analysis of Airline Pilots Subjective Feedback to Human Autonomy Teaming in a Reduced Crew Environment

Mathew Cover[1(✉)], Chris Reichlen[1], Michael Matessa[2],
and Thomas Schnell[1]

[1] University of Iowa, Iowa City, IA 52242, USA
mathew-cover@uiowa.edu
[2] Rockwell Collins, Cedar Rapids, IA 52498, USA

Abstract. Building upon several software tools developed in the last decade for ground-based flight-following operations, a new use for the software is being applied to the cockpit environment of NextGen operations. Supporting automation of several tasks and streamlining information flow to the pilot via a tablet interface, software and datalink capabilities (Controller-pilot data link communications, CPDLC) were evaluated in a multi-location study. This paper provides an overview of the project and analyzes the subjective feedback from line pilots participants who evaluated this new capability. Workload and situation awareness metrics were compared across the conditions and scenarios presented to the pilots in this study. Overall, the pilot participants were subjectively in favor of the Human Autonomy Teaming (HAT) software tool suite and procedures they experienced and utilized in this study of a NextGen environment.

Keywords: NextGen · Automation · Workload

1 Introduction

1.1 Project Background

This project has origins in the mid-2000's, when researchers started to focus on the development of distributed simulation tools that enabled interaction between ground controllers and aircrew in flight simulators over long distances. Research examining these interactions was accomplished by geographically distributed teams occupying multiple facilities, alleviating the need for participating researchers to be co-located [1]. The present study followed a similar approach by leveraging existing distributed simulation capabilities at the University of Iowa Operator Performance Lab (OPL) and the California State University Long Beach Center for Human Factors in Advanced Aeronautics Technologies (CHAAT). This federation of simulations involved inter-action of aircrew flying the Boeing 737-800 simulator at OPL with the simulation manager, confederate dispatcher, and air traffic controller at CHAAT [2]. The simulator at OPL is equipped with data recording capabilities that are specialized to assess new technologies, emphasizing aviation-specific human factors considerations. The ground station was based at California State University – Long Beach in the CHAAT lab.

S. Yamamoto and H. Mori (Eds.): HIMI 2018, LNCS 10905, pp. 359–368, 2018.
https://doi.org/10.1007/978-3-319-92046-7_31

1.2   Study Outline

The Human Autonomy Teaming (HAT) software tool suite, installed on a Surface Pro tablet in the aircraft simulator (Fig. 1), provided pilots with semi-automated electronic checklist, audio/voice cues and commands, as well as provided alternate airport destinations in cases of emergencies.

Fig. 1. Simulator configuration: Boeing 737-800 cockpit with HAT software running on surface pro located left of the captain seat

One of the features of the HAT software tool suite was a traffic display developed at NASA called the Traffic Situation Display (TSD). The TSD would allow a pilot to see their own aircraft with respect to nearby airports and other traffic in the area. A control panel to the left of the TSD would allow a pilot to zoom in and out as well as pan around the display to view other regions as necessary to facilitate any required or desired planning by the pilot. A screenshot of the tablet on the TSD page is shown in Fig. 2.

A large portion of the HAT tool visual interface consisted of software developed by NASA called the Autonomous Constrained Flight Planner (ACFP). This tool was previously developed and tested in ground-based flight-following operations for controllers on the ground supporting multiple aircraft under their watch [3]. The ACFP in the HAT condition provided input for the pilot to make adjustments to criteria of how the algorithm selects alternate airports/runways. These 'weights' are located on sliders as seen in Fig. 3. A pilot can move the weights along the slider controls to adjust what criteria should receive priority when the automation selects top alternate airport candidates to present to the pilot.

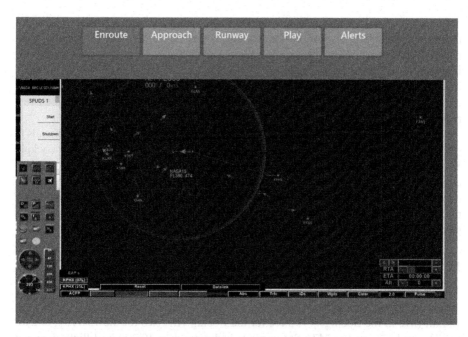

Fig. 2. Traffic Situation Display (TSD)

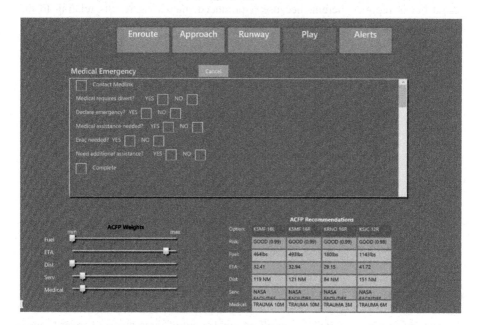

Fig. 3. Slider controls can be found in the bottom left corner of this screenshot, allowing the pilot to adjust the importance of different criteria in ACFP

The automated checklist feature of HAT supported the study aircrew in accomplishing pre-selected tasks for off-nominal events that would normally be accomplished manually. One goal of automating these tasks is to decrease the overall workload of the flight crew, allowing more attention to be devoted to aircraft control. A potential long-term outcome of transitioning tasks, normally accomplished by human crewmembers, to automation may be a reduction in the number of crew members required in the cockpit.

The HAT tool also provided the pilots with a mechanism to use voice commands to navigate the application, thus freeing their hands to perform other tasks in the cockpit. Since voice interaction is not a standardized feature in cockpits today, a laminated quick reference card was provided to pilots to use as reference should they need to consult it at any time during the simulations. This voice command feature enabled the pilot to call up approach plates and airport maps of various destinations as well as switch between screens, or tabs.

1.3   Participant Demographics

Twelve pilots participated in the study. All twelve were active (not retired) and at the time were flying for a major US airline. All had current qualification type in Boeing aircraft (five in the Boeing 777, three for the Boeing 787, two for the 757/767, one for the 737, and one for the 747). Seven pilots had more than 10,000 h of flight time and five had less than 10,000 h. Pilot flight experience was not a part of the experimental design but is reported herein, because it produced interesting results relating to the impact of experience (see Sect. 3) (Fig. 4).

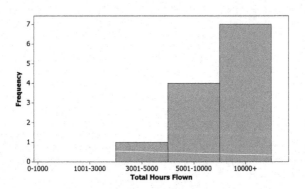

Fig. 4. Total hours flown as line pilot

2   Scenarios and Conditions

A variety of scenarios containing an off-nominal event of three levels of severity were designed to stimulate the pilots to use the HAT tool and procedures. The three levels of severity were: Light (L), Moderate (M), and Severe (S). A table enumerating the six different non-normal events and their designated level of severity are shown in Table 1 below. Typically, a scenario lasted for 15–20 min, although depending on pilot action, a few went longer than 30 min.

Table 1. Off-nominal events

Severity level	Off-nominal event
Light – 1 (L1)	Weather radar failure
Light – 2 (L2)	Anti-skid inoperable
Moderate – 1 (M1)	Windshield overheat
Moderate – 2 (M2)	Airport weather
Severe – 1 (S1)	Wheel well fire
Severe – 2 (S2)	Medical emergency

The scenarios were varied such that no pilot saw the same scenario more than once. Additionally, the scenarios were counterbalanced so that all appeared an equal number of times, both with and without the HAT tool to allow comparisons of subjective workload and situation awareness ratings. Each scenario appeared in the HAT and No-HAT conditions six times. Pilot participants experienced all six scenarios over the course of their day in the simulator (three in the morning and three in the afternoon). For each pilot, the scenarios were grouped by HAT condition, meaning they completed the three HAT and the three No-HAT scenarios consecutively (i.e. there was no alternating between conditions).

Pilots 1, 3, 5, 7, 9, and 11 completed three scenarios with the HAT condition first and then three scenarios without HAT. Pilots 2, 4, 6, 8, 10, and 12 completed three scenarios without HAT first and then three scenarios with HAT. This alternating order between subjects was designed help minimize impact order and learning effects. The experimental matrix for the study can be found in Fig. 5.

HAT First

PPT1	HAT			NoHAT		
	L1	M1	S1	M2	S2	L2
PPT3	HAT			No HAT		
	L2	M2	S2	S1	L1	M1
PPT5	HAT			No HAT		
	M1	S1	L1	S2	L2	M2
PPT7	HAT			No HAT		
	M2	S2	L2	L1	M1	S1
PPT9	HAT			NoHAT		
	S1	L1	M1	L2	M2	S2
PPT11	HAT			No HAT		
	S2	L2	M2	M1	S1	L1

No HAT First

PPT2	NoHAT			HAT		
	M1	S1	L1	L2	M2	S2
PPT4	NoHAT			HAT		
	S2	L2	M2	L1	M1	S1
PPT6	NoHAT			HAT		
	S1	L1	M1	M2	S2	L2
PPT8	NoHAT			HAT		
	L2	M2	S2	M1	S1	L1
PPT10	NoHAT			HAT		
	L1	M1	S1	S2	L2	M2
PPT12	NoHAT			HAT		
	M2	S2	L2	S1	L1	M1

Fig. 5. Experimental run matrix for all 12 pilot participants (PPT) in this study

3 Subjective Responses

For every simulator scenario, pilots completed a NASA Task-Load Index (TLX) [4] and 10-Dimensional Situation Awareness Rating Technique (SART) [5] questionnaires to capture their subjective workload and situational awareness ratings. Questionnaires were administered with paper and pen immediately following scenario completion.

A General Linear Model (GLM) Analysis of Variance (ANOVA) was performed on the TLX and SART scores to investigate effects of condition and scenario on these variables. Condition (HAT vs. No HAT) had no significant effect on TLX ratings ($F_{1,54} = 2.19$, p = 0.145) or SART scores ($F_{1,53} = 0.23$, p = 0.630), while scenario (L1-2, M1-2, S1-2) showed a significant main effect on TLX ratings ($F_{5,54} = 7.24$, p < 0.005). Post-hoc Tukey pairwise comparisons showed that the L1 scenario produced significantly lower scores than the S1 (t = 4.284, p < 0.005) and S2 (t = 3.269, p = 0.022) scenarios and the L2 scenario produced significantly lower scores S1 scenario (t = 3.221, p = 0.025). The M1 scenario produced lower scores than the S1 (t = 4.622, p < 0.005) and S2 (t = 3.608, p = 0.008) scenarios, while the M2 scenario produced lower scores than the S1 (t = 3.978, p < 0.005) and S2 (t = 2.963, p = 0.049) scenarios. The effect of scenario was not of practical significance, as this was by design. It did, however, validate that scenario difficulty level generally met the intent of the experiment.

As discussed previously, post-hoc analysis suggested experience was an important consideration in interpreting the results. Splitting the pilots into two groups, there were five pilots who had fewer than 10,000 h and seven pilots who had more than 10,000 h of total flight time. A review of the data using cumulative histograms indicated that workload and situation awareness may have shown more variation resulting from the presence or absence of HAT when considering overall flying experience. Due to the imbalance of experience levels, a non-repeated measures GLM ANOVA, considering the effects of condition and scenario, was performed to investigate the impact of total flying time on TLX and SART scores. Results indicated that total flying time had a significant effect on TLX score ($F_{1,64} = 5.22$, p = 0.026). A post-hoc Tukey pairwise comparison showed that the group with higher total flying time had significantly lower scores overall (t = −2.285, p = 0.0257). The ANOVA did not show a significant effect of total flying time on SART score ($F_{1,63} = 1.98$, p = 0.165).

Figures 6, 7, 8 and 9 show boxplots and empirical CDFs of SART and TLX scores broken down by condition and experience groups. A qualitative examination of these

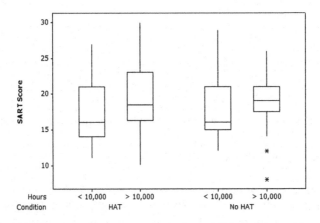

Fig. 6. SART scores broken down by condition (HAT vs. No HAT) and number of hours flown (more than vs. less than 10,000 h)

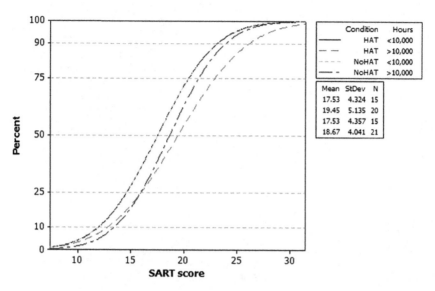

Fig. 7. Empirical CDF of SART scores broken down by condition (HAT vs. No HAT) and number of hours flown (more than vs. less than 10,000 h)

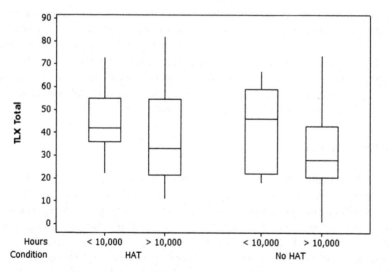

Fig. 8. TLX scores broken down by condition (HAT vs. No HAT) and number of hours flown (more than vs. less than 10,000 h)

graphs indicates several interesting findings. In Figs. 6 and 7, the differences in SART are slightly more prevalent in the high total flight time group than the low flight time group. The HAT condition, within the higher flight time group, produced the highest SART scores. As Fig. 7 shows, this difference is most evident in the higher percentiles of SART score (75th and above).

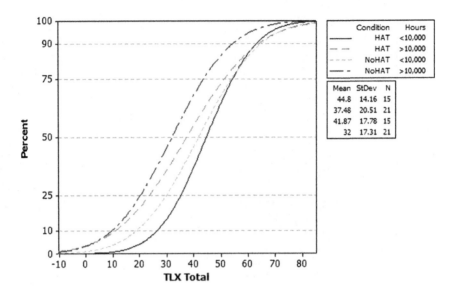

Fig. 9. Empirical CDF of TLX total score broken down by condition (HAT vs. No HAT) and number of hours flown (more than vs. less than 10,000 h)

Workload using the NASATLX (Figs. 8 and 9), on the other hand, indicated different effects of condition on the two experience groups. Overall, TLX scores were lower for the group with higher flight time. Within this group, TLX scores were highest in the HAT condition. The separation was greatest when TLX scores were highest (i.e. 75th – 90th percentiles). Within the low experience group, the HAT condition also produced generally higher TLX scores. Contrary to the group with higher flight time, there was greater separation in the lower percentiles (10th – 25th).

4 Questionnaire

Pilots completed additional questionnaires (developed by the research team) after both HAT and non-HAT scenarios designed to elicit information from them regarding various workload/understanding aspects of the tool and procedures. There was a number scale that inquired about specific components and overall feelings about the ACFP tool within HAT. Pilots circled the number on the scale where they felt most comfortable in agreement with those statements listed in the questionnaire.

There was also a final questionnaire administered after both conditions of scenarios had been experienced by the pilots. Several of these questions asked for a preference of the condition when considering the six different scenarios that were experienced by the pilot in the simulator. To determine whether the preferences for HAT or No-HAT were statistically significant from no preference, we ran single sample t tests against the mean value of 5 (No Preference). The preference was overwhelming in favor of the HAT condition (present) in four of the five questions asked. In the remaining question (Q2, about situational awareness), the difference was non-significant. The table highlighting these questions can be found in Table 2 and Fig. 10.

Table 2. Statistical significance of Questions 1–5 of the final questionnaire comparing HAT and no HAT conditions

Question	Mean (SD)	p
Q1: With regard to keeping up with operationally important tasks, I preferred completing the tasks in	7.3 (2.1)	0.0031
Q2: My situational awareness was better with	6.6 (2.7)	~
Q3: My workload was lower with	7.3 (1.5)	0.0002
Q4: My ability to integrate information from a variety of sources was better with	7.7 (1.6)	<0.0001
Q5: My overall efficiency was better with	6.9 (1.9)	0.0047

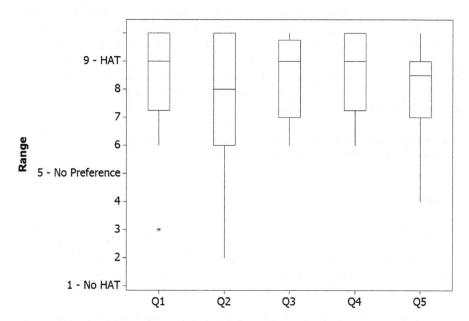

Fig. 10. Questions 1 through 5 responses on the final questionnaire across all subjects (N = 12)

5   Pilot Suggestions

Pilots also provided a number of suggestions for improvement of the HAT tool for future use. Pilot feedback was considered to be critically important as they are the end users of this tool. Among the suggestions pilots provided was increasing the reliability of the voice recognition to use natural language rather than a strict grammar set.

An implementation of the TSD screen with convective weather integrated was also a requested feature from the pilot participants. This would help paint a more comprehensive picture for the pilots to understate the situation they are in and why the ACFP may be providing the recommendations that it is. Not having an illustration of the convective weather would sometimes lead pilots into questioning why a proposed route/destination would avoid what would otherwise appear to be clear airspace.

There was also some interaction enhancements requested from pilots such as a pinch-to-zoom feature for the TSD as this is now commonplace among most tablet applications in use today. Another interaction request was to have the HAT tool suite automatically switch to the Approach or Runway tab if an airport chart was summoned by voice command and the display was not already on that tab.

6   Conclusions

Overall, the pilot's subjective feedback in the SART, TLX, and comments provided in questionnaires indicate that the HAT tool suite and procedures may be useful cockpit tools to supplement current day operations. Many pilots commented on its usefulness as a confirmation to actions that the pilots were going to take in several of the non-normal events. An analysis of the TLX and SART scores show that there isn't any significant detriment to workload or situational awareness with the presence of HAT in the cockpit. However, the effect of experience may warrant further evaluation. The subjective data analysis also suggests that pilots would welcome the software on an EFB or perhaps even certified avionics in the future.

Acknowledgements. This work was partially supported by the NASA Single Pilot Understanding through Distributed Simulation NRA #NNA14AB39C

References

1. Strybel, T.Z., Canton, R., Battiste, V., Johnson, W., Vu, K.P.L.: Recommendations for conducting real-time human-in- the-loop simulations over the internet. In: Proceedings of the Human Factors and Ergonomics Society Annual Meeting, vol. 50, no. 1, pp. 146–150. Sage Publication, Los Angeles (2006)
2. Matessa, et al.: Using Distributed Simulation to Investigate Human-Autonomy Teaming (current issue)
3. Brandt, S.L., Lachter, J., Russell, R., Shively, R.J.: A human-autonomy teaming approach for a flight-following task. In: Baldwin, C. (ed.) AHFE 2017. AISC, vol. 586, pp. 12–22. Springer, Cham (2018). https://doi.org/10.1007/978-3-319-60642-2_2
4. Hart, S.G.: NASA-task load index (NASA-TLX); 20 years later. In: Proceedings of the Human Factors and Ergonomics Society Annual Meeting, October 2006, vol. 50, no. 9, pp. 904–908. Sage Publication, Los Angeles (2006)
5. Selcon, S.J., Taylor, R.M.: Evaluation of the situational awareness rating technique (SART) as a tool for aircrew systems design. In: AGARD, Situational Awareness in Aerospace Operations, 8 p. (1990). (SEE N 90-28972 23-53)

Integration of an Exocentric Orthogonal Coplanar 360 Degree Top View in a Head Worn See-Through Display Supporting Obstacle Awareness for Helicopter Operations

Lars Ebrecht[✉], Johannes M. Ernst, Hans-Ullrich Döhler,
and Sven Schmerwitz

German Aerospace Center (DLR), Institute of Flight Guidance,
Lilienthalplatz 7, 38108 Braunschweig, Germany
lars.ebrecht@dlr.de
http://www.dlr.de/fl/en

Abstract. The objective was the development of an HMI for helicopter obstacle awareness and warning systems in order to improve the situational and spatial awareness as well as the workload of helicopter pilots. The related work concerning obstacle awareness and warning systems, situational awareness, orthogonal coplanar and perspective representations plus previous work done by DLR was depicted and discussed. The two main aspects of the developed HMI concept were explained, i.e., the combination of the exocentric orthogonal coplanar top view with the egocentric perspective view, and secondly three ways for the integration of the obstacle awareness display inside a head-worn see-through display. The developed HMI concept was applied to two helicopter offshore operations and its specific obstacle situation. The first operation is a hoist operation at the lower access point of an offshore wind turbine. The second regards the landing operation on an offshore platform. From a technical point of view, especially concerning available sensor technologies, helicopter might be fitted with obstacle awareness systems in future. The HMI design is still under investigation in order to support the pilot in a holistic and balanced way.

Keywords: Situational awareness · Spatial awareness
2D/3D representations · Helicopter offshore operations
Human machine interface · Multimodal men machine interaction
Cockpit display systems · Augmented reality

1 Introduction

Helicopter offshore operations are conducted in rough environments under adverse conditions, e.g., strong winds or limited visual conditions. Under these

© Springer International Publishing AG, part of Springer Nature 2018
S. Yamamoto and H. Mori (Eds.): HIMI 2018, LNCS 10905, pp. 369–382, 2018.
https://doi.org/10.1007/978-3-319-92046-7_32

conditions helicopter pilots have to operate very close to obstacles, like wind turbine poles and blades, towers or cranes. Even though specific operations ensure safety as much as possible, hoisting near a wind turbine or landing on an offshore platform remain quite challenging operations.

Helicopter operations in the proximity of obstacles are more or less daily business for helicopter pilots, equal to on- or offshore. Nevertheless, obstacles represent a serious issue concerning operational hazards and accidents [1]:

"Operational Safety Issues:
Helicopter Obstacle See and Avoid: Obstacle collisions are the second most common accident outcome in this domain, making obstacle see and avoid one of the key safety issues. This involves the provision of the best equipment and strategies to help flight crew maintain safe clearance from obstacles during take-off and landing."

When operating in the vicinity of obstacles, pilots are forced to tackle different problematic aspects. Firstly, they have to estimate the distance to an obstacle, in particular between an obstacle and the helicopter main rotor. The situation becomes more difficult when obstacles get or already are in the back of a helicopter, out of pilot's field of view. This might end up in a tail rotor strike. Accordingly, pilots need a clear view and understanding of the overall obstacle situation, as well. Apart from that, pilots have to manage the helicopter systems, have to conduct specific procedures, and have to be prepared for degraded situations.

1.1 Objective

Some helicopters offer helmet mounted displays (HMD) or more generally head worn see-through displays. HWDs combine and relate additional computed information with the reality. HWDs consider the head orientation of the pilot, which offers various possibilities for the HMI creation. Hence, HWDs allow pilots keep looking out of the cockpit window, paying attention to the surroundings, especially when operating in the vicinity of obstacles. Nevertheless, pilot's head orientation and field of view is limited.

This contribution introduces a new display concept, which combines the native egocentric perspective with an exocentric orthogonal coplanar 360 degree top view in a head-worn see-through display (HWD). The egocentric perspective comprises the natural view of the pilot to surroundings plus the indication of the primary flight information. The primary flight information comprises all mandatory information, the pilot needs to continue flying any time, e.g., the present attitude, altitude, heading, air speed, ground speed and vertical speed of the helicopter. Accordingly, the exocentric orthogonal coplanar 360 degree top view has been added as inset beside the primary flight symbolics. Thus, pilots should be able to catch the overall obstacle situation as well as being able to determine the distances to each obstacle while looking out and keep flying.

Figure 1 depicts the DLR's primary flight information symbolics for HWD. Pilots see all the green drawn contents in addition to the reality [2]. The background shows the offshore wind park Alpha Ventus, which is located in the North Sea [3]. The research offshore platform FINO1 is placed in front of the picture. Figure 2 shows a HWD integrated in the DLR's simulation and evaluation environment, the Generic Experimental Cockpit (GECO). Outside the cockpit the offshore wind park Alpha Ventus is depicted, too.

Fig. 1. DLRs primary flight information symbolics for HWD [2] (Color figure online)

Fig. 2. See-through HWD in DLR's Generic Experimental Cockpit [4]

On the one hand, the extension of an exocentric top view to the egocentric perspective seems to be promising in order to combine the pros of perspective with the pros of a coplanar view. On the other hand, the combination of the top view with the egocentric view and furthermore in a see-through HWD may confuse, cause additional workload or attentional tunneling effects. Consequently, the main question related to the developed display concept is, if the display concept is suitable to provide situational and spatial awareness as well as reducing the workload of helicopter pilots operating in the vicinity of obstacles for safer operations.

1.2   Use Cases

The first evaluation of the developed HMI concept regards two target operations: a winching operation at the lower access point of an offshore wind turbine and an offshore platform landing.

Figure 3 depicts the situation of the first target operation, the hoist operation at the lower access point of an offshore wind turbine. Usually, offshore wind turbines have two access points, an upper one at the top, close to the wind turbine's head and blades plus a second one, at a lower level, a few meters above the sea surface. In the majority of cases, the upper one is used. Only if materials have to be lowered or if injured persons have to be picked up, the lower one is used. The lower access point bears two main challenges. The first results from the fact that pilots can not hover the helicopter straight above the target position. Due to the main rotor size, pilots have to keep a certain distance in order to avoid a collision of the main rotor with the wind turbine pole. The wind turbine rotor is stopped cross to the wind direction at this time. The second issue is that a helicopter pilot is not able to observe the wind turbine pole. Because winches being mounted on the side of a helicopter, pilots have to hover abeam the wind turbine looking to the side with approximately 90 degree offset to the cockpit and helicopter orientation.

Figure 4 illustrates the second target operation, landing on a fixed offshore platform. Similar to the previous case, helicopter pilots operate in the vicinity of high obstacles, like towers, cranes or wind turbine poles, e.g., on construction ships waiting for their installation. When approaching a landing position, the helicopter always faces the wind direction. The landing position and the obstacles have a fixed orientation. Hence, the landing position on an offshore platform may be either before, abeam or behind an obstacle (refer Fig. 4a–c) depending on the wind direction. Among these, case (c) is the most challenging one. When the obstacle comes behind the helicopter, it will be out of pilots field of view. Apart from that, offshore platforms can have more than one obstacle. Thus, in reality you have combinations of the depicted cases (a), (b) and (c). In comparison to that, in the first case the hover position is not fixed. The wind turbine is always abeam to the side of a helicopter. The position can be all around the wind turbine pole due to the fact that the lower access point offers a 360 degree access. Accordingly, the approach to the hoist position of use case one is always the same (equal to Fig. 4b).

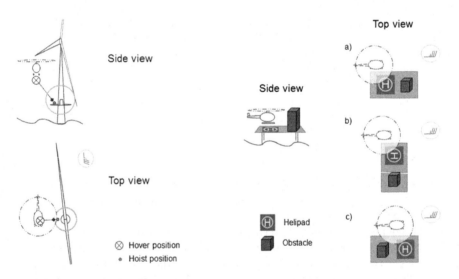

Fig. 3. Hoist operation at a wind turbine

Fig. 4. Offshore platform landing

1.3 Project Context

The work of the contribution represents an outcome of the project "Development of powerful and efficient Avionic-Platforms for Fixed and Rotary Wing Aircraft" (AVATAR). The joint project comprises industrial partners and research institutions. It is funded by the German Federal Ministry of Economics and Energy in the frame of the national Aeronautical Research Program V "LuFo V".

1.4 Contents

The following Sect. 2 describes related work, concerning helicopter obstacle awareness and warning systems, the pros and cons of orthogonal and perspective representations as well as preliminary work done by DLR. Section 3 introduces the two main aspects of the developed HMI concept, this means the extension of the egocentric perspective view by an exocentric orthogonal view plus the integration of the 360 degree top view in the HWD. Section 4 depicts the developed HMI concept regarding the two target operations. Section 5 comprises the conclusions.

2 Related Work

Present helicopters are equipped with several flight information and warning systems. So far, obstacle awareness and warning systems (OAWS) do not belong to the helicopter equipment. As described in the previous section, it is highly desirable assisting helicopter pilots with a system detecting obstacles nearby. Furthermore, an OAWS should provide a proper HMI in order to support pilot's situational and spatial awareness.

2.1 Helicopter Obstacle Awareness and Warning Systems

Until now, two promising helicopter OAWS had been prototyped and evaluated, one by Agusta Westland and the other by Airbus Helicopters. In 2014 M. Brunetti from Agusta Westland presented a novel obstacle proximity Lidar system (OLPS) [5]. The system used three Lidar sensors detecting obstacles 360 degrees around the helicopter, with a range of up to 25 m. Each of the three sensors covered approximately 210 degrees with an accuracy of circa 10 cm and 0.25 degrees. The sensors had been mounted below the helicopter main rotor, one in front looking ahead and the other two with 120 degrees to the left and right. One year after Waanders et al. from Airbus Helicopters presented a competitive Rotor Strike Alerting System (RSAS) [6]. This system used commercial of-the-shelf radar sensors from the automotive domain. Four sensors covered 360 degrees. Each sensor covered 100 degrees with an accuracy of 10 cm and less than 10 degrees azimuth up to 80 m. Each sensor was mounted below the main rotor every 90 degrees beginning at 45 degree offset from the helicopter front. Both sensor systems had been evidenced being able to detect obstacles properly in order to realize an OAWS. Hence, from a technological point of view, helicopters might be fitted with sensors enabling an OAWS in future.

2.2   Human Machine Interface of Obstacle Awareness and Warning Systems

The mentioned OAWS offer a display concept for head down displays (HDD), presented on one of the panel mounted cockpit displays or on an additional display beside the cockpit displays. Both display concepts used an orthogonal coplanar 360 degree top view. Agusta Westland developed a 360 degree × "25 m" polar grid [5], while Airbus Helicopters implemented three concentric circles representing three distances and alerting levels [6]. The latter display concept offered a 360 degree circle divided by sectors with 9 degrees. A more or less filled sector indicated the distance to the shortest detected obstacle. According to the three mentioned alerting levels, the color green, yellow or red were used, if a distance is below the alerting levels, i.e., 35 m, 15 m and 5 m to the helicopter main or tail rotor. In comparison to Airbus Helicopters, Agusta Westland draw the detected obstacle outline over the grid. The polar grid of Agusta Westland's display concept used 15 m, 10 m and 5 m distance circles in relation to the helicopter main and tail rotor. As a special feature, Agusta Westland used a beam highlighting the shortest obstacle distance and direction.

Beside the panel mounted head down display concepts of the aforementioned OAWS, Agusta Westland and Airbus Helicopters applied a multimodal HMI using audio. Agusta Westland implemented a variable frequency tone and vocal announcements, i.e., warning and caution, while Airbus Helicopters evaluated discrete tones, indicating the distance to the closest obstacle. Apart from that, research does also investigate other human senses as HMI, e.g., audio or tactile cues. For instance, you can design a multimodal HMI using audio in addition to or instead of any display in order to put the pilots attention to the closest obstacle [7]. You can also imagine tactile cues in addition or instead of visual information presentation, e.g., soft stops, vibrations or directed ticks to the helicopter controls, in order to prevent pilots flying further in the direction of an obstacle [8]. However, so far multimodal HMIs seems to be best practice to emphasize one information, for example, relating to the most critical obstacle.

From the authors' point view, the best way to gain the overall spatial obstacle situation on the pilots side, seems to be a visual figure. Hence, the authors are looking forward to evaluate the potentials of head worn display concepts, as a basic visual display concept as OAWS HMI. Anyway, further investigation concerning multimodal HMIs, providing an optimal holistic balanced HMI for OAWS, will follow.

2.3   Situational Awareness

Situational awareness means the understanding of what happened, what happens plus what may happen based on the consideration of all corresponding aspects [9]. Situational awareness comprises the perception of information, understanding and processing the perceived information in order to tackle the current and next situation [10]. Further, situational awareness is the product of different subitems, i.e., system and mode, operational, task, and spatial awareness [11,12].

The system and mode awareness addresses the understanding of features, functionality and behavior of the technical system in use. It also includes the awareness concerning the level of automation, i.e., knowing the available modes as well as being aware of the present mode of operation of any assistance system. In this context for instance, the flight attitude, the mode of the autopilot, the hover or landing assistance and the obstacle awareness and warning system. The operational and task awareness comprise all procedures as well as corresponding actions that have to be conducted by the pilot, e.g., the hoist procedure or the landing procedure for offshore platforms. The spatial awareness considers close and far surroundings. In the present case, the flight path, the sea surface, wind orientation, humidity and sight, wind turbines, installation ships or offshore platforms.

Even though the presented display concept for HWD primarily addresses the spatial awareness, the other previous mentioned aspects of the situational awareness can not be omitted. Hence, one very important issue of the evaluation is, to evidence a proper and balanced generation of the spatial and situational awareness without attentional tunneling effects or an increased workload due to visual clutter.

2.4   Orthogonal vs. Perspective Representations

Regarding spatial awareness, one has to respect the properties of perspective and orthogonal coplanar representations regarding their effects to the visual perception.

Perspective representations depict the natural stereoscopic view of human beings. They fit very well to our three-dimensional mental model and imagination of surroundings. On the other side the perspective causes the line of sight ambiguity [13,14] and includes the depth compression [15]. Both effects prevent an accurate estimation of distances, heights and orientations, which represents a serious issue. In addition, human beings underestimate distances under good visual conditions and vice versa. The estimation of distances and heights can be assisted by the application of grids [16–18]. Nevertheless, the negative effects of perspective depictions cannot be compensated completely.

In comparison to that, an orthogonal, coplanar representation properly supports the determination of distances and orientations. Unfortunately, coplanar representations do not provide a proper three-dimensional picture. Terrain elevation, i.e., mountains and valleys, may be indicated using color codes. Heights of obstacles are textually annotated. You have to read and to compare numbers in order to know which obstacle has which height. A direct visual comparison per se of heights of different objects causes some efforts. Hence, perceiving heights is more or less difficult and matching coplanar representations with the reality as well. Due to the fact that coplanar representations by itself differ very much from the natural view, users have to become familiar with its usage.

In conclusion, orthogonal coplanar representations do not support a 3D mental model of surrounding objects properly and perspective views hinder the determination of distances and orientations. Consequently, neither a perspective view nor an orthogonal representation for itself will assist the spatial awareness sufficiently. Both views already are utilized in cockpits. Synthetic vision primary flight displays provide a perspective view and synthetic vision navigation displays an exocentric orthogonal coplanar top view (as 120 degree arc or 360 degree circle).

2.5   Previous Work

The very first step related to the presented concept was made by the development of so called "virtual aircraft-fixed cockpit instruments" (AFCI) [19,20]. The AFCI intend to provide a more flexible and customizable HMI using conventional cockpit instruments as virtual representatives in a HWD besides the primary HWD symbolics (see Fig. 5). The AFCI include a Primary Flight Display (PFD), the first of two insets of the lower half in Fig. 5, a Navigation Display, to the right of the PFD inset. Furthermore, an airport arrival chart is shown below the two aforementioned insets. The position of the insets are related to the helicopter airframe. Hence, if the pilot moves his head then the instruments remain on their airframe related position and may get out of the display area of the HWD. Further, the AFCI concept offered the possibility to change the placement and configuration of the depicted virtual instruments. After entering a changing mode, the pilot is able to use a cursor to grab one of the previously mentioned displays and replace or resize it. The cursor is bound to the center of the HWD and follows the head orientation (see Fig. 6).

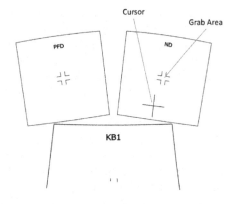

Fig. 5. Virtual aircraft-fixed cockpit instruments [19]

Fig. 6. AFCI interaction concept [19]

3   Method

The developed display concept comprises two key aspects. The first is related to the combination of the egocentric perspective view of a HWD with the exocentric orthogonal coplanar 360 degree top view. This mainly addresses the primary motivation, i.e., the generation of a proper situational and spatial awareness concerning obstacles close to helicopters plus the flight state. The second aspect regards the integration of the inset within the HWD beside the primary HWD symbolics, in other words the placement of the inset. This aspect concerns potential display clutter and the amount of displayed information.

3.1 Extension of the Egocentric Perspective View by an Exocentric Orthogonal View

According to the aforementioned explanations in Sects. 1.1 and 2.4, an orthogonal coplanar 360 degree top view has been implemented as inset beside the primary flight information symbolics in the egocentric view of the HMD (see Fig. 7). The inset shows the aircraft in the center of a compass rose. The compass rose contains a polar coordinate system with 12 sectors. Sector arcs depict the available space to surrounding obstacles within each sector. The space behind the obstacle is indicated by further sector arcs.

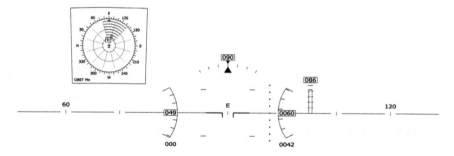

Fig. 7. Orthogonal coplanar 360 degree top view in a HWD depicting surrounding obstacles (Color figure online)

The display is comparable to the filled sectors of the Airbus Helicopters OAWS display. Alternatively, only one arc may be shown indicating the distance to an obstacle, but this might be insufficient to be perceptible by the pilot. Depending on the resolution and accuracy of the selected sensor system in use, it is also possible to draw the outlines of detected obstacles like the Agusta Westland obstacle display. However, providing too much details might not be helpful. Figure 7 depicts the implemented display concept as black-and-white image.

3.2 Integration of the 360 Degree Top View in an HWD

As shown before in Fig. 7, the obstacle situation display is placed beside the primary flight information symbolics. On the one hand, the display concept should provide a complete situational awareness, i.e., concerning all the different aspects as described in Sect. 2.3. On the other hand, in order to provide a balanced HMI it must be evaluated, how the obstacle situation is used by a pilot. This means, do pilots use it more or less simultaneously together with the primary flight information, i.e., switching frequently by eye movement in between or do pilots want to have more distance to the primary flight symbolics in order to be able to manage the amount of information as well as the focus on the different information moving their head.

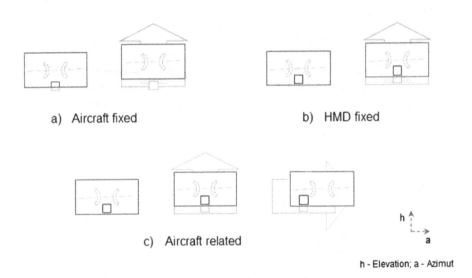

a) Aircraft fixed   b) HMD fixed

c) Aircraft related

h - Elevation; a - Azimut

Fig. 8. Integration concept for the 360 degree top view inset in the HWD

Hence, three different inset orientations had been implemented for the evaluation: an aircraft fixed, an HWD fixed and an aircraft related orientation. The first option implements a virtual aircraft fixed instrument. The inset is fixed to a certain position related to the helicopter airframe, like a real panel mounted cockpit instrument displayed on a LCD or like a head-up display. Depending on the head orientation, the inset can be inside or outside of the field of view and display area of the HWD (see Fig. 8a). In the second layout, the inset has also a fixed position. In comparison to the first layout, the inset is bound to the display area of the HWD. Accordingly, the inset remains visible all the time at same position inside the HWD, even if pilots move their head (refer to Fig. 8b). The third option represents a combination of the first and second option. The inset is located at a certain position related to the helicopter airframe, too. Whenever this position is within the display area of the HWD according to the head orientation, then the inset is displayed and remains at this position. Otherwise the inset is shown at the border of the HWD display area, pointing in the direction of the aircraft related position of the inset. Thus, pilots may focus on it as well as they are guided to it when changing the line of sight from the primary symbolics to the obstacle situation display (Fig. 8c).

4   Results

The concept is implemented using the programming language C and OpenGL and standard PC hardware. After the concept study, the implementation will be ported to an industrial hardware platform for future avionic systems. The display concept is applied to the helmet mounted see-through display JedEye^{TM} (see Fig. 2). The JedEye^{TM} is an industrial high performance prototype, developed by Elbit Systems Ltd. The main features of this monochrome green, binocular optical see-through HMD are its wide FOV (80 × 40 degree with 60 degree horizontal overlap) together with a high display resolution (2200 × 1200 px in total; 1920 × 1200 px per eye). The magnetic 400 Hz head tracker has an accuracy of 0.25 degree.

Figure 9 illustrates the obstacle situation and awareness display in use concerning the hoist operation at the lower access point of an offshore wind turbine. On the right side of the image, you see the base of the wind turbine AV2 [3]. In the background, other wind turbines of the offshore wind park Alpha Ventus and the offshore research platform FINO1 are visible. The scene is rendered by the game engine Unity [21]. The developed HWD symbolics superpose the scene. On the left, you see the exocentric orthogonal coplanar 360 degree view, enabling the pilot to observe the location and distance to the wind turbine pole, while the helicopter is approaching the target hover position abeam the wind turbine pole. The current flight state is displayed by 2D symbolics in the center within the egocentric perspective view.

Figure 10 shows the developed HMI concept applied to a landing on the research offshore platform FINO1 [22]. The helipad is depicted in front at the lower border of the picture. Furthermore, you see a lattice tower to the right. This tower is 101 m high above the sea and 5 meters beside the helipad border. The helipad is 25 m above the sea surface. Other wind turbines of Alpha Ventus are placed behind in the background. Similar to the previous case, the HWD symbolics overlay the scene. The obstacle is indicated in the inset in parallel to the landing information in the center.

The figures demonstrate the potentials of HWD as well as its drawbacks, e.g., the brightness and contrast of the symbolics to the background or the interaction of the symbolics with the outside world. As discussed, the intents and potentials of the developed display concept has to evidenced in a first concept study. This study is planned for the next months. Helicopter pilots will test the HWD HMI in combination with the two target operations in DLRs simulation and evaluation environment.

5   Conclusion

The contribution presented a new HMI concept for head-worn see-through displays featuring helicopter obstacle awareness and warning systems. The need of obstacle awareness and warning systems for helicopter operations as well as the need for the integration of an exocentric orthogonal coplanar 360 degree top view

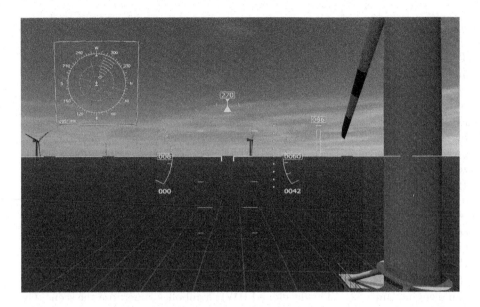

Fig. 9. HWD OAWS in case of conducting a hoist operation at the lower access point of an offshore wind turbine

Fig. 10. HWD OAWS in case of landing on the research offshore platform FINO1

beside the native perspective view in a head worn see-through display had been motivated. Besides, the background concerning helicopter obstacle awareness and warning systems, situational awareness and orthogonal as well as perspective representations has been pointed out and discussed. Three different options are presented in order to investigate the potential use of the integrated obstacle display beside the primary flight symbolics by the pilot. Finally, the concept was applied and presented to two helicopter offshore operations.

References

1. European Aviation Safety Agency (EASA): Annual Safety Review 2017 (2017). https://www.easa.europa.eu/document-library/general-publications/annual-safe ty-review-2017#group-easa-download
2. Doehler, H.-U.: Improving visual-conformal displays for helicopter guidance. SPIE Newsroom. International Society for Optics and Photonics (2013). https://doi.org/ 10.1117/2.1201310.005162. ISSN: 1818-2259
3. Offshore Wind Park Alpha Ventus (2018). https://www.alpha-ventus.de/english/
4. German Aerospace Center (DLR), Institute of Flight Guidance: Generic Experimental Cockpit (GECO) (2018). http://www.dlr.de/fl/desktopdefault.aspx/tabid-1964/1601_read-3009/
5. Brunetti, M.: The guardian project: reasons, concept and advantages of a novel obstacle proximity LIDAR system. In: Proceedings of the 40th European Rotorcraft Forum 2014, 2–5 September 2014, Southampton, UK (2014). ISBN: 9781510802568
6. Waanders, T., et al.: Helicopter rotorstrike alerting system. In: Proceedings of the 41st European Rotorcraft Forum 2015, 1–4 September 2015, Munich, Germany (2015). ISBN: 9781510819832
7. Niermann, C.A.: Potential of 3D audio as human-computer interface in future aircraft. In: Harris, D. (ed.) EPCE 2016. LNCS (LNAI), vol. 9736, pp. 429–438. Springer, Cham (2016). https://doi.org/10.1007/978-3-319-40030-3_42. ISBN 978-331940029-7. ISSN 03029743
8. Muellhaeuser, M.: Tactile cuing with active cyclic stick for helicopter obstacle avoidance: development and pilot acceptance. CEAS Aeronaut. J. (2017). https://doi.org/10.1007/s13272-017-0271-2. ISSN 1869-5582
9. Endsley, M.R.: Toward a theory of situation awareness in dynamic systems. Hum. Factor **37**(1), 32–64 (1995)
10. Endsley, M.R., Jones, D.J.: Designing for Situation Awareness: An Approach to User-Centered Design, 2nd edn. CRC Press, Boca Raton (2004)
11. Wickens, C.D.: Situation awareness and workload in aviation. Curr. Dir. Psychol. Sci. **11**(4), 128–133 (2002)
12. Wickens, C.D. Spatial awareness biases. Technical report No. ARL-02-6/NASA-02-4, University of Illinois, Aviation Research Lab (2002)
13. Wickens, C.D.: The when and how of using 2-D and 3-D displays for operational tasks. Proc. Hum. Factors Ergon. Soc. Annu. Meet. **44**(21), 3–403 (2000)
14. Wickens, C.D., Thomas, L.C., Young, R.: Frames of reference for the display of battlefield information: judgment-display dependencies. Hum. Factors: J. Hum. Factors Ergon. Soc. **42**(4), 660–675 (2000)

15. Wickens, C.D., Hollands, J.G., Banbury, S., Parasuraman, R.: Engineering Psychology and Human Performance, 4th edn. Pearson Education, Upper Saddle River (2013)

16. Domini, F., Shah, R., Caudek, C.: Do we perceive a flattened world on the monitor screen? Acta Psychol. **138**(3), 359–366 (2011)

17. Bolton, M.L., Bass, E.J.: Using relative position and temporal judgments to identify biases in spatial awareness for synthetic vision systems. Int. J. Aviat. Psychol. **18**(2), 183–206 (2008)

18. Bolton, M.L., Bass, E.J., Comstock, J.R.: Using relative position and temporal judgments to assess the effects of texture and field of view on spatial awareness for synthetic vision systems displays. Proc. Hum. Factors Soc. **50**(1), 71–75 (2006)

19. Doehler, H.-U., Ernst, J.M., Lueken, T.: Virtual aircraft-fixed cockpit instruments. In: Degraded Visual Environments: Enhanced, Synthetic, and External Vision Solutions 2015, vol. 9471, p. 94710E (2015)

20. Ernst, J.M., Doehler, H.-U., Schmerwitz, S.: A concept for a virtual flight deck shown on an HMD. In: ISI/SCOPUS, vol. 9839(10), Seiten 983909-1 (2016). SPIE Press. SPIE Defense + Commercial Sensing 2016, 17–21 April 2016, Baltimore. https://doi.org/10.1117/12.2224933

21. Unity Technologies (2018). https://unity3d.com/

22. Offshore Research Platform FINO1 (2018). http://www.fino1.de/en/

Evaluating User Interfaces Supporting Change Detection in Aerial Images and Aerial Image Sequences

Jutta Hild[1(✉)], Günter Saur[1], Patrick Petersen[2], Michael Voit[1],
Elisabeth Peinsipp-Byma[1], and Jürgen Beyerer[1,2]

[1] Fraunhofer Institute of Optronics, System Technologies and Image
Exploitation (IOSB), Karlsruhe, Germany
jutta.hild@iosb.fraunhofer.de
[2] Vision and Fusion Laboratory, Karlsruhe Institute of Technology (KIT),
Karlsruhe, Germany

Abstract. Change detection in images taken from the same scene at different times is an important subtask in domains like remote sensing, medical diagnosis, or video surveillance. As human attention is limited, support by computing systems might be beneficial. In this contribution, the benefit of optimized image presentation and the availability of a change mask computed by an automated change detection algorithm is evaluated. In a user study, twelve participants performed change detection in different types of aerial images and aerial image sequences, using parallel side-by-side or alternating flicker image presentation, and performing with and without a change mask. The results show better change detection performance (higher hit rates, shorter completion time, less perceived workload) using the alternating flicker image presentation for the large majority of data sets. With an automated change mask available, the participants' hit rates increase even more, up to 95% for image pairs and up to 84% for image sequence pairs.

Keywords: Change detection · Human observer · Change mask
User interface · User study

1 Introduction

Image interpretation occurs in various domains, for example in remote sensing, video surveillance, radiology, or driver assistance systems. Detecting changes in images taken from the same scene at different points of time is an important subtask. In remote sensing, aerial photos of the same geographical region are compared, for example in order to detect changes in urban development [1, 2]. In radiology, MRI images of the same body region are compared, for example, to detect image changes for disease assessment [3, 4]. In dynamic environments, driver assistance systems must detect critical changes of environments [5], or in video surveillance [6], changes with respect to a certain static background have to be detected.

A human observer is able to perform such tasks quite well due to both visual and cognitive capabilities [7]. Particularly, the latter allows assessing visually distinguished

© Springer International Publishing AG, part of Springer Nature 2018
S. Yamamoto and H. Mori (Eds.): HIMI 2018, LNCS 10905, pp. 383–402, 2018.
https://doi.org/10.1007/978-3-319-92046-7_33

differences in the images as task-relevant changes due to the observer's expert knowledge. However, change blindness – not to notice that a change occurred – is a perceptual phenomenon indicating that human attention is limited [8]. Hence, support for the human observer is required and might be very beneficial. However, designing appropriate support might be challenging as stated for example by Parasuraman et al. [9].

In this contribution, two approaches for support of the human observer are investigated. One is optimized image presentation, aiming to make change perception for the human observer as easy as possible. The other one is assistance by automated change detection algorithms providing a change mask visualizing all automatically detected changes. In a user study, we evaluate the impact of the two approaches on human change detection performance when analyzing either aerial images or aerial image sequences. The remainder of this chapter describes characteristics of the two approaches and their consideration in the user study. Section 2 describes the methodology of the user study, Sect. 3 provides the results, and Sect. 4 the conclusion.

1.1 Image Presentation for Easy Perception of Changes

When doing change detection, the human observer scans the images serially, and systematically checks locations for task-relevant changes. The straightforward arrangement of two images is aligning them side-by-side, which is possible either in a printed version on a physical desktop, or on a screen. Using such parallel presentation requires repositioning of the observer's eyes for each comparison at least once. Commercial systems like, for example, ERDAS Imagine provides side-by-side parallel presentation, in the following named "Parallel". However, it also provides presenting the images alternately, aligned at the same position on screen, in the following named "Flicker". Using Flicker presentation, the comparison of corresponding image areas does not require repositioning of the observer's eyes. Furthermore, with aligned images, Flicker induces apparent motion for changes. As a result, new or vanishing objects flash and, hence, may be easier to detect; but numerous (flashing) changes might overload the observer's visual system. In order to get insight on human performance using Parallel or Flicker image presentation both are evaluated in the user study.

1.2 Automated Change Detection in Aerial Imagery

Providing automated change detection is the first of four design principles St. John and Smallman proposed in order to support a human observer "maintaining and recovering situation awareness" [10]. In recent years, there has been much progress in this research domain [11]. Changes in aerial imagery from two flight missions (a previous and a current one) are detected automatically in three steps:

1. Selection of two images of the observed scene area.
2. Adaption of the image pair to each other geometrically (alignment) and radiometrically.
3. Detection of changes by computing a change mask.

The main challenge consists in distinguishing between relevant changes for a certain task (e.g., persons, vehicles) and non-relevant changes (e.g., varying brightness, shadows, imaging artifacts, 3D parallax, and vegetation). In recent years, approaches of automated change detection applied to UAV video have been published which combine several image differencing methods. In this contribution, a change detection algorithm introduced by Saur and Krüger [12] is used. Using feature-based detection, directed change masks can be computed which distinguish between new and vanished objects. Such change masks might provide the human observer with a means for verification of their own change detection results. In order to get insight how change mask availability would influence human change detection performance, change masks are considered in the user study, too.

2 Methodology

Performing change detection, essentially, a binary classification task has to be accomplished. It is crucial to detect all task-relevant changes ("true positives"), and not to miss any of them ("false negatives"); furthermore, it is crucial not to include any not-task-relevant changes ("false positives"), but to disregard those as "true negatives". Besides avoiding errors, it is also of interest to complete a change detection task in a short time. Hence, in the user study, the task instruction for the participants is to detect and frame changes in the aerial images and aerial image sequences, respectively, as fast as possible. In the following, the experimental design is described in detail.

2.1 Independent and Dependent Variables

As outlined above in the introduction, the user study aims to determine the impact of the two user interface design aspects *Image presentation* and *Change mask* to human change detection performance. Hence, there are two 2-level factors (independent variables) modified during the experiment: *Image presentation* with levels *Parallel* and *Flicker*, and *Change mask* with levels *Provided* and *Not Provided*. Combination results in four experimental conditions to evaluate:

- Parallel + Change mask not provided (in the following abbreviated *Par*, Fig. 1)
- Parallel + Change mask provided (*ParCM*, Fig. 2)
- Flicker + Change mask not provided (*Fli*, Figs. 3 and 4)
- Flicker + Change mask provided (*FliCM*, Fig. 5)

Figures 1, 2, 3, 4 and 5 show for one exemplary aerial image pair how images are presented using the experimental system (see Sect. 2.4). Flicker presentation *Fli* is provided by displaying the image of Figs. 3 and 4 alternately. Figure 5 shows the *FliCM* condition in an exemplary way only for Image 1.

Considered dependent variables describing effectiveness are *hit rate* (aka *recall*)

$$P(\text{Detected Change}|\text{Actual Change}) = \frac{t_p}{t_p + f_n} \tag{1}$$

Fig. 1. Screenshot example for *Par:* Parallel presentation of an aerial image pair, left: image 1, right: image 2.

Fig. 2. Screenshot example for *ParCM:* Parallel presentation with change mask provided on both images.

Fig. 3. Screenshot example for *Fli:* Flicker presentation of image 1 from the aerial image pair.

Fig. 4. Screenshot example for *Fli:* Flicker presentation of image 2 from the aerial image pair.

Fig. 5. Screenshot example for *FliCM:* Flicker presentation with change mask provided (exemplary for image 1 from the aerial image pair).

and *precision*

$$P(Actual\ change|Detected\ Change) = \frac{t_p}{t_p + f_p} \tag{2}$$

with t_p representing task-relevant changes correctly framed, f_n representing not-task-relevant changes correctly not framed, and f_p representing not-task-relevant changes incorrectly framed.

Further considered dependent variables are efficiency as *task completion time*, and user *workload* measured using the NASA-TLX questionnaire [13, 14].

2.2 Test Tasks

Visual Stimuli (Image Data) and Task Instructions for Single Trials. The four experimental conditions are evaluated using four different data sets. The three data sets A to C use pairs of aerial images; data set D uses pairs of aerial image sequences. In order to cover different image types occurring in image analysis, data sets A and B use orthophotos derived from data material purchased form the *Agency for Digitization, High-Speed Internet and Surveying (Bavaria, Germany)*, and data sets C and D use UAS imagery used by Saur and Krüger [12] for investigation of automated change detection. Furthermore, the data sets differ in terms of the amount of present not-task-relevant changes like shadows.

- Data set A, in the following named *Ortho,* contains pairs of orthophotos captured several years apart (Fig. 6); task instructions tell to frame changes observed for "vehicles", "grave changes in infrastructure and vegetation (new/missing buildings, streets, electricity pylons, pieces of forest, bank building)", and "structural modifications on buildings (roofs, solar panels)".
- Data set B, in the following named *OrthoFake,* consists of image pairs of an original orthophoto and a manually modified fake version (objects added/retouched, some image noise added; Fig. 7, also Fig. 1); task instructions tell to frame changes observed for "vehicles (cars, tractors, boats)", or "all object types (e.g., vehicles, containers, trees, hay bales, buildings)".
- Data set C, *UASimage,* contains UAS image pairs captured some minutes up to several hours apart (Fig. 8); task instructions tell to frame changes observed for "vehicles", "cars", "trucks", "persons", or "persons on the street".
- Data set D, *UASimSeq,* consists of pairs of UAS video sequences captured several hours apart (Fig. 9); all pairs have a duration of about 4 s. Task instructions told framing changes observed for "persons", or "parking vehicles/cars/trucks".

Table 1 shows the characteristics of the data sets including the numbers of task-relevant changes. The differing numbers of trials issued from the fact that there was less UAS image material than orthophoto material available. Furthermore, the image content of the UAS image material provided less options for changes which could be precisely phrased in a task instruction. In total, there were 68 test tasks, 58 single image pairs, and 10 image sequence pairs.

Fig. 6. Data set A *Ortho* image pair as provided in Parallel presentation by the experimental system. Task instruction: please frame observed large changes in buildings (answer: one roof, one solar panel, and two complete new/missing buildings).

Fig. 7. Data set B *OrthoFake* image pair. Task instruction: please frame observed changes in buildings and vehicles (answer: one new/missing building, two boats, one group of cars).

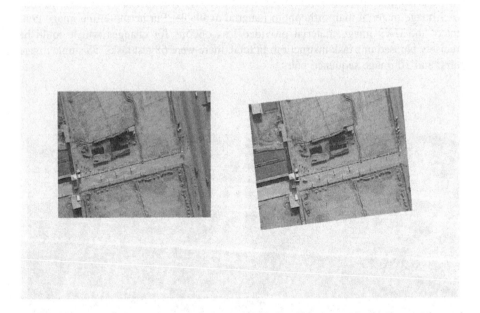

Fig. 8. Data set C *UASimage* image pair as provided in Parallel presentation by the experimental system. Task instruction: frame observed changes in (new/missing) parking cars (answer: five changes).

Presentation and Utilization of Change Mask. There are two ways the observer may include a change mask into their change detection procedure:

- Looking for changes autonomously and utilizing the change mask afterwards for confirmation of the determined result.
- Displaying the change mask from the beginning and verifying all locations, the change mask marks as changes.

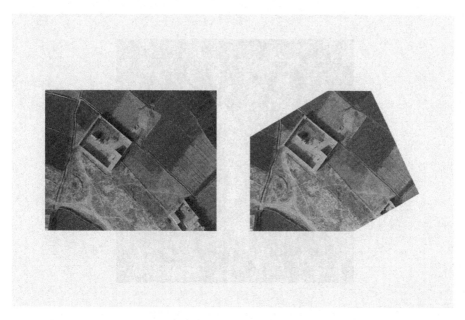

Fig. 9. Data set D *UASimSeq* image pair as provided in Parallel presentation by the experimental system. Task instruction: please frame observed changes in persons (answer: three changes, one of them currently visible).

Table 1. Data sets characteristics.

Data sets	Trials	Number of task-relevant changes			Amount of not-task-relevant changes
		Total	Single trials	Mean (SD)	
			Max	Min	
A *Ortho*	16	65	8	1	4 (2)	Plenty (shadows, vegetation)
B *OrthoFake*	24	132	10	2	5 (2)	Almost none (manually added noise only)
C *UASimage*	18	48	5	1	3 (1)	Few (shadows)
D *UASimSeq*	10	34	9	1	3 (3)	Several (shadows, moving objects)

While the second method might be useful in certain situations, the first method appears to be the more general method with the advantage of the observer considering the image data unprejudiced. Hence, this was the method applied in the user study.

For image pair data sets *OrthoFake* and *UASimage*[1], participants were advised to first carefully search the images for changes and frame them (Fig. 10). When feeling

---

[1] For data set *Ortho,* change masks provided very many false positives due to shadows and vegetation and were hence not included in the user study.

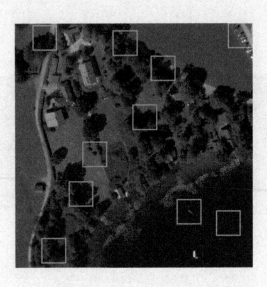

Fig. 10. *Fli* presentation of image 1 from the *OrthoFake* aerial image pair (see Fig. 3) with framed changes.

confident having found all changes or just having inspected the images thoroughly, they ought to check their results by superimposing the change mask using a functional key (see Fig. 11 for *FliCM*, Fig. 12 for *ParCM*).

For the image sequence data sets *UASimSeq*, using a functional key for superimposing a change mask to verify detected changes does not work as due to scene dynamics there is not enough time left. Hence, each image sequence is presented twice to the participants: the first time without change mask, and the second time immediately after with the change mask provided as a continuously displayed red transparent overlay (see Fig. 13 for *ParCM*, Figs. 14 and 15 for *FliCM*). Participants now use the change mask in the second way outlined above: they perceive the change mask's change suggestions and confirm them by framing, or ignore them. However, due to the short image sequence duration and the relatively few changes, participants may be able to recall what they just framed in the condition without change mask. Hence, they basically utilize the change mask for confirmation of their own result, again, similar to the first way outlined above.

The change mask for each image pair or image sequence pair was calculated using the algorithm described in Sect. 1.2 [12].

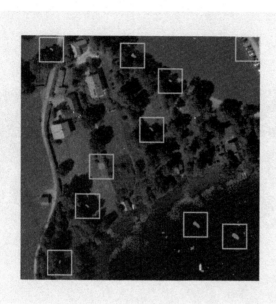

Fig. 11. *FliCM* presentation of *OrthoFake* aerial image 1 with framed changes and superimposed change mask (see Fig. 5).

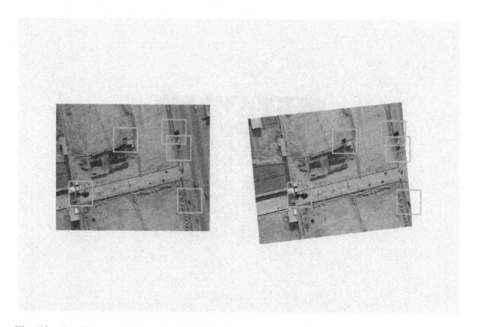

Fig. 12. *ParCM* presentation of a *UASimage* image pair with framed changes (Task instruction: changes in parking cars) and superimposed change mask (see Fig. 8).

Fig. 13. *ParCM* presentation of *UASimSeq* pair with superimposed change mask (currently one change visible, see Fig. 9) (Color figure online)

Fig. 14. *FliCM* presentation of image 1 of a *UASimSeq* pair (Fig. 9, left) with superimposed change mask and frame (currently one task-relevant change visible). (Color figure online)

Fig. 15. *FliCM* presentation of image 2 of the *UASimSeq* pair (Fig. 9, right) on top of image 1 with superimposed change mask (currently one task-relevant change visible). (Color figure online)

2.3   Participants

Twelve subjects (3 female, 9 male, average age 27 (SD 6) years), all non-expert image analysts, participated in the user study. All had normal or corrected to normal vision (4 using glasses, 1 using contract lenses).

All participants completed each type of test data set using both Parallel and Flicker image presentation. To ensure that each participant is presented only once with each pair of images/image sequences, test trials of each test data set were divided into two equal parts I and II. Thus, each participant completed both for Parallel and Flicker presentation 8 *Ortho* trials, 12 *OrthoFake* trials, 9 *UASimage* trials, and 5 *UASimSeq* trials. Participants performed the two presentation types immediately after another; within each data set, half of the participants first performed using Parallel, the other half started using Flicker. Six participants used test sets I with Parallel presentation, and test sets II with Flicker presentation, the other six participants used test sets II with Parallel, and test sets I with Flicker. Test data set order varied using a Latin Square, resulting in four different orders, each completed by three participants:

- *Ortho – OrthoFake – UASimage – UASimSeq*
- *OrthoFake – UASimage – UASimSeq – Ortho*
- *UASimage – UASimSeq – Ortho – OrthoFake*
- *UASimSeq – Ortho – OrthoFake – UASimage*

2.4   Apparatus

For presentation of the test tasks, an experimental system was implemented (Java application). It provides both Parallel and Flicker presentation for image pairs and for image sequence pairs. For presentation, a 24 in screen with a resolution of 1920×1200 pixels was used. Participants were sitting with a distance of 65 cm to the screen.

Parallel presentation displays the *Ortho* and *OrthoFake* images centered and covering (width x height) $20.9° \times 20.9°$ of visual angle; Flicker presentation is also centered and covers $23.5° \times 23.5°$. Switching between images can be performed manually using a functional key, however, in the user study, it was accomplished automatically every 600 ms. This decision was made to avoid additional manual load.

UASimage and *UASimSeq* pairs are displayed centered covering $17.1° \times 13.6°$ of visual angle for both Parallel and Flicker presentation. Image 1 is displayed aligned with the screen (see for example Figs. 12(left), 13(left), 14, and 15 where image 1 is partially covered by image 2). Image 2 is displayed geometrically adapted to image 1 (see Fig. 12 for Parallel, Fig. 16 for Flicker). For *UASimSeq* data, image 2 is in addition trimmed showing only the overlapping parts with image 1 (see Figs. 13(right) and 15). This was done as the dynamic input of the *UASimSeq* pairs is quite loading for the visual system of the observer, particularly, for Flicker with the automated image switching every 600 ms. Only the overlapping image parts are relevant for change detection, so trimming helps reducing the dynamics by cutting off irrelevant image parts. As all *UASimSeq* pairs were presented with both Flicker and Parallel, the trimmed version was used for both presentations.

The sizes of the target changes cover for *Ortho* between $0.5° \times 0.26°$ of visual angle for the smallest targets and $7° \times 3.5°$ for the largest targets; for *OrthoFake* sizes cover between $0.18° \times 0.18°$ and $2.2° \times 2.2°$; for *UASimage* as well as for *UASimSeq* sizes cover between $0.18° \times 0.18°$ to $2.2° \times 0.44°$.

Framing is accomplished by a point-select-operation using the gaze + key press interaction technique, performing a selection operation (NumPad-ENTER-key) on gaze pointing at the change location; frame size is $2.4° \times 2.4°$ as a compromise for all occurring target change sizes. This interaction technique is used as it had allowed significantly faster while equally accurate performance than mouse input for moving target selection (framing) in videos [15], which happens for the data set D *UASimSeq*. Gaze pointing was provided using a Tobii X60 remote eye-tracking device. The manufacturer reports a 60 Hz sampling rate, a typical accuracy of $0.5°$ (with head movements additional $0.2°$), and a head movement box of 44 cm $\times$ 22 cm (width height) at a distance of 70 cm from the monitor; hence, no chinrest was used (as for an observer in practice). In order to reduce the measurement uncertainty arising from technical issues of gaze estimation in the eye-tracker as well as from physiological issues of the human eye, the raw gaze data provided by the eye-tracker is processed using the algorithm introduced by Kumar et al. [16]. Additional evaluation of mouse input was dispensed as this would have exceeded the recommended session duration.

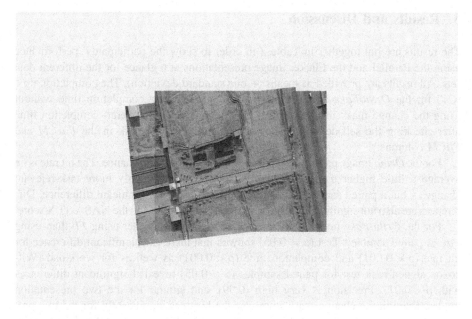

Fig. 16. Flicker presentation of *UASimage* pairs, image 1 aligned with screen, image 2 (currently underlying) geometrically adapted to image 1.

2.5   Procedure

After a short introduction about the character of the test tasks, the participants performed a standard 9-point calibration for the Tobii X60 eye-tracking device. The participants repeated calibration until the average measurement uncertainty on screen was 1° of visual angle or less.

After that, the participants completed the four test data sets with Parallel and Flicker presentation. Before each test data set, the participants completed a short training using an image (sequence) pair, which was not part of the test data to get familiar with the interaction. Each trial started with a 5-s display of the task instruction provided in large keys on the screen, followed by the display of the images or image sequences. In case of the image pairs, the participants first looked for changes, and afterwards checked their result superimposing the change mask using a functional key. In case of the image sequence pairs, they first performed the trial without change mask, and right after with change mask (cf. explanations in Sect. 2.2). After each test data set, the participant answered the NASA-TLX questionnaire in order to rate their subjectively perceived workload. They rated the six subscales mental demand, physical demand, temporal demand, performance, effort, and frustration on a 21-point rating scale (the best rating is 0, the worst rating is 100, the scale uses steps of five).

Finally, the participant answered the ISO 9241-9 questionnaire [17] assessing the gaze interaction used for change framing.

3   Results and Discussion

The results are put together in Table 2 in order to show the participants' performance using the Parallel and the Flicker image presentations at a glance for the different data sets. All results are provided as means (+ one standard deviation). The completion time (CT) for the *OrthoFake* and *UASimage* data sets give the completion time without using the change mask in the *Par* and *Fli* columns, and the overall completion time after checking the self-determined result using the change mask in the *ParCM* and *FliCM* columns.

For the *Ortho* image pairs, participants showed similar performance. The hit rate is on average a little higher using *Fli* – participants detected slightly more task-relevant changes – but a paired sample t-Test ($\alpha = 0.05$) showed no significant difference. Differences are also not significant for precision, completion time and the NASA-TLX score.

For the *OrthoFake* image pairs, participants performed better using *Fli* than using *Par*. A paired sample t-Test ($\alpha = 0.05$) showed that there is a significant difference for hit rate ($p < 0.001$) and completion time ($p < 0.001$) as well as for workload (Wilcoxon signed-rank test for paired samples ($\alpha = 0.05$) revealed significant differences with $p < 0.01$). Precision is very high (0.99), and similar for the two presentation modes. Providing a change mask improves the hit rates significantly for the two presentation modes. The improvement is particularly high when using Parallel presentation. However, the participants performed significantly better ($p < 0.01$) using *Fli* (Flicker without a change mask) than using *ParCM* (Parallel with a change mask). Using *FliCM*, the participants achieved a very high hit rate and precision with less than half of the completion time (on average) required using *ParCM*. The subjective workload was also much lower for Flicker presentation.

For the *UASimage* image pairs, participants also performed better using the Flicker presentation. Paired sample t-Tests ($\alpha = 0.05$) showed significant difference for hit rate (*Fli-Par* $p < 0.01$, *FliCM-ParCM* $p < 0.01$), for precision (*Fli-Par* $p < 0.05$), and for completion time ($p < 0.001$); a Wilcoxon signed-rank test for paired samples ($\alpha = 0.05$) revealed significant differences for perceived workload with $p < 0.01$. Providing a change mask improves the hit rates significantly for the two presentation modes (*Par-ParCM* $p < 0.001$, *Fli-FliCM* $p < 0.001$). Using *FliCM*, the participants achieved a very high hit rate and precision with a significantly shorter completion time and perceived workload.

For the *UASimSeq* image sequence pairs, participants performed best using *FliCM*. Looking at *FliCM* and *ParCM*, precision is significantly better for *FliCM* ($p < 0.05$); the difference for the hit rates is not significant ($p = 0.1089$), but better on average for *FliCM*. For both presentation types, the change mask conditions allowed significantly better hit rates (*Fli-FliCM* $p < 0.01$; *Par-ParCM* $p < 0.05$) and precision (*Fli-FliCM* $p < 0.001$; *Par-ParCM* $p < 0.001$). Considering workload, a Wilcoxon signed-rank test for paired samples ($\alpha = 0.05$) revealed significant differences for perceived workload with $p < 0.01$.

In summary, the participants achieved for the three datasets *OrthoFake*, *UASimage* and *UASimSeq* better performance using the Flicker presentation than using the Parallel presentation. They achieved not only higher hit rate and precision, but achieved those

Table 2. Results.

Data Set	Measures	*Par*	*Fli*	*ParCM*	*FliCM*
Ortho	Hit rate	0.80 (0.25)	0.85 (0.26)		
	Precision	0.87 (0.23)	0.83 (0.26)		
	CT in sec	50 (25)	50 (30)		
	Workload	36 (25)	34 (23)		
OrthoFake	Hit rate	0.45 (0.28)	0.88 (0.17)	0.82 (0.20)	0.94 (0.13)
	Precision	0.99 (0.05)	0.99 (0.04)	0.99 (0.04)	0.99 (0.04)
	CT in sec	48 (24)	21 (10)	71 (30)	31 (23)
	Workload	43 (28)	17 (17)		
UASimage	Hit rate	0.67 (0.40)	0.80 (0.37)	0.86 (0.31)	0.95 (0.16)
	Precision	0.83 (0.31)	0.90 (0.27)	0.87 (0.25)	0.90 (0.23)
	CT in sec	27 (18)	20 (14)	39 (23)	27 (18)
	Workload	35 (25)	21 (18)		
UASimSeq	Hit rate	0.51 (0.43)	0.56 (0.40)	0.76 (0.34)	0.84 (0.24)
	Precision	0.67 (0.43)	0.78 (0.37)	0.80 (0.30)	0.92 (0.20)
	Workload	60 (27)	49 (27)		

results in considerably shorter completion time and with less perceived workload. Providing a change mask improved hit rate and precision substantially for both presentation modes. Obviously, the less required gaze repositioning together with the apparent motion induced by the Flicker presentation had a beneficial impact on the overall performance.

For the data set *Ortho,* the participants performed similar with the two presentation modes, and the perceived workload was similar, too. As mentioned before in Table 1, the other three data sets comprised only few to several not-task-relevant changes; the *Ortho* data set comprised plenty not-task-relevant changes, making it more difficult to find the relevant changes within the plenty. Using the Flicker presentation, the not-task-relevant changes have a distracting impact on the observer's attention. Hence, the Flicker presentation might not have the same beneficial effect as for cases with less not-task-relevant changes. However, all participants voted for Flicker presentation as their favorite for all data sets.

Furthermore, all participants appreciated the change mask very much. They assessed the visual presentation (opaque red for single images, transparent red for image sequences) to be well perceivable and very beneficial for their overall change detection performance.

The participants also appreciated the gaze interaction technique gaze + key press for framing the changes. Figure 17 shows the results of the ISO 9241-411 questionnaire, where several features have to be rated on a 7-point-scale (1: very bad, 7 very good). All features got a good rating. The little lower value for accuracy might come from the fact that users are used to pixel-level mouse accuracy. However, eye-tracker accuracy is not on a pixel-level (see Sect. 2.4) due to measurement uncertainty of the eye-tracker, and in addition, framing targets in an image sequence is challenging and typically not possible on pixel-level either. Overall, gaze input might be an alternative for framing changes when performing change detection using a computer system.

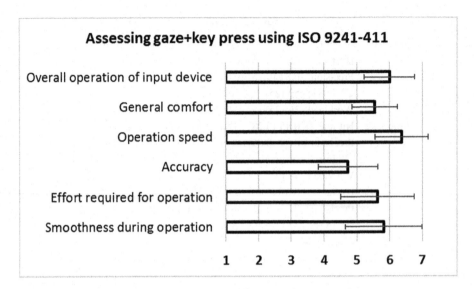

Fig. 17. Results of the assessment of the gaze input technique used for change framing.

4 Conclusion

Change detection in images taken from the same scene at different times is an important subtask in various domains. As human attention is limited, human observers require support in order to increase their overall change detection performance. The presented user study investigated how two system features would affect change detection performance:

- Image presentation (Flicker: alternating and aligned at the same position on the screen versus Parallel side-by-side), and
- Availability of change mask calculated for the image pair by an automated algorithm.

The results show that the twelve participants performed similar with Flicker and Parallel when looking for changes in aerial image pairs comprising plenty of not-task-relevant changes; average hit rates (precision) were 85% (83%) for Flicker versus 80% (87%) for Parallel.

For image pairs containing few to several not-task-relevant changes, the participants performed substantially better using Flicker presentation. They achieved significantly higher hit rates (up to 88%) and higher precision (up to 99%), they required substantially shorter completion time (up to 50 percent less on average), and they perceived less workload. Providing the participants in addition with a change mask calculated by an automated change detection algorithm improved the overall change detection performance again substantially, enabling hit rates for Flicker up to 95% on average (up to 86% for Parallel).

For image sequence pairs, the participants performed only slightly better using Flicker presentation if no change mask was provided; achieved hit rate was 56% on average (versus 51% for Parallel), precision was 78% (versus 67% for Parallel). Providing a change mask improved hit rate and precision again substantially: for Flicker, the hit rate improved to 84% on average (precision to 90%); for Parallel, the hit rate improved to 76% (precision to 80%).

To conclude, using Flicker image presentation and providing a change mask, participants achieved the best change detection performance, with a hit rate of up to 95% for image pairs and up to 84% for image sequence pairs.

References

1. Bouziani, M., Goïta, K., He, D.C.: Automatic change detection of buildings in urban environment from very high spatial resolution images using existing geodatabase and prior knowledge. ISPRS J. Photogram. Remote Sens. **65**(1), 143–153 (2010)
2. Leichtle, T., Geiß, C., Wurm, M., Lakes, T., Taubenböck, H.: Unsupervised change detection in VHR remote sensing imagery–an object-based clustering approach in a dynamic urban environment. Int. J. Appl. Earth Obs. Geoinf. **54**, 15–27 (2017)
3. Doi, K.: Diagnostic imaging over the last 50 years: research and development in medical imaging science and technology. Phys. Med. Biol. **51**(13), R5 (2006)
4. Bosc, M., Heitz, F., Armspach, J.P., Namer, I., Gounot, D., Rumbach, L.: Automatic change detection in multimodal serial MRI: application to multiple sclerosis lesion evolution. Neuroimage **20**(2), 643–656 (2003)
5. Fang, C.Y., Chen, S.W., Fuh, C.S.: Automatic change detection of driving environments in a vision-based driver assistance system. IEEE Trans. Neural Netw. **14**(3), 646–657 (2003)
6. Hodgetts, H.M., Vachon, F., Chamberland, C., Tremblay, S.: See no evil: cognitive challenges of security surveillance and monitoring. J. Appl. Res. Mem. Cognit. **6**(3), 230–243 (2017)
7. Spotorno, S., Faure, S.: Change detection in complex scenes: hemispheric contribution and the role of perceptual and semantic factors. Perception **40**(1), 5–22 (2011)
8. Styles, E.: The Psychology of Attention. Psychology Press, New York (2006)
9. Parasuraman, R., Cosenzo, K.A., De Visser, E.: Adaptive automation for human supervision of multiple uninhabited vehicles: effects on change detection, situation awareness, and mental workload. Mil. Psychol. **21**(2), 270 (2009)
10. John, M.S., Smallman, H.S.: Staying up to speed: four design principles for maintaining and recovering situation awareness. J. Cognit. Eng. Decis. Mak. **2**(2), 118–139 (2008)
11. Hussain, M., Chen, D., Cheng, A., Wei, H., Stanley, D.: Change detection from remotely sensed images: from pixel-based to object-based approaches. ISPRS J. Photogram. Remote Sens. **80**, 91–106 (2013)
12. Saur, G., Krüger, W.: Change detection in UAV video mosaics combining a feature based approach and extended image differencing. IPRS Int. Arch. Photogram. Remote Sens. Spat. Inf. Sci. **XLI-B7**, 557–562 (2016)
13. Hart, S.G.: NASA-task load index (NASA-TLX); 20 years later. In: Proceedings of the Human Factors and Ergonomics Society Annual Meeting, vol. 50, no. 9, pp. 904–908. Sage Publications, Los Angeles (2006)
14. NASA TLX Homepage. https://humansystems.arc.nasa.gov/groups/TLX/downloads/TLXScale.pdf. Accessed 05 Feb 2018

15. Hild, J., Kühnle, C., Beyerer, J.: Gaze-based moving target acquisition in real-time full motion video. In: Proceedings of the Ninth Biennial ACM Symposium on Eye Tracking Research & Applications, pp. 241–244. ACM, New York (2016)
16. Kumar, M., Klingner, J., Puranik, R., Winograd, T., Paepcke, A.: Improving the accuracy of gaze input for interaction. In: Proceedings of the 2008 Symposium on Eye Tracking Research & Applications, pp. 65–68. ACM, New York (2008)
17. ISO: 9241–411 Ergonomics of human-system interaction–Part 411: Evaluation methods for the design of physical input devices. International Organization for Standardization (2012)

Towards Autonomous Weapons Movement on an Aircraft Carrier: Autonomous Swarm Parking

James Hing[(⊠)], Kyle Hart, and Ari Goodman

Naval Air Warfare Center – Aircraft Division, Lakehurst, NJ, USA
{james.hing,kyle.m.hart,ari.b.goodman}@navy.mil

Abstract. To maintain weapons throughput on Nimitz class aircraft carriers, aviation ordnance is transported from the magazines to the flight deck and loaded onto aircraft in a process called "Strike Up". This time and labor intensive process requires multiple sailors to push weapon skids through the aircraft carrier to staging area on the flight deck.

Augmenting sailor tasking through the use of robotic equipment is one method to improve sortie generation rates (launching of aircraft), optimize manpower, and lower risk to sailors. Seen in Fig. 1, weapons skids spend extended time parked in various locations along their routes such as on elevators or in portions of hangar decks. This work presents improvements to the authors' previous work [1] to develop a Human Machine Interface (HMI) and the appropriate control methods toward supervisory control for parking multiple robotic skids in a cluttered and dynamic environment.

The HMI consists of four parts: (1) the user interface, (2) automated definition and conflict free assignment of parking goals within a user defined parking boundary, (3) collision free navigation with multiple differential drive robotic vehicles, and (4) an external infrastructure-free approach to localization and mapping.

Distribution Statement A – Approved for public release; distribution is unlimited, as submitted under NAVAIR Public Release Authorization YY-2018–28.

Keywords: Non-holonomic · Artificial potential fields · Autonomous parking

1 Introduction

High weapons throughput for Nimitz class aircraft carriers is one of many important steps to maintaining a sufficient rate of aircraft launches (sortie generation rates). To help maintain this throughput, aviation ordnance is transported from the magazines to the flight deck and loaded onto aircraft. This process, called "Strike Up", can take a significant amount of time as sailors push weapons skids through a circuitous route from the magazines to a staging area on the flight deck called the "Bomb Farm". Multiple sailors are needed to move a single heavy loaded skid, and as Fig. 1 shows, at any moment, there can be many skids on deck.

The aircraft carrier deck is not only busy with weapons movement, but it is also a dangerous environment. Sailors are continually completing tasks in close proximity to

This is a U.S. government work and its text is not subject to copyright protection in the United States; however, its text may be subject to foreign copyright protection 2018
S. Yamamoto and H. Mori (Eds.): HIMI 2018, LNCS 10905, pp. 403–418, 2018.
https://doi.org/10.1007/978-3-319-92046-7_34

Fig. 1. Weapon skids on elevator (Navy.mil Photos)

aircraft launching, recovering, and taxiing around the deck. The close proximity to aircraft puts anyone on deck at risk of bodily injury, whether it is from manual labor, loud noises, or accidents.

Optimizing and augmenting sailor tasking through the use of robotic equipment is one method of lowering the risk of bodily harm to the sailors. Specific to weapons transport, robotic transporters have been touted as a way of helping to improve the sortie generation rate and optimize manpower. An example advantage of an autonomous transport system would be for a single operator to control multiple skids all at once. In this context, we refer to multiple skids (more than three) as a swarm. In previous works on swarm control, little focus has been on the action of autonomously parking multiple systems in a particular formation. Seen in Fig. 1, weapon skids spend much time parked in various locations along their routes such as on elevators or in portions of hangar decks. This work presents improvements on previous work [1] to develop a Human Machine Interface (HMI) and the appropriate control methods to enable supervisory control for parking multiple skids in a cluttered and dynamic environment. This system could reduce the time required for a single operator or multiple operators to move the skids and setup in or exit from areas such as elevators and storage areas.

The HMI consists of four parts: (1) Automatically defining parking goal configurations for each weapon skid within a boundary, (2) the control methods for moving multiple non-holonomic weapon skids, (3) a user interface, and (4) a localization and obstacle detection method for the robotic systems.

The rest of this paper is organized as follows: Sect. 2 provides a brief overview of related work in this area. Section 3 describes the four parts of the HMI system. Section 4 presents simulation and hardware test results. Section 5 concludes the paper with a discussion and future work.

2   Related Works

Formation control of multiple robotic vehicles (swarms) is a very active area of research. Many different strategies have been developed for controlling formations of swarms under different scenarios such as movement of a formation in a corridor or amongst obstacles. A majority of these control strategies can be categorized as either leader-follower [2], behavior-based [3], or virtual structure approach [4]. Multiple works have utilized a potential field approach to maintain the formation of a swarm of non-holonomic robots while moving to a target location [5–7]. Those works focused on maintaining a consistent formation during movement or maintaining a consistent formation on a predefined contour line. Multiple works have focused on mobile robotic platforms during parking [8, 9], but those works did not address issues of parking of multiple non-holonomic robotic platforms at the same time in close proximity to one another.

The example of weapons movement on a carrier demonstrates a Navy specific scenario where parking occurs multiple times during transport and there are multiple skids. While the works listed above present strategies for the movement of a swarm formation from point A to point B, they are not focused on the parking of formations. A parking task involves identifying safe spaces within an area enclosed by a contour. In this work, "safe" means clear of obstacles. A contour could be a physical construct such as walls or painted lines on the ground, or a virtual construct such as an operator defined virtual boundary. Ekanayake and Pathirana [10] developed a scalable control algorithm to navigate a group of mobile robots into a predefined shape and spread them inside while avoiding inter-member collisions. However, each robot was treated as an omnidirectional point mass with each robot having the same mass and mobility. Their work also did not consider dynamic obstacles within the environment.

This work builds off of previous work presented by the authors in [1]. In [1], a convex optimization approach is presented for parking multiple heterogeneous weapon skids within a user defined convex boundary. In this new work, methods for assigning parking goals are presented that enable automated parking assignment in non-convex boundaries. The previous work assigned robotic vehicles to optimized goals sequentially which can lead to robots blocking each other from getting to their assigned goals when operating in tight spaces. This new work presents a method for adaptively reassigning robotic vehicles to appropriate goals using the Hungarian algorithm with a formation driven cost function. We also present a path planning and navigation control method that overcomes some of the local minimum challenges listed in [1]. Part of the new navigation method includes a modified Reciprocal Velocity Obstacle approach that uses assigned priorities, enabling those robotic vehicles with higher priorities to move more freely through a formation. The previous interface has been revamped and all operations are now conducted on a Ubuntu computer using the Robot Operating System (ROS) toolset [11]. Notably, there is no longer a need for an overhead camera to extract positions of each robotic vehicle as localization and mapping are all performed onboard the robot using onboard sensors.

3   Autonomous Skids Parking System

Improvements on the work in [1] are implemented to address limitations. Changes to the four components of the HMI system are highlighted in the following sections.

3.1   User Interface

The two panels of the user interface are shown in Figs. 2 and 3. The goal of the user interface is to enable control of one sailor the ability to select assets in one location of the ship and to command them to move to other locations on the ship in a supervisory function. This means that the operator is not directly controlling the robots to move forward, left, right, etc. but is instead commanding end goals and then monitoring the robotic vehicles as they autonomously drive to the goals. The interface was generated using toolsets available as part of the Robot Operating System (ROS) on Ubuntu. In particular, rqt and RViz tools were used to generate the interface and visualization.

The main functionality of the interface is to give the operator situational awareness of the current pose of each of the robotic vehicles, the map of the environment, the location of detected obstacles, and awareness of the planned path for each robotic vehicle. An example of the interface can be seen in Fig. 3. The operator can select robots either by clicking on them individually on the "map view" or by dragging a selection boundary around the robots of interest. The parking boundary is then selected by the operator by drawing a selection boundary on the "map view". The operator can select if they want the goals automatically distributed using a Stacking, Convex Optimization, or Potential Field method. The operator can also individually assign goals by selecting a single robot and then clicking on the desired parking goal.

After a central controller assigns the goals, each robot's controller plans the path to the goal and commands the robot to follow that path. At any moment, an operator can cancel the paths of all robots by selecting the emergency stop "eSTOP" button or they can select each individual robot to stop. The operator can also command a new goal. Additionally, the operator can take control of any robot and drive it directly via a joystick/game pad.

3.2   Parking Goals Within Non-convex Boundaries

In [1], the authors present a convex optimization method for assigning goals for weapon skids within a user defined convex boundary. The convex boundary requirement significantly limits the operator's parking options. In the case of parking skids on an aircraft carrier, there are scenarios where parking within non-convex boundaries would be advantageous. For example it is beneficial to park around obstacles. To enable this capability, a stacking method and a potential field method is implemented for assigning goals within non-convex boundaries.

Each of the following methods use a boundary defined by the operator which is discretized into a series of position coordinates. Inputs to the method are, the number of vehicles to park, two dimensional bounding boxes of the vehicles, and desired end orientation. Currently, the desired orientation is an input rather than an output of the system for reasons described in Sect. 3.2.2.

3.2.1   Stacking Parking Method

The stacking method is a simple method of moving from the left most location of the boundary and working along the rows and columns of the boundary to place the goals. Each goal is chosen if there is no overlap between the goals and any previously placed

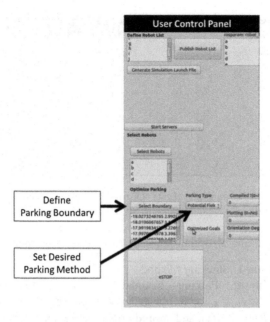

Fig. 2. User control panel of the HMI. Enables user selection of robots, selection of parking boundary, selection of parking method, and emergency stop.

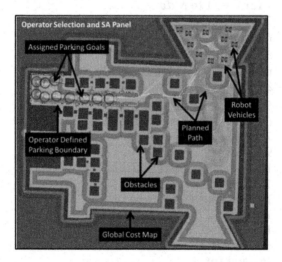

Fig. 3. Selection and situational awareness panel of the HMI. Showing example of group of robot vehicle's path plans to assigned parking goals within a user defined boundary. Environment is a mock representation of a section of a crowded hangar bay. (Color figure online)

goal (with a safety margin) such that a skid occupying the goal location would not collide with another skid. An example of the stacking method within a user defined boundary can be seen in the left side of Fig. 4.

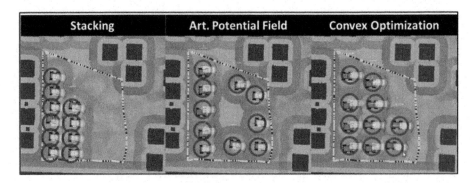

Fig. 4. Parking goal distributions. (Left) Distribution using stacking method, (Middle) Distribution using artificial potential fields, (Right) Distribution using convex optimization

The stacking parking algorithm proceeds as follows (Algorithm 1).

Algorithm 1: Stacking Method
Input: Parking boundary converted to polygon $Poly_{boundary}$, Robot geometries converted to polygon, $Poly_{robot}(i)$, and desired orientation, θ for N robots. **Output:** World reference parking goals for each robot $p_{Goal}(i)$
1:**for** $i \leftarrow 1$ **to** N **do**   2:**for** $stepX \leftarrow$ $\min(Poly_{boundary})$ **to** $\max(Poly_{boundary})$ 3:**for** $stepY \leftarrow$ $\min(Poly_{boundary})$ **to** $\max(Poly_{boundary})$ 4: **if** $(stepX, stepY)$ inside $Poly_{boundary}$ 5: **for** $j \leftarrow 1$ **to** $(i - 1)$ 6:   **if** $(stepX, stepY)$ inside $Poly_{robot}(j)$ 7:   continue to next iteration   9:   **else**   10: $p_{Goal}(i) = (stepX, stepY, \theta)$ 11:   **end if**   12: **end for**   13: **else**   14:   continue to next iteration   15: **end if**   16: **end for**   17: **end for**   18:**end for**

3.2.2 Artificial Potential Field Method

The artificial potential field (APF) method takes the stacking method a step farther. Starting from the parking results of the stacking method, the APF method further distributes the parking goals within the boundary by minimizing the total energy of the parking goals within the defined boundary as defined by

$$U_{Total}(q) = \sum U_{boundary}(q) + \sum U_{pGoal}(q) \tag{1}$$

where U_{Total} is the potential field around the centroid of the parking goal, q. U_{pGoal} is the potential function of the parking goal centroids $pGoal$. Each defined parking goal (centroid) is treated as a "floating" point. Each point on the discretized parking boundary, and other $pGoals$ act as repulsive potentials to the floating parking points whose potentials are defined by

$$U_{boundary} = \begin{cases} 0 & q > dist\ thresh \\ dist(q, boundary)^{-1} & q \leq dist\ thresh \end{cases} \tag{2}$$

$$U_{pGoal} = dist(q, p, Goal)^{-1} \tag{3}$$

where $U_{boundary}$ is the potential function of all the discretized points along the user defined parking boundary and it only has value when q is within a defined distance threshold, *dist thresh*. *dist* is the Euclidean distance function.

The induced force driving the location of each parking goal centroids is defined by

$$F_{PGoal} = \nabla U(q) \tag{4}$$

After a few iterations, the parking goals move away from the boundary points and each other. With enough time, the parking goals settle into a minimum energy point and stop moving. It is important to note that the minimum energy point could be a local minimum and not the global minimum. After settling, these points are determined to be the parking goals within the user defined boundary as seen in Fig. 4.

In practice, the APF method can loop through many iterations before all the spots settle into a local minimum. Usually close to a local minimum, the parking spots will only making small adjustments between iterations. To prevent long wait times, a set number of iterations is used as a threshold and the parking goals are set at the end of the iterations if a local minimum is not achieved. It would also be reasonable to set the threshold to be triggered when the change in overall energy between iterations is less than a set value.

We investigated the use of the vertices on the geometric shape of the vehicles to use as repulsive potentials. In this case, the sum of the forces on each vertex contributes to the resultant centroid force acting on the floating point parking goal. While this did achieve a minimum energy state, the resulting parking goals for the vehicles give the appearance of disorder as seen in Fig. 5. Instead, we chose to plan on a formation basis with constant orientation to increase operator's confidence.

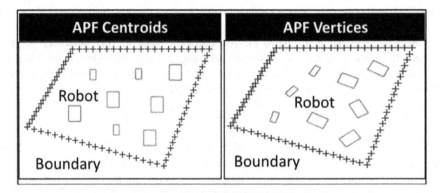

Fig. 5. Comparison between distributing parking goals using artificial potential fields via robot centroids (left) vs. robot geometric vertices (right). The robots are of varying sizes.

3.3   Adaptive Goal Assignment Using Hungarian Algorithm with Formation Driven Cost Function

In [1], final parking goals are assigned sequentially to the robot swarm, regardless of the location of those robots with respect to the parking goals. For example, parking goal One would be assigned to robot vehicle One, parking goal Two would be assigned to robot vehicle Two, and so on. This leads to inefficient movement of the robotic group within the parking boundary because no attention is paid to the amount of movement that the assigned parking goal requires from each robotic vehicle. Many times this means that there is considerable amount of avoidance maneuvers taking place to keep robots from crashing into each other within the parking boundary. Also, this leads to scenarios where one robot will reach its parking goal and end up blocking all other robots from reaching their assigned parking goals as seen in Fig. 6. To

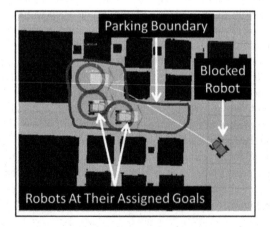

Fig. 6. Example of parked robots blocking other robots from reaching their assigned goals.

alleviate this problem, we implemented an adaptive goal assignment method that uses the Hungarian algorithm to assign the proper parking goals for the robotic vehicles.

The Hungarian algorithm is a combinatorial optimization algorithm that solves the problem of assigning m agents to n tasks [12]. It minimizes the overall costs in assignment based on a user defined cost function. In our case, the agents are the robotic vehicles and the tasks are the parking goals.

Optimization cost functions can include a precomputed estimate of distance travelled, time taken for each robot to reach the various parking goals, and/or Euclidian distance. For example, an intuitive computationally cheap approach would be to choose Euclidean distance such that the parking assignments are set based on minimizing the total distance of the group of robots from the goals. However, there are still very basic scenarios where this approach can lead to a robot blocking another robot from reaching its assigned goal, even when recalculating new assignments at every time step. A simple example of this is shown in Fig. 7.

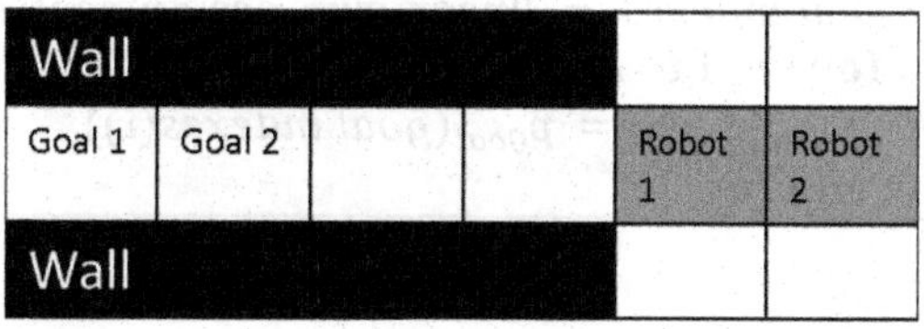

Wall					
Goal 1	Goal 2			Robot 1	Robot 2
Wall					

Fig. 7. Hallway example where the total cost (Euclidean Distance) for Robot 1 to Goal 2 and Robot 2 to Goal 1 is identical to the Robot 1 to Goal 1 and Robot 2 to Goal 2. It is arbitrary which goal the Hungarian Algorithm will assign with a Euclidian distance cost function, leading to cases when robots are blocked.

The cost function used in this work is based on minimizing the cost to move the group of robots from the current shape of the formation to the final shape of the parked goals at a current instant in time. It is important to note that this cost function is not dependent on the real world distances between the robots and actual parking goals. Instead, the centroid of the parking formation as a whole is superimposed on the centroid of the current robot formation. It is the Euclidean distances of the robots to these superimposed parking goals that is used in the cost function for the Hungarian algorithm goal assignments. At each instant in time, goal assignments are calculated and sent to the robotic group. This cost function alleviates the problem of robots getting to their assigned goals and blocking others from passing through.

The goal assignment algorithm proceeds as follows (Algorithm 2).

Algorithm 2: Parking Goal Assignment Method
Input: N parking goals, $p_{Goal}(i)$, N robot positions $q(i)$. A means to calculate linear sum assignment (Hungarian Algorithm) given calculated cost matrix, $cost$. **Output:** $p_{Goal\_NEW}(i)$ arranged in proper order for robots, q
1: $offset = mean(p_{Goal}) - mean(q)$ 2: $q_{offset} = q + offset$ 3: **for** $i \leftarrow 1$ **to** N **do**   2: **for** $j \leftarrow 1$ **to** N **do**   3: $diff = p_{Goal}(j) - q_{offset}(i)$ 4: $cost(i,j) = norm(diff)$ 5: **end for**   6: **end for**   7: $goal\ indexes = linear\_sum\_assignment(cost)$ 8: **for** $i \leftarrow 1$ **to** N **do**   9: $p_{Goal_{NEW}}(i) = p_{Goal}(goal\ indexes(i))$ 10: **end for**

3.4 Navigation with Path Planner and Reciprocal Velocity Obstacles Using Priorities

In [1], navigation is achieved through the use of an artificial potential field frame work to drive a non-holonomic robotic vehicle (differential drive system). The system works well in many cases to drive the robotic vehicles to their goals while avoiding each other. It also has the added benefit of enabling a robotic skid to drive through and out of formations. However, the downside to this method is that it only works in scenarios where a local minimum is not reached. For example, Fig. 8 shows two robots swapping positions within two formations. Potential fields will not succeed as the robots will get stuck in the corners of the rooms due to a local minimum condition.

To address this drawback, we used a path planner based on Dijkstra's algorithm [13] to plan the route for each robot to their assigned parking goals. The path planner takes in the map of the environment, the goal (i.e. assigned parking goal), and all detected obstacles, and then calculates an optimal path to the goal. Parts of robots detected by other robots are treated as detected obstacles. The potential fields approach from [1] is then used to drive the robotic vehicle to follow the path generated by the Dijkstra's algorithm as seen in Fig. 3 by the green lines extending from each robot to the assigned parking goal.

3.4.1 Reciprocal Velocity Obstacles Using Priorities

A Reciprocal Velocity Obstacle (RVO) approach using priorities is implemented to enable a robotic system to move through existing formations. An example of an RVO

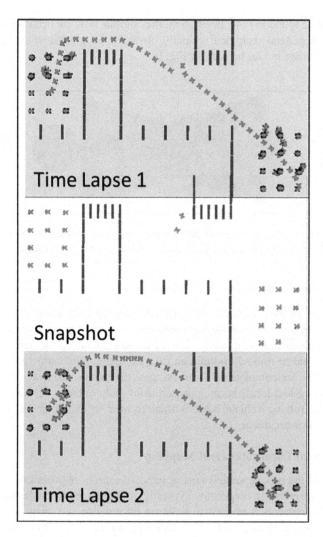

Fig. 8. Time lapse images of two robots swapping places within parking formations using RVO.

implementation is shown in Fig. 8. The reason for this capability is to handle scenarios when a single weapon skid within a formation (e.g. a skid in the middle) is needed to go elsewhere. This capability will also allow for skids or sailors to walk through the parked formation without having to move each skid individually. The skid needing to leave the formation, or needing to get through the formation, is given a high priority (i.e. more authority to occupy a space) and other skids move out of the way.

RVO is a technique that has been widely used for safe navigation among moving obstacles [14]. In very simple terms, each moving robot looks at other robot positions, velocities, and geometric shape, and uses that information to determine the best velocity to avoid a collision. The description and mathematical formulation of the RVO algorithm can be found in [14]. In addition, they outline how the behaviors extend with priority, called general reciprocal velocity obstacles. A graphic showing a velocity obstacle determination can be seen in Fig. 9.

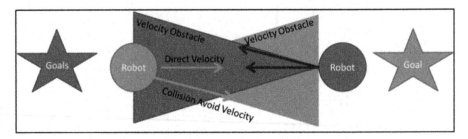

Fig. 9. Robots (circles) trying to get to their assigned goal (star). Velocity resulting from RVO is the closest velocity to the direct velocity without entering the velocity obstacles.

3.5   Robot and Obstacle Localization

In [1], positions of the robot and obstacles are extracted from images obtained from a top down camera viewing the operational environment. On an aircraft carrier, the weapon skids traverse through many areas of the ship including the hangar bay and on the deck. It would be difficult to build an external imaging infrastructure onto the ship that would allow for complete visual tracking of the robotic assets at all times. In this work, a mapping and localization approach using only onboard sensors (i.e. LiDAR, camera) to the robotic vehicle itself is implemented to eliminate the restrictions of needing a top down camera.

3.5.1   Onboard Localization and Mapping

Localization of the robotic vehicles using onboard sensors requires information about the system's operating environment. Typically, a simultaneous localization and mapping (SLAM) approach is utilized to build up information (i.e. map) about the operational environment. In this work, the SLAM approach called "GMapping" developed by [15] is used to build a map of the operational environment for the robotic systems. For weapons movement on the aircraft carrier, it is not unreasonable to believe that a rough blueprint of the various areas of the ship could be fed to the robotic skids prior to operations, thereby lessening the workload of SLAM. Along with this, the robotic skids can be continuously running GMapping to update the maps should major obstacles change (e.g. cargo pallets placed or moved).

If a general map of the operational environment exists, the robotic skids only need to localize within that map. An algorithm called Adaptive Monte Carlo Localization (AMCL) is used to localize the robotic vehicles. AMCL uses a particle filter to track the pose of the robot against the known map as described in [16].

Obstacles within the environment are detected by each robotic vehicle's onboard sensors. This obstacle information is used to update a robot's local map. A central controller uses each robot's location to manage parking assignments and sends those assignments back to the appropriate robots. Each robot then uses its local map, which includes detected obstacles, to plan appropriate paths to the assigned goal.

3.5.2 Shared Mapping for Common Reference Frame

A common reference frame for all information being received is an important aspect of the central controller to coordinate the movement of a swarm. It will be a challenge on an aircraft carrier to guarantee the exact known location of every robot at boot up, making it difficult to define a common reference frame. When a map of the operational environment exists, the reference frame of the map is then shared by all robots that are localizing themselves within that map. However, in scenarios where a map of the environment does not exist prior to operations, a map needs to be created by the robotic systems. Using GMapping, a local reference frame to the robot creating the map is used. The central controller then needs to take this local reference frame and make it a global reference frame shared by all the robotic vehicles.

There are a few approaches that could be used to generate a global shared reference frame. A couple examples are: (1) Robotic vehicles traversing the same areas can use AMCL to localize within the map created by another robot. (2) Robotic vehicles can visually see each other from afar and can use visual features to create a commonly shared reference frame.

4 Results and Discussion

The new interface and control methods were tested in hardware using two robotic vehicles and in simulation using multiple simulated robotic vehicles. The simulation setup was demonstrated at the Department of Defense Lab day held at the Pentagon where visitors were able to select simulated robots and command them to park in user defined boundaries. The simulated environment, shown in Fig. 3 is a generalized representation of a crowded hangar bay with staging areas and elevators for the robotic skids to park on. With minimal instruction, visitors were able to use the interface with ease and were able to successfully park the simulated robots in many different types of user defined boundaries without collisions and without inter-robot obstruction. The hardware tests used two robots due to laboratory resources limitations but the operator was able to command the robotic systems around a pre-mapped environment using the developed interface without the use of any overhead camera system. An example of the hardware test is shown in Fig. 10.

RVO with priorities enabled multiple differential drive robotic vehicles to move through and into formations. An example simulation result can be seen in Fig. 8 where two robots swap locations within parked formations. It was also tested in hardware with two robots passing by each other without colliding as shown in Fig. 11. The bottom starting robot is given a higher priority than the top starting robot. The top robot takes a larger avoidance route because of this. RVO shows promise as being a good method for performing local obstacle avoidance. The RVO implementation, as written in this work, grows linearly with respect to the number of robots.

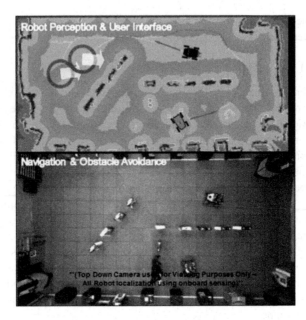

Fig. 10. Hardware demonstration of user interface and robot control. Two robots are autonomously navigating to the assigned goals.

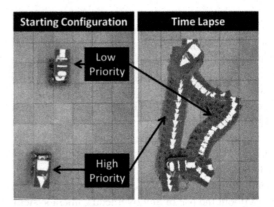

Fig. 11. Hardware demonstration of RVO. Two robots commanded to pass each other. One with higher priority that the other. (Right) Time lapse showing lower priority robot taking larger avoidance route due to RVO with priorities.

5   Conclusions and Future Work

This work presents improvements over the previous HMI and artificial potential field approach to parking multiple autonomous differential drive skids. The improvements enabled the control of multiple robotic systems by a single operator in a wider range of environment scenarios and without the need for an overhead camera to extract pose information of the robotic systems.

Now that the majority of the functionality of the operator interface and robotic control system is in place, one of the next steps will be to conduct a formal evaluation of operator performance when controlling multiple robotic vehicles using the interface. There is also technical work needed to enable the control methods to work for operating robotic skids with different steering mechanism (e.g. Ackerman steering). The goal assignment methodology in this work requires that the geometry of all parking spots be the same size and defined by the largest bounding box of the robotic vehicles in the commanded group. This enables the adaptive shifting of parking assignments without concern that a larger robot could be assigned to a space that it can't fit into. Future work can investigate goal assignments that take into account varying robotic vehicle sizes (addressed by our APF methodology) and the appropriate sequence at which they should be filled.

In general, there are many additional elements of the autonomous vehicle system that need to be further developed before an autonomous weapon skid is fielded. As stated in [17], there are aspects of the human system interface, monitoring and diagnosis, planning and decision, sensing and perception, and networking and collaboration that have to be addressed at the vehicle management system level, the mission management system level, and the command and control system level to field a safe and successful autonomous system. This work touched on much of the vehicle management system level and a small bit in the mission management system side. There are many other elements that need to be addressed in future work to transition this technology.

References

1. Hing, J., Boczar, R., Hart, K.: Parking autonomous skids. In: Yamamoto, S. (ed.) HCI 2015. LNCS, vol. 9173, pp. 557–568. Springer, Cham (2015). https://doi.org/10.1007/978-3-319-20618-9_55
2. Cowan, N., Shakerina, O., Vidal, R., Sastry, S.: Vision-based follow-the-leader. In: IEEE/RJS International Conference on Intelligent Robots and Systems (2003)
3. Scharf, D.P., Hadaegh, F.Y., Ploen, S.R.: A survey of space formation flying guidance and control (part 2). In: American Control Conference, Boston, MA (2004)
4. Ren, W., Beard, R.W.: A decentralized scheme for spacecraft formation flying via the virtual structure approach. In: American Control Conference, Denver (2003)
5. Lawton, J., Beard, R., Young, B.: A decentralized approach to formation maneuvers. IEEE Trans. Robot. Autom. 19(6), 933–941 (2003)
6. Liang, Y., Lee, H.-H.: Decentralized formation control and obstacle avoidance for multiple robots with nonholonomic constraints. In: American Control Conference, Minneapolis, MN (2006)
7. Elkaim, G., Kelbley, R.: A lightweight formation control methodology for a swarm of nonholonomic vehicles. In: IEEE Aerospace Conference, Big Sky, MT (2006)
8. Kondak, K., Hommel, G.: Computation of time optimal movements for autonomous parking of non-holonomic mobile platforms. In: International Conference on Robotics and Automation, Seoul, Korea (2001)

9. Masaki, H., Kangzhi, L.: Automatic parking benchmark problem: experimental comparison of nonholonomic control methods. In: Chinese Control Conference, Hunan, China (2007)
10. Ekanayake, S., Pathirana, P.: Formations of robotic swarm: an artificial force based approach. Int. J. Adv. Rob. Syst. $7(3)$, 173–190 (2010)
11. Quigley, M., Conley, K., Gerkey, B., Faust, J., Foote, T., Leibs, J., Wheeler, R., Ng, A.: ROS: an open-source robot operating system. In: ICRA Workshop on Open Source Software (2009)
12. Kuhn, H.: The Hungarian method for the assignment problem. Nav. Res. Logist. Q. **2**, 83–97 (1955)
13. Dijkstra, E.W.: A note on two problems in connexion with graphs. Numer. Math. **1**, 269–271 (1959)
14. Van den Berg, J., Lin, M., Manocha, D.: Reciprocal velocity obstacles for real-time multi-agent navigation. In: Proceedings of the IEEE International Conference on Robotics and Automation (ICRA) (2008)
15. Grisetti, G., Stachniss, C., Burgard, W.: Improved techniques for grid mapping with rao-blackwellized particle filters. IEEE Trans. Rob. **23**, 34–46 (2007)
16. Fox, D.: KLD-sampling: adaptive particle filters. In: Advances in Neural Information Processing Systems, vol. 14, pp. 713–720 (2001)
17. Naval Studies Board: Autonomous Vehicles in Support of Naval Operations (2005)

Monitor System for Remotely Small Vessel Navigating

Masaki Kondo[1(✉)], Ruri Shoji[1], Koichi Miyake[1], Tadasuke Furuya[1],
Kohta Ohshima[1], Etsuro Shimizu[1], Masaaki Inaishi[1],
and Masaki Nakagawa[2]

[1] Tokyo University of Marine Science and Technology,
2-1-6 Etchujima Koto-ku, Tokyo, Japan
mkondo@kaiyodai.ac.jp
[2] Tokyo University of Agriculture and Technology,
2-24-16 Nakacho, Koganei City, Tokyo, Japan

Abstract. In this research, we examine the necessary information for remotely vessel navigating and the display system for the mariner in a re-mote place. We also conduct experiments using small vessels. In remotely vessel navigating, it is important to send all information on the hull and vessel around to vessel operator at a remote location with the lowest possible delay. Since river traffic is being researched, it is targeted for remote ship navigating of small vessels in urban rivers. In this research, we examined the monitor for the ship operator for remotely vessel navigating of small vessels, and conducted the remotely vessel navigating experiments. Although it is the minimum necessary function for remotely vessel navigating, it's possible to navigate. It is necessary to show a lot of information to the operator for safe navigation, such as distance to distant vessels, obstacles, weather, oceanic conditions etc. and consider to avoid data traffic.

Keywords: Ship navigation · Remote navigating

1 Introduction

In this research, we examine the necessary information for remotely vessel navigating and the display system for the mariner in a remote place. We also conduct experiments using small vessels.

In remotely vessel navigating, it is important to send all information on the hull and vessel around to vessel operator at a remote location with the lowest possible delay. Rolls Royce [1] does research and verification on autonomous vessel and remotely vessel navigating in emergency, promoting the development of remote vessel navigating in MUNIN project [2]. The accident proportion of small vessels is high due to human error [3]. Since river traffic is being researched, it is targeted for remote ship navigating of small vessels in urban rivers.

© Springer International Publishing AG, part of Springer Nature 2018
S. Yamamoto and H. Mori (Eds.): HIMI 2018, LNCS 10905, pp. 419–428, 2018.
https://doi.org/10.1007/978-3-319-92046-7_35

2 Remote Navigating of Small Vessels

For remotely small vessel navigating (Fig. 1), vessel navigating function, communication function, and emergency response function are necessary. navigating and
monitoring techniques are required for the marine vessel navigating function. To
prevent accidents, monitoring is very important. In general, a marine vessel operator
navigates while checking the surrounding environment such as height, distance of a
pier, speed of another ship, weather and oceanographic information. In the case of
remotely navigating, these information are also necessary. However, if the amount of
data to be transmitted increases, the problems that communication delays and data
losses happen. All of data must be synchronized. Therefore, it is necessary to research
on information transportation and display method for the remote operator.

In this research, we set up an environmental display computer for processing
information of monitoring, a controlling computer for autopilot on-board. Also, a
display computer, a control computer and a communicating computer are installed on
the remote navigate side. We conduct remotely vessel navigating experiment by
controlling the control computer at a remote place.

Fig. 1. Small vessel (inside)

3 Remotely Navigating System

For navigating, the autopilot system was used. This system has a generally purpose
computer that can communicate with Transmission Control Protocol (TCP). From the
control computer inside "RAICHO I", information of vessel is transmitted to the control
computer on the remotely navigating side, and the control commands are sent from the
control computer on the remotely vessel navigating side. At the same time, the image is

delivered from the environment display computer by User Datagram Protocol (UDP). A total of 5 people boarded "RAICHO I", a person holding a small vessel license who can be a captain in the event of an emergency, one manager of the system and three guards. The pipeline of the information display is shown in Fig. 2. Experiments were conducted twice, changing information display method, remotely vessel operator.

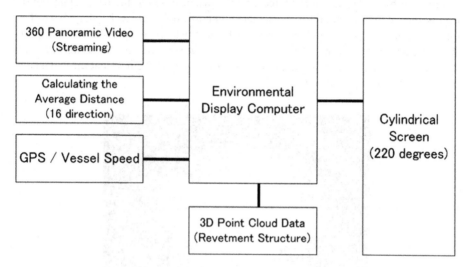

Fig. 2. Pipeline of the information display

4   Monitor for Remote Operator

Vessel navigating is different from driving a car. In the case of a car, the road, lane and direction are determined. Also, in places where the roads cross, traffic is controlled with signals. Although in a water area where small vessels sail, the direction of is determined, it is not as precise as a car. Also, when overtaking, it's important to decide where can be sail to according to the size and speed of the vessel. In some cases, it meanders to avoid the pulling wave. The situation of surroundings must be watched constantly, so that the monitor is more important than using in car.

Navigating such a vessel from a remote place, the same information as actual navigating that includes navigate information and monitor information are required. Information of navigating is "vessel speed" indicating the speed of ship, "bow direction" indicating the direction that the vessel is facing, and "panoramic streaming video", the monitor information which can overlook all directions, so that the captain can navigate a small vessel. Comparing to visual observation, the distance to objects (bridge, berthing boat) is hard to be transmitted from the image. Therefore, in addition to the necessary items for normal vessel navigating, a method for displaying distances to obstacles such as bridge piers and berths will be researched. Also, for low speed communication, the virtual environment map of the place where the vessel is located is displayed.

5   Distance Display to Object

Two cameras were used to calculate the distance from the vessel to the surroundings, fitting with fisheye lens (360° × 240°). It takes a lot of time to calculate the distance of all the pixels in the range that can be shot with camera (Fig. 3). Also, it is predicted that errors will increase due to the shake of vessel. Therefore, dividing the whole circumference by 16 and calculating the average distance within the rectangular region, the distance to the surroundings is calculated with low delay. It can be judged in a short time by indicating the calculation result whether distance is short in color (Fig. 4). While acquiring the image of the whole circumference, the calculated distance was overwrote. The distance between the two cameras is set to 95 cm so that the distance of 40 m ahead can be calculated.

Fig. 3. Captured image from fisheye camera

Fig. 4. Display average distance (Color figure online)

6 Low Speed Communication

Assuming a case where it is difficult to receive video due to deterioration of the communication environment, based on the position information sent from the vessel, a 3-dimensional space map based on point cloud data (Fig. 5) created in advance with a stereo camera is displayed on the marine vessel side screen. During low-speed communication, video distribution with a large amount of data is not performed, only autopilot command and vessel position information are transmitted. From the position information, the position in the virtual environment is obtained.

Fig. 5. Point cloud data (river side)

7 Screen Evaluation Experiment for Distance Display

It is difficult to judge the distance to the obstacle from the image displaying on the screen, and whether both the steering angle and the output are appropriate, although the distance to the surrounding obstacles is displayed (Fig. 6). When navigating inside the vessel, it is easy to grasp the size and movement of the hull, but in remotely ship navigating it is difficult to grasp the surrounding environment even if the panoramic video was displayed, the evaluation of the operator is feeling uneasy as a result. In this experiment, it is not a situation where it is easy to remotely vessel navigate. It is a guide to steer and display that can grasp the movement of the hull which is necessary. Based on the results of these experiments, we reexamined the display of bearing lines, which aligns the shape of the cylindrical screen on the remote navigating side (220° in the horizontal direction) and the horizontal angle of view of the image (Fig. 7).

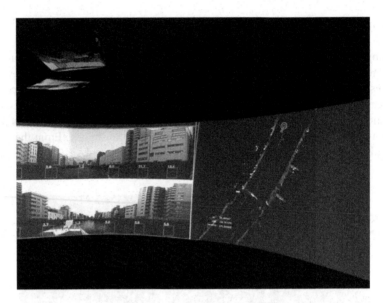

Fig. 6. Screen evaluation experiment

Fig. 7. The cylindrical screen

8 Screen Evaluation Experiment Displaying Vessel Navigating Support Information

We conducted a re-experiment in the same water area. From the previous result, the new display system (Fig. 8) was changed as follows.

- Draw a line indicating the bow direction and stern direction
- Draw bearing line every 30° so that can easily specify the steering angle (Fig. 9)
- Vessel speed displayed on screen.

The distance was feeling easy to grasp because of the display on the large screen overall, and the vessel speed indication was a reference for vessel navigating. Especially the image of the stern was a good evaluation for judging the position of the hull when passing through a bridge or narrow water channel.

Based on the previous experimental results, we added information such as azimuth on the image so that it is easy to grasp the movement of the hull, so it was a stable remotely vessel navigating. It is necessary to research on vessel navigating support functions such as adjustment of shooting equipment, future indications to be considered, predicted moving position, indication of water depth, but we think that it's a good result.

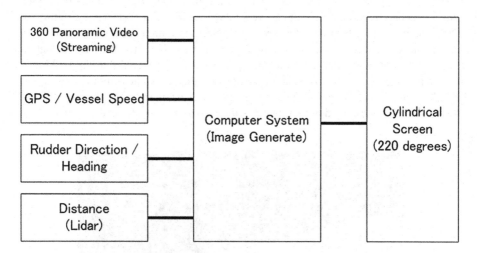

Fig. 8. New pipeline of the display system configuration

Fig. 9. Display system

Fig. 10. Panoramic images (bow/stern)

In the display of the surrounding conditions of the vessel, it is thought that visibility of the operator himself influenced the evaluation. In a narrow waterway sailing by a small vessel, it is necessary to secure the rearward visibility as well as the bow direction because the distance to other vessels such as the same navigation line and anti-navigation line is short. In accordance with the angle of 220° of the cylindrical screen watched by the operator, it can avoid the collision with obstacles by cutting out 220° centering on the bow and stern from the whole circumference image, reducing the movement of the head by displaying it all in two steps (Fig. 10). We believe that we can provide sufficient information to the operator. We considered that the shaking of the capture device influenced the operator, and this time the capture device was installed on the 3-axis stabilizer at remotely vessel navigating. At the time of hull shake, since the capture device moves in the horizontal direction, the center of the

Fig. 11. Simple Lidar

Fig. 12. The experiment

screen and the bow direction didn't overlap, affecting the designation of the steering angle. It is necessary to take measures to make the bow appear in the captured image.

We researched on the 3-dimensional space map created beforehand based on position information of hull in case of narrow communication band instead of video. If there is no difference in the environment at the time of creating the 3-dimensional space map and the navigation, sailing is possible based on the map, but navigation information on other vessels can't be acquired so that collision can't be avoided. Therefore, we used Simple Lidar which can acquire distance information faster than the stereo distance calculation using the capture device, and verified whether the positional relation of surrounding obstacles can be displayed at a high speed (Fig. 11). Figure 12 shows the state of the experiment. Since it is possible to acquire the distance if it is in the range up to 30 m, it is useful when low speed navigation in a narrow channel.

9 Conclusions

In this research, we examined the monitor for the ship operator for remote controlling of small vessels, and conducted the remote vessel controlling experiments. Although it is the minimum necessary function for remote vessel controlling, it's possible to control. It is necessary to show the operator a lot of information for safe navigation, such as distance to other vessels, obstacles, weather, oceanic conditions, etc., and consider processing to reduce traffic. Especially, we have to pay attention to "the height" in the river. It is necessary to obtain the height information during navigation because the height of the water surface changes according to passage of time. However, we emotionally judge whether the vessel can pass through under the bridge as usual, because we can stereo graphically look at the bridge. The ship operator hesitates because he can not feel the depth and the height of the bridge even though he recognizes the bridge itself on the screen. We examine making use of 3D-panoraminc video and Head Mount Display as the solution. This is thought to be beneficial to operate vessels when they get to the shore. We also reexamine and enhance the security of remote vessel controlling system based on the result of the experiment.

References

1. Rolls-Royce. http://www.rolls-royce.com/. Accessed 9 Feb 2018
2. MUNIN Project. http://www.unmanned-ship.org/munin/. Accessed 9 Feb 2018
3. Japan Coast Guard: Current situation and countermeasures of marine accidents in Heisei 26. http://www.kaiho.mlit.go.jp/info/kouhou/h27/k20150318/k150318-2.pdf. Accessed 1 Feb 2018

The "Watch" Support System for Ship Navigation

Masaki Kondo[1]([✉]), Ruri Shoji[1], Koichi Miyake[1], Ting Zhang[1],
Tadasuke Furuya[1], Kohta Ohshima[1], Masaaki Inaishi[1],
and Masaki Nakagawa[2]

[1] Tokyo University of Marine Science and Technology,
2-1-6 Etchujima Koto-ku, Tokyo, Japan
mkondo@kaiyodai.ac.jp
[2] Tokyo University of Agriculture and Technology, 2-24-16 Nakacho,
Koganei City, Tokyo, Japan

Abstract. Floating and non-floating objects such as other ships, buoys and so on must be alarmed before becoming obstacles for ship navigations. In this research, we have aimed to predict obstacles around a ship from maritime navigation images using an image recognition method and display them effectively to its operator. Faster R-CNN was used as detection method. We prepared a dataset composed of three categories for training and testing machine learning. We enumerated parameter values to obtain the best detection rate of obstacles by CNN. Then, we employed the best set of parameters for further experiments. The results are summarized as follows: (1) the detection rate of buoys is about 55 [%]; (2) large ships are sometimes mistaken for small boats. It remains to improve the detection rate and to decrease misclassifications; (3) the detection rate of small boats with distance of about 3 nautical mile(nm) from the ship is 86 [%], the detection rate of buoys with distance of about 2 [nm] from the ship is 100 [%].

Keywords: Ship navigation · Image processing · Faster R-CNN

1 Background and Purpose

AIS (Automatic Identification System) [2] is obliged to be mounted on all passenger ships and vessels over 300 gross tons engaged in international voyages as well as vessels over 500 gross tonnage even not being engaged in international voyages by the SOLAS convention (The International Convention for the Safety of Life at Sea). The installation has been completed from 2002 to 2008. With AIS, it becomes possible to acquire the navigation information such as positions of other ships and their paths to predict their future courses. Researches have been made on integrating navigation information into a superimposed display where navigation information on other vessels captured by vision, radar, automatic collision avoidance aid device ARPA (Automatic Radar Plotting Aid), AIS and so on are integrated and displayed.

S. Yamamoto and H. Mori (Eds.): HIMI 2018, LNCS 10905, pp. 429–440, 2018.
https://doi.org/10.1007/978-3-319-92046-7_36

NT-NAV was proposed by Hayuma et al. in 2003 [3, 4]. INT-NAV obtains information from multiple sources and integrates radar information, AIS information, Obstacle Zone by Target (OZT) on the image captured by a camera installed on the ship in order to shorten the time taken for information acquisition processing of an arbitrary ship. We gained the same or higher evaluation about the easiness of obtaining various information and superiority of guard work by doing the performance evaluation in shipbuilding simulator in 2007 [5] and in real sea area in 2008 [6] in comparison with the conventional ship maneuvering method. INT-NAV has the problem that the display range depends on the angle of view of the video camera and it is difficult to obtain full circumference navigation information of the ship.

As a method to provide visual information and radar ARPA information simultaneously, the visual recognition support equipment was developed by Hikida et al. in 2009 [7, 8]. The visual recognition support apparatus installs the HUD before the compass and supports the visual inspection by superimposing the radar echo and the ARPA information on the HUD. Due to the angle of view of the HUD, the field of view of the visual recognition device is only in the front, and it is difficult to obtain information on the ship behind, so a normal radar display was attached [9]. Furthermore, a mechanism capable of rotating the HUD around the compass was added to solve the problem of the limited field angle. Evaluation was carried out on this visual recognition support apparatus by a ship maneuvering simulator and actual ship experiments [10]. As a result, it has been confirmed that the work amount and information acquisition time are reduced as compared with the case of using the radar, and the work precision is kept equivalent to the case of using the radar. When ship's body shakes, however the problem is that the actual ship viewed differs from the display of direction information.

On the other hand, many small vessels such as fishing vessels are rarely equipped with AIS as compared with large vessels being equipped with a variety of navigational aids and can only be recognized by visual or video cameras, so it is difficult to predict their future courses. In 2014, 75% of maritime accidents were made by small vessels [1] in Japan. For this reason, studies to identify vessels from images have been conducted in recent years [11–14]. These studies are aimed at identifying and tracking vessels, but identifying various obstacles on the sea such as buoys have not been done.

This research aims to develop a system for small vessels which detects and displays obstacles around a ship from maritime navigation images. Obstacles such as other vessels and buoys on the sea are extracted from maritime environmental images captured by the system and classified into three categories. The classifier is trained by Faster R-CNN [15] which is one of image recognition methods. The classifier is tuned by changing parameters of Faster R-CNN. Then, we evaluate detection rates by the classifiers with different parameter values. We display obstacles and examine the accuracy of obstacle detection by using the best classifier and maritime navigation images.

The system that displays information such as AIS has already been made. Obstacles are displayed as shown in Fig. 1. Our system detects and adds obstacles such as small fishing vessels and buoys not equipped with AIS, which have been recognized only by visual inspection, and can recognize many obstacles on the sea. By showing the presence of obstacles to the operator using our system, we think it will be useful for ship navigation and marine accident reduction.

Fig. 1. Display example

2 AIS Information and Camera Image

The server installed in the SHIOJIMARU laboratory collects data from various vessel navigation instruments in the Ship of Tokyo University of Marine Science and Technology (SHIOJIMARU) (total length: 49.63 [m], width: 10.0 [m]). In this research, these data are received by a small server installed on the ship and transferred to a remote place. AIS information is distributed in the signal format [16] defined by NMEA (National Maritime Electronics Association). Various information about ships can be obtained by decoding the AIS information. Its example is shown in Fig. 2.

```
AIVDM,1,1,,,1BK2<nhU1wawjBRD?IH@i@`2UUS9,U*U8
AIVDM,1,1,,,16K2`o@000b0HGDD>oSJibH22500,0*44
AIVDM,1,1,,,16U9rj0001b0:FRD@:0rdbh22500,0*0C
AIVDM,1,1,,,1815?J0018awm72D7Vp0H0F0080W,0*2D
AIVDM,1,1,B,17qFQp0000aw@IpDCDMbIQv22@0W,0*7F
AIVDM,1,1,,,16K2UAPOisawj00D;MSUUIR42500,0*59
AIVDM,1,1,,,16K2Pg0001awF=tD@UQan1J42@0j,0*1F
AIVDM,1,1,,,16K2dU0OhIawJiJDBB;GdU<40<5M,0*3E
AIVDM,1,1,,,A@4757QAvOagH2JdO=h`4gse4VhIwQTB<ACusA4r,0*2D
AIVDM,1,1,,,16LhwP?001b0I@:D>nhWKQv>0500,0*32
AIVDM,1,1,,,15D`>n001@awou<DGKRmEID20<<G,0*48
AIVDM,11,,B,16KJd=@01mawhHtD<:Q<db9n0D4P,0*7A
AIVDM,1,1,,,15DHVn001qawRV>D@j:oCUn20@12,0*7B
AIVDO,1,1,,,16K8@QA1@0awFSn>DL:7V:00500,0*4B
AIVDM,1,1,B,16SIj:0028awe@0D7k7v6L20D2n,0*5B
AIVDM,1,1,,A,362@4PPB>awUqODAPOU9I0401f1,0*18
AIVDM,1,1,,,16K2;B@01gawSMOD=TSEbIP40H17,0*10
AIVDM,1,1,B,16K2?u001gbO>`LDD8fpM6f42817,0*0C
AIVDM,11,,B,16KVq7@01uawbbnD=4atUJ:<0<50,0*02
AIVDM,1,1,,,36K8@NPP@>aw?n0D@4kQ=5>620o1,0*02
AIVDM,1,1,,,A@4757QAvOagH2JdO>D`50gF1CQHwcP07hkvc@=D,0*36
```

Fig. 2. AIS information

Information included in AIS is classified into four types: static information, dynamic information, navigation related information, and navigation safety related information. Static information includes ship identification number MMSI (Maritime Mobile Service Identity), ship name, call sign, total length and width, ship type and so on. Dynamic information includes ship position, navigational condition, ground course, ground speed, turning rate, navigation status and so on. Navigational information includes drafts, destinations, loads and so on. Navigation safety information is a text message containing voyages or weather warnings that each ship can arbitrarily create as needed. A list of information included in AIS is shown in Table 1.

Table 1. Classification of AIS

Static information	MMSI number, IMO number, ship name, ship type, call sign, full length, width, antenna position
Dynamic information	Ship position (latitude and longitude), ground speed, ground course, azimuth direction, turning rate, voyage status, UTC
Voyage related information	Destination, load, draft, estimated time of arrival
Navigation related safety information	Safety short text message

Also, a camera is installed in the upper part of the mast of the ship, and it is possible to capture the state of the surrounding ships on the deck of SHIOJIMARU. In this research, we convert the serial signal flowing from the AIS transceiver installed in the ship and transmit it to the server in the ship via TCP (Transmission Control Protocol) communication. From there, AIS information is transmitted by UDP (User Datagram Protocol) communication to the server on the information display side installed at a remote place. Images captured by the camera are also transmitted to the server on the information display side via the inboard LAN. The transmitted camera images and AIS information are superimposed on the large screen installed on the Etchujima campus at Tokyo University of Marine Science and Technology.

3   Maritime Environment Images

On September 21, 22, and October 6, 2017, we used SHIOJIMARU to photograph Maritime environment images for deep learning. We captured obstacles at sea from various angles using three digital cameras from various shipboards. The resolution of three digital cameras are 3,216 × 2,136 [pixel], 2,592 × 1,728 [pixel], 3,216 × 2,136 [pixel]. Also on 5th October 2017, we captured obstacles at sea near the Tokyo Bay URAGA Waterway Route from the land. We also used another digital camera of 3,216 × 2,136 [pixels]. 7,553 acquired images are employed for the experiments.

First, we extract the areas of obstacles from maritime environment images. Secondly, we classify the extracted regions into three categories of large ships, small boats and buoys. Examples of maritime environment images and extracted obstacle areas are shown in Fig. 3.

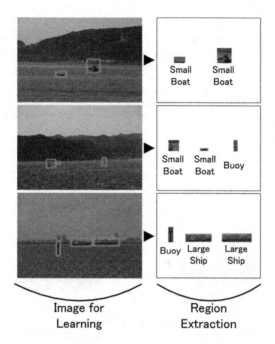

Fig. 3. Image example

4 Machine Learning

We employ Faster R-CNN which is one of the image recognition methods and is a fast object detection algorithm based by deep learning. We use the library Caffe [17] which is a framework provided by Berkeley Vision and Learning Center (BVLC) for classification and object detection by Faster R-CNN. We first extract features from maritime environment images input using six layers of convolution neural networks and create a feature map. A window scans the created feature map and outputs the object candidate regions. Detection windows called anchors are used for object detection. The window scans the feature map and outputs some object anomaly regions judged to have a high possibility of obstacles by fitting several preset anchors. The obtained object candidate regions are again processed by a convolution process and classified into the three categories. By repeating learning, the classifier capable of detecting obstacles from maritime environment images is created. The flow of learning and detection of Faster R-CNN is shown in Fig. 4.

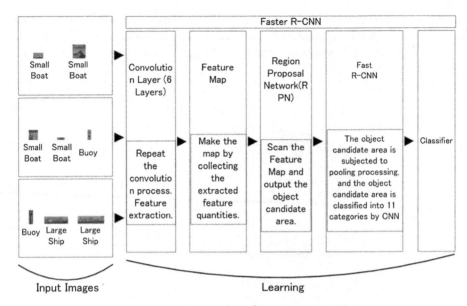

(a). Learning

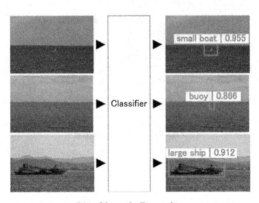

(b). Obstacle Detection

Fig. 4. Faster R-CNN

5   Maritime Navigation Images

Maritime navigation images were taken for evaluation on 7th February 2017 by a video camera installed at SHIOJIMARU. These images capture the surroundings of SHIO-JIMARU navigating over the sea. Maritime navigation images are composed of the length of 13 min and 3 s with about 30 frames per 1 [s], thus the total 23,490 frames with the screen size 1,920 × 1,080 [pixel]. The obstacles in maritime navigation images are three large ships, one small boat and one buoy. We extracted image frames from the images at 1 [s] intervals, and performed obstacle detection by Faster R-CNN. An example of maritime navigation images is shown in Fig. 5.

Fig. 5. Maritime navigation images example

6 Obstacle Detection Experiment

From 793 extracted image frames, large ships, small boats and buoys were detected by Faster R-CNN. Its detection rate is defined by the following Eq. (1).

$$Detection\ Rate\ [\%] = \frac{Number\ of\ Frames\ that\ Detected\ Obstacles}{Number\ of\ Frames\ that\ Obstacle\ is\ Captured} \quad (1)$$

In this research, the experiment was conducted by changing parameters of Faster R-CNN. In Faster R-CNN, when the size of the input image is large, memory errors of the computer occurs. Also, Faster R-CNN resizes the input image before learning and object detection in order to unify the sizes of different input images. In this research, we determine the best classifier by changing the image size, the anchor size and the number of learning iterations. The image size was set to $1{,}920 \times 1{,}080$ [pixels], $2{,}000 \times 1{,}200$ [pixels], $1{,}500 \times 900$ [pixels], $1{,}000 \times 600$ [pixels]. The anchor size was set to 9 types, 15 types, 18 types, 30 types. The number of learning iterations was 400 thousand times. Then, the detection rate was obtained.

The relationship between the anchor size and the detection rate is shown in Table 2. The image size and the number of learning iterations were unified at $1{,}000 \times 600$ [pixels] and 300 thousand times. The 9 kinds of the anchor size and the 18 kinds produce the better detection rate than others. In this research, we conduct the subsequent experiments with the 9 kinds of the anchor size.

Table 2. Anchor size and detection rate

Anchor size [type]	Detection rate [%]				
	Larege ship1	Larege ship2	Larege ship3	Small boat	Buoy
9	10.24	3.73	0.00	29.35	53.59
15	13.39	1.49	0.00	24.38	53.11
18	25.98	7.46	0.00	21.89	50.72
30	16.54	3.73	0.00	5.47	39.71

The relationship between the image size and the detection rate is shown in Table 3. The anchor size and the number of learning iterations were unified at 9 kinds and 300 thousand times. The detection rate was good in the case of 1,920 × 1,080 [pixel]. Since the size of maritime navigation images used in the experiment is also 1,920 × 1,080 [pixel], the image size is fixed to 1,920 × 1,080 [pixel] in the subsequent experiments.

Table 3. Image size and detection rate

Image size [pixel]	Detection rate [%]				
	Larege ship1	Larege ship2	Larege ship3	Small boat	Buoy
2,000 × 1,200	7.87	7.46	0.00	21.89	54.07
1,920 × 1,080	11.81	22.39	0.00	29.35	55.98
1,500 × 900	4.72	0.00	0.00	25.87	55.98
1,000 × 600	10.24	3.73	0.00	29.35	53.59

Finally, the relationship between the number of learning iterations and the detection rate is shown in Table 4. The image size and the anchor size were unified at 1,920 × 1,080 [pixel] and 9 kinds.

Table 4. Number of learning iterations and detection rate

		Number of learning iterations [times]			
		200,000	250,000	300,000	350,000
Detection rate [%]	Larege ship1	11.81	11.81	11.81	11.81
	Larege ship2	22.39	22.39	22.39	22.39
	Larege ship3	0.00	0.00	0.00	0.00
	Small boat	29.70	29.35	29.35	29.35
	Buoy	55.71	55.98	55.98	55.98

Regarding the detection rate, the same result was obtained for more than 250 thousand times. Looking at the detection rate of 250 thousand times, the highest detection rate was 55.98 [%] of buoys. Although the detection rate got lower overall, we think that it was affected by the distance to our ship and encountering situation.

On the sea, obstacles are captured in various shapes and sizes in the image, so that the detection rate was damaged. In the case of the same obstacle, it is considered that the detection rate differs depending on the front, the side, and the back in the image. Also, it is difficult to distinguish obstacles that are far from our ship because they are small in the image. Misclassification is one of the reasons why the detection rate of large ships is lower than that of buoys and small boats. Since there were many cases when large ships were misclassified as small boats, the detection rate of large ships was low. The results are shown in Table 5. Here, parameters of Faster R-CNN are set as follows: the image size being 1,920 × 1,080 [pixel], 9 kinds of anchor size employed and the number of learning iterations as 250 thousand times.

Table 5. Detection Rate shown in Confusion Matrix

	Classified category			
	Larege ship	Small boat	Buoy	Not detected
Larege ship1	11.81	75.59	0.00	12.60
Larege ship2	22.39	39.55	0.00	38.06
Larege ship3	0.00	53.23	0.00	46.77
Small boat	0.00	29.35	0.00	70.65
Buoy	0.00	0.00	55.98	44.02

The misclassification rate of large ships as small boats was about 40 [%] or more in all three classes of ships. In maritime environment images, large ships are captured large if they are close but small if they are far away. Therefore, distinguishing images of large ships captured far away and those of small boats becomes very hard. An example of detecting obstacles is shown in Fig. 6.

Fig. 6. Example of obstacle detection

7   Display to the Operator

A navigation system must inform the operator of the existence of obstacles at the distance that the operator is made possible the collision avoidance maneuvering. Particularly, the distance is very important from small ships and buoys not equipped with AIS. In the experiment, three large ships, one small boat, and one buoy were captured. Because it existed in the heading direction of the ship, the distance to the ship was shortened with the passage of time. The distance to the ship and the detection rate are summarized in Table 6.

Table 6. Detection rate by distance

		Detection rate [%]				
		Larege ship1	Larege ship2	Larege ship3	Small boat	Buoy
Distance [nm]	4–5	0.00	-	0.00	0.86	-
	3–4	3.84	11.39	-	0.00	-
	2–3	22.22	38.18	-	86.56	17.86
	∼2	32.00	-	-	-	100.00

The detection rate of large ships was 21 [%] if they were within 4 [nm] and 33 [%] if they were within 3 [nm]. Because the detection rate of small boat was 86 [%] if it is within 3 [nm] and 68 [%] if it is within 4 [nm], most of small boat can be recognized if it is within 3 [nm]. Buoy who was 56 [%] within 3 [nm] became the detection rate of 100 [%] within 2 [nm].

From this experiment, it was confirmed that it is possible to detect and display the AIS non-equipped obstacles of the distance of about 2 [nm] from the ship and display the AIS mounted ships. Therefore, it became possible to show the existence of obstacles which is the main cause of the accident to the navigation support.

In the future it is necessary to make the experiments with the variety of situations, such as backlight and bad weather condition. Also, this research is used by a marine vessel maneuver in a remote place. By a proposed method, it is possible to display obstacles that are difficult to understand in images. Images display in a remote place are affected by the communication situation. Images may be interrupted. Even if images are temporarily interrupted, it may be possible to navigate while avoiding obstacles by sending the position and classification of all obstacles.

8   Conclusion

In this research, we made a classifier that classifies obstacles in captured maritime environment images into three categories using Faster R-CNN. Experiments were carried out on maritime navigation images. The kinds and positions of the detected obstacles are superimposed on the display system as shown in Fig. 1. In addition to

AIS information, the operator can acquire information on obstacles not equipped with AIS.

The results are summarized as follows:

(1) We created data set of three categories of obstacles in maritime environment images that can be used for machine learning.
(2) We detected obstacles using Faster R-CNN. We examined the detection rate by changing image size, anchor size and the number of times of learning which are parameters of Faster R-CNN.
(3) We carried out the obstacle detection experiment using maritime navigation images. The detection rate was about 55 [%] for buoys.
(4) Large ships were sometimes mistaken for small boats. We will consider how to classify them.
(5) The detection rate of small boats with distance of about 3 [nm] from the ship is 86 [%], the detection rate of buoys with distance of about 2 [nm] from the ship is 100 [%].

Future work is to detect targets and movements affecting our own ship, and to study the method to raise prediction accuracy and notify them to the operator effectively. Parameters of Faster R-CNN and category classification of obstacles are related to the detection accuracy. In the case of the same obstacle, we consider to divide categories by the size of the obstacle shown in the image. We will also verify the relationship between the detection rate and the number of data used for learning, as well as the relationship with the shooting locations. We will increase maritime environmental images and maritime navigation images and proceed with these verifications.

References

1. Japan Coast Guard: Current situation and countermeasures of marine accidents in Heisei 26. http://www.kaiho.mlit.go.jp/info/kouhou/h27/k20150318/k150318-2.pdf. Accessed 1 Feb 2018
2. Japan Coast Guard: Compulsory aircraft equipped with AIS and existing ship installation date list. http://www.kaiho.mlit.go.jp/06kanku/onomichi/b_infomation/b_3_ais/tousai.htm. Accessed 1 Feb 2018
3. Hayuma, I., Takahiko, F., Junji, F., Yuichiro, O.: Study on integration and display of navigation information. Jpn. Navig. Assoc. Pap. **109**, 133–140 (2003)
4. Hikida, K., Fukudo, J., Okazaki, T.: Development and evaluation of a new safety navigation support system INT-NAV: advanced safe navigation support system integrated display of landscape image and navigation information, Technical Report of the Institute of Electronics Information and Communication Engineers, SSS Safety, vol. 106, no. 311, pp. 21–24 (2006)
5. Junji, F., Jyoutema, I., Yuichiro, O., Akio, A.: Production of integrated information display apparatus and evaluation of its basic performance. In: Japan Navigation Association Proceedings, vol. 116, pp. 27–30 (2007)
6. Junji, F., Hayuma, I., Takaaki, M., Hiroaki, F.: Evaluation of real area of integrated information display apparatus. Jpn. Navig. Assoc. Pap. **118**, 73–81 (2008)

7. Hikida, K., Mitsuto, N., Fukuto, J., Yoshimura, K.: Development of radar target acquisition and recognition support equipment by visual observation. Jpn. Navig. Assoc. Pap. **121**, 7–12 (2009)

8. Hikida, K.: Aiming to Reduce the Burden of Lookout Work - Development of Visual Recognition Support Equipment-, Ship and Sea Science, pp. 16–18 (2009)

9. Hikida, K., Fukudo, J., Numano, M., Inaoka, T., Nakato, M.: Development of visual recognition support equipment. Jpn. Navig. Assoc. Pap. **122**, 7–13 (2010)

10. Hikida, K., Yoshimura, K., Fukudo, J., Numano, M.: Evaluation of visual recognition support equipment. Jpn. Navig. Assoc. Pap. **125**, 1–8 (2011)

11. Yohei, M.: Ship image recognition using HOG features. Jpn. Navig. Assoc. Pap. **129**, 105–112 (2013)

12. Nobuo, K., Junji, F.: Motion recognition by interframe difference using sea observation images. Jpn. Navig. Assoc. Pap. **113**, 107–113 (2005)

13. Masatoshi, S., Tachihiro, Z., Masato, H., Masaki, O.: Ship extraction from navigation environmental time series images by HSV analysis, vol. 116, pp. 69–76 (2007)

14. Tianhui, Z., Masatoshi, S., Masaki, O.: Real-time ship extraction method from navigation environmental time series images. Jpn. Navig. Assoc. Pap. **117**, 175–182 (2007)

15. Ren, S., et al.: Faster R-CNN: towards real-time object detection with region proposal networks. Neural Inf. Processing. Syst. 1–10 (2015)

16. Raymond, E.S.: AIVDM/AIVDO protocol decoding, http://catb.org/gpsd/AIVDM.html. Accessed 1 Feb 2018

17. Caffe|Deep Learning Framework. http://caffe.berkeleyvision.org/. Accessed 1 Feb 2018

Discussion on the Application of Active Side Stick on Civil Aircraft

Xianxue Li[⊠], Baofeng Li, and Haiyan Liu

Shanghai Aircraft Design and Research Institute, No. 5188 JinKe Road,
PuDong New District, Shanghai 201210, China
lixianxue@comac.cc

Abstract. With the maturation of fly by wire technology, side stick has been used in aircraft flight control and active side stick is attracting more and more attention to achieve better HMI performance. In this paper, the development and application of side stick technology is discussed. From the research, we make a suggestion that the COMAC should consider apply some new technologies such as active side sticks to maintain competitiveness in the future.

Keywords: Active · Sidestick · Civil aircraft

1 Introduction

A side stick or sidestick controller is an aircraft control column (or joystick) that is located on the side console of the pilot, usually on the right hand side, or outboard on a two-seat flightdeck, just as shown in Fig. 1. Side stick is divided into three basic types: mechanical side stick, passive side stick and active side stick.

It's a long history for the development of side stick. The earliest Wright Flyer used a single-axis side stick to control the pitch of the aircraft and White's most early designs used the same control type which is the mechanical side stick. With the maturation of fly-by-wire technology, side stick began to be used in aircraft flight control widely. The side stick was first applied in military aircraft [1]. In the 1970s, the American F-16 aircraft firstly adopted the fly-by-wire system and side stick as the primary control and bomber B-1B also adopted side stick later. From then on, side stick re-attracted people's attention. Until 1980s, the first fly-by-wire civil aircraft with side stick as the primary control appeared, which was Airbus A320 series. After that, side stick was used as mail control equipment on the following series of civil aircraft of Airbus. However, Boeing still uses the traditional control column as the primary control equipment from B-707 to B-787 series. The side stick at this stage is called passive stick which lacks of control feedback from the aircraft or the other pilot. To solve this problem, active side stick came up which pilots can feel the force and visual feedback through the active side stick to have better situation awareness.

© Springer International Publishing AG, part of Springer Nature 2018
S. Yamamoto and H. Mori (Eds.): HIMI 2018, LNCS 10905, pp. 441–449, 2018.
https://doi.org/10.1007/978-3-319-92046-7_37

Fig. 1. Typical side stick layout

2   Passive Side Stick

The principal reason for using side stick is ergonomics. Removing the control yokes allows for larger flight displays in the latest cockpits and pilots can move closer to the instrument panels, allowing use of touchscreens. Besides Airbus, other aircraft company are also inclined to apply the side stick as the primary flight control, like Dassault, Embraer, Bell, Sikorsky, Comac, Irkut and Sukhoi.

At present, the side sticks used in civil aircraft such as Airbus series and Gulfstream series are passive side sticks, that is to say, the pilots can not directly feel the force feedback and displacement of surface through the side stick. For the fly by wire aircraft, it has been the mainstream to use side stick instead of traditional control column. Compared with the mechanical manipulation of the traditional control column, side stick has many advantages, however there are some disadvantages. The passive side stick is fixed by spring without information feedback, and the left side stick does not have mechanical or electrical linkage with the right side stick. When the autopilot works, the side stick is locked in neutral position. The displacement of the side stick is input by pilot's operation and transfer to the angular velocity command, and then compares it with the true angular rate feedback signal which the error is calculated by the control law and finally the operational signal generates. During above operation, the pilot can only feel the gradient force produced by the spring, the inertia and friction produced by the mechanism. The command of side stick (displacement command) is processed by the flight control computer to control the control surface and the pilot does not get any direct feedback through side stick from the control surface.

The advantages of a passive side stick over the traditional center control column are as follows:

(1) Side stick is conducive to manipulation of the front controller side.

Stick device unit is more simplified, the side stick device unit is located on the outside console and its length and displacement are much shorter than the traditional center control column, and multiple controllers can be integrated on one side stick. The traditional control column not only widens the distance between the pilot and the dashboard, but also blocks pilot's normal operation of the front control panel. The removal of the central control column actually precludes operational obstacles in front of the pilot. Coupled with the ability to move the seat forward, it will undoubtedly greatly facilitate pilot's flight control of the front controller.

(2) Side stick can improve the visibility of the front display.

The center control column obstructs the front lower dashboard. This blockage can be worse when the pilot holds and maneuvers the center control column. After removing the central control column and by using side stick, the visibility of the pilots will suddenly clear. The adverse effects on the visibility of the front monitor are eliminated.

(3) It can improve the operation performance.

With a side stick control, the pilot can control the flight more precisely and effortlessly and by using side stick there is more space which allows the pilots not only lift their feet on both feet, but also extend the activities of the lower limbs freely to increase flight comfort.

(4) Removing the central control column and other devices, in fact, increases the available space at the front of the cockpit.

On the other hand, the replacement of a conventional side stick reduces the amount of space available for the pilot to complete the maneuvering of the aircraft. This provides more possibilities for designing aircraft cockpit with more equipments.

(5) It can reduce the control of the system weight, compared with the steering column, the side bar on the cabin space requirements greatly reduced.
(6) The weight of primary flight control system is reduced to enhance airline company's economic income.
(7) It can reduce maintenance costs, reduce flight workload, improve aircraft handling quality, facilitate pilots access, facilitate the front controller and so on.

At the same time, passive side sticks also have some disadvantages:

(1) Unable to change hands to operate the aircraft.

Since the side stick is located at the outside of the console, the left pilot can only operate the side stick with his left hand and the right pilot can only operate the side stick with his right hand. As a result, it is difficult for pilots to alternately control the side stick with both hands, which is a major disadvantage of the side-stick piloting device.

(2) It is difficult for the pilot to perceive the steering effect directly.

With the fly-by-wire aircraft, there is no longer a conventional mechanical connection between the pilot and the controlled surface of the aircraft. When the pilot maneuvered the aircraft, he could no longer directly feel the reaction force after the controlled movement of the aircraft, so it was difficult to correctly sense the effect of the maneuver and make corrections in time.

(3) Feedback information is not easy to distinguish.

Compared with the traditional center control column, the displacement of the side stick is usually small. Whether visually or tactile, it is difficult for pilots to feel the effect or effect of their own control action from slight changes in the position of the side stick. When the pilot explicitly discerns the displacement of the side stick, the actual control input may have been excessive.

The pilot's perception of the change in position of the side stick is usually better than the resolution of the force. Therefore, the use of pure power type side stick will increase the difficulty of pilots perceived control effect. Persistent force in the case of people, the size of the force changes will become less and less sensitive, which is the inherent tactile adaptation. Tactile adaptation also increases the difficulty of pilots correctly sensing maneuvering effects without any other clues.

(4) Passive sidebar because there is no linkage, will increase the difficulty of training drivers.

The biggest drawback of the "passive" sides tick now used in civil aircraft is the lack of control feedback from the aircraft or the other pilot.

However the transfer of "active inceptor" technology to the commercial sector from the military is helping to overcome that objection. Active side stick that provide tactile and visual feedback in response to pilot and autopilot commands are used in the F-35 and CH-53K heavy-lift helicopter.

3   Active Side Stick

As early as 10 years ago, many aircraft manufacturers and equipment manufacturers have studied the active side stick and got some progress. Until now, Gulfstream G500, MC-21 and other commercial aircraft have adopted the active side stick and is in the process of airworthiness certification. It is also reported that the Boeing and Airbus are also working on the research of active side stick, and they also consider to adopt active side stick as the primary flight control system on their next generation aircraft.

In addition that the active side stick has the same visual notification about operating authority and aural notification about dual manipulation just like the passive side stick, variable gradient force feedback is provided since the electric motor is used instead of the spring, and the linkage function between left side stick and right side stick is also realized to acknowledge the pilot if there is dual manipulation. Besides, the active side stick also moves along with the control surface when autopilot is engaged. The active side stick can also receive the feedback from the aircraft control surface (such as the aerodynamic load on the rudder surface) which is different from the passive side stick with the spring that can only provide fixed force. It applies the force and displacement feedback to the side stick through actuators, so that the pilot can feel the state of the aircraft more realistically [2].

Compared with passive side stick, there are many advantages for active side stick.

- The coupling linkage between the left and right side stick makes the relationship between captain and copilot more suitable for human factors, so that the two drivers feel more direct interaction with each other.
- Variable force gradient gives more flexibility to operate the aircraft.
- The feeling of change in force allows the pilot to have a direct feeling of envelope limitations (such as soft tilt limits, speed limits, etc.)
- It can provide soft stop or hard stop.
- The anti-drive of the autopilot can provide more direct control information for the pilot.

4 Application of Active Side Stick

Active side stick are mostly used in military aircraft, such as F-35 and CH-53K heavy-lift helicopter. Meanwhile, some civil manufacturer also adopt active side stick, such as Gulfstream's G500 and G600 business jets, and Russian's MC-21 aircraft. Their active side stick are mainly from several famous suppliers like BAE, UTAS, Sagem and Stirling.

4.1 Sagem's Active Side Stick

In 2014, Sagem introduced a prototype of its new active side stick unit for civil airplanes and helicopters. This active side stick was demonstrated in a flight simulator featuring a full glass cockpit at NBAA business aviation trade show and exhibition in Orlando, Florida (USA).

This side stick controller features very high dispatch reliability, a robust design that stands up to all types of contingencies and an optimized architecture supporting the real-time adjustment of force feel laws. Based on its stick shaker and stick pusher function and the fine synchronization between all side stick units, this system greatly facilitates pilot/co-pilot coordination and the management of stressful situations.

4.2 BAE's Active Side Stick

BAE has developed a commercial active side stick with its own investment. The commercial version of the active side stick is dual-duplex, using dissimilar processors so there are no common failure modes between channels. Side stick force and position-sensing is quadruplex redundant to meet certification requirements.

All of the hardware and software are packaged within a single line-replaceable unit that interfaces with the aircraft's flight control computers, while dedicated electrical links connect the pilot and copilot inceptors. This ensures both side sticks move

446   X. Li et al.

together in response to both pilot and autopilot commands, providing crew situational awareness equivalent to conventional pilot controls.

BAE's active side stick has been used on Gulfstream's business jets G500 and G600, shown in Fig. 2. Besides, Embraer's KC-390 also adopts BAE's active side stick.

Just like active side stick's characteristics, BAE's active side sticks have the characteristics that breakout forces, force displacement gradients and soft stops in each axis.

Fig. 2. BAE's active side stick

4.3 UTAS's Active Side Stick

France-based subsidiary of UTC Aerospace Systems (UTAS) designed a active side stick which can couple the motions of the left-seat and right-seat controllers. The side stick also provides back driven feedback based on the aircraft's actions, so more pressure is required to make inputs that cause higher manoeuvring loads.

UTAS's active side stick was adopted on Russia's MC-21 which was firstly used on a large commercial aircraft.

4.4 Stirling's Active Side Stick

Stirling has pioneered active control technology since the early 1990s. Stirling's family of active controls are feature-rich, highly reconfigurable and suitable for single or dual (linked) cockpit configurations. Stirling's range of active products includes side sticks,

Table 1. Stirling's active side sticks

Name	Feature	Image
Next Generation Inceptor	• Configurable Feel, Force and Dynamic Characteristics • No Software Linked Dual Cockpit Controls • Custom Grip	
Compact Stick	• Configurable Feel, Force and Dynamic Characteristics • Software Linked Dual Cockpit Controls • Custom Grip	
Cyclic/High Force Stick	• Configurable Feel, Force and Dynamic Characteristics • Software Linked Dual Cockpit Controls • Custom Grip	

throttles, collectives, cyclics and pedals. These controls are extremely compact, fully active and benefit from low acquisition and through-life costs. It has several active side stick and supplies the active sticks and throttles for helicopter flight and the new F-35 pilot training simulators (Table 1).

5   Application Analysis of Active Side Stick

Although there has no civil aircraft equipped with the active side stick that gets airworthiness certification delivered, some commercial aircrafts such as Gulfstream G500 and MC-21 have used the active side stick and are working on the process of airworthiness certification. Besides, Airbus, Boeing and many equipment manufacturers also have done much research on active side stick for the next generation of aircraft. As a new type of aircraft for the next twenty years, the wide body aircraft by COMAC should also consider apply some new technologies such as active side sticks to maintain competitiveness in the future.

However, there are some potential risks for active side stick in airworthiness certification and human factors which need special attention. In order to maintain the competitiveness in the future, the active side stick technology should be studied in detail in the early stage of the aircraft development.

5.1   Control Law Analysis of Active Side Stick

The control law's difference of the flight control system between active side stick passive side stick is very large. To avoid too much change, the control law of active side stick should develop on the basis of the passive side stick control law. However, the change of manipulation manner with the active side stick is still very obvious.

It's very significant for the effect on pilots' control philosophy and the design of new active side sticks.

5.2   Airworthiness Certification Analysis

One of the important issues with active sidebar aircraft is airworthiness certification. There is no active side stick that has got airworthiness certificated on civil aircraft until now. Therefore, it needs more effort to establish a dedicated airworthiness clause by aircraft manufacturers and the airworthiness certification authority. Airworthiness certification should focus on the following questions:

- Numerical requirements for the stick force and force gradient,
- Cross-coupling certification between captain and co-pilot side stick;
- Back-driven certification;
- Control law certification.

5.3 Human Factors Analysis

(1) Force distribution on both sides of the side stick.

Most people are dextromanual, so the force feedback of the side stick should be properly defined to meet both left hand and right hand, then the left and right pilots can comfortably operate the side stick with either left or right hand.

(2) Button on the side stick should be rearranged.

The trim button may be added on the active side stick or other buttons be cancelled, so the layout of the buttons on the side stick need rearranged according to the importance.

(3) The force gradient need be newly defined.

The manner of safety protection is different between passive side stick and active side stick, and it's needed to remind pilots the current state of the aircraft through the force change.

(4) Force superposition from pitch and roll.

The relationship between the movement of active side stick and surface of the aircraft should be studied, including the force and displacement of active side stick from left, right, forward and afterward movement.

6 Conclusions

In summary, the use of active side stick on civil aircraft is technically feasible, but there is some certain risks in airworthiness certification. In order to maintain the competitiveness in the future, a detailed study of the active side stick technology should be conducted in the early stages of aircraft development. According to the actual situation, even though passive side sticks are adopted, the interface of active side stick should also be reserved to consider the application of active side stick on the improved generation aircraft.

According to the investigation from the pilots, the demand of linkage function of the side sticks is the most urgent, since it can effectively solve the problem of dual input. Therefore, in the application of active side stick, the linkage function of the side stick can be realized first, then the function of variable gradient force with the speed and the stroke, and finally the follow up function of side stick.

References

1. Duanqin, X., Xiaochao, G., et al.: Discuss on the advantages and disadvantages of aircraft side stick and its improvement. Chin. J. Ergon. **12**(1), 36–38 (2016)
2. Hegg, J.W., Smith, M.P., Yount, L., et al.: Features of active sidestick controllers. In: Proceedings of 13th AIAA/IEEE Digital Avionics Systems Conference, pp. 305–308. Institute of Electrical and Electronics Engineers, Inc., New York (1994)

Testing Human-Autonomy Teaming Concepts on a Global Positioning System Interface

Ricky Russell[(✉)]

San José State University, San José, CA 95192, USA
ricky.russell@sjsu.edu

Abstract. The field of Human-Autonomy Teaming (HAT) aims to understand how humans and automated systems can work together to optimize performance. Tenets of HAT [13] have been established from aviation based research. The current study explored the generalizability of these tenets to the automotive navigation domain. Current Apple Maps user interface (UI) and Global Positioning System (GPS) routes were tested against HAT enhanced Apple Maps UIs and GPS routes. The HAT enhanced versions were designed to improve the communication between the human and automation regarding goal accomplishment and awareness of the automation's processing. Participants reported the HAT enhanced UIs were a significant improvement over their current GPS systems and nearly all reported they would use the HAT enhanced UIs over their current systems. The systems were rated similarly on all other measures indicating that both systems supported their ability to understand the routes and make the necessary decisions. Data indicate a larger sample size is needed to achieve significance for some of these variables. Overall, the subjective data collected indicated preference for the HAT enhanced GPS UI lending support to the generalizability of HAT tenets to domains outside of aviation.

Keywords: Automation · Global Positioning Systems
Human-Autonomy Teaming · Human factors · Transparency

1 Introduction

Increasingly complex automated systems are becoming more capable of handling tasks that would traditionally be handled by an operator. Automation brings advantages such as performing complex calculations instantly and gathering information from multiple sources simultaneously. Unfortunately, an automated system's capabilities are constrained by its programming and the brittle nature of computing [11]. A human operator on the other hand, can think abstractly and gather information from outside of the system. The field of Human-Autonomy Teaming (HAT) is focused on understanding how human operators and automated systems can work together to optimize performance.

Shively et al. [13] have identified three tenets of HAT. The tenets include Bi-Directional Communication, which allows the operator and automation to have a clear understanding of mission goals and the ability to clearly convey when the goals

© Springer International Publishing AG, part of Springer Nature 2018
S. Yamamoto and H. Mori (Eds.): HIMI 2018, LNCS 10905, pp. 450–464, 2018.
https://doi.org/10.1007/978-3-319-92046-7_38

are not being met (and provide alternate options if possible), an Operator Directed Interface allowing for the operator to dynamically allocate their tasks, and Transparency, allowing the operator to understand what the automation is doing and why. The tenets have thus far been tested on an automated flight re-routing tool [12]. A study comparing the flight re-routing tool against a version of the tool enhanced with HAT showed that participants preferred the enhanced HAT version of the tool [9]. The participants reported that HAT improved the automations ability to handle unusual situations, was more helpful in making aircraft diversion decisions, and made their task management easier. The tenets are assumed to be generalizable to other domains such as photography and automotive navigation systems [11].

Since automotive navigation systems are common, free, and have a large population of users, the commonly used Global Positioning Systems (GPS) Waze, Google Maps, and Apple Maps were evaluated. A thorough front-end analysis of the GPS systems was conducted to identify if HAT was present and how HAT could be better implemented. Each of the tenets were already in use across the GPS platforms, to some degree (Table 1). The current study tested those tenets identified as "Low" HAT versions of Bi-Directional Communication and Transparency against enhanced, "High" HAT prototypes. This new "High HAT" system was designed around the current HAT concepts of Situation Awareness-based Agent Transparency [10], Feed-forward/Feedback loop [10], HAT design patterns [13], and propositions for implementing HAT on an automotive navigation system [11]. Separate elements of the enhanced HAT interface reflect different HAT tenets (Bi-Directional Communication, and Transparency), allowing for us to manipulate enhancements in these areas independently.

Table 1. HAT Tenets currently seen in common GPS applications

Transparency	ETA	Color coded traffic	Driver updates (accidents, road hazards, police)	Event notifications, accident notifications delaying normal drive time
Operator directed	Speeding notification – operator specifies when/how to receive alert	Plays (food, gas, coffee)	Sync your calendar for trip reminders	
Bi-directional communication	Provides route to operator's destination	Planned drive notifications from Calendar	Propose reroute with justification (save time)	ETA changes in real time
	Shared language (consistent with driver language)	Receiving speeding notifications	Ability to report hazards, police, accidents	Can change route options to get adjusted routes

2 Method

2.1 Prototype

The mobile prototype used for testing was built using Axure RP 8 Team Edition (version 8.1.0.3372). The user interface (UI) was designed to resemble Apple Maps. Screenshots of Apple Maps routes [1–8] were used for the visual design and the map background. All test configurations use real Apple Maps settings and the Low HAT conditions use real Apple Maps route screenshots (Fig. 1). The enhanced HAT configurations stay consistent with the Apple Maps UI but include enhanced HAT features (Fig. 2). To build these configurations blank Apple Maps screenshots were used and route information was overlaid on top of the map screenshots.

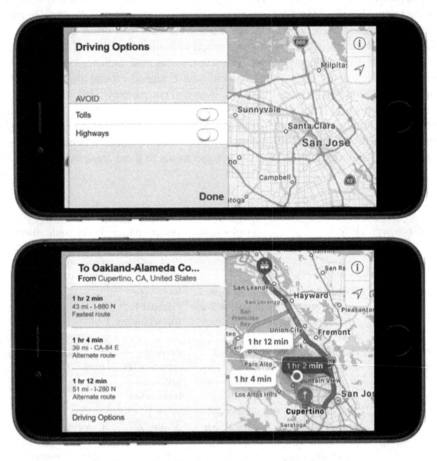

Fig. 1. Low HAT Apple maps configuration

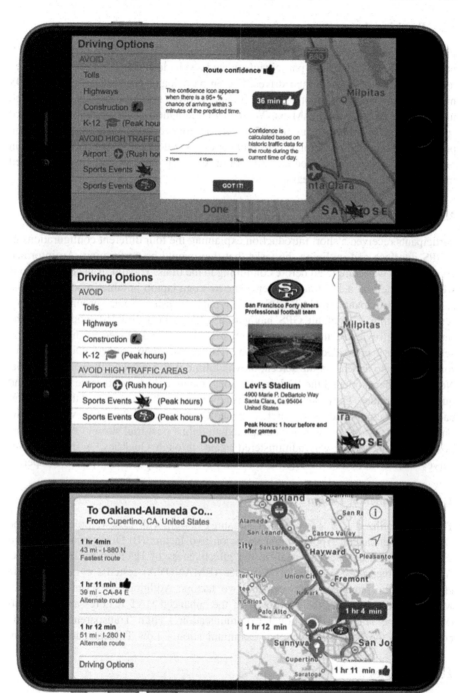

Fig. 2. Apple maps with enhanced bi-directional communication and transparency

2.2 Participants

Fifteen San José State University students (Mean age = 25 years old, range = 20–49) participated in this study. All participants owned a vehicle and 14/15 reported that they used a mobile GPS to plan their drives. Eleven participants reported using Apple iOS mobile systems and four reported using Google Android systems. Participants reported using Apple Maps, Google Maps, Waze, and MapQuest. One participant reported spending over 120 min per day in his/her vehicle, four reported spending 90–120 min per day, two reported spending 60–90 min per day, three reported spending 30–60 min per day, and five reported spending 30 min or less each day. All participants reported traffic as moderate to very heavy.

2.3 Procedure

Participants received a short introduction explaining the four different configurations of a GPS interface and outlining the tasks that they would be performing. An informed consent form was completed before continuing to the trials. The four trial configurations were completed on a 13″ touch screen Lenovo Yoga laptop in tablet mode. Each trial started the participant at a HAT GPS landing page. Pressing 'Setup' brought them to a screen depicting the default GPS navigation settings in a disabled mode. Once the settings were reviewed the participants pressed 'Done' and moved to the navigation screen where a preloaded destination was. Participants then pressed 'Directions' for route options to appear on the screen, at which time they evaluated the routes provide. Next, participants opened the 'Driving Options' menu, adjusted two route settings, and queried new routes. They then evaluated the new routes, which completed the trial. After each trial was completed the researcher opened Qualtrics in a different browser window on the laptop and the participants completed an online post-trial survey. A final online post-simulation survey was administered again, through Qualtrics. A demographics survey with a debriefing message at the end was then completed via pen and paper.

2.4 Experimental Design

This study utilized a 2 × 2 repeated-measures design with Low (Apple Maps) and High HAT (enhanced Apple Maps) levels on two factors (Bi-Directional Communication and Transparency) to analyze the effectiveness of HAT on a GPS interface. A repeated-measures Analysis of Variance (RMANOVA) tested for the two main effects and the interaction between the two factors. Additional descriptive statistics were collected regarding the usefulness of the enhanced HAT features and user preference of the High Bi-Directional Communication + High Transparency condition compared to the Low Bi-Directional Communication + Low Transparency condition.

2.5 Variables

Independent Variables

Transparent Design. The Low Transparency conditions contained transparent elements found in common GPS systems (color-coded traffic, ETA, and route attributes)

(Table 1). The High Transparency conditions contained the Low Transparency elements and additional 'Route Confidence' icons that appeared next to some route ETAs on the map and among the corresponding route's attributes in the navigation menu (Fig. 2). Upon initial setup, a lightbox appeared on the mobile screen explaining why and when a 'Route Confidence' icon would appear and was required reading before proceeding with the trial.

Bi-directional Communication. The Low Bi-Directional Communication conditions contained the bi-directional communication elements present in Apple Maps systems that allowed the user to input their mission goal (get to a destination), then adjust the mission goal (change the navigation settings to 'Avoid tolls' and 'Avoid highways') to receive more suitable options for achieving their mission goal. The High Bi-Directional Communication conditions contained all the elements present in the Low Bi-Directional Communication conditions, with the addition of five navigation settings that would allow the user to avoid common causes of traffic. Information about when to expect these areas to cause traffic was displayed in the 'Driving Options' menu, and the user was required view where high traffic areas were during their initial setup experience.

Dependent Variables. At the end of each trial a six item questionnaire (Table 2) was administered online via Qualtrics. All items were on a scale from 0 = Completely Disagree to 50 = Neutral to 100 = Completely agree. As an overall effectiveness measure, overall means of the six questions from each condition were calculated and analyzed with a RMANOVA.

Table 2. Post-trial survey questions.

Q1	Improvement	"This interface is an improvement over my current system." (Accompanied by pictures of the interface used in the trial)
Q2	Transparency	"I had enough information to choose the best route available"
Q3	Transparency	"I understand why it gave me the available routes"
Q4	Bi-directional communication	"I could modify the routes to suite my needs"
Q5	Bi-directional communication	"I could give the automation appropriate information to generate suitable routes"
Q6	Trust	"If I were to choose a route from the available options I would trust it"

Additional Measures. At the end of the study a final, six item survey (Table 3) regarding the High HAT tenets' usefulness and participants' desire to use them was administered online through Qualtrics. Questions one through three were on a scale from 0 = Completely Useless to 50 = Neutral to 100 = Completely Useful, questions four and five were on a scale from 0 = Never to 50 = Neutral to 100 = Always, and question six was a 'Yes' or 'No' question.

Table 3. Post-simulation survey questions.

Q1	High BC settings usefulness	"How useful were these settings?" (Accompanied by a picture of the High BC Settings)
Q2	Transparency usefulness	"How useful was the confidence?" (Accompanied by pictures of the confidence lightbox and maps with the icon)
Q3	High BC re-routing usefulness	"How useful was the ability to adjust routing?" (Accompanied by pictures of the High BC rerouting sequence)
Q4	Bi-directional communication willingness to use	"I would use this feature if it were available on a GPS" (Accompanied by pictures of the High BC features)
Q5	Transparency willingness to use	"I would use this feature if it were available on a GPS" (Accompanied by pictures of the confidence lightbox and maps with the icon)
Q6	High BC + High T willingness to use	"I would use HAT GPS over my current GPS system" (Accompanied by pictures of all the High HAT features)

Control Variables. Participant differences were controlled for by blocking on replicates. Route recommendations impacted by any level of HAT tenets were controlled for by keeping the route order standard within each block, and randomizing the order of HAT levels by block. To control for traffic variability the route traffic was kept minimal. To ensure the interactions participants had with the prototype would be consistent with what is commonly used a mobile Apple iOS prototype platform was chosen. Apple Maps was chosen as the visual design to control for users rating a new interface's visual design that contains the same features and route information as Apple Maps as an improvement or downgrade compared to their current GPS system.

3   Results

The HAT tenets Bi-Directional Communication and Transparency across two levels (Low and High) were analyzed for their impacts on UI Improvement, Transparency, Bi-Directional Communication, and Trust using a 2×2 RMANOVA with an alpha level of .05. Overall Effectiveness was analyzed by computing the mean scores from each condition for questions one through six with a RMANOVA. Additional measures were reported as descriptive statistics.

3.1   Interface Improvement

There was a main effect for Bi-Directional Communication (BC) on participants' rating of UI improvement scores (Fig. 3). The High BC conditions were rated as a significant UI improvement over the Low BC conditions (High BC $M = 89.07$, SD = 13.73; Low BC $M = 78.14$, $SD = 19.13$), $F (1, 14) = 8.90$, $p = .01$, $d = -0.67$. There was a main effect for Transparency (T) on participants' rating of UI improvement scores

(Fig. 4). The High T conditions were rated as a significant UI improvement over the Low T conditions (High T M = 88.57, SD = 11.40; Low T M = 78.64, SD = 21.46), F (1, 14) = 8.48, p = .01, d = −0.60. There was not a significant interaction effect (Fig. 5) for BC × T (Low BC × Low T M = 71.60, SD = 23.56; Low BC × High T M = 84.67, SD 14.69; High BC × Low T M = 85.67, SD = 19.35; High BC × High T M = 92.47, SD = 8.11), F (1, 14) = 0.63, p = .44.

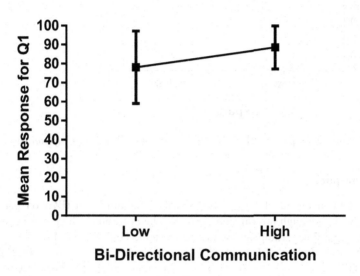

Fig. 3. Main effect plot of bi-directional communication for "This interface is an improvement over my current system."

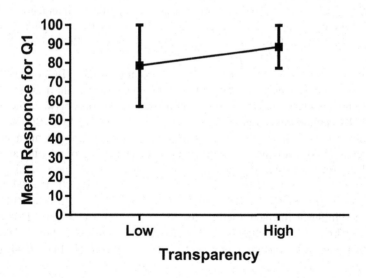

Fig. 4. Main effect plot of Transparency for "This interface is an improvement over my current system."

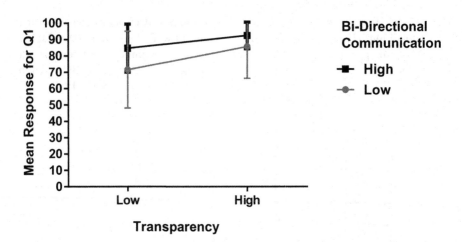

Fig. 5. Interaction plot for "This interface is an improvement over my current system."

3.2   Transparency

There was no main effect for Bi-Directional Communication (BC) on participants' rating of having enough information to choose the best route. There was no significant difference regarding participants' scores between the High BC and Low BC conditions (High BC $M = 92.93$, $SD = 11.72$; Low BC $M = 91.63$, $SD = 14.91$), F (1, 14) = 0.03, $p = .87$, $d = -0.10$. There was no main effect for Transparency (T) on participants' rating of having enough information to choose the best route. There was no significant difference regarding participants' scores between the High T and Low T conditions (High T $M = 92.60$, $SD = 12.56$; Low T $M = 91.97$, $SD = 14.08$), F (1, 14) = 0.06, $p = .81$, $d = -0.05$. There was not a significant interaction effect for BC × T (Low BC × Low T $M = 91.07$, $SD = 17.71$; Low BC × High T $M = 92.20$, SD 12.12; High BC × Low T $M = 92.87$, $SD = 10.45$; High BC × High T $M = 93$, $SD = 12.99$), F (1, 14) = 0.03, $p = .87$.

There was no main effect for Bi-Directional Communication (BC) on participants' rating of their understanding of why the automation provided the routes displayed. There was no significant difference regarding participants' scores between the High BC and Low BC conditions (High BC $M = 93.63$, $SD = 12.44$; Low BC M = 88.47, SD = 19.88), F (1, 14) = 1.90, $p = .19$, $d = -0.32$. There was no main effect for Transparency (T) on participants' rating of their understanding of why the automation provided the routes displayed. There was no significant difference regarding participants' scores between the High T and Low T conditions (High T $M = 92.23$, $SD = 14.06$; Low T $M = 89.87$, $SD = 18.26$), F (1, 14) = 0.71, $p = .42$, $d = -0.15$. There was not a significant interaction effect for BC × T (Low BC × Low T $M = 86.33$, $SD = 23.58$; Low BC × High T $M = 90.60$, SD 16.18; High BC × Low T $M = 93.40$, $SD = 12.94$; High BC × High T $M = 93.87$, $SD = 11.94$), F (1, 14) = 0.44, $p = .52$.

3.3   Bi-directional Communication

There was no main effect for Bi-Directional Communication (BC) on participants' rating of having the ability to modify the routes to suit their needs. There was no significant difference regarding participants' scores between the High BC and Low BC conditions (High BC $M = 91.70$, $SD = 15.33$; Low BC $M = 89.17$, $SD = 19.55$), F (1, 14) = 0.87, $p = .37$, $d = -0.15$. There was no main effect for Transparency (T) on participants' rating of having the ability to modify the routes to suit their needs. There was no significant difference regarding participants' scores between the High T and Low T conditions (High T $M = 88.77$, $SD = 22.56$; Low T $M = 91.97$, $SD = 12.31$), F (1, 14) = 0.76, $p = .40$, $d = -0.19$. There was not a significant interaction effect for BC × T (Low BC × Low T $M = 89.47$, $SD = 15.50$; Low BC × High T $M = 89.87$, SD 23.60; High BC × Low T $M = 94.73$, $SD = 9.13$; High BC × High T $M = 88.67$, $SD = 21.52$), F (1, 14) = 1.99, $p = .18$.

There was no main effect for Bi-Directional Communication (BC) on participants' rating of having the ability to give the automation appropriate information to generate suitable routes. There was no significant difference regarding participants' scores between the High BC and Low BC conditions (High BC $M = 91.77$, $SD = 12.02$; Low BC $M = 86.77$, $SD = 22.75$), F (1, 14) = 1.48, $p = .24$, $d = -0.29$. There was no main effect for Transparency (T) on participants' rating of having the ability to give the automation appropriate information to generate suitable routes. There was no significant difference regarding participants' scores between the High T and Low T conditions (High T $M = 89.37$, $SD = 18.64$; Low T $M = 89.17$, $SD = 16.14$), F (1, 14) = 0.01, $p = .94$, $d = -0.01$. There was not a significant interaction effect for BC × T (Low BC × Low T $M = 85.33$, $SD = 21.79$; Low BC × High T $M = 88.20$, SD 23.71; High BC × Low T $M = 93.00$, $SD = 10.48$; High BC × High T $M = 93.53$, $SD = 13.56$), F (1, 14) = 1.38, $p = .26$.

3.4   Trust

There was no main effect for Bi-Directional Communication (BC) on participants' rating of whether they would trust a route from the available options (if they were to choose one). There was no significant difference regarding participants' scores between the High BC and Low BC conditions (High BC $M = 93.10$, $SD = 11.27$; Low BC $M = 91.67$, $SD = 12.58$), F (1, 14) = 0.02, $p = .64$, $d = -0.12$. There was no main effect for Transparency (T) on participants' rating of whether they would trust a route from the available options (if they were to choose one). There was no significant difference regarding participants' scores between the High T and Low T conditions (High T $M = 93.20$, $SD = 11.07$; Low T $M = 91.57$, $SD = 12.78$), F (1, 14) = 0.51, $p = .49$, $d = -0.14$. There was not a significant interaction effect for BC × T (Low BC × Low T $M = 92.07$, $SD = 11.09$; Low BC × High T $M = 91.27$, SD 14.08; High BC × Low T $M = 91.07$, $SD = 14.46$; High BC × High T $M = 95.13$, $SD = 8.07$), F (1, 14) = 1.38, $p = .26$.

3.5   Overall Effectiveness

Overall means from Questions 1–6 from each condition were calculated and analyzed with a RMANOVA. There was no main effect for Bi-Directional Communication (BC) on overall effectiveness across the means from each condition. There was no significant difference regarding participants' scores between the High BC and Low BC conditions (High BC $M = 92.03$, $SD = 10.81$; Low BC $M = 87.64$, $SD = 15.05$), F (1, 14) $= 2.62$, $p = .13$, $d = -0.34$. There was no main effect for Transparency (T) on participants' rating of whether they would trust a route from the available options (if they were to choose one). There was no significant difference regarding participants' scores between the High T and Low T conditions (High T $M = 90.79$, $SD = 12.79$; Low T $M = 88.88$, $SD = 13.07$), F (1, 14) $= 1.48$, $p = .37$, $d = -0.24$. There was not a significant interaction effect for BC $\times$ T (Low BC $\times$ Low T $M = 85.98$, $SD = 15.88$; Low BC $\times$ High T $M = 89.30$, SD 14.23; High BC $\times$ Low T $M = 91.79$, $SD = 10.27$; High BC $\times$ High T $M = 92.28$, $SD = 11.35$), F (1, 14) $= 0.87$, $p = .37$.

3.6   Additional Measures

Tenet Usefulness. Participants were asked to rate how useful the individual High HAT tenets were, on a 0 = Completely Useless to 100 = Completely Useful scale (Fig. 6).

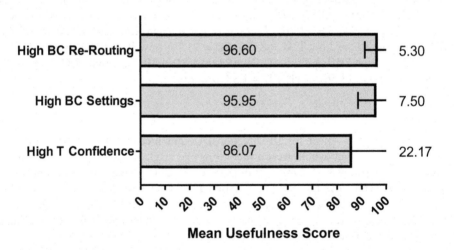

Fig. 6. Mean usefulness scores and standard deviations of HAT tenets.

Desire to Use the Tenets. Participants were asked to rate if they would use the individual High HAT tenets if they were available on a GPS, on a 0 = Never to a 100 = Always scale (Fig. 7).

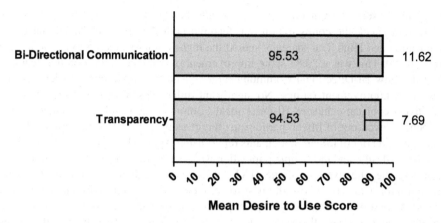

Fig. 7. Mean scores and standard deviations for desire to use the individual High HAT tenets.

Desire to Use the HAT GPS Interface. Participants were asked to respond 'Yes' or 'No' to the statement "I would use HAT GPS over my current GPS system" (Fig. 8).

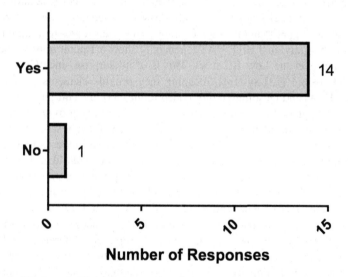

Fig. 8. Total 'Yes' and 'No' responses for desire to use the HAT GPS interface over participants' current GPS system.

4 Discussion

The results show that by enhancing the HAT factors present on the current Apple Maps UI users rated the UI as a significant improvement over their current GPS systems. Both factors (HAT tenets) manipulated showed a significant UI improvement rating over users' current GPS system. While there was not a significant two-way interaction effect, the configuration that featured both the High Bi-Directional Communication and High Transparency scored the highest improvement ratings on a scale from 0–100. This was a 20.87 score improvement over the configuration that featured both the Low Bi-Directional Communication and Low Transparency, which had the lowest UI improvement ratings. No significant differences were found regarding the effectiveness of the enhanced Bi-Directional Communication or Transparency factors.

Since eleven out of fifteen participants report using Apple iOS platforms (iPhone, iPad) and nine out of fifteen participants report using Apple Maps, it is safe to conclude that the UI design and the sample population were appropriate. The lack of significance regarding the effectiveness of the HAT tenets from participants' ratings of having enough information to choose the best route, their understanding why the automation provided the routes, their ability to give the automation the appropriate information to generate suitable routes, their ability to modify the routes, and their trust in the routes (if they were to choose one) could be attributed to the low sample size for this pilot study, the low amount of traffic depicted on the routes, or because the participants did not have to choose the 'best' route. This suggests that a larger sample size is needed for the effect sizes we found.

An interesting observation was found when evaluating the individual High HAT tenet usefulness and desire to use scores. Participant mean scores regarding the usefulness of the individual High HAT tenets show the High BC ratings to be approximately 10 points higher than the High T ratings. The mean scores regarding participant desire to use the individual High HAT tenets only show a 1 point improvement for the High BC tenet over the Low BC tenet. This is of no surprise since GPS systems are already SAT 1+2+3 [12] systems, meaning they provide elements such as a future predictions (ETA) and reasoning (color coded traffic). For the current study the High T condition was upgraded to a SAT 1+2+3+U [12] system (by providing uncertainty in the form of route confidence). While the confidence did not score as high as the Bi-Directional Communication elements regarding its usefulness, participants still reported a high desire to use it. This suggests that participants liked having the confidence rating and may find it more useful if the traffic present on the maps is increased.

The user's preference for the enhanced HAT interfaces is consistent with previous research regarding a HAT tenet enhanced aviation dispatch display system [11]. While this study does not provide enough evidence to state conclusions based on the utility of the enhanced HAT tenets, participants have now reported preference for HAT tenet enhanced systems over non-enhanced systems in both the aviation dispatch and automotive navigation domains, supporting the assumption that HAT tenets are generalizable to domains outside of aviation [13].

Future HAT tenet research may replicate this study but increase the amount of traffic present on the routes. This adjustment would be intended to address a lack of utility for the HAT features among the present study. For example, if there is little to no traffic depicted on the original routes, users may have less of a reason to rely on a confidence rating because there is no reason to question the predicted ETA. Or, if there is only traffic near an obstacle or high traffic event, users may not find it necessary to adjust route settings and take a less direct route that avoids the traffic area. Another possible method adjustment is to shift the user task from simply evaluating the routes and reroutes to finding and choosing the best route available for the destination. This would provide the researchers an opportunity to rate the quality of the routes provided and manipulate them. This would allow the researchers to measure if there is a difference in user's ability to find the 'best' route across the different HAT tenet configurations.

Acknowledgments. I would like to thank the Human-Autonomy Teaming Lab at NASA Ames Research Center for their guidance on appropriately implementing the HAT tenets in the prototype GPS design, their ongoing feedback which helped shape the study, and the knowledge that they have shared with me. I would like to thank my project advisor Dr. Sean Laraway for his ongoing support of this project and for the opportunities that he has provided me. I would like to thank Dr. Susan Syncerski for her assistance with recruiting participants and providing the lab space used for the study. I would also like to thank Dr. Daniel Rosenberg for teaching me interaction design, without which I would not have been able to build these prototypes.

References

1. Apple: 6 screenshots of the Apple Maps directions (3 original and 3 with the driving options turned on) from NASA Ames Research Center, Moffett Field, CA to The Bay Fish 'N Chips, Sunnyvale, CA (n.d.). Accessed 19 Dec 2017
2. Apple: Screenshot of Apple Maps map area between NASA Ames Research Center, Moffett Field, CA and The Bay Fish 'N Chips, Sunnyvale, CA (n.d.). Accessed 19 Dec 2017
3. Apple: 6 screenshots of the Apple Maps directions (3 original and 3 with the driving options turned on) from 824 Blair Ave, Sunnyvale, CA 94087 to Oakland Coliseum, Oakland, CA (n.d.). Accessed 04 Jan 2018
4. Apple: Screenshot of Apple Maps map area between Sunnyvale, CA and Oakland, CA (n.d.). Accessed 04 Jan 2018
5. Apple: 5 screenshots of the Apple Maps directions (2 original and 3 with the driving options turned on) from San Jose State University, San Jose, CA to American Barbell Clubs, Santa Clara, CA (n.d.). Accessed 05 Jan 2018
6. Apple: Screenshot of Apple Maps map Area between San Jose State University, San Jose, CA and American Barbell Clubs, Santa Clara, CA (n.d.). Accessed 05 Jan 2018
7. Apple: 6 screenshots of the Apple Maps directions (3 original and 3 with the driving options turned on) from Red Rock Café, Mountain View, CA to the San Jose Museum of Art, San Jose, CA (n.d.). Accessed 05 Jan 2018
8. Apple: Screenshot of Apple Maps map Area between Red Rock Café, Mountain View, CA to the San Jose Museum of Art, San Jose, CA (n.d.). Accessed 05 Jan 2018
9. Brandt, S.L., Lachter, J., Russell, R., Shively, R.J.: A human-autonomy teaming approach for a flight-following task. In: Baldwin, C. (ed.) AHFE 2017. AISC, vol. 586, pp. 12–22. Springer, Cham (2018). https://doi.org/10.1007/978-3-319-60642-2_2
10. Chen, J.Y.C., Lakhmani, S.G., Stowers, K., Selkowitz, A.R., Wright, J.L., Barnes, M.: Situation awareness-based transparency and human-autonomy teaming effectiveness. Theor. Issues Ergon. Sci. **19**(3), 259–282 (2018). https://doi.org/10.1080/1463922X.2017.1315750
11. Lachter, J., Brandt, S.L., Sadler, G., Shively, R.J.: Beyond point design: general pattern to specific implementations. In: Baldwin, C. (ed.) AHFE 2017. AISC, vol. 586, pp. 34–45. Springer, Cham (2018). https://doi.org/10.1007/978-3-319-60642-2_4

12. Meuleau, N., Plaunt, C., Smith, D.E., Smith, T.B.: An emergency landing planner for damaged aircraft. In: Proceedings of the Twenty-First Innovative Application of Artificial Intelligence Conference, pp. 114–121 (2009)
13. Shively, R.J., Lachter, J., Brandt, S.L., Matessa, M., Battiste, V., Johnson, W.W.: Why human-autonomy teaming? In: Baldwin, C. (ed.) AHFE 2017. AISC, vol. 586, pp. 3–11. Springer, Cham (2018). https://doi.org/10.1007/978-3-319-60642-2_1

Effectiveness of Human Autonomy Teaming in Cockpit Applications

Thomas Z. Strybel[1(✉)], Jillian Keeler[1], Vanui Barakezyan[1],
Armando Alvarez[1], Natassia Mattoon[1], Kim-Phuong L. Vu[1],
and Vernol Battiste[2]

[1] Department of Psychology, California State University Long Beach,
Long Beach, CA 90840, USA
{thomas.strybel,kim.vu}@csulb.edu,
jill.keeler@student.csulb.edu,
vanuibarakezyan@yahoo.com,
mandoalva9@gmail.com, nat.mattoon@gmail.com
[2] San Jose State University Foundation, Moffett Field, CA 94035, USA
vernol.battiste@nasa.com

Abstract. Single pilot and/or remotely piloted operations are becoming feasible in the national airspace system because of advances in autonomous systems, and the development of Human-Autonomy Teams (HAT). We compared a recommender tool for cockpit applications installed with HAT tools or No HAT tools, using simulations of off-nominal events varying in severity. Pilots on average spent more time with the tool when the HAT features were present, but there was considerable variability between pilots in tool usage. However, greater time spent using the tool was associated with lower subjective workload (NASA TLX).

Keywords: Human-Autonomy Teaming · Workload · Situation awareness

1 Introduction

The development of autonomous systems has been ongoing for some time in the aviation, rail, medical and automotive industries, among others. It is unlikely that these systems will completely replace the human element in the system in the near term; therefore, designers are focused on automation systems that serve as team members or partners with human operators, a concept known as "Human Autonomy Teaming (HAT)." HAT represents a significant shift in the view of automation as a simple replacement for human functions. Nowhere is this change more apparent than in the aviation industry. In the 20th century automation served to replace skilled crew members. New jet engines eliminated the need for the flight engineer, advanced navigation aids (e.g. VOR, INS) eliminated the need for a navigation officer, and improvements in radio communication eliminated the need for communications officer. As a result, the standard five-person crew (two pilots, a navigator, radio operator and flight engineer) was reduced to just two pilots [1]. It is not surprising then, that some thought and development starting at the end of the 20th century is considering further reductions in crew size.

© Springer International Publishing AG, part of Springer Nature 2018
S. Yamamoto and H. Mori (Eds.): HIMI 2018, LNCS 10905, pp. 465–476, 2018.
https://doi.org/10.1007/978-3-319-92046-7_39

1.1 Human Autonomy Teaming (HAT)

Reduced Crew Operations (RCO) or Single Pilot Operations, however, will not be feasible without support from advanced autonomous systems in combination with ground support [2, 3]. The Human Automation Teaming Laboratory at NASA Ames Research Center has been evaluating HAT autonomous systems for assisting pilots and dispatchers in these environments. In a program of automation support development and simulation evaluations, the HAT Laboratory has evaluated tools for a ground dispatcher assisting pilots dealing with off-nominal events and cockpit tools to assist reduced crews [4, 5]. At the same time, a conceptual model for HAT is being developed that consists of three tenets (for details see [3, 4, 6]).

Transparency. The human operator must understand the intent and reasoning of the autonomous agent, and determine the factors used by the agent in arriving at a solution or recommendation. The operator must have knowledge of the general logic being used by the agent, and have an accurate mental model of its functioning. At the same time, the autonomous agent must understand the preferences, attitudes and states of the human team members. This latter specification may be one of the more challenging aspects facing designers of autonomous-agent crew members.

Bi-directional Communications. Fast and effective communications between humans and autonomous agents is essential for effective HAT. Communications will establish shared knowledge of team goals, current status and errors, either human or autonomy. Effective communication also requires that the human crew effectively and accurately direct the autonomous agent, and override its decisions, if necessary.

Operator-Directed Interface. The risk of automation failure can override the benefits of an automation agent, so it is important the operator be able to allocate tasks depending on the current situation. This also serves to keep the operator in the loop.

1.2 Assessing the Effectiveness of HAT

The effectiveness of HAT will be determined by additional factors, based on previous research in the area of human-automation interaction, and human-human team performance. Therefore, evaluations of the HAT designs must include assessments at several system levels, and assess system and operator performance. Operator factors that will determine HAT effectiveness include the following:

Trust. Effective HAT will depend on the degree of human trust in the autonomous agent. Trust is a complex state that is similar to but not exactly the same as human-human trust. Human crew members must place themselves in a position of uncertainty and vulnerability with respect to the autonomous agent, in the expectation that the agent is doing what is supposed to do, or communicate why it is unable to do so. Trust must be appropriately calibrated in order to avoid negative consequences of over trust (i.e., complacency) and under trust (i.e., workload; [7]).

Workload. When automation agents become an integral team member, it is critical that they not increase the workload of the human team members (e.g., [8]). Excessive workload may be produced by difficulty understanding the agent's current reasoning, awkward communication procedures, and lack of trust in the agent.

Situation Awareness. In addition to awareness of the goals, tasks, systems and environments, human crew members must be aware of the current status of the automated agent, and the automated agent must be aware of the current state of human crew members [9, 10].

Individual Differences. Human team members can have varied skills that impact how they collaborate and accomplish mission goals. Human operators may also have different attitudes toward an automated agent as crew member. It is important therefore that the agent know about these differences and take them into account when interacting with them [9].

1.3   NASA's HAT Demonstrations

NASA's HAT Laboratory has an ongoing program of design and evaluation of HAT tools for aerospace applications [4]. In 2016, collaboration tools for ground operators supporting RCO were developed based on the HAT tenets, and these were evaluated in a simulation demonstration [11, 12]. The following tools were developed based on the HAT tenets.

Plays. Using plays ensured that the HAT agent was operator directed. The ground operator initiated automation procedures from a set of plays in order to establish system goals and clarify roles and responsibilities between the automation agent and human operator. These plays were called in response to off-nominal events, and served to activate the Autonomous Constrained Flight Planner System (ACFP) and an electronic checklist that showed tasks allocated to either human or autonomous agent.

ACFP. The major tool developed on HAT principles, the ACFP determines alternative flight plans based on a number of factors such as weather, services, and fuel. Transparency was promoted by displaying the values of each factor used to determine the recommendations, and their relative weights in the decision. Bi-Directional communications was enabled by displaying the weight values as sliders that could be adjusted. The operator could then request new recommendations.

Traffic Situation Display. This display provided additional transparency. The TSD is a 3-dimensional display of traffic in the area surrounding the currently serviced aircraft. Weather, turbulence, ATIS at the destination and other potential divert options could be provided at the operator's request.

Voice I/O. Voice commands for selecting plays or requesting information, enhanced the principle of bi-directional communication. Voice also announced to the operator the current activity of the automation agent.

The simulation evaluation was based on a small sample. Nevertheless preliminary results indicated that workload was lower with HAT tools compared to No-Hat tools. Operators took more time to uplink revised flight plans in the HAT condition, even when no adjustments were made to the recommendations of the autonomous recommender system [12]. Participants rated the HAT condition more favorably than the no-HAT condition: diversion recommendations were rated more acceptable, and confidence in the recommendations was higher. Moreover, HAT displays were preferred for keeping up with operationally important issues, ensuring situation awareness, integrating information, and reducing workload [11].

1.4   Present Investigation

The tools in Brandt et al. [11] were modified and installed on a tablet workstation, in order to provide them to line pilots. A simulation test was conducted with line pilots in a distributed simulation network that compared pilot performance, behavior and subjective responses to autonomous agents based on HAT vs. No-HAT principles. The present paper reports how pilots used these automated tools to deal with off-nominal events.

2   Method

2.1   Participants

Twelve ATP participants participated in this simulation. All were line pilots (2 Captains and 10 First Officers). Eleven participants had over 5000 h of line experience (one with 3000–5000 h) and 10 had over 3000 h of glass cockpit experience. For additional details on the sample, see [13].

2.2   Apparatus

A distributed simulation network was established between University of Iowa and California State University Long Beach, with support from NASA Ames Research Center and Rockwell-Collins, Inc. Pilots flew a Boeing 737 motion-base simulator located in the Operations Performance Laboratory (OPL) at University of Iowa. Confederate dispatchers and air traffic controllers were located in the Center for Human Factors in Advanced Aeronautics Technologies (CHAAT) at California State University, Long Beach (CSULB) which also housed servers for the HAT tools. The simulation network was made possible with NASA's MultiAircraft System (MACS), and ADRS along with additional tools for generating flight diversion recommendations, displaying automated checklists, and a Cockpit Display of Traffic Information (CDTI). Voice communications between the pilot, ground support and simulation personnel were accomplished via TeamSpeak™ software. For additional details of the distributed simulation configuration, see [5].

HAT tools were installed on a Microsoft Surface Prime tablet that was mounted on the left wall of the cockpit. The tablet contained separate pages that provided functions or information based on the flight phase: Enroute, Approach, Runway, Play and Alerts. These could be selected by touch or voice commands. The Approach and Runway pages provided charts and information regarding airports, runways, etc. Alerts served to initiate most off nominal events found in the simulation. When an alert occurred, the Alerts button would turn orange and the specific alert would be listed on the page in red or orange, depending on severity.

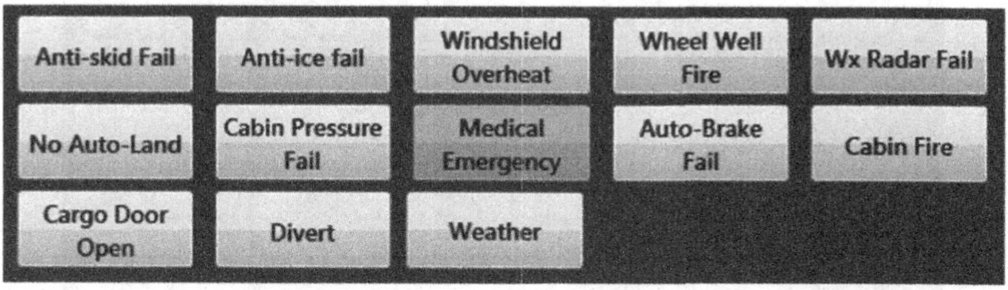

Anti-skid Fail	Anti-ice fail	Windshield Overheat	Wheel Well Fire	Wx Radar Fail
No Auto-Land	Cabin Pressure Fail	Medical Emergency	Auto-Brake Fail	Cabin Fire
Cargo Door Open	Divert	Weather		

Fig. 1. Panel of plays available to the pilot which could be activated by either touch or voice.

The pilot acknowledged the alert, and utilized the Plays and Enroute pages to resolve the event. The information displayed depended on the automation condition (HAT or No HAT). The Plays page contained a set of plays that corresponded to one of the off-nominal events. The pilot would call the play, either via touch input (see Fig. 1) or voice commands. When a play was activated, the information on the page changed depending on the automation condition.

The Autonomous Constrained Flight Planner (ACFP) is a flight-planning recommender system for assisting pilots in generating and evaluating routes. The ACFP shows a table of airports based on four diversion recommendations. In the No-HAT condition, these four alternative routes are displayed, but no information as to rationale used to arrive at the recommendations was shown. In effect, if the pilot did not like any of the recommendations, he or she would have to generate a diversion flight plan without assistance from the ACFP. In the HAT condition, the ACFP provided the recommended airports and the basis of its reasoning, as shown in Fig. 2. In addition to a rating of risk, the factors used in the decision-making process and their values were displayed in tabular form. The factor weight values were shown above the table as sliders. The pilot could generate a new set of recommendations by adjusting the relative weights (moving the sliders) and requesting a new set of options.

In the HAT Condition, an electronic checklist was also displayed based on the play that had been called. The checklists were based on the QRH manual, but included steps that would be performed by the automated agent. In the No-HAT Condition, traditional paper-based checklists were provided in the QRH manual located in the center console of the cockpit. For both conditions additional paper documentation was provided on voice commands, ATIS and in the Medical Emergency event, Medlink.

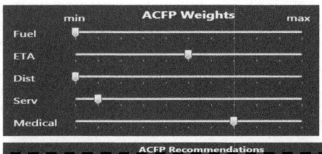

Fig. 2. ACFP display with factor weights in the HAT condition. In the No-HAT condition, only the top "Option" row is displayed.

2.3   Experimental Design and Procedure

Each pilot flew six, 12–15-min scenarios, three in the HAT Condition and three in the No-HAT Condition, with the order counterbalanced. Within each condition, the scenarios varied in the severity of the off-nominal event, as shown in Table 1. The off-nominal event began roughly 4 min into the scenario. The event was signaled by an alert on the tablet. Once the pilot acknowledged the alert, the plays were displayed and the pilot could select the play corresponding to the event. This brought up the ACFP along with the Automated Checklist and Weighting factors in the HAT Condition. In the No-HAT Condition, the pilot would find the appropriate checklist in the QRH manual and begin working through it. In the Medical Emergency event, the event was alerted by a confederate experimenter serving as flight attendant and MedLink. Pilots also communicated via voice with air traffic control and ground dispatch for additional information regarding weather and alternative airports. All flight plan changes required clearance from air traffic control.

At the end of each scenario, pilots completed the NASA TLX workload questionnaire [14], Situation Awareness Rating Technique [15] and a questionnaire asking about the usefulness of the tools. After all scenarios were completed for one condition, the pilots completed a questionnaire that asked about the tools just used, and a Trust in Automation scale [16] (for results of the questionnaires, see [13]). After both conditions were completed, pilots were debriefed as to their preference for HAT agents and other concerns regarding the tools and simulations.

Table 1. Specific off-nominal events for each event severity level

Severity level	Event
High	Medical
	Wheel well fire
Moderate	Airport weather
	Windshield overheat
Low	Weather radar failure
	Anti-skid inoperable

The design was repeated measures with the factors automation condition (HAT vs. No HAT) and event severity (Low, Moderate and High). Here we report on the extent to which pilots utilized the information on the tablet relative to information on the instrument panel, and paper documents, for HAT vs. No Hat conditions. We recorded the eye gaze of the pilots in each scenario. Cameras were located on the tablet, instrument panel and on the right panel next to the first officer seat. The amount of time gazing at each major source of information, (tablet, instrument panel and documents) was determined by examining the position of the participant's sclera relative to the information source. Tablet gaze time was measured as time looking at the left-side panel of the cockpit. Instruments gaze time was the time spent looking forward at the cockpit instruments including time for adjusting flight parameters. Documents gaze time was the time looking for and reading documents that were originally located in the center console, but often ended up on the pilots lap. It also included time to query the confederate flight attendant in some scenarios. Finally, other interactions with the tablet were recorded, such as the frequency of weight adjustments in the HAT Condition, and which, if any, of the alternative recommendations were selected. We analyzed NASA TLX workload and SART situation awareness scores for each automation condition and event severity.

3   Results

3.1   Eye Gaze

Repeated measures ANOVAs were performed on the total time to resolve each off-nominal event with the factors Automation Condition (HAT vs. No HAT) and Event Severity (low, moderate, high). All effects were non-significant, indicating that the Automation Condition did not affect the time to resolve the event. As shown in the top row of Table 2, resolution times for HAT and No HAT Conditions were nearly identical. ANOVAs were also run on the time spent gazing at each source of information (tablet, instrument panel and documents). Because individual gaze times are related to total event times, we converted the gaze times to the proportion of time spent on each information source for each event, and ran repeated measures ANOVAs on these proportions as well.

Table 2. Mean gaze times and proportion of gaze times: HAT vs. No HAT

Event, gaze times and proportion of gaze	HAT	No HAT	p
Mean (SEM) time to resolve off-nominal event (s)	295.5 (13.7)	301.2 (16.1)	.80
Mean (SEM) gaze time - Tablet (s)	137.8 (10.0)	102.4 (12.4)	**.002**
Mean (SEM) gaze time - Instruments (s)	125.7 (10.3)	134.9 (9.6)	.323
Mean (SEM) gaze time - Documents (s)	30.6 (5.3)	58.5 (9.7)	**.012**
Proportion time spent on Tablet (%)	46.6 (2.7)	34.4 (3.4)	**.003**
Proportion time spent on Instruments (%)	43.1 (2.7)	46.7 (2.7)	.25
Proportion time spent on Documents (%)	9.9 (1.5)	18.6 (2.8)	**.007**

As shown in Table 2, in the HAT condition, participants spent significantly more time fixated on the tablet $F(1, 10) = 169.04$; $p < .003$, and significantly less time looking at written documentation, $F(1, 10) = 9.14$; $p < .013$. The time spent on instruments panel was non-significant, however. More time on documents was required in the No-HAT condition because checklists were paper-based, and pilots would have to find the QRH, locate the correct QRH, and follow the checklist. In the HAT condition, electronic checklists were displayed on the Tablet. When gaze times were converted to proportions of time, similar results were obtained. Pilots in the HAT condition spend roughly 47% of the event-resolution time focused on the tablet, and only 10% of the time on documents, compared with the No-HAT condition in which 34% of the event-resolution time was spent looking on the tablet, and 18% on documents.

The proportion of time spent on each display also depended on Event Severity. Significant main effects of Event Severity were obtained for proportion of time on tablet, $F(2, 20) = 4.24$; $p < .029$, and proportion of time on instruments, $F(2, 20) = 4.24$; $p < .029$, but not on documentation. As shown in Fig. 3, relatively more time was spent on the tablet and less time on instruments for moderately severe events. In fact post hoc analysis determined that the difference between each measure were non-significant for low and high severity events. For moderately severe events, pilots spend more time looking at the tablet and less time looking at instruments. All interactions between condition and event severity were non-significant.

3.2   Workload and Situation Awareness

Repeated measures ANOVAs were also run on the post-scenario measures of workload (NASA TLX) and situation awareness (SART). For both measures, the effects of Automation Condition were non-significant. For workload only, a significant main effect of Event Severity was obtained, $F(2, 20) = 16.14$; $p < .0001$. As shown in

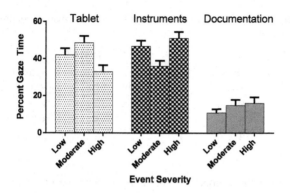

Fig. 3. Percentage gaze time for each information source as a function of event severity

Fig. 4, workload scores were highest for the high-severity events, with no difference in workload for low- and moderate-severity events. TLX scores for high-severity events (M = 46.48, SEM = 9.585) were on average 15 points higher compared with moderate- and low-severity events (M = 32.65, SEM = 7.09; M = 32.49, SEM = 7.26 for moderate and low severity events, respectively), and this was confirmed with post hoc comparisons. Figure 4 also shows that across all levels of Event Severity, there were no differences in TLX scores between HAT and No-Hat conditions. Event Severity and Automation Condition did not significantly affect SART scores.

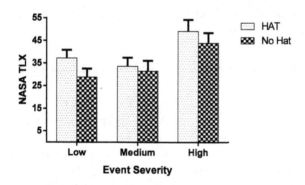

Fig. 4. NASA TLX scores for HAT and No HAT conditions as a function of event severity

3.3   Other Measures of HAT Interactions

There was considerable variability between pilots in the frequency of interactions with ACFP. For example, five of the twelve pilots never adjusted the weights in any HAT scenario; two pilots adjusted weights in only one scenario. This means that 7 of 12 pilots had little or no interactions with factor adjustments. Moreover, slider use depended on Event Severity: three pilots adjusted the weights in the low-severity condition, six pilots in the moderate severity condition and four pilots in the high-severity conditions. Note that three pilots adjusted weights at all severity levels.

474   T. Z. Strybel et al.

Table 3. Number of resolutions accepted by rank of risk: automation condition (HAT vs. No HAT) and event severity

	HAT			No HAT		
	Ranked first	Ranked second	Ranked third - fourth	Ranked first	Ranked second	Ranked third - fourth
Low	11	0	1	10	1	1
Moderate	9	3	0	5	3	1
High	4	3	1	6	2	1

We also counted the number of ACFP resolutions accepted by pilots based on automation condition and severity level, and the rank of the recommendation. As shown in Table 3, most pilots accepted the top ranked recommendation for low-severity events, but as severity increased, fewer top-ranked recommendations were accepted. In fact, for the high-severity events, only 4 pilots accepted the highest-ranked solution in the HAT Condition, and 6 in the No-HAT Condition. For the high-severity condition four pilots in the HAT condition, and 3 in the No-HAT condition, rejected all recommendations.

The lack of effect of automation condition on workload and situation awareness may have been due to differences between pilots in the amount HAT-tool interactions because pilots received minimal training on these tools. To investigate this possibility, we correlated the relative proportion of time spent on each display with workload and situation awareness measures separately for HAT and No-HAT Conditions. These correlations and are shown along with significance values in Table 4. Significant correlations are shown in bold.

Table 4. Correlation between percent time on tablet, NASA TLX and SART. Significance level in parentheses.

	NASA TLX		SART	
	HAT	No HAT	HAT	No HAT
Percent tablet time	**-.35** (.05*)	-.18 (.34)	**.41** (.02*)	.13 (.49)
Percent instrument time	.25 (.16)	.13 (.51)	**-.35** (.04*)	-.11 (.58)
Percent document time	.15 (.40)	.11 (.56)	-.07 (.698)	-.06 (.74)

The proportion of time spent on the tablet was significantly and negatively correlated with TLX scores, meaning that when a greater proportion of time was spent with the HAT tools subjective workload was lower. Moreover, time spent on the Tablet was significantly and positively correlated with SART scores, indicating that more time on the tablet produced higher levels of subjective situation awareness. In the No-HAT Condition, time on tablet was unrelated to workload and situation awareness. However, TLX and SART scores were highly correlated with one another; low workload was associated with high SART situation awareness. Consequently, we computed semi-partial correlations between proportion of time on the tablet with TLX and SART. The semi-partial correlation between TLX and tablet time was reduced to −.28, and was marginally significant. The semi-partial correlation between SART and tablet time was reduced to .21, which was non-significant. In sum, when pilots spent more time interacting with the HAT tools, they reported lower subjective workload.

4 Discussion

These preliminary results from an investigation of automation tools developed based on HAT tenets can be summarized as follows. First, HAT-designed tools did not affect the time to resolve off–nominal events despite the fact that pilots spent more time on the tablet in the HAT condition, and less time on cockpit instruments. Moreover, pilot workload and situation awareness on the average was unchanged by HAT vs. No-HAT tools. On the one hand, this suggests that HAT did not increase workload, which is a good thing. On the other hand, HAT did not improve pilot situation awareness, which is surprising given the additional information provided in the HAT Automation Condition. One possible reason for these lack of difference may be due to the sensitivity of the instruments themselves, suggesting that changes in workload and situation awareness may be too subtle to be detected by subjective instruments.

Another possible explanation may lie in the variability between pilots in their use of the HAT-designed automation. For example, providing pilots with opportunities for modifying the weights used to arrive at flight diversions was intended to promote the HAT tenet of bi-directional communication. However, five pilots never used the sliders, and two only adjusted the weights once. This could mean that sliders are ineffective for bi-directional communications. Moreover, when the severity of the event was high, pilots were less likely to accept the recommendations of the ACFP. Perhaps the improved transparency of this tool becomes less important when rapid decisions are required as in the case of medical emergencies or wheel well fires. In other words, pilots varied in the use of the tools. This is shown most clearly in the correlational analysis of time on tablet, workload and situation awareness. Greater time spent on the tablet was related to lower subjective workload. At this point, we are unable to determine why some pilots made more use of the tablet in HAT conditions than others, but clearly this factor must be considered when designing HAT agents. As pointed out by Chen et al. [9] individual differences in attentional control, spatial ability and gaming experience affect how operators interact with autonomous robots. It is possible that these differences played a role in the use of our HAT agent.

Acknowledgements. This research was supported by the NASA cooperative agreement #NNA14AB39C "Single Pilot Understanding through Distributed Simulation (SPUDS)," R. J. Shively, Technical Monitor.

References

1. Fadden, D.M., Morton, P.M., Taylor, R.W., Lindberg, T.: First-Hand Evolution of the 2-Person Crew Jet Transport Flight Deck (2015). http://ethw.org/First-Hand:Evolution_of_the_2-Person_Crew_Jet_Transport_Flight_Deck
2. Norman, R.A.: Economic Opportunities and Technological Challenges for Reduced Crew Operations. The Boeing Company (2007)
3. Shively, R.J., Lachter, J., Brandt, S.L., Matessa, M., Battiste, V., Johnson, W.W.: Why human-autonomy teaming? In: Baldwin, C. (ed.) AHFE 2017. AISC, vol. 586, pp. 3–11. Springer, Cham (2018). https://doi.org/10.1007/978-3-319-60642-2_1
4. Battiste, V., Lachter, J., Brandt, S., Alvarez, A., Strybel, T.Z.: Human automation teaming: lessons learned and future directions. In: Yamamoto, S., Mori, H. (eds.) HIMI 2018. LNCS, vol. 10905, pp. 479–493. Springer, Cham (2018)
5. Matessa, M., Vu, K.-P.L., Strybel, T.Z., Battiste, V., Schnell, T., Cover, M.: Using distributed simulation to investigate human-autonomy teaming. In: Yamamoto, S., Mori, H. (eds.) HIMI 2018. LNCS, vol. 10905, pp. 541–550 (2018)
6. Lachter, J., Brandt, S.L., Sadler, G., Shively, R.J.: Beyond point design: general pattern to specific implementations. In: Baldwin, C. (ed.) AHFE 2017. AISC, vol. 586, pp. 34–45. Springer, Cham (2018). https://doi.org/10.1007/978-3-319-60642-2_4
7. Parasuraman, R., Sheridan, T.B., Wickens, C.D.: Humans: still vital after all these years of automation. Hum. Factors 50, 511–520 (2008). Golden Anniversary Special Issue
8. Cummings, M.L., Stimpson, A., Clamann, M.: Functional requirements for onboard intelligent automation in single pilot operations. In: AIAA 2016, p. 1652 (2016)
9. Chen, J.Y.C., Barnes, M.J.: Human-agent teaming for multi-robot control: a literature review (ARL-TR-6328). Aberdeen Proving Grounds, MD: Human Research and Engineering Directorate (2013)
10. Endsley, M.R.: From here to autonomy: Lessons learned from human–automation research. Hum. Factors 59, 5–27 (2016)
11. Brandt, S.L., Lachter, J., Russell, R., Shively, R.J.: A human-autonomy teaming approach for a flight-following task. In: Baldwin, C. (ed.) AHFE 2017. AISC, vol. 586, pp. 12–22. Springer, Cham (2018). https://doi.org/10.1007/978-3-319-60642-2_2
12. Strybel, T.Z., et al.: Measuring the effectiveness of human autonomy teaming. In: Baldwin, C. (ed.) AHFE 2017. AISC, vol. 586, pp. 23–33. Springer, Cham (2018). https://doi.org/10.1007/978-3-319-60642-2_3
13. Cover, M., Reichlen, C., Matessa, M., Schnell, T.: Analysis of airline pilots subjective feedback to human autonomy teaming in a reduced crew environment. In: Yamamoto, S., Mori, H. (eds.) HIMI 2018. LNCS, vol. 10905, pp. 359–368 (2018)
14. Battiste, V., Bortolussi, M.: Transport pilot workload: a comparison of two subjective techniques. In: Proceedings of the Human Factors and Ergonomics Society Annual Meeting, vol. 32, no. 2, pp. 150–154. SAGE Publications (1988)
15. Taylor, R.M.: Situational awareness rating technique (SART): the development of a tool for aircrew systems design. In: Situational Awareness in Aerospace Operations (AGARD-CP-478), pp. 3/1–3/17. NATO-AGARD, Neuilly Sur Seine (1990)
16. Jian, J., Bisantz, A., Drury, C.: Foundations for an empirically determined scale of trust in automated systems. Int. J. Cogn. Ergon. 4, 53–71 (2000)

Intelligent Systems

Human-Automation Teaming: Lessons Learned and Future Directions

Vernol Battiste[1]([envelope]), Joel Lachter[2], Summer Brandt[1],
Armando Alvarez[3], Thomas Z. Strybel[3], and Kim-Phuong L. Vu[3]

[1] San Jose State University Foundation, Moffett Field, CA 94035, USA
{vernol.battiste, summer.l.brandt}@nasa.gov
[2] NASA Ames Research Center, Moffett Field, CA 94035, USA
joel.lachter@nasa.gov
[3] California State University Long Beach, Long Beach, CA 90840, USA
mandoalva9@gmail.com,
{thomas.strybel, kim.vu}@csulb.edu

Abstract. Full autonomy seems to be the goal for system developers in almost every area of the economy. However, as we move from automated systems to autonomous systems, designers have needed to insert humans to oversee automation that has traditionally been brittle or incomplete. This creates its own problems as the operator is usually out of the loop when the automation hands over problems that it cannot handle. To better handle these situations, it has been proposed that we develop human automation teams that have shared goals and objectives to support task performance. This paper first summarizes a body of research to develop ground station automation support for single pilot transport operations. Then the paper will describe an initial model of Human Automation Teaming (HAT) which has three elements: transparency, bi-directional communications, and human-directed execution. Transparency in our model is a method for giving insight into the reasoning behind automated recommendations and actions, bi-directional communication allows the operator to communicate directly with the automation, and finally the automation defers execution to the human. The model was implemented through a number of features on an electronic flight bag (EFB) which are described in the paper. The EFB was installed in a mid-fidelity flight simulator and used by 12 airline pilots to support diversion decisions during off-nominal flight scenarios. Pilots reported that working with the HAT automation made diversion decisions easier and reduced their workload. They also reported that the information provided about diversion airports was similar to what they would receive from ground dispatch, thus making coordination with dispatch easier and less time consuming. These HAT features engender more trust in the automation when appropriate, and less when not, allowing improved supervision of automated functions by flight crews.

Keywords: Autonomous systems · Human-autonomy teaming

S. Yamamoto and H. Mori (Eds.): HIMI 2018, LNCS 10905, pp. 479–493, 2018.
https://doi.org/10.1007/978-3-319-92046-7_40

1 Introduction

In every area of the economy there are plans to move from manual (human controlled systems) to autonomous (no human required systems). As the technology needed to support this rapid movement has improved, almost on a daily basis, there is greater recognition that human oversight of these systems will be needed in the near future. For example, when automakers and robot designers use the term autonomy they generally mean: autonomy within a limited range of functions or for a broad range of functions with human oversight. Before proceeding with the discussion of Human Autonomy Teaming, we would like to offer a few definitions of autonomous from Dictionary.com [1]:

> **Government.** a. self-governing; independent; subject to its own laws only. b. pertaining to an autonomy or a self-governing community.
> **Business.** Having autonomy; not subject to control from outside; independent: a subsidiary that functions as an autonomous unit.
> **(of a vehicle)** navigated and maneuvered by a computer without a need for human control or intervention under a range of driving situations and conditions: an autonomous vehicle.

These definitions clearly describe systems that have both the ability and freedom to make independent judgments. However, some of our most advanced systems – Waymo's self-driving car, Tesla's auto-pilot – still require human oversight. For example, current "autonomous" cars have significant problems dealing with traffic when it is directed by people (e.g., flagmen or police officers) and with static objects in the roadway [2, 3]; thus, the need to team up autonomous systems with humans to improve overall system safety and efficiency.

People working with automation, even when that automation has a certain level of autonomy, does not equate to human autonomy teaming. HAT requires that there be some level of cooperation and coordination in achieving goals. This paper tells the story of how our research at NASA in support of work on single pilot operations (SPO) and reduced crew operations (RCO) came to incorporate HAT. The goal of that research was to explore the possibility of reducing the crew complement on commercial flight decks from two pilots to one. Based on task analysis, a concept of operations was developed that called for automation and a ground operator (similar to a dispatcher) to support the single pilot. Our initial prototype ground stations provided an ability to coordinate with a human ground operator, and provided (increasing levels of) automation. As we included more automation, our research participants expressed distrust of the automation and uncertainty about the rational for the suggestions recommended by the automation. This led us to begin work to make the automation act more like a teammate.

After a brief discussion of our pre-HAT work, this paper will present our initial vision for HAT, followed by a discussion of our HAT implementation to support an advanced airline dispatcher ground station and a final implementation of HAT tools on the flight deck. The majority of data reported in this paper will be flight dispatcher and commercial transport pilot ratings and their comments on the usability and acceptability of the HAT tools.

2 Pre-HAT SPO/RCO Work

2.1 Technical Interchange Meeting

NASA began its work on SPO by convening a technical interchange meeting (SPO TIM) to discuss the feasibility of SPO [4]. Two types of challenges resulting from the removal of the second pilot were often mentioned: workload and redundancy (see also [5]). The consensus of attendees was that to make SPO feasible, workload needed to be reduced to a level where a single pilot could handle it. Also, and perhaps more important, removing the second pilot raises issues about how to replicate the redundancy they currently provide which is required for certification and flight safety. The group converged on two approaches to the workload and redundancy problem: onboard automation or external support from other people.

2.2 Experiment 1: Together Versus Apart

In our first experiment, we evaluated the effect of crews working together, versus being in separate locations, on crew communications and workload (see Fig. 1), as suggested by Thomas Sheridan at the SPO TIM [4, 6]. In this study flight deck automation replicated that found on current transport category aircraft. Ten two-pilot crews flew both together and apart – at separate redundant ground stations – while resolving off-nominal diversion scenarios.

Lessons Learned. In this experiment we found that while control manipulations can be acknowledged non-verbally in two-pilot operations, acknowledgement may be forgotten or require extensive radio use. Additionally, there is a risk of shared situation awareness (SA) being reduced when pilots are physically separated. Pilots appeared to have increased uncertainty about roles and responsibilities (e.g., Do I have the aircraft or do you?), uncertainty about control manipulation (e.g., Are you entering the altitude?) and uncertainty about completed actions (e.g., Did you put that in the CDU?).

Based on these results and additional feedback from our pilots, we developed tools to facilitate remote collaboration – Crew Resource Management (CRM) Tools. These tools were then implemented in our ground station and evaluated in the next experiment.

Fig. 1. Pilots flew together on the left and captain and first officer separated on right.

2.3   Experiment 2: Higher Fidelity with CRM Tool Manipulation

In our second experiment 18 two-pilot crews flew high-workload off-nominal scenarios that required diversions [7]. However, this time with CRM indicators we developed to show roles and responsibilities, shared charts, shared flight deck displays and video that allowed the pilots to see each other (see Fig. 2). As in the first experiment, crews flew side-by-side in a baseline configuration, (this time in a high-fidelity full motion simulator) and separated. In the separate condition the captain remained on the flight deck and the first officer flew a prototype ground operator station that incorporated aspects of both a flight deck and an airline dispatch station. To assist in planning diverts, the ground station was equipped with a rerouting tool incorporating a previously developed NASA technology, the Emergency Landing Planner (ELP; [8, 9]), which assessed the suitability of airports near the aircraft and returned recommendations for which airport would make the best divert. This tool also provided routing information to the selected airport. A simple dispatcher task to reroute aircraft around convective weather was introduced.

Lessons Learned. Data from this second experiment was generally positive for our shared tools (CRM indicators, video, flight deck displays, and shared charts) although pilots had multiple suggestions for improvement. A communication analysis showed that crews spent more time communicating, shared more decision-relevant information and were more responsive to each other when CRM indicators were available, suggesting these tools directed crewmembers' attention to their joint responsibility for safe decision-making [10]. We also found that when the captain requested assistance from the ground dispatcher, the dispatcher focused on that aircraft and stopped performing

Fig. 2. SPO II ground station. CRM indicators circled on the right and video of the cockpit circled on the left.

the rerouting task. We concluded that a ground operator working off-nominal aircraft should be relieved from servicing other aircraft. This procedure is similar to current practice in Airline Operations Centers: dispatchers often hand off their nominal aircraft to other dispatchers and give one-on-one support to aircraft that need to divert. We refer to this one-on-one mode of operation as dedicated assistance.

2.4   Experiment 3: Investigation of Situation Awareness Issues

In the third study we tested two concepts of operation. If SPO was to be considered for implementation, a ground operator must give dedicated assistance to aircraft in high workload or off-nominal situations. However, in order for SPO to be cost effective, the ground operator must handle more than one aircraft. In this third study we evaluated two ground station concepts of operation:

Specialist, in which the ground operator only performs dispatch functions and hands the aircraft to a separate person (pilot) who provides dedicated assistance to the aircraft when needed; and

Hybrid, in which the ground operator performs dispatcher functions and, when needed can hand off all other aircraft and provide functions during dedicated assistance.

The CRM tools and the ELP [8] were similar to those in the previous experiment (see Fig. 3) [5]. In this experiment thirty-five commercial airline pilots participated. In the hybrid condition a ground operator (the participant pilot) acted as a dispatcher until one troubled aircraft (a confederate pilot) had an off-nominal situation, at which time the dispatcher entered dedicated support; assuming the role of first officer for that flight and handing off the other aircraft. Varying the level of interaction the ground operator had with both the "to-be-troubled" aircraft and with the airspace in general, prior to dedicated support, allowed us to look at the effects of this initial exposure on performance. In the specialist condition, the participant pilot was simply handed the troubled aircraft with a brief message (e.g., "Sir, flight 123 needs dedicated assistance") without prior exposure to either the flight or other environmental conditions such as the weather.

Lessons Learned. We found no performance difference between our two ground station support concepts - hybrid and specialist. This suggests that with the tools provided participants could gain sufficient SA to perform the task relatively quickly. From a concept of operations perspective, it suggests that the decision of whether to have ground pilots waiting to takeover distressed aircraft or increase training -cost and time- for flight-followers could be made on economic grounds.

Overall, participants found the ground station tools (Information on the aircraft control list (ACL), shared charts, the traffic situation display (TSD) with ELP recommendations, and CRM indicators) to be useful. Of particular interest were their impressions of the ACL. Pilots reported that the ACL improved their SA. One pilot commented, "I would like to see a lot more info on the ACL. I really liked the concept."

Fig. 3. SPO III dispatch ground station: (a) flight deck displays for the selected aircraft; (b) TSD, ACL with ELP recommendations; and (c) CRM tools and sharable charts.

2.5   Experiment 4: Monitoring Multiple Aircraft

The previous three experiments focused on the ability of a ground-based flight-follower to perform piloting duties, sometimes helping to manage a single-piloted aircraft under high workload and off-nominal conditions. This study examined the ability of this flight follower to work with a fleet of aircraft. These flight-followers could not actually control the aircraft as they could in the previous studies, however, with additional automation they did perform some of the functions normally associated with the pilot not flying/monitoring in a two-person crew.

In order to facilitate the increased monitoring task, a new Aircraft Monitoring and Management System (AMMS) was introduced. This system gathered data from various sources (e.g., monitoring weather data, ATC clearances, aircraft position, and EICAS alerts) and placed prioritized alerts on a redesigned ACL when threats were detected (see Fig. 4) [5]. The route replanning tool, presented to the left of the TSD, used in Experiment 3 was augmented to display ATIS at the destination airport as well as indicate which of a number of risk factors were present in any potential divert location [12, 13]. Operators could request ratings for airports that were not recommended by the tool and could adjust the weighting of various factors going into the recommendation. The modified tool was renamed the Autonomous Constrained Flight Planner, ACFP. Five certified dispatchers and five commercial airline pilots participated in the build one evaluation. Participants ran two one hour-long scenarios. Each scenario required participants to make approximately six diversions using the ground station tools.

Lessons Learned. The dispatchers and pilots were very positive about the ground station. Specifically, they agreed that the automation and displays did a good job of

Fig. 4. Build 1 ground station. Bottom center, ACL, augmented with timeline, alerting information; above the ACL is the TSD; to the left is flight controls and displays for the selected aircraft in read-only mode; on the right is CONUS map and charts.

integrating information. They found that the alerts reduced the workload of the monitoring task. They also found the ACFP route replanning tool useful; "The ACFP is outstanding... We like to be able to verify stuff, so what is really cool is you guys have that ability, you don't just blindly trust, you can verify by literally looking at the ATIS and say, 'Ah! I think that is pretty accurate'. However, they also had significant issues with risk ratings. One participant reported, "I was not always sure what the tool was prioritizing: weather, distance, or time. [Because of this] I skewed my decisions more toward a personal judgment".

Voice recognition and voice synthesis technologies were used to support both the ability to perform some functions by voice and to receive briefings from the ground station. However, our system lacked robustness and thus was not fully utilized by the operator. It also did not show the proper etiquette, speaking over the operator and pilot. We also found that dispatchers and pilots differed in their attitude toward the concept of enhanced ground support. While dispatchers were eager for the additional information and tools at the ground station, pilots on the other hand were more cautious about interruptions from the ground.

3   Our Concept for HAT

Based on these initial studies it was clear that the automation tools which were designed and implemented in the ground station were helpful in performing the flight-following task. Thus, we continued to work with dispatchers and pilots to develop more automation. However, there were issues noted with respect to transparency and trust in the automation. Thus, in the next series of studies we began to integrate new collaborative decision making technologies [14–16], which we will collectively refer to as human-autonomy teaming or HAT.

3.1   Why HAT?

HAT attempts to address a long standing issue with automation: while engineers attempt to develop systems for as many foreseeable conditions as possible, these systems inevitably end up in conditions they cannot handle. Sometimes this is because the engineers could not find a way to handle the situation. Typically in these cases the manual will explicitly call for the human operator to take control (e.g., the autopilot shutting off on Air France 447). In other cases, the engineers simply did not foresee the conditions. In either case, the human operators suddenly find themselves in tricky off-nominal conditions, often with little understanding of how they got there [17].

To overcome these issues, we sought to develop a framework for HAT in which automation could be treated as a teammate. Over the last 40 years, aviation has developed a model for good teamwork referred to as Crew Resource Management, or CRM. Our initial HAT framework focused on three design tenets inspired by CRM: transparency, bi-directional communication (including a shared language), and operator directed execution [16].

3.2   HAT Tenets

Transparency: Good CRM between humans requires team members to understand what the others are doing and why. When teaming with automation, intention is often less intuitively obvious, so transparency about reasoning is necessary. Transparency of the automation has to do with whether its functioning is easily understood by operators. Operators must have knowledge of the general logic of how it works so that they can develop accurate mental models of its functioning, and be able to discern what mode the automation is in [17]. In the case of early fly-by-wire aeronautics systems, for example, test pilots placed little trust in the automation because the functioning was obscure to them [18].

Bi-directional Communication: Good CRM between humans requires people with different information to enter a dialog about how best to achieve their goals. This implies explicit discussion of goals (as opposed to intent inferencing), as well as confidence, and rationale. To facilitate this dialogue a shared language or "phraseology" is needed to improve communication efficiency. This dialogue can be initiated with plays called by the operator. The play is an adaptable system of assigning specific tasks prior to a mission based on delegated agreements that can be invoked by the human.

Operator Directed Execution: Good CRM requires someone to be responsible for final decisions and that decisions should be explicit. Through the use of "plays" this responsibility is ascribe to the human, and will continue to be for the near future. This does not mean that the automation can never act autonomously. Through the use of plays operators can still delegate tasks to automation, but only the human can execute the final action. However, we argue that automation should be adaptable. Goals, operating modes and levels of automation should change at operator direction or based on prior agreements.

3.3   HAT Agent

The HAT tenets described above give us general guidelines for implementing HAT. An important (and, to date, unanswered) question is the degree to which specific implementations can be used across multiple kinds of automation. That is, can we develop a "HAT Agent" that would add teaming capabilities to a variety of automation? This HAT agent could encapsulate a number of important teamwork functions such as maintaining a goal structure, coping with counterfactual "what if" questions, and understanding when to interrupt an ongoing task. It might also provide interfaces for HAT interactions such as cooperative decision-making and calling, modifying, and monitoring plays (a type of share plan of action, see [15]). A sketch of such an agent is presented in Fig. 5 [16].

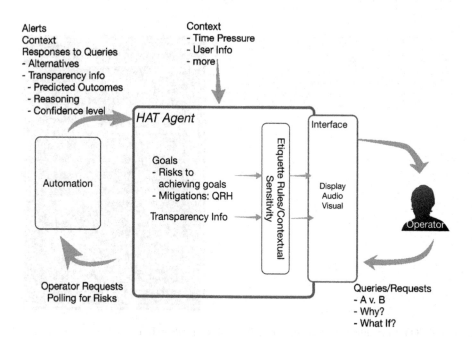

Fig. 5. Initial model of HAT interactions.

4   Implementing HAT for RCO

In our initial implementation of HAT based on CRM principles, we developed an agent that only mimics intelligence because the knowledge that it presents is instantiated by our programmers and not learned through an interaction with the real world. However, as discussed in the next section, we attempted to imbue our ground station with the HAT principles outlined above.

4.1   Experiment 5: HAT no HAT

Experiment 5 was based on the HAT tenants outlined above and a human automation teaming approach was taken to the design of ground station automation.

The interface was implemented using the playbook approach to set goals and manage roles and responsibilities between the operator and the automation [15]. It provided 13 different plays the operator could call to address off-nominal airspace and system simulation events. When the operator selects a play, the ACFP is initiated with preset weights, and the corresponding play checklist appears on the display identifying shared operator tasks in white and automation tasks in blue (see Fig. 6) [19]. As per our tenets, the operator was always responsible for executing any recommendations.

For Bi-Directional Communication, weights were preset for each play and presented in slider bars (top of Fig. 7). The operator was able to negotiate with the system by altering these weights to what the operator considered appropriate for the situation. The operator can perform "what if exploration" by changing the weights to see how the divert recommendations are affected. Using the example shown in Fig. 7, if the

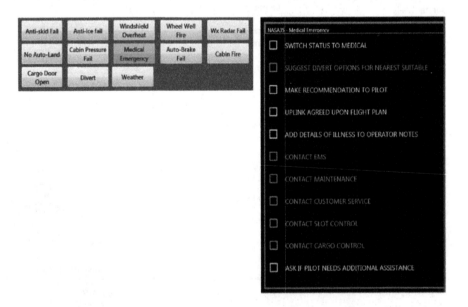

Fig. 6. Operator directed interface for calling plays in the HAT condition and associated checklist of roles and responsibilities. (Color figure online)

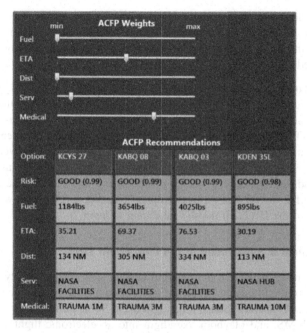

ACFP Weights			
Fuel	min ▮———————————— max		
ETA	————————▮————		
Dist	▮————————————		
Serv	—▮—————————————		
Medical	—————————▮————		

ACFP Recommendations			
Option:	KCYS 27	KABQ 08	KABQ 03	KDEN 35L
Risk:	GOOD (0.99)	GOOD (0.99)	GOOD (0.99)	GOOD (0.98)
Fuel:	1184lbs	3654lbs	4025lbs	895lbs
ETA:	35.21	69.37	76.53	30.19
Dist:	134 NM	305 NM	334 NM	113 NM
Serv:	NASA FACILITIES	NASA FACILITIES	NASA FACILITIES	NASA HUB
Medical:	TRAUMA 1M	TRAUMA 3M	TRAUMA 3M	TRAUMA 10M

Fig. 7. Transparency and bi-directional communication in the HAT condition implemented by ACFP recommendations (on bottom) and weights (on top).

operator decided that estimated time of arrival (ETA) to the airport was a higher priority than available services, the operator could adjust the ACFP weights and find new recommendations.

To address the significant task of monitoring 30 aircraft, an aircraft monitoring and messaging system (AMMS) was implemented in the ACL. The AMMS alerted the dispatcher to any non-normal events associated with the aircraft:

Weather along the current cleared path
Deviations from the current cleared path – both track and altitude
Adverse event at the destination airport that would render the airport unusable (weather minimums, airport closures, etc.)
Any system problems on the aircraft.

Previous research indicates that autonomous cooperation between robots can improve performance of human operators [20] and improve team performance. The idea of autonomous agents reporting problems to a central authority (call center) was proposed by Xu in 2012 [21]. Google maintains a call center to oversee its self-driving cars. The ACL coupled with the AMMS reduces the monitoring task of our ground station operator. The AMMS with its access to the information listed above allows the system to diagnose any problem and alert the operator, freeing up resources which can be used to service additional aircraft.

During this study four flight dispatchers and two pilots participated. After 3.5 h of training on the ground station, they managed the flight-following task during two

50 min scenarios, with and without HAT tools. During a scenario they managed approximately 30 aircraft, and worked with our pseudo-pilots to complete six diversions. During the study we collected subjective and objective data; only the subjective data is reported.

Lessons Learned. In a study comparing ground stations with and without HAT, ground station operators (both dispatchers and pilots) preferred the ground station with HAT over the station without HAT features. They reported that the ground station with integrated HAT features (ACFP and AMMS) were preferred for keeping up with operationally important issues. Workload in the HAT condition was lower, as measured by both subjective rating and eye-gaze duration data. Participants agreed that the automation and displays did a good job of integrating information, and they liked the new HAT interface to the ACFP (for example, "The sliders, I thought, were pretty well done."; "I loved the HAT…It doesn't take long to learn.").

4.2   Experiment 6: Integration of HAT on Flight Deck

Since the ground dispatch and the onboard captain share responsibility for the safety and efficiency of the flight and both must consent on any flight deviations, a clear next step was to install the HAT tools on the flight deck. So in the final study, we integrated the ACFP, AMMS, and playbook paradigm into an electronic flight bag (EFB) and installed it on the flight deck (see Fig. 8). Twelve airline transport pilots participated in our flight deck assessment of HAT tools which were presented on an EFB. Each pilot flew three off-nominal, 15 min scenarios in both the HAT and no HAT conditions. In each condition scenario difficulty varied – high, medium and low.

Fig. 8. Flight deck HAT setup: (A) EFB for interacting with HAT features.

Lessons Learned. We found no differences for HAT ratings on situation awareness, workload or trust. However, participants showed a significant preference for HAT over No HAT conditions. Moreover, as with any emerging technology, the participants provided suggestions for improving the HAT agent. These suggestions included a better voice interface that uses natural language, better labeling of anchor points on our slider tools, and suggestions for providing pilots with additional information.

5   Conclusions

This paper describes a line of research whose goal was to explore the feasibility and acceptability of single pilot operations for commercial transport aircraft and the development of a human autonomy teaming approach to automation which supported the single pilot and flight dispatcher. One of the significant impediments to SPO was the loss of nonverbal cues when crews were not co-located. To mitigate this problem crews communicated more often and openly discussed roles and responsibility. Normally the roles and responsibilities are decided by the captain prior to a flight or during a flight they both will hand off responsibility as needed with just a nod or a single utterance – I got the stick. To remediate this loss of nonverbal cues we developed the CRM tools which allowed the team to quickly assess current roles and responsibilities. Data from the first two studies suggested that the suite of tools introduced and empirically tested to address CRM challenges stemming from non-co-located crews was generally useful although pilots had multiple suggestions for improvement. Another impediment was the importance of SA prior to providing dedicated support to a single pilot aircraft. This issue had significant implications for how quickly the single pilot could expect the needed dedicated support and consequently the number of piloted needed to provide dedicated support. In the third SPO study we found no difference between our two operational concepts – hybrid and specialist. We concluded from this that if the ground station displays present the environmental and systems data which are important to gaining overall situation awareness of the vehicle needing dedicated support, either concept would be feasible. The data from this study showed that with appropriate displays, ground operators can jump in and provide assistance, even if they are coming from a place where they have minimal situation awareness. Lastly, the final three studies suggest moving to a human autonomy teaming concept reduced the need to continuously monitor individual aircraft. With the HAT tools, when a problem arose on any particular aircraft, the ground operator would be immediately alerted and could call a play which immediately provided resolution alternatives. Additionally, some tasks could be handed off to the automation, reducing task workload.

In the future we plan to continue the development of our HAT agent, giving it some adaptive capabilities, and the ability to learn from its environment. However, we will be mindful of Miller and Parasuraman's [15] caution about the technical and philosophical issues with adaptive systems – by their nature they usurp delegation authority from the human. Finally, we plan to evaluate the use of HAT concepts and tools in our future work on Urban Air Mobility, which seeks to safely and efficiently move cargo and passenger in urban areas.

Acknowledgements. This work was partially supported by the NASA Single Pilot Understanding through Distributed Simulation NRA #NNA14AB39C.

References

1. Dictionary.com. http://www.dictionary.com/browse/autonomous. Accessed 21 Feb 2018
2. ODI Resume Report from NHTSA Office of Defects Investigation on Investigation PE 16-007 concerning Tesla automatic vehicle control systems. (See ODI report PE 16-007) (2017)
3. National Transportation Safety Board Collision Between a Car Operating With Automated Vehicle Control Systems and a Tractor-Semitrailer Truck Near Williston, Florida, 7 May 2016. Highway Accident Report NTSB/HAR-17/02, Washington, D.C. (2017)
4. Comerford, D., Brandt, S.L., Lachter, J, Wu, S.C., Mogford, R., Battiste, V., Johnson, W.W.: NASA's Single Pilot Operations Technical Interchange Meeting: Proceedings and Findings. NASA-CP- 2013-216513. NASA Ames Research Center, Moffett Field (2012)
5. Lachter, J., Brandt, S.L., Battiste, V., Matessa, M., Johnson, W.W.: Enhanced ground support: lessons from work on reduced crew operations. Cogn. Technol. Work **19**(2–3), 279–288 (2017)
6. Lachter, J., Battiste, V., Matessa, M., Dao, Q.V., Koteskey, R., Johnson, W.W.: Toward single pilot operations: the impact of the loss of non-verbal communication on the flight deck. In: Proceedings of the International Conference on Human–Computer Interaction in Aerospace, HCI-Aero 2014, Santa Clara, California, 30 July–01 August 2014. ACM, New York (2014a). Article No. 29
7. Lachter, J., Brandt, S.L., Battiste, V., Ligda, S., Matessa, M., Johnson, W.W.: Toward single pilot operations: developing a ground station. In: Proceedings of the International Conference on Human–Computer Interaction in Aerospace, HCIAero 2014, Santa Clara, California, 30 July–01 August 2014. ACM, New York (2014b). Article No. 19
8. Meuleau, N., Plaunt, C., Smith, D., Smith, C.: Emergency landing planner for damaged aircraft. In: Proceedings of the 21st Innovative Applications of Artificial Intelligence Conference, pp. 114–121 (2009)
9. Meuleau, N., Neukom, C., Plaunt, C., Smith, D., Smith, T.: The emergency landing planner experiment. In: Scheduling and Planning Applications Workshop (SPARK), ICAPS-2011 (2011)
10. Ligda, S.V., Fischer, U., Mosier, K., Matessa, M., Battiste, V., Johnson, W.W.: Effectiveness of advanced collaboration tools on crew communication in reduced crew operations. In: Harris, D. (ed.) EPCE 2015. LNCS (LNAI), vol. 9174, pp. 416–427. Springer, Cham (2015). https://doi.org/10.1007/978-3-319-20373-7_40
11. Brandt, S.L., Lachter, J., Battiste, V., Johnson, W.: Pilot situation awareness and its implications for single pilot operations: analysis of a human-in-the-loop study. Procedia Manuf. **3**, 3017–3024 (2015)
12. Lyons, J.B., Sadler, G.G., Koltai, K., Battiste, H., Ho, N.T., Hoffmann, L.C., Smith, D., Johnson, W., Shively, R.: Shaping trust through transparent design: theoretical and experimental guidelines. In: Proceedings of the 7th Annual International Conference on Applied Human Factors and Ergonomics, Orlando, Florida (2016)
13. Sadler, G., Battiste, H., Johnson, W., Ho, N., Lyons, J., Hoffmann, L., Smith, D., Shively, R.: Effects of transparency on pilot trust and acceptance in the autonomous constraints flight planner. In: Proceedings of the 35th Annual Digital Avionics Systems Conference, Sacramento, California (2016)

14. Chen, J.Y.C., Barnes, M.J.: Supervisory control of multiple robots in dynamic tasking environments. Ergonomics **55**(9), 1043–1058 (2012)
15. Miller, C.A., Parasuraman, R.: Designing for flexible interaction between humans and automation. Hum. Factors **49**, 57–75 (2007)
16. Shively, R.J., Lachter, J., Brandt, S.L., Matessa, M., Battiste, V., Johnson, W.W.: Why human-autonomy teaming? In: Baldwin, C. (ed.) AHFE 2017. AISC, vol. 586, pp. 3–11. Springer, Cham (2018). https://doi.org/10.1007/978-3-319-60642-2_1
17. Endsley, M.R., Kiris, E.: The out-of-the-loop performance problem and levels of control in automation. Hum. Factors **37**, 381–394 (1995)
18. Mindell, D.: Digital Apollo: Human and Machine in Spaceflight. MIT Press, Cambridge (2008)
19. Brandt, S.L., Lachter, J., Russell, R., Shively, R.J.: A human-autonomy teaming approach for a flight-following task. In: Baldwin, C. (ed.) AHFE 2017. AISC, vol. 586, pp. 12–22. Springer, Cham (2018). https://doi.org/10.1007/978-3-319-60642-2_2
20. Lewis, M., Wang, H., Chien, S., Velagapudi, P., Scerri, P., Sycara, K.: Choosing autonomy modes for multirobot search. Hum. Factors **52**, 225–233 (2010)
21. Xu, Y., Dai, T., Sycara, K., Lewis, M.: A mechanism design model to enhance performance in human-multirobot teams. In: Proceedings of the Annual Human Agent Robot Teamwork Workshop, Boston, MA, 5 March 2012
22. Strybel, T.Z., et al.: Measuring the effectiveness of human autonomy teaming. In: Baldwin, C. (ed.) AHFE 2017. AISC, vol. 586, pp. 23–33. Springer, Cham (2018). https://doi.org/10.1007/978-3-319-60642-2_3

On Measuring Cognition and Cognitive Augmentation

Ron Fulbright$^{(\boxtimes)}$

University of South Carolina Upstate, Spartanburg, SC, USA
rfulbright@uscupstate.edu

Abstract. We are at the beginning of a new age in which artificial entities will perform significant amounts of high-level cognitive processing rivaling and even surpassing human thinking. The future belongs to those who can best collaborate with artificial cognitive entities achieving a high degree of cognitive augmentation. However, we currently lack theoretically grounded fundamental metrics able to describe human or artificial cognition much less augmented and combined cognition. How do we measure thinking, cognition, information, and knowledge in an implementation-independent way? How can we tell how much thinking an artificial entity does and how much is done by a human? How can we measure the combined and possible even emergent effect of humans working together with intelligent artificial entities? These are some of the challenges for researchers in this field. We first define a cognitive process as the transformation of data, information, knowledge, and wisdom. We then review several existing and emerging information metrics based on entropy, processing effort, quantum physics, emergent capacity, and human concept learning. We then discuss how these fail to answer the above questions and provide guidelines for future research.

Keywords: Information theory · Representational information
Cognitive work · Cognitive power · Cognitive augmentation
Cognitive systems · Cognitive computing

1 Introduction

Until now, humans have had to do all of the thinking. However, we are at the beginning of a new era in human history in which artificial entities, we sometimes call "cogs" or "AIs," will perform greater amounts of high-level cognition rivaling or surpassing that of humans. The new era will see human cognitive performance augmented by working with such artificial entities. Although true machine intelligence has been predicted for decades, so far technology has fallen short of the grand vision. However, some impressive milestones have been achieved recently.

Chess was once considered the Holy Grail of artificial intelligence. However, in 1997, IBM's Deep Blue, defeated the reigning human chess champion [1]. Deep Blue was an expensive, specially-built computer system taking many years to develop. Now, even the simplest handheld electronic devices run Chess programs able to defeat all but the most advanced human players. Human chess players are achieving higher ratings than ever before by working with chess computers to refine their game. In this way,

© Springer International Publishing AG, part of Springer Nature 2018
S. Yamamoto and H. Mori (Eds.): HIMI 2018, LNCS 10905, pp. 494–507, 2018.
https://doi.org/10.1007/978-3-319-92046-7_41

human chess players are already cognitively augmented. In 2011, a cognitive system built by IBM, called Watson, defeated the two most successful human champions of all time in the game of *Jeopardy!* [2–4]. Watson communicated in natural-language, demonstrated multi-level distributed reasoning, competed in real-time over being the first to ring the "buzzer," and engaged strategic wagering. In 2016, Google's AlphaGo defeated the reigning world champion in Go, a game vastly more complex than Chess [5, 6]. In 2017, a version called AlphaGo Zero learned how to play Go by playing games with itself and not relying on any data from human games [7]. AlphaGo Zero exceeded the capabilities of AlphaGo in only three days. Also in 2017, a generalized version of the learning algorithm called AlphaZero was developed capable of learning any game. After only a few hours of self-training, AlphaZero achieved expert-level performance in the games of Chess, Go, and Shogi [8]. These herald a new type of artificial entity, one able to achieve, in a short amount of time, expert-level performance in a domain without special knowledge engineering or human input.

Of course, these systems are not built to just play games. Watson and AlphaGo represent a new kind of computer system built as a platform for a new kind of application [9, 10]. For example, since 2011, IBM has been actively commercializing Watson technology to serve (and in many ways create) the emerging multi-billion dollar cognitive computing market. The Cognitive Business Solutions group consults with companies to create cogs. The Watson Health group's focus is to commercialize Watson technology for the health sector [11–14]. In her keynote address at the 2016 Consumer Electronics Show, Chairwoman, President, and CEO of IBM Ginni Rometty announced more than 500 partnerships with companies and organizations across 17 industries each building new applications and services utilizing cognitive computing technology based on Watson [16]. Many of these systems currently under development are intended for use by the average person.

In the coming age, many of us will encounter cognitive systems first via our handheld electronics. Also once deemed a pinnacle of artificial intelligence, natural language understanding and synthesis is now built into our voice-activated personal assistants such as Apple's Siri, Microsoft's Cortana, Google Now, Facebook's M, and Amazon Echo's Alexa [17–21]. The artificial entities we communicate with will rapidly increase in sophistication and cognitive ability. As cogs become able to perform higher-order cognitive processing, human-cog partnerships of the future will go far beyond what is possible today. Cogs will be able to consume vast quantities of unstructured data and information and deeply reason to arrive at novel conclusions and revelations, as well as, or better than, any human expert. Cogs will then become colleagues, co-workers, and confidants instead of tools. Because cogs will interact with us in natural language and be able to converse with us at human levels, humans will form relationships with cogs much like we do with friends, fellow workers, and family members. Fulbright has suggested such systems may very well lead to the democratization of expertise in much the same way the Internet has democratized information [22].

But what exactly is happening here? Obviously, these systems are processing information and generating information and knowledge that did not previously exist. However, do we have a way to measure how much knowledge has been created? Can we determine how much and by what quality the information has been altered? Do we have a way to compare one system's cognitive ability with another? How can we compare artificial cognition with human cognition? We predict humans will become cognitively augmented but how will we know when that happens? We do not yet have the metrics available to analyze cognition in this way. The purpose of this paper is to review some existing and emerging information and cognition metrics and suggest paths forward.

2 The Cognitive Process

The knowledge management and information science fields view data, information, knowledge, and wisdom (DIKW) as a hierarchy based on value as shown in Fig. 1 [23]. Information is processed data, knowledge is processed information, and wisdom is processed knowledge. Each level is of a higher value than the level below it because of the processing involved.

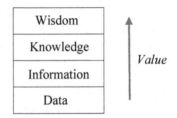

Fig. 1. The DIKW hierarchy.

At the most fundamental level, a cognitive process transforms data, information, or knowledge, generically referred to as *information stock*, as depicted in Fig. 2.

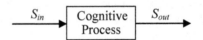

Fig. 2. A cognitive process as a transformation of information stock.

By viewing cognitive processing this way, the path to measuring cognition becomes a question of how to measure the effect a cognitive process has on the information stock and how to account the resources expended in performing the transformation. To measure this, we seek one or more characteristics altered by a cognitive process. Are there existing metrics we can use?

3 Previous Work

3.1 Entropy and Information Theory

There is a long history of entropic measures of information. In physics, particularly statistical mechanics and thermodynamics, such metrics are associated with the concept of order and disorder. In the mid-1800s, Clausius coined the term *entropy* to describe the amount of heat energy dissipated across the system boundary ultimately leading to the Second Law of Thermodynamics. In the late 1800s, Boltzmann related thermo-dynamic entropy of a system, S, to the number of equally likely arrangements (states) W, where k is Boltzman's constant [24]

$$S = k \ln W. \tag{1}$$

Boltzmann's entropy is a measure of the amount of disorder in a system. In 1929, Leo Szilard was one of the first to examine the connection between thermodynamic entropy and information by analyzing the decrease in entropy in a thought experiment called Maxwell's Demon [25]. Szilard reasoned the reduction of entropy is compen-sated by a gain of information:

$$S = k \sum_i p_i \ln p_i, \tag{2}$$

where p_i is the probability of the ith event, outcome, or state in the system. In 1944, Erwin Schrodinger wondered how biological systems (highly ordered systems) can become so structured, in apparent violation of the Second Law of Thermodynamics and realized the organism increases its order by decreasing the order of the environment [26]:

$$S = k \ln D \tag{3a}$$

Schrodinger calls - S, the negative entropy (or negentropy), a measure of order in a system. Leon Brillouin refined the idea and described living systems as importing and storing negentropy [27]. The ideas of Schrodinger, Szilard, and Brillouin involve a flow of information from one entity to another and use entropy to measure the flow.

The field of *information theory* was inspired by entropic measures. Ralph Hartley defined the information content, H, of a message of N symbols chosen from an alphabet of S symbols as [28]

$$H = \log S^N = N \log S \tag{4}$$

Since S^N messages are possible, one can view the number of messages as the number of possible arrangements or states and therefore see the connection to ther-modynamic entropy equations. Hartley's equation represents a measure of disorder in probability distribution across the possible messages, Hartley equates the measure of disorder with the information content of a message. In 1948, Claude Shannon devel-oped the basis for what has become known as *information theory* [29–31]. Shannon's equation for entropy, H, is

$$H = -K \sum_{i=1}^{v} p(i) \log_2 p(i), \tag{5}$$

where p(i) is the probability of the ith symbol in a set of v symbols and K, is an arbitrary constant enabling the equation to yield any desired units. Shannon, as did Hartley, equates order/disorder and information content. Shannon entropy relies on the uncertainty of what the next symbol might be. The more unpredictable it is, the higher the Shannon information content for that symbol. The information content, I, of a message consisting of m symbols is

$$I = mH = -mK \sum_{i=1}^{m} p(i) \log_2 p(i) \tag{6}$$

Shannon's information theory has been used extensively for several decades and several other metrics have evolved from it including: *joint entropy* (measuring the uncertainty of a set of independent variables), *conditional entropy, mutual information* (measuring related information), *relative entropy* (measuring shared information), and a generalization of entropy metrics called *Renyi entropy*:

$$\text{Joint Entropy } H(X, Y) = -\sum_x \sum_y p(x, y) \log_2 p(x, y) \tag{7}$$

$$\text{Conditional Entropy } H(X|Y) = -\sum_{x,y} p(x, y) \log_2 p(x|y) \tag{8}$$

$$\text{Mutual Information } I(X; Y) = -\sum_{x,y} p(x, y) \log_2 \frac{p(x, y)}{p(x)p(y)} \tag{9}$$

$$\text{Relative Entropy } D_{KL}(p(X)\|q(X)) = \sum_x p(x) \log_2 \frac{p(x)}{q(x)} \tag{10}$$

$$\text{Renyi Entropy } H_\alpha(X) = \frac{1}{1 - \alpha} \log_2 \left(\sum_{i=1}^{n} p_i^\alpha \right) \tag{11}$$

where $p(x, y)$ is the probability of the two occurring together, and $p(x|y)$ is the probability of x given y. Different values of α in Renyi entropy yields other entropic metrics. $\alpha = 0$ yields Hartley entropy (Eq. 4). $\alpha = 1$ yields Shannon entropy (Eq. 5).

3.2   Algorithmic Information Content

In the 1960's, Ray Solomonoff, Gregory Chaitin, Andrey Kolomogorov and others developed the concept of *algorithmic information theory* (Kolmogorov-Chaitin complexity) as a measure of information [32–35]. The algorithmic information content, I, of a string of symbols, w, is defined as the size of the minimal program, s, running on the universal Turing machine generating the string

$$I(w) = |s|, \tag{12}$$

where the vertical bars indicate the length, or size, of the program s. This measure of information concerns the complexity of a data structure as measured by the amount of effort required to produce it. A string with regular patterns can be "compressed" and produced with fewer steps than a string of random symbols which requires a verbatim listing symbol by symbol. Like the entropic measures described above, this description equates order/disorder to information content, although in a different manner by focusing on the computational resources required.

3.3   Information Physics and Digital Physics

Starting with Szilard and Maxwell's Demon, described above, many have identified a deep connection between information and physical reality even describing information as having a physical manifestation in the universe. Shannon credited Szilard's work as his starting point in the 1950s. In 1967, Konrad Zuse suggested the universe itself is a computational structure, a notion now known as *digital physics*. Edward Fredkin was an early pioneer of digital physics maintaining all physical processes in nature are forms of computation or information processing. Rolf Landauer, for example, stated "information is physical" [36]. Stephen Wolfram has concluded the universe is digital in nature and can be described as simple programs [37]. Seth Lloyd has proposed the universe is a quantum computer and everything in it is "chunks of information." Lloyd maintains that merely by existing, physical systems register information and by evolving over time transform and process that information [38].

Digital physics permits us to state fundamental limits of information storage and computation. Since a bit of information requires a system to be in a particular state, the total number of bits a system can encode is limited by the total number of possible states. Furthermore, for information to be transformed, a system must move from one state to another. The Margolus/Levitin theorem implies the total number of elementary operations a system can perform per second is limited by its energy:

$$\#ops/sec \leq \frac{2E}{\pi\hbar} \tag{13}$$

where E is the system's average energy above the ground state and $\hbar$ is Planck's reduced constant [39]. The total number of bits available for a system to process is limited by its entropy:

$$\#bits \leq \frac{S}{k \ln 2} \tag{14}$$

where S is the system's thermodynamic entropy and k is Boltzmann's constant. The speed information can be moved from place to place is limited by the speed of light. Therefore, the maximum rate at which information can be moved in and out of a system with size R is

$$rate \approx \frac{cS}{kR} \tag{15}$$

attained by taking all the information S/kB ln2 in the system and moving it outward at the speed of light.

3.4   Emergent Capacity

While previous measures equated information content directly with entropy, in 1990, Tom Stonier suggested a more complex, exponential, relationship between entropy, S, and information, I [40–42]:

$$I = I_0 e^{-S/K} \tag{16}$$

where K is Boltzmann's constant, S is Shannon entropy, and I_0 is the amount of information in the system at zero entropy. Stonier maintained information content is in some way dependent on the *structure* present in a system and uses Shannon's entropy to provide the measure of that structure.

In 2002, Ron Fulbright explored the idea of information being an emergent property evolving from underlying complexity in a system [43]. Research in cellular automata and artificial life has shown emergence requires some randomness in the system. Systems that are too structured are not dynamic enough to allow structures to evolve. Systems that are too random are too dynamic to allow evolving structures to persist. Wolfram identified four classes of systems relating how dynamic evolving structures versus how random the system is [44]:

Class I: Patterns quickly degenerate into a homogeneous state.
Class II: Simple static or periodic structures evolve.
Class III: Increasingly random and chaotic patterns evolve.
Class IV: Persistent complex structures evolve.

Also studying cellular automata, Chris Langton demonstrated a phase change phenomenon by varying the randomness of cellular automata evolution and therefore identifying the ideal amount of randomness, λ_c, at which maximal emergent behavior evolves [45]. Langton discovered the ideal amount of randomness to be an intermediate value. Figure 3 shows the relationship between Langton's results and Wolfram's classes.

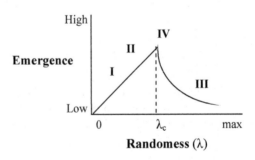

Fig. 3. Wolfram classes superimposed on Langton's phase change showing maximal emergent behavior occurring at an intermediate level of randomness.

Fulbright uses normalized entropy to measure a system's randomness and proposes the following equation for the capacity for emergent behavior by a system

$$I = m e^{\eta log_2 \frac{1}{\eta}} \tag{17}$$

where $\eta = S/S_{max}$ and m is the size of the system [43].

3.5   Representational Information Theory

Renaldo Vigo has recently proposed a new kind of information theory, *generalized representation information theory* (GRIT) [46, 47]. Key to GRIT is how humans extract concepts (known as categories) from some given information (known as categorical stimuli). Elements in a categorical stimulus may vary from one another to various degrees and human concept extraction is based on the detection of patterns. Vigo's *generalized invariance structure theory* (GIST) maintains it is easier to extract a concept from information with less variance (more similarity between elements) than it is from information with a more variance (less similarity between elements). Values calculated with GIST formulae agree with empirical evidence from human trials lending further credence to the theory. GRIT defines concept-learning difficulty, called *structural complexity*, as

$$\psi(F) = |F| e^{-k\phi^2(F)} \tag{18}$$

where F is a well-defined category, ϕ is a measure of the categorical invariance of F, $|F|$ is the cardinality (size) of the category, and k is a scaling parameter accounting for categories with different number of dimensions (e.g. $k = 2/D$ where D is the number of dimensions of category F). Subjective representational information is then the change in structural complexity due to some transformation as given by

$$h_s(F \rightarrow F') = \frac{\psi(F') - \psi(F)}{\psi(F)} = G \tag{19}$$

where $F' \subseteq F$ and F is transformed into F'. Later, we refer to this as *cognitive gain, G,* achieved by the transformation. Representational information, h_S, is defined as the percentage change in structural complexity effected by the transformation and ranges from -1 to $+1$. Positive values of h_S indicate the learnability of the concept has gotten more difficult while negative values indicate the learnability of the concept has gotten easier. The representational information associated with any portion of the original can be calculated even down to the single element. Elements lowering structural complexity are more valuable than elements raising structural complexity.

Like Stonier and Fulbright's metric, this measure relates the value of information to the structure of the ensemble. However, Stonier and Fulbright's measures rely on entropic measures of patterns whereas in GRIT, patterns are measured by invariance.

3.6   Cognitive Augmentation Metrics

Building on GRIT, Fulbright has proposed several metrics useful for describing situations in which human cognition is augmented by the cognition of an artificial entity.

Cognitive work, is an accounting of all changes in structural complexity caused by a transformation of some arbitrary information, S, keeping in mind some intermediate results may be deleted and therefore not represented in the output, ψ_{lost} [48]. Cognitive work of a cognitive process is given by

$$W = |\psi(S_{out}) - \psi(S_{in})| + \psi_{lost} \tag{20}$$

Cognitive work is a measure of the total effort expended in the execution of a cognitive process. It is important to note it requires both representational information (Eq. 19) and cognitive work to characterize a cognitive process.

When humans (H) and artificial entities, or cogs, (C) work together, each are responsible for some amount of change in representational information (cognitive gain (G)) and each expend a certain amount of cognitive work (W):

$$W_H = \sum_i W_H^i \quad G_H = \sum_j G_H^j$$
$$W_C = \sum_i W_C^i \quad G_C = \sum_j G_C^j \tag{21}$$

and the total amount by the ensemble is

$$W^* = W_H + W_C \quad G^* = G_H + G_C \tag{22}$$

Given that we can calculate the individual cognitive contributions, it is natural to compare their efforts. In fact, doing so yields the *augmentation factor*, A^+:

$$A_W^+ = \frac{W_C}{W_H} \quad A_G^+ = \frac{G_C}{G_H} \tag{23}$$

Note humans working alone without the aid of artificial entities are not augmented at all and have an $A^+ = 0$. If humans are performing more cognitive work than artificial entities, $A^+ < 1$. This is the world in which we have been living so far. However, when cogs start performing more cognitive work than humans, $A^+ > 1$ with no upward bound. That is the age that is coming.

Fulbright defines other efficiency metrics by comparing cognitive gain and cognitive work to each other and to other parameters such as time, t, and energy, E:

$$\text{Cognitive Efficiency:} \; \xi = \frac{G}{W} \tag{24}$$

$$\text{Cognitive Power:} \; P_G = \frac{G}{t} \quad P_W = \frac{W}{t} \tag{25}$$

$$\text{Cognitive Density:} \; D_G = \frac{G}{E} \quad D_W = \frac{W}{E} \tag{26}$$

4 Discussion

Figure 2 represents cognition as the transformation of data, information, and knowledge (information stock) from a lower-value form to a higher-value form. If we can define a set of metrics to measure what the cognitive process does to the information stock and describe the effort/resources expended by the cognitive process, we could measure cognition to some degree. How well do the existing "information metrics" discussed above measure cognition and cognitive effort?

Information physics and digital physics metrics describe the universe at the quantum level. For the matter of measuring cognition and its effect on information stock, these metrics are too fine-grained. Imagine trying to assess the overall strategy of a sports team's performance by taking voltage measurements of individual neurons in the player's legs. We seek metrics of cognition working at the macro level where concepts, meaning, and semantics are the important, emergent, characteristics to measure.

Entropic-based metrics and algorithmic-based information content measures depend on the randomness (probability distribution) of the information stock. Does a cognitive process either increase or reduce the randomness of the information stock and if it does, does that increase the value of the information stock? Certainly, some cognitive processes seek to decrease randomness. We can imagine an alphabetization, or any other sorting algorithm, producing a much-less random output than its input. However, entropic measures and algorithmic measures attribute less information content to the ordered output than it does to the non-ordered input, so the metric operates opposite from what we expect. Furthermore, what about cognitive processes that do something other than change the order of the information? So, while entropic metrics measure a certain characteristic of information stock, they seem to fall short of a comprehensive measure of all cognitive processes.

The emergent-based metrics hold promise because they do not equate information content directly with randomness/order, rather they imply information content is the emergent result of structure having enough complexity to support a certain amount of information. However, Wolfram, Langton, and Fulbright studied cellular automata. The idea of information, knowledge, and wisdom being the emergent result of processing of lower-level information stock is appealing but the metrics need to be evolved so they calculate real and intuitive values for human-types of information stock. More research along these lines is encouraged.

Representational information theory is promising because it ties information content to human comprehension (learnability of a concept from categorical stimuli). If we seek to measure human-level cognition and processing of human-type of information stock, then a cognition metric should involve the human component. No other metric discussed here includes human understanding and meaning. However, it remains to be shown how rigorous statements about the *value* of information can be made using these metrics because they rely on *learnability* as the key quality and *invariance* as the key characteristic. Do these speak to value? Like the order/disorder discussion above for entropic metrics, some cognitive processes increase the learnability of a concept but other cognitive processes do not. Furthermore, some information is more important because of its implications and just measuring learnability is insufficient. For example, one can image two different categorical stimuli each with the same amount of invariance. Are these two of the same informational value? One could be samples of pizza toppings with the idea of identifying the best pizza while the other one may identify the type of cancer a patient has. Isn't the latter more valuable than the former? However, representational information metrics, nor any other metric, are not able to distinguish the two.

The current version of the cognitive augmentation metrics described above are based on representational information theory. However, they could be based on any other metric that measures the effect cognitive processing has on the information stock. One line of future research could seek to establish different fundamental metrics to underlie the cognitive augmentation metrics. Another line of future research could be to develop entirely new cognitive augmentation metrics not yet envisioned. Cognitive augmentation metrics have been shown to be able to differentiate between human and artificial cognition, they have not yet been used to solve an important problem in cognitive augmentation nor have they been used to predict results that could be tested in future research. We encourage researchers to both employ cognitive augmentation metrics in their research and seek to ground the theory in empirical reality.

A failure of all metrics discussed here is none speak to the *level* of cognition. Human cognition has been studied for decades and different levels of cognition have been defined. Bloom's Taxonomy is a famous hierarchy relating different kinds of cognition as shown below from easiest (remember) to hardest (create) [49] (Fig. 4).

- **Remember** (Recognize, Recall)
- **Understand** (Interpret, Exemplify, Classify/Categorize, Summarize, Infer/Deduce, Compare, Explain)
- **Apply**   (Execute, Implement, Calculate)
- **Analyze**   (Differentiate, Organize, Attribute)
- **Evaluate** (Check, Critique)
- **Create**   (Generate, Plan, Produce)

Fig. 4. Bloom's Taxonomy expresses different levels of human cognitive processing

Each of these is a cognitive process effecting a transformation of information stock but the amount of effort involved increases dramatically as one goes down the list. Most cognitive systems and AIs today execute only the first few levels of Bloom's Taxonomy but cogs will quickly move into the higher-ordered types of processes. However, none of the metrics discussed here can distinguish between "easy" processing and "hard" processing. This is a critical line of inquiry for future research.

5   Conclusion

A new age is coming in which human cognition will be augmented by collaborating with artificial entities capable of high-level cognition. However, we do not yet have theoretically-grounded and empirically-grounded metrics to describe human or artificial cognition. For many years, the author thought a single metric could be developed to measure cognitive processes. However, current thinking is no single metric is possible and a family of metrics will have to be developed and verified empirically to fully characterize cognition. We have discussed several existing metrics from physics, information theory, and emergence theory. These metrics have deficiencies but each measure a certain characteristic of data, information, knowledge, and wisdom (information stock) such as order/disorder and learnability.

The family of metrics envisioned might very well employ the existing metrics discussed here but, future research must identify *all* important characteristics of information stock and devise ways to measure these characteristics. Researchers must not focus only on physical characteristics of the information stock. Things like value of information, importance, and emotional effect of information are human-oriented quantities possibly with subjective valuations. Researchers must include these and other human-centered issues in cognitive metrics. Finally, not all cognition is equal. We see this in human development. As a child ages, their cognitive abilities become more sophisticated, climbing the Bloom's Taxonomy hierarchy. As the coming cognitive systems age unfolds, artificial entities will master these higher-level cognitive processes. Researchers must devise metrics that consider level of cognition.

References

1. IBM: Deep Blue (2018). http://www-03.ibm.com/ibm/history/ibm100/us/en/icons/deepblue/. Accessed Feb 2018
2. Jackson, J.: IBM Watson Vanquishes Human Jeopardy Foes (2011). http://www.pcworld. com/article/219893/ibm_watson_vanquishes_human_jeopardy_foes.html. Accessed May 2015
3. Ferrucci, D.A.: Introduction to "This is Watson". IBM J. Res. Dev. **56**(3/4), 1:1–1:15 (2012)
4. Ferrucci, D., Brown, E., Chu-Carroll, J., Fan, J., Gondek, D., Kalyanpur, A., Lally, A., Murdock, J.W., Nyberg, E., Prager, J., Schlaefer, N., Welty, C.: Building Watson: an overview of the DeepQA project. AI Mag. **31**(3), 59–79 (2010)
5. Silver, D., et al.: Mastering the game of Go with deep neural networks and tree search. Nature **529**, 484–489 (2016)
6. DeepMind: The story of AlphaGo so far (2018a). https://deepmind.com/research/alphago/. Accessed Feb 2018
7. DeepMind: AlphaGo Zero: learning from scratch (2018b). https://deepmind.com/blog/alphago-zero-learning-scratch/. Accessed Feb 2018
8. ChessBase: AlphaZero: Comparing Orangutans and Apples (2018). https://en.chessbase.com/post/alpha-zero-comparing-orang-utans-and-apples. Accessed Feb 2018
9. Wladawsky-Berger, I.: The era of augmented cognition. Wall Str. J.: CIO Rep. (2013). http://blogs.wsj.com/cio/2013/06/28/the-era-of-augmented-cognition/. Accessed May 2015
10. Isaacson, W.: The Innovators: How a Group of Hackers, Geniuses, and Geeks Created the Digital Revolution. Simon & Schuster, New York (2014)
11. IBM: IBM Forms New Watson Group to Meet Growing Demand for Cognitive Innovations (2014). https://www03.ibm.com/press/us/en/pressrelease/42867.wss. Accessed May 2015
12. IBM: IBM Launches Industry's First Consulting Practice Dedicated to Cognitive Business (2015a). https://www-03.ibm.com/press/us/en/pressrelease/47785.wss. Accessed Nov 2015

13. IBM: Watson Health (2015b). http://www.ibm.com/smarterplanet/us/en/ibmwatson/health/. Accessed Nov 2015
14. Sweeney, C.: Tech leader brings Wellville initiative to Lake County. North Bay Bus. J. (2015). http://www.northbaybusinessjournal.com/northbay/lakecounty/4293852-181/tech-leader-brings-wellville-initiative#page=0#kXTgUCrErV81oRDk.97. Accessed Nov 2015
15. Gugliocciello, G., Doda, G.: IBM Watson Ecosystem Opens for Business in India (2016). https://www-03.ibm.com/press/us/en/pressrelease/48949.wss. Accessed March 2016
16. Rometty, G.: CES 2016 Keynote Address (2016). https://www.youtube.com/watch?v=VEq-W-4iLYU
17. Apple: Siri (2015). http://www.apple.com/ios/siri/. Accessed Nov 2015
18. Microsoft: What is Cortana? (2015). http://windows.microsoft.com/en-us/windows-10/getstarted-what-is-cortana. Accessed Nov 2015
19. Google: Google Now: What is it? (2015). https://www.google.com/landing/now/#whatisit. Accessed Nov 2015
20. Hempel, J.: Facebook Launches M, Its Bold Answer to Siri and Cortana (2015). http://www.wired.com/2015/08/facebook-launches-m-new-kind-virtual-assistant/. Accessed Nov 2015
21. Colon, A., Greenwald, M.: Amazon Echo (2015). http://www.pcmag.com/article2/0,2817,2476678,00.asp. Accessed Nov 2015
22. Fulbright, R.: How personal cognitive augmentation will lead to the democratization of expertise. In: Fourth Annual Conference on Advances in Cognitive Systems, Evanston, IL, June 2016 (2016). http://www.cogsys.org/posters/2016. Accessed Jan 2017
23. Ackoff, R.: From data to wisdom. J. Appl. Syst. Anal. **16**, 3–9 (1989)
24. Jaynes, E.T.: Gibbs vs Boltzmann entropies. Am. J. Phys. **33**, 391–398 (1965)
25. Szilard, L.: On the decrease of entropy in a thermodynamic system by the intervention of intelligent beings. Behav. Sci. **9**, 301–3104 (1964)
26. Schrodinger, E.: What is Life?. Cambridge University Press, Cambridge (1944)
27. Brillouin, L.: Physical entropy and information. J. Appl. Phys. **22**, 338–343 (1951)
28. Hartley, R.V.L.: Transmission of information. Bell Syst. Tech. J. **7**, 535–563 (1928)
29. Weaver, W., Shannon, C.E.: The Mathematical Theory of Communication. University of Illinois Press, Urbana (1949)
30. Pierce, J.R.: An Introduction to Information Theory, 2nd edn. Dover Publications, New York (1980)
31. Goldman, S.: Information Theory. Dover Publications, New York (1953)
32. Chaitin, G.J.: On the length of programs for computing finite binary sequences. J. Assoc. Comput. Mach. **13**, 547–569 (1966)
33. Chaitin, G.J.: Algorithmic information theory. IBM J. Res. Dev. **21**, 350–359, 496 (1977)
34. Kolmogorov, A.N.: Three approaches to the quantitative definition of information. Probl. Inf. Transm. **1**, 1–17 (1965)
35. Solomonoff, R.J.: A formal theory of inductive inference. Inf. Control **7**, 1–22, 224–254 (1964)
36. Landauer, R.: Dissipation and noise immunity in computation and communication. Nature **335**, 779–784 (1988)
37. Wolfram, S.: A New Kind of Science. Wolfram Media, Champaign (2002)
38. Lloyd, S.: Ultimate physical limits to computation. Nature **406**, 1047–1054 (2000)
39. Margolus, N., Levitin, L.B.: The maximum speed of dynamical evolution. In: Toffoli, T., Biafore, M., Leao, J. (eds.) PhysComp 1996. NECSI, Boston (1996). Physica D **120**, 188–195 (1998)
40. Stonier, T.: Information and the Internal Structure of Universe. Springer, London (1990). https://doi.org/10.1007/978-1-4471-3265-3

41. Stonier, T.: Beyond Information: The Natural History of Intelligence. Springer, London (1992). https://doi.org/10.1007/978-1-4471-1835-0
42. Stonier, T.: Information and Meaning: An Evolutionary Perspective. Springer, Berlin (1997). https://doi.org/10.1007/978-1-4471-0977-8
43. Fulbright, R.: Information domain modeling of emergent systems. Technical report CSCE 2002-014, May 2002, Department of Computer Science and Engineering, University of South Carolina, Columbia, SC (2002)
44. Wolfram, S.: Universality and complexity in cellular automata. Physica D **10**, 1–35 (1984)
45. Langton, C.G.: Computation at the edge of chaos. Physica D **42**, 12–37 (1990)
46. Vigo, R.: Complexity over uncertainty in generalized representational information theory (GRIT). Information **4**, 1–30 (2013)
47. Vigo, R.: Mathematical Principles of Human Conceptual Behavior. Psychology Press, New York (2015). ISBN 978-0-415-71436-5
48. Fulbright, R.: Cognitive augmentation metrics using representational information theory. In: Schmorrow, Dylan D., Fidopiastis, Cali M. (eds.) AC 2017. LNCS (LNAI), vol. 10285, pp. 36–55. Springer, Cham (2017). https://doi.org/10.1007/978-3-319-58625-0_3
49. Bloom, B.S., Engelhart, M.D., Furst, E.J., Hill, W.H., Krathwohl, D.R.: Taxonomy of educational objectives: the classification of educational goals. In: Handbook I: Cognitive Domain. David McKay Company, New York (1956)

Framework to Develop Artificial Intelligent Autonomous Operating System for Nuclear Power Plants

Jae Min Kim and Seung Jun Lee[✉]

Ulsan National Institute of Science and Technology,
50, UNIST-gil, Ulsan 44919, Republic of Korea
{jaemink,sjlee420}@unist.ac.kr

Abstract. As artificial intelligent (AI) technology has been dramatically developed, various industries have been challenged to apply it. In a view of nuclear power plants (NPP), it seems that AI technology applies to NPPs at the last because NPPs are required the most stringent level of regulatory guideline for safety. To overcome it, AI technology should be applied incrementally into the NPPs rather than all at once. According to the unintended shutdown records during startup and shutdown operation from 1997 to 2017 in Korea, it is reported that human errors accounts for 40% of the total. This is because operators feel heavy burden to monitor hundreds of parameters for a long time of operating time. Also, there are lots of startup and shutdown operating history that can be used for correcting the data from the NPP simulator. Therefore, this work proposes a framework to develop AI automatic operating system for startup and shutdown operations of NPPs. Operating procedures of startup and shutdown operations are categorized. In addition, AI technologies will be introduced to find out the most suitable learning algorithm. It is expected that economic loss from human error during startup and shutdown operation will be reduced as AI system developed.

Keywords: Artificial intelligent · Startup and shutdown operation
Operating procedure

1 Introduction

As data processing and artificial intelligence (AI) technologies have been dramatically improved, automation researches have been carried out in various fields. For a long time, driving on complex city streets has been regarded that only humans can do. However, it becomes one of the successful outcomes of AI technologies now.

Although nuclear power plants (NPPs) are one of the safety-critical infrastructures, NPPs also have been partly automated and even higher level of automation has been investigated.

Figure 1 shows typical operation modes and their automation levels in PWR plants [1]. In startup and shutdown modes, shown as heat-up and cool-down stages respectively on Fig. 1, completely manual operations are conducted. Heat-up and cool-down operation modes are a process that is likely to cause high workload and human error because the operator needs to check lots of plant parameters and control many components in

© Springer International Publishing AG, part of Springer Nature 2018
S. Yamamoto and H. Mori (Eds.): HIMI 2018, LNCS 10905, pp. 508–517, 2018.
https://doi.org/10.1007/978-3-319-92046-7_42

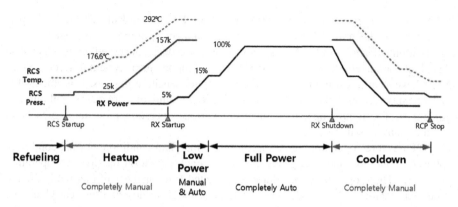

Fig. 1. Typical operation mode and automation level in PWR

accordance with the change of the plant power during the operation. This operational environment may cause high probability of human errors. Actually, there have been records about unintentional trip during startup and shutdown operation for 1997 to 2017 from Operational Performance Information System for Nuclear Power Plant developed by Korea Institute of Nuclear Safety, which is a regulatory body in Korea. From the records, human errors account for about 40% of the total trip history. Once NPP is tripped, tremendous economic loss occurs. NPPs are designed to shut down in a hazardous moment taking account of huge economic loss. In general, these factors make operators feel a heavy burden to shut down the plant by their mistake. If a new system that can reduce human error is introduced into a NPP, then it will also be economically beneficial.

Therefore, this work proposes an automated startup and shutdown system for an NPP to minimize possible human operator errors. To develop an automation framework, task analysis of startup and shutdown procedures, investigation of various deep learning techniques, training data collection, and the performance evaluation and validation are necessary. As the beginning of the work, the task analysis and some suggestions related with deep learning techniques will be introduced in this paper.

2 Special Characteristics of Nuclear Power Plant Operation

2.1 The Need for Operating Procedures of a NPP

In Korea, pressurized water reactor (PWR) is the representative type of NPP operating currently. PWR has two main systems divided into primary system and secondary one. Primary system consists of nuclear core, steam generator, and reactor coolant system with various safety system. The reactor is cooled with pressurized water so that water does not be boiled during the operation. In general, since one steam generator forms one loop with hot leg and cold legs, it is referred to as a loop instead of a steam generator. Secondary system consists of turbine, generator, and condenser. Primary system produces thermal energy by nuclear fission reaction and it is transferred into the secondary side to generate electricity. As well as two systems, there are too many components to be controlled or monitored.

NPPs are operated on main control room (MCR) and worksite where local operators work. To keep the NPP safe, hundreds of parameters are monitored whether they satisfy limits of safety. In MCR, operator group make decisions based on the plant state that plant parameters show on and control target components or system if needed. Since NPPs have enormous numbers of complicated and sensitive devices, it is impossible that operators control the plant with full understanding of parameters without any guidelines. To support operators, there are various operating procedures present to prepare many possible cases.

NPP has three operation modes according to the plant condition; normal condition, abnormal condition and emergency condition. Each condition has its own operating procedure. These operating procedures are resulted from analysis of thermal hydraulic code and reactor analysis code. Since it is not enough to describe the real plant sate from thermal hydraulic code and reactor core analysis code, the results get reliable continuously by comparing with empirical data.

During most of operating time, the plant is on the normal state. Once something goes wrong, it is not determined as abnormal state. Rather than, the plant parameters are checked according to entry conditions listed in the abnormal operation procedures. Operator should select a proper procedure in accordance with combination of alarms and symptoms. Once the plant state enters abnormal condition, operators try to restore normal condition from current state. If it fails, the plant state goes to emergency stage and reactor core must be tripped immediately. The emergency operation procedure is needed in this moment to shut-down the plant in safe.

2.2 Startup and Shutdown Operation

Startup and shutdown operating procedures are corresponding to the normal operation mode. Reference operating procedures are based on OPR1000, developed in Korea. Startup operation covers cold shutdown to hot standby mode and shutdown operation has the reverse process. Reactivity maintains subcriticality during whole processes. Cold leg temperature is under 99 °C at cold shutdown state and over 177 °C at hot standby condition. It is known that operating time takes about 20 h to startup or shutdown the plant.

Operators follow instructions on the procedure and decide whether the task should be done or not. The conditions depend on plant operating parameters such as pressure and water level of the pressurizer, water level and pressure of the steam generators and temperature of the reactor. Startup and shutdown operating procedures also provide such parameters informing the plant state. As the beginning of the research, it is important to analyze startup and shutdown operating procedure to find out key parameters that indicate the plant state.

2.3 Classification of Operating Procedures of Startup and Shutdown Operations

The procedures consist of 6 sections. The first part declares the purpose of the procedure. In the second part includes reference documents. The third section introduces cautions and limitations during process to keep the plant state safe. It is important part for training AI because these limitations will become indicators when to stop. The forth part is about initial conditions. The fifth part contains the procedure and the last section is an appendix.

While some instructions are simple and straightforward, some instructions require continuous monitoring, dynamic situation assessment, and decision making based on knowledge and experience in the fifth part. Therefore, startup and shutdown operation cannot be automated easily with rule-based code for the reasons above. Those ambiguous statements and tasks in operating procedure that should be determined by skilled operators who have worked long time in this field. To make matter worse, in Korea, the policy to reduce dependency on NPP might lead to lack of startup and shutdown operating experiences for the new operators.

The types of procedure instructions are categorized into two condition judgments and three controls: simple judgment, complex judgment, simple check, simple control, and dynamic control. Each type needs different methods for automation. In a statement, judgement play the role of the entry condition to conduct indicated controls.

1. *Simple judgement.* Simply judge by comparing with given numbers on the procedures. i.e. if reactor coolant temperature reaches 146 °C, trip shutdown cooling system pump manually and arrange the system in safety injection mode.
2. *Complex judgement.* The entry condition is ambiguous or complicated because operator should refer to other system operating procedure. i.e. if needed, conduct 'long term secondary system purification operation' according to system-26 'condenser system' or system-27 'condenser purification system' operating procedure to improve the secondary system part of steam generator.
3. *Simple check.* Check the component or system state without action. i.e. Check the level of steam generator is maintained 30 to 50% by startup feedwater pump or auxiliary feedwater pump.
4. *Simple control.* Statements can be done by simple action like binary control. i.e. Close safety injection tank isolation valves at main control room.
5. *Dynamic control.* Control components or system according to other operating procedure to meet the demanded states. i.e. if reactor coolant system pressure reaches under 52.7 kg/cm^{2}A, make safety injection tank pressure to 21.1 kg/cm^2 by following system-03 'safety injection system'.

Judgement 2 and control 5 are hard to be implemented with rule-based commands. These commands need more information that is not indicated on the procedure. AI algorithm should be able to take a proper decision like skilled human operator do.

According to this classification method, the result of classified startup and shutdown operating procedures of OPR1000 is shown in Table 1.

Since the procedure is designed for the real plant and the simulator which is NPP has, there is a possibility to have more target components and systems in real plant. The operating procedures of APR1400, which is the latest type of PWR in Korea, will be analyzed with the same manner. But the number of targets and types of statements will be similar or increasing because APR1400 is originated from OPR1000. That is, more than 200 or 300 targets will be treated when AI algorithm is trained. Strategies to select the deep learning algorithm and how to train it should consider that the data will continue to accumulate for 20 h, the average operating time for startup and shutdown.

Table 1. Categorizing statements of startup and shutdown operating procedures

Startup operation procedure		Cold shutdown to hot standby		
Minimum # of target comp. and sys.		309		
	Judgement		Control	
Type	1	2	3	4	5
# of instructions	20	31	44	22	33
Shutdown operation procedure		Hot standby to cold shutdown		
Minimum # of target comp. and sys.		194		
	Judgement		Control	
Type	1	2	3	4	5
# of instructions	33	12	11	33	34

3 Autonomous Operation Model

3.1 Deep Learning Algorithm for Automatic Operation

There is a research to develop an automation system for emergency operating procedures using fuzzy colored Petri net [2]. However, startup and shutdown operator procedures have different forms with emergency operating procedures and more dynamic situations should be considered because plant power and status of various component vary continuously. While conventional neural networks have required large amounts of hand-labelled training data, reinforcement learning algorithms must be able to learn from a scalar reward signal that is frequently sparse, noisy and delayed [3]. In addition, considering characteristics of NPPs, learning algorithms should not assume that data sets are independent.

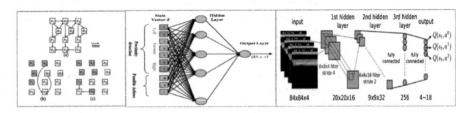

Fig. 2. Bayesian reinforcement learning (RL) algorithms (left), artificial neuron network RL algorithm (center), and deep RL (right)

Therefore, to find out the most appropriate learning algorithm for startup and shutdown operations, various methods were investigated as shown in Fig. 2 [3–5]. From the review, it is evaluated that guided reinforcement learning approach is appropriate for startup and shutdown operations. This is because we do not have data that can clearly distinguish between good and bad and there is too much information to be considered. In the future, the outcome of this work would be a group of AI models with multiple layers of hidden layers.

One of the difficulties to apply learning methods to NPP operation is to set score for determining a performed action is appropriate or not. While games have obvious goal (high score), NPP status cannot be easily scored with a specific number. Therefore, there is need to gather the factors that can be used as a score to describe the plant state well. For example, the difference between the target range and the current value of significant plant parameters can be candidates for scoring. Also, operating time could be another option. As far as the operating limits allow, the variables are changed as quickly as possible to finish the entire process earlier than before. Future works will focus on verification of these candidates.

3.2   Learning Data Generation Using Full Scope Simulator

Lack of data is one of the main causes which make it difficult to apply deep learning algorithms to nuclear field. For many decades, especially only a few historical data of unanticipated reactor trip have been reported. Even though NPPs have long history, it is not enough to train AI algorithms. Therefore, it is not avoidable to use not only empirical data but also the data obtained from a simulator for training.

Simulator includes hydraulic analysis code such as RELAP and MARS-KS, and reactor core analysis code such as RAST-K and NESTLE neutronic codes to describe the real NPP. To make a result of the simulator as real as possible, empirical data adjust the result. Therefore, even if raw data from system codes are insufficient, simulator results gets better reliability constantly.

Moreover, unlike other accident scenarios, startup and shutdown operation, the target operation modes of this work, have lots of operating records through overhaul period or refueling issue. Considering these advantages, it is appropriate to develop automatic system from these operation modes.

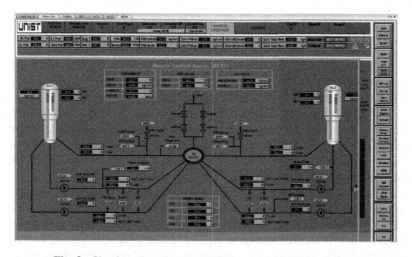

Fig. 3. Simulator based on 1400MWe pressurized water reactor

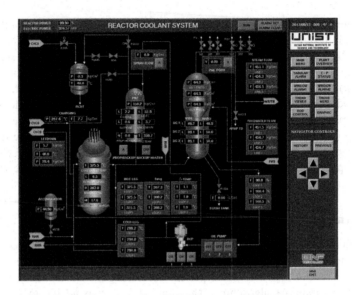

Fig. 4. Simulator based on 998MWe pressurized water reactor

In this work, two simulators will be used to make operating data. One simulator developed by Western Service Corporation models 1400MWe PWR as shown in Fig. 3. It will be mainly used because it has the similar structure with APR1400. In addition, since it allows to be run by a script, it is big beneficial to data production. The other simulator developed by Korea Atomic Energy Research Institute models 998MWe PWR as shown in Fig. 4. One of advantages is that the simulator has its own operating procedure. Unfortunately, it does not have script operating function yet.

A scenario file composed of a bundle of script commands can be input in advance to run the 1400MWe PWR simulator. The most common and simplest form of script is as follows.

$$[+]delaytime\,[ObjectName]$$

At the first, declare the time when the command line happens. Then write down the object with desired state. For example, '+5 var hsACHIS0268C.iswPos = 0' means that the switch of variable hsACHIS0268C will change its position as 0 (close state) after 5 s. With this function, it is possible to enter numerical values directly as well as simple components operations. However, there is a problem that naming methods of the procedure, the simulator, and the system code, used for script, are all different. For this reason, it takes a lot of time to match their names. Therefore, it must be done in advance before data production stage.

Later, to train the selected algorithm with as many data as possible, an automatic training data generation system was developed connected to 1400 MWe PWR simulator. Data production is big issue because a long operation time is required, and even if the operation time is reduced through simulator function, the reliability of the data may be deteriorated. The automatic data generation system should cover those problems.

4 Framework of Startup and Shutdown Operation Procedure

4.1 Gradual Approach to Full-Autonomous AI System

The AI autonomous operating system can be divided into three level depending on the degree of utilization of the operating procedure because the procedures do not use all the parameters of the plant. As developing a new system, it is necessary to decide what level it will be made. Three levels of AI system can be categorized as follows.

Level1: autonomous system based on operating procedure only.

This is the method to make the tasks automate only appeared in the operating procedure. It is the simplest approach, however, there is no chance to operate the plant more efficiently than the existing procedure does. Only the order of statements or time to take an action could be changed. Therefore, reducing the time it takes to operate will be an indicator of a better system.

Level2: autonomous system considering plant parameters based on the operating procedure.

This level basically follows the procedure but makes it possible to consider other related systems. For example, if water level of the pressurizer should be lower, the system will consider not only the components suggested by the procedure but also any components which affects it. From this level, even if the same procedure is performed, the variables to be handled might be different. Likewise, it is possible to develop a new instruction by the AI system.

Level3: full autonomous system without learning the operating procedure.

This level guarantees the maximum feasibilities of all controllable parameters without the procedure. Obviously, this level will be the most difficult to be implemented. However, once it succeeds in development, it is expected that the new system will produce a procedure that is completely different from the existing one. As Alpha-Go has made its own strategies, it might consist of something people have not considered ever.

In fact, the operating procedure has high reliability because it is made by experts of a NPP operation and revised continuously. Nevertheless, AI can discover a new way of operation that no one else can think of. The higher level of the system, the greater the likelihood of new discoveries. Obviously, higher level is hard to implement. There are hundreds of controllable components and thousands of parameters to be monitored. Moreover, these parameters make data continuously for about 20 h. That is, there are almost infinite cases to be learned.

Therefore, as feasibility study, the objective of this work is to create a level 1 AI model first. If it succeeds, level of AI model will then be stepped up to the higher level.

4.2 Identification of Input Data and End State for AI Algorithm Training

To train AI algorithm, it is important to identify input data type and end state. Input data types can be classified from the fifth part of the operating procedures. It is needed to distinguish which components or systems can be operated because as mentioned in Sect. 2, there are lots of ambiguous statements that needs complex thinking process. End states can be defined by using the third and sixth parts of the procedures which deal with operating and variable limits. AI algorithm should stop the simulator when it meets end states. These termination conditions will check the outcome of operations by AI model.

It feels easy to think so far, however, the procedure suggests only the range of variables. Limitation changes dynamically as the plant state differs. In a data set, a single variable has no meaning. The input variables should be defined considering the combination of the variables. Similarly, termination conditions should be determined in the same way.

In addition, input data type and end state should be change according to the level of the AI system to be developed. The higher level is, the more complex the variables become. In the level 3 approach, problem is not only the almost infinite number of cases but also termination conditions to be set.

This work cannot be done in the present state because it will be better to declare input data and end state while learning the algorithm, rather than defining them when organizing data.

5   Concluding Remarks

In this paper, framework to develop AI autonomous operating system for nuclear power plants is suggested. Since NPP is one of the industries where safety is the highest priority, such automatic system should be applied gradually with thorough verifications. Once startup and shutdown operations succeed to be operated by the AI algorithm, it will be a stepping stone for other operating conditions such as abnormal and emergency conditions.

Considering unique characteristics of NPPs, simple guided machine learning or reinforcement learning has limitation. Rather than, this work suggests guided reinforcement learning for developing level 1 AI model which is based on the present operating procedures. If it is successful to apply to the most sensitive subject, NPP, AI-based autonomous technology is expected to apply to other industrial sectors as well.

Acknowledgments. This work was supported by the Korea Institute of Energy Technology Evaluation and Planning (KETEP) and the Ministry of Trade, Industry & Energy (MOTIE) of the Republic of Korea (No. 20171510102040)

References

1. Seong, P.H., Kang, H.G., Na, M.G., Kim, J.H., Heo, G.Y., Jung, Y.S.: Advanced MMIS toward substantial reduction in human errors in NPPs. Nucl. Eng. Technol. **45**(2), 125–140 (2013)
2. Lee, S.J., Seong, P.H.: Development of automated operating procedure system using fuzzy colored petri nets for nuclear power plants. Ann. Nucl. Energy **31**, 849–869 (2004)

3. Mnih, V., Kavukcuoglu, K., Silver, D., Graves, A., Antonoglou, I., Wierstra, D., Riedmiller, M.: Playing Atari with deep reinforcement learning. arXiv preprint arXiv:1312.5602 (2013)
4. Vlassis, N., Ghavamzadeh, M., Mannor, S., Poupart, P.: Bayesian reinforcement learning. In: Wiering, M., van Otterlo, M. (eds.) Reinforcement Learning, pp. 359–386. Springer, Heidelberg (2012). https://doi.org/10.1007/978-3-642-27645-3_11
5. Hatem, M., Abdessemed, F.: Simulation of the navigation of a mobile robot by the Q-learning using artificial neuron networks. In: CIIA (2009)

Embodiment Support Systems: Extending the DEAR Causal Inference Framework Through Application to Naturalistic Environments and Inclusion Within a Decision Support System

Ryan A. Kirk[1](✉) and David A. Kirk[2]

[1] Kirk Enterprises, Seattle, WA 98106, USA
info@ryankirk.info
[2] Consulpack LLC, Minneapolis, MN 55448, USA
dave.kirkl@comcaset.net

Abstract. Historically decision support systems (DSSs) have struggled to augment users in complex, non-linear scenarios. Such environments occur anytime feedback loops contribute to the resultant states of independent variables. These types of environments are becoming both more common and more pervasive in everyday life. Digital platforms are becoming increasingly complex. Recommendation systems are becoming a part of everyday life. These authors have focused previous efforts upon first outlining statistical techniques useful for nonlinear systems and more recently upon incorporating such techniques within a decision making framework. This system is designed to provide deep insight into complex non-linear environments and augment the user's effectiveness in managing risk. This paper will focus upon the application and extension of the DEAR framework to the inherent complexity of naturalistic environments. It will offer an example of an embodiment of this approach into a DSS as a part of an embodiment support system (ESS). Finally, it will advocate the creation of hybrid DSS/ESS architectures.

Keywords: Business integration · Decision support systems
Intelligent systems · Knowledge management · Nonlinear systems

1 Background

Two common ways that science advances are through methods and through theory [1]. Research can advance through coming up with innovative methods for testing phenomenon. Frameworks can advance as new ways of thinking about past experimental results offers an alternative lens with which to create new theories. However, there is another way; advances in instruments can lead to new discoveries which in turn could lead both to new methods and to new theories.

Advances in instrumentation used to constitute new types of apparatuses. While this is still true, software offers several alternative techniques for improving existing apparatuses. Software can increase the speed, sensitivity, scale, or statistical inference offered by existing instruments. Everyday devices now connected devices offer new

© Springer International Publishing AG, part of Springer Nature 2018
S. Yamamoto and H. Mori (Eds.): HIMI 2018, LNCS 10905, pp. 518–530, 2018.
https://doi.org/10.1007/978-3-319-92046-7_43

insights into everyday behaviors. Global networks of sensors offer increased situational awareness and increased statistical power for climatological research. User interfaces and intelligent services offer windows through which knowledge workers can explore these new worlds.

Knowledge workers are an important part of modern, service-oriented culture. These individuals are tasked with helping guide important, fast-paced decisions within both private and public sectors. Technology used to facilitate decision making is increasingly relying upon intelligent systems to either partially or to fully automate the decision-making process [2, 3]. These interactive learning processes involve humans as part of a decision-making feedback loop. Humans' judgement serve to guide future machine judgement and vice versa [4].

The human-machine feedback loop has existed since the dawn of computing. The key premise is that ideal aspects of each help to augment the combined capabilities of both. In order for these types of technologies to be effective they must scaffold human capabilities. There are several ways this can take place. For example they can increase awareness, inference, attribution, or the overall decision-making process. This approach works well and improvements in DSSs.

As intelligent systems become increasingly integrated within our everyday lives the primary orientation of DSSs changes from being interface-driven to becoming service-driven. In these new environments humans no longer fettered by the capabilities of visual perception. Speech assisted technologies bring enablement for more interactive environments. Augmented reality allows digital artifacts to enter the everyday. Social networks allow machines to rely upon real-time crowd consensus. While this changes the type of interactions in the human-machine feedback loops it does not change the central premise that mutual augmentation is the most effective strategy for increasing combined capabilities.

2 Cognitive Extension

Human-Computer Interaction as a field is interested in applying the tools and techniques from the research sciences to the creation of technologies or processes that better integrate with the human experience. While traditionally concerned with human factors or usability design, increasingly and recently this field has focused upon the usability of information more broadly as it relate to facilitating decision-making [5]. Intelligent systems are becoming more common. While such systems may behave autonomously, their effectiveness (at least currently) is still quintessentially linked to their ability to enhance or to extend human activities.

Intelligent systems often utilize a form of statistics called machine learning. Machine learning is often a form of applied statistical methods whose goal is not to test for significance of variation but rather to classify which hypothesized group is most likely given the information present. (i.e. we are not asking if A is different than B but rather whether A belongs to [B, C, D, ... ,]). A challenge for such systems historically has been coming up with approaches that generalize to unforeseen circumstances [6–10].

A similar challenge is in coming up with models that are grounded [2, 6, 9, 10]. Grounded approaches are those whose results we can explain based upon an understanding of the objects and the objective function. In cases where the definition of the object is subjective (e.g. the presence of emotional states) then models have to first determine what states exist before classification can occur. While grounding models is effective, researchers struggle to understand why some grounded models are better than others. Classification error alone is a fragile technique for evaluating model efficacy. A theory for why this occurs relates to the ability of the models to generalize to unseen examples.

Representative models are grounded models that function through encoding some properties of the attributes into the model itself [6, 9, 10]. For example, a representative model of color might encode the dimensions of color perception. However, while a model built using the dimensions of color perception that humans use would be grounded, fewer dimensions might be able to more efficiently capture this variation. The challenge in this later case is arguing that the compressive dimension is still representative.

Research suggest that grounded models benefits from first pre-processing the input. By first creating a natural ontology and later mapping this to an external ontology models are more likely to be able to statistically differentiate along meaningful axes [9, 10]. Thus, grounded models contain representations that are naturally separable based upon observed data. This is in contrast to ontological-backed models that may or may not be directly grounded. Taking this further, representational models would be those grounded models that also contain perceptual dimensions that are indicative of the conceptual landscape. Here perception implies that the model encodes the external world into a parsimonious and natural internal representation.

While specific on the need for such encodings, there is flexibility in the mechanism for such encodings; these encodings could direct, connectionist, or something else entirely. Techniques such as the use of covariance, concomitance, correlation, or convolution have worked well in the past. The important distinction is that such models do not encode artificial features even if they are useful for classification purposes. Once built, the incorporation of these models built upon information from instruments into decision-making frameworks extends our cognition.

2.1 DEAR Framework

The DEAR framework operates in four stages: detect aberrant or nonlinear behavior, evaluate for causality, assess risk, and recommend action [11]. This approach is based upon decision-making theory from operations research, risk management from public and private institutions, and statistical inference from chaos theory. The detection technique is a combination of classical statistical inference combined with methods for detecting nonlinearity. The core technique for evaluating causal relationships is based upon the convergent cross-mapping (CCM) technique proposed by Sughihara but has later been extended by other researchers [12–15]. This technique quantifies biases in pairwise relationships using a matrix of time-embedded offsets [15–17]. The risk assessment phase uses the optimal Kelly Criterion analysis to determine the relative risk of various outcomes with a goal towards system optimization [18]. The recommendations then assemble the output of the risk assessment and compare and contrast them with alternative uses of an institution's resources.

The DEAR framework is based upon the results of experimentation and application to multiple domains. These authors have already successfully applied the DEAR framework to financial, social, and epidemic scenarios. They have applied the CCM component of this framework to physical systems. Naturalistic systems such as these domains represent areas of study that originally motivated chaos theory as well as areas that would substantially benefit from DSSs focused upon non-linearity. As with any system, understanding the underlying relationship governing the observable variables within such systems is based on discerning cause and effect relationships. However, unlike limited test environments, naturalistic systems also represent the unpredictable characteristics inherent in the real world. These systems exhibit complexity, nonlinearity, and multivariate interactions. As a process DEAR could be built directly into an interface as a part of a DSS or as a part of broader service-oriented solution. Either way this approach is inherently sociotechnical since the end result requires inclusion of relativistic criteria such as an institution's objective function and/or alternative uses of institutional resources.

The four stages of the DEAR framework are: detect nonlinear abnormalities (D), evaluate causality (E), assess risk (A), and recommend action (R) [11]. Taken together these stages help to enhance cognition while reducing cognitive bias. Detection helps facilitate attention and perception while minimizing extraneous. Evaluation helps with hypothesis formation. Assessment helps with attribution and with evaluation of alternatives. Recommendation aids prospective memory and minimizes unnecessary actions. If this process is included in a service as a part of an intelligent system then the process has the added advantage that it should also serve to maximize the ratio of exploitative versus exploratory behaviors. This would result in more efficient form of agency.

2.2 Instruments

Instruments facilitate cognition when they enable just out of reach activities. These activities could be attentional, perceptual, procedural, attributional, etc. Perceptual enhancement is perhaps the clearest example of how instruments can enhance cognition. The ability to observe electromagnetic radiation in frequencies outside of the visible spectra is a clear example of where instruments enhance human cognition.

A recent development in the field of astronomy has come from an unexpected source. As high definition televisions (HD TV) became more common, broadcast standards for such devices also had to improve. With the advent of ATSC 1.0 it became possible to broadcast HD TV signals across radio wave broadband. In order to be effective this technology required that each TV have a receiver. Due to commercialization and mass production these receivers became increasingly affordable. Today these receivers cost less than 20 USD and are capable of receiving anywhere from 15 kHz–2.5 MHz. Fortunately for astronomers these frequencies are also characteristic of interplanetary emissions. This has led to the development of software defined radio (SDR) so that individuals can use these devices for a multitude of purposes.

SDR is similar to trends taking place in other, similar fields. For example in telecommunications the software defined networking (SDN) is enhancing the ability of humans to manage the quality of service for large networks of interconnected devices.

In industrial processes the industrial internet of things (IoT) allows real time awareness and troubleshooting of various forms of production. What these technologies have in common is their use of software and their use of a collection of sensors. These technologies also often make use of wireless media for transmission and receiving.

Modern instruments also have a sense of agency that enables certain types of insights that were not previously possible. In these scenarios devices can perform several types of tasks. They can behave passively when they simply observe broadcast traffic. They engage in remote sensing when they use telemetry to observe behavior of other devices. Finally, they can engage in remote control when they use telecommands. This combination of behaviors when combined with large arrays of sensors allows a level of awareness and control beyond what single machine, human interactions offers.

2.3   Experimental Context

Astronomy is a field where humans have always relied upon formal logic to perform inferences for phenomenon beyond what meets the naked eye. Early observations of distant planets relied upon statistics to mediate subtle individual differences in perception. This way of thinking allowed for increased precision in observation and in turn led to the formation of more advanced theories of planetary mechanics. Over time telescopes became increasingly advanced and it is now prohibitively difficult to increase the capabilities of individual telescopes.

Astronomy benefits from the use of SDRs since they substantially increase the number of signals simultaneously analyzed. The very large array (VLA) telescope is a famous example of a radio telescope that has uses multiple independent observations to enhance overall capabilities. The low relative cost combined with the potential for highly distributed deployment promises to allow SDRs to crowdsource the development of useful radio arrays. The challenge that arrays such as the VLA face comes in correlating the outputs of the sensors into an integrated sense of awareness. The hope is that tbe conjoint application of the DEAR framework with these new instruments will yield new capabilities for astronomy.

As with any natural phenomenon, radio wave emission from Jupiter certainly vary. They range from disorganized noise to strikingly organized "songs". The later types are typically characterized as being either L-Bursts or S-Bursts denoting whether their duration is either long or short [19–21]. The immediate cause for these emissions is thought to be patterns in the emission of high energy particles from the planet. The mechanism responsible for these emissions is a combination of the mixture properties of the planet as well as the planet's interaction with the Jovian system.

The exact causes of Jupiter's magnetic fields to form is still an active area of research. While it is impossible to directly interact with the planet, astronomers have some ideas as to what causes these emissions. They believe that Jupiter acts as a particle accelerator. Similar to how particle accelerators work in laboratories on Earth, the accelerators on Jupiter contain a conduit, a reservoir, and magnetic fields. Later for any number of reasons a strong magnetic field interacts with these reservoirs and pulls particles outwards into a band of emitted radiation. This creates a broad spectrum pattern detectable via radio telescopes.

These emissions form patterns within certain bandwidths of radio spectrum. These patterns are quite differentiable from the background noise of the typical Jovian radiation [20, 21]. So much so that astronomers describe these patterns as the "songs of Jupiter". These patterns represent synchronization taking place in the emissions. Non-relativistic, cycltronic acceleration causes radiation that forms longer patterns called L-Bursts. Faster, relativistic, syncotronic acceleration causes radiation that forms shorter patterns called S-Bursts. The difference in duration is due to the increased speed of the particles and this translates to changes in duration of these wave patterns. However, the wavelengths remain similar such that observers can detect them using similar bandwidth receivers.

In many cases, the cause of these magnetic fields is well studied and fall into four categories: Io-A, Io-B, non-Io-A, and non-Io-B. While other moons also interact with Jupiter's fields, none compare in magnitude of interaction to that of Io and its position along the Central Meridian Longitude (CML) as perceived by Earth-based telescopes [20, 21]. It strongly interacts with Jupiter's magnetic field as it processes it orbit. Depending upon its position relative to Earth the emission pattern varies slightly. However, in some cases the causes are still unknown. In these cases the emissions may be due to natural interaction between these reservoirs and Jupiter's magnetic field.

While the causes of these burst phenomenon are not conclusively known, there are some hypotheses. S-Bursts might emanate from deeper within the planet where higher energy sources exist. Conversely the L-Bursts might emanate from along the gassy outer surface of the planet. A primary interest to the authors of this paper is whether these two phenomenon might interact with each other. Could the presence of S-Bursts somehow be causing the emergence of the L-Bursts to take place?

2.4 Experimental Apparatus

Just because an instrument is capable of enhancing perception does not mean it is free from bias. As with the conventional dimensions of human perception, enhanced dimensions will also have certain innate biases. For example, an instrument observing the radio waves resultant from some astronomical phenomenon might have no information loss. However, it may also have distortion due to red-shift or it may suffer information loss due to modulation. As with any medium, there is also interference. So how can we know whether we are observing one phenomenon versus other phenomenon?

As described above the context of this study will focus upon increasing human understanding of Juptier's radio emissions. With something such as Jupiter we have the advantage of being able to use conventional perception to observe Jupiter via telescopes using the visible spectrum. However, what if the phenomenon does not engage in visible light or is too distant to be enhanced using lenses? In such cases the degrees of separation between human perception and instrumentation increases. As this separation further and further it becomes increasingly important that our approach to cognitive enhancement is one that is both representative and grounded. Higher frequency emissions that cause visible light are difficult to receive by the same means with which we receive radio waves. This means that instruments often use different techniques to observe different spectrums of light. In such cases how is it possible to maintain confidence in comparisons across media? It may be easier to, for example, use modulation to up-convert the frequency of radio waves to that of visible light than vice versa.

This experiment used SDR technology to monitor Jupiter's radio emissions. To properly observe interplanetary emissions, researchers used antennas intended for receiving radio frequencies in the 20 MHz range. In addition, the antenna chosen should have provided some gain. This was useful since it helped to differentiate against terrestrial and interplanetary noise. SDR receivers require the presence of a computer to translate input into intelligible time series signals. SDR receivers also typically require some sort of console software useful for interfacing with the SDR drivers. The combination of computer and console allowed operators to adjust SDR settings during observation. The computer once equipped with storage media also allowed researchers to be able to record these time series values into files useful for subsequent analysis.

2.5   Method

This was an observational study. The goal of the study was to determine whether there is a causal relationship between the S-Burst and L-Burst activity taking place in the Jovian system. This study observed samples of these behaviors taking place and then analyze them using the DEAR framework to determine whether there are causal relationships present within these emission patterns. Once armed with the results of the DEAR analysis researchers then used this to inform their subsequent investigative behaviors of the Jovian system. This combination of approaches was used in an effort to test the effectiveness of a feedback approach that combined this decision support framework with an iterative research project.

The detect phase of the DEAR framework focuses upon detecting signal abnormalities. Since space is vast and it is difficult to ubiquitously observe the appropriate signals, the researchers had to narrow the focus down to time periods of substantial interest. Through studying the historic patterns of the Jovian system, researchers were able to determine when and where to listen for S-Burst and L-Burst activity. The historic Io-A and Io-B interactions result in fairly predictable intervals during which time these emissions are easier to detect. Through listening to several of these periods of activity researchers were able to detect examples of these patterns. However, these examples of S-Burst and L-Burst activity were quite noisy and were difficult to analyze even though they were taking place in the correct bandwidth during the correct periods of time. For this reason, researchers instead extracted samples of S-Burst and L-Burst activity during a known Io-B type activity from 3/10/2002 via online archives [19].

Once the samples were collected, researchers then used various noise filters and signal amplification techniques within the SDR console. The duration of S-Burst and L-Burst range in the seconds and for this reason researchers prioritized obtaining high quality signal samples even if the duration of the samples are quite short. To further increase success of analysis researchers tried to obtain files with a sample rate in the millisecond range. In order to compare files from different sources with each other, down-sampling took place on the file with the short sample interval so that the two files had the same sample interval. Researchers used Audacity to perform this sampling. The amplitudes were composited into a text file for CCM analysis.

The evaluation phase of the DEAR framework focuses upon assessing the presence of causal relationships between variables of interest. Upon processing the input files researchers used Sugihara's CCM process to examine causal relationships between S-Burst and L-Burst activity.

The assessment phase of the DEAR framework examines the risk involved in maintaining a policy given the presence of causal relationships. In this context the risk relates to the opportunity cost of using the radio telescope to examine the Jovian system when it could be used to explore other regions and epochs. To assess this risk researchers would use Kelly's original equation to determine the relative risk that researchers would not obtain useful information in the Jovian region as compared to the default probability of obtaining information in other regions.

Finally, the recommendation phase of the DEAR framework would issue policy recommendations for researchers operating the radio telescope based upon the results of the previous causal inference and related risk assessment. In this case the recommendation would either be to continue evaluating the Jovian system or to continue exploring other regions.

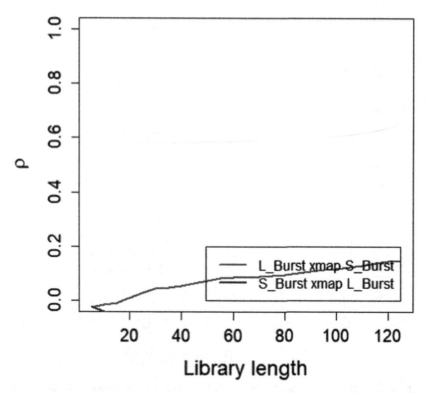

Fig. 1. The CCM xmap results for Jupiter S-Burst and L-Burst activity for library lengths <0,120> . The blue, top line shows L-Burst xmap S-Burst and the red, bottom line shows S-Burst xmap L-Burst. The blue, top line suggests weak evidence for the case that S-Burst may influence L-Burst behavior. (Color figure online)

3 Results

Researchers ran CCM on an Io-B Storm containing both S-Burst and L-Burst activity. The result of this analysis showed only weak evidence to suggest a causal relationship between these two phenomenon. The S-Burst xmap L-Burst showed none to negative causality as the library length increased (Fig. 1). The correlation of causality for L-Burst xmap S-Burst approached 0.2 which suggests weak evidence for the case of S-Bursts influencing L-Bursts. However, even this weak trend disappears when examining longer library lengths (Fig. 2).

Since the evaluation phase did not yield evidence for causality, the assessment and recommendation phases of experiment were much shorter. With no return on investment the assessment phase would indicate to continue exploration elsewhere. However, the weak evidence for causality causes the researchers to also avoid excluding this possibility from future searches. For this reason, the policy recommendation would be to continue natural exploration including occasionally revisiting the Jovian system.

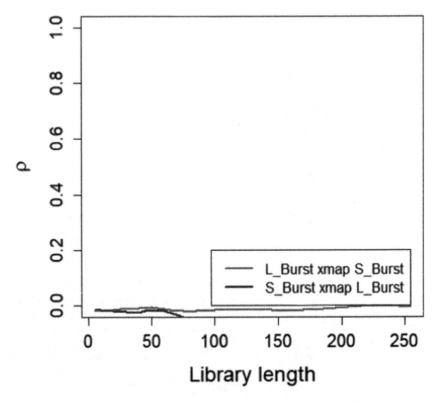

Fig. 2. The CCM xmap results for Jupiter S-Burst and L-Burst activity for library lengths <0,120>. The blue, top line shows L-Burst xmap S-Burst and the red, bottom line shows S-Burst xmap L-Burst. There is no evidence for causality present. (Color figure online)

4 Conclusions

There is not strong evidence for causation between S-Bursts and L-Bursts. However, this is not surprising since there is also not strong evidence for this being the case. The, the combination of the use of the DEAR framework in conjunction with an iterative testing approach made it possible for the researchers to inform their investigative process. Nonetheless, this research also unveiled new challenges that this context present.

The challenges faced in this experiment are characteristic both of the particular context as well as of applied decision making. In general humans struggle to make decisions in contexts that are either too complex or in contexts where the time and space dimensions are substantially different than those in everyday scenarios. Astronomy is an area of study where the distances as well as the time scales offer substantial challenges. The vastness of space make omniscience unfeasible. The rapid timescales of EMF combined with the long timescales of cosmological objects makes causal inference nearly impossible without the aid of DSSs.

These limitations stem from the inability of human perceptual dimensions to create useful conceptualizations which in turn impacts human attribution. For this reason, DSS for these contexts require augmenting human embodiment as well as human cognition. Human embodiment is finite and restricted by our innate sensorimotor apparatuses and the related perceptual dimensions. In contrast, extended mind is a concept that refers to the ability to augment human cognition through extending our perceptual awareness using information systems and, usually, visualization apparatuses. Extended embodiment would be a similar concept extended into to human embodiment.

Artificial Intelligence (AI) is a technology that enables machines to perform certain automated actions where the likelihood of a given action to take place is based upon that actions past success [22, 23]. The success of an action depends upon the objective function the AI uses. While objective functions vary quite a bit from application to application they generally focus upon increasing the prevalence of a desired outcome and/or upon the minimization of an undesirable outcome. The desirability of an outcome can be gauged based upon human experience. In so doing this approach extends human embodiment by incorporating its quintessential objectives into synthetic systems. Thus, the two systems continue to collaborate via sharing common goals.

Science in general, and Astronomy in particular, are great places for researchers to focus upon building embodiment support systems (ESSs). The vastness of space both in terms of time and physical distance make this area of research a practical paradigm within which to develop DSS/ESS hybrid systems.

5 Next Steps

The future of DSSs will include a hybrid DSS/ESS. Such systems will extend AI within the operations of the systems in order to support extended embodiment [24, 25]. The DEAR framework will become a part of a larger DSS/ESS framework in which it will continue to serve its role as a routine useful for assessing causality. Similar to the way that DSSs facilitate cognition, this system will focus upon extending the human embodiment into new situated contexts.

The long-term goal of such a system will be to build a synthetic mental model upon which a decision making cadence can operate. Effective perceptual dimensions will be essential for this system to be able to form concepts. As this system discovers concepts it will incorporate them into the model. Upon initial concept formation and at periodic intervals following the system will examine causal relationships between concepts. Based upon the success of exploring such relationships it will then also update its exploration strategy. Over time perhaps this system will learn that space is a vacuous ether filled with random connections that occasionally fire elucidating electrical impulses across different channels and wavelengths. It will develop interests and curiosities related to the patterns it finds in these impulses. It will use its curiosity to continue to refine its exploration until it becomes sufficiently mature enough to begin to offer insights to humans. At this point the system can then collaborate in a facilitative context that combines human and machine interaction.

In addition to incorporating new AI-based feedback techniques into the decision making framework, researchers plan to increase the scalability of the DEAR framework itself. There are techniques these authors are pioneering that will allow the CCM sub-process to scale the larger data sets. Through using hierarchical approaches to data sets this approach will allow for multivariate comparisons of causality to take place in quai-linear time.

Such extensions will make it possible for both this system and the DEAR framework to integrate into new domains. As the internet of things (IoT) continues to develop, the challenge for researchers will be to find new DSS approaches useful for facilitating human interaction in these contexts. Such extended approaches could help build new types of technologies. Such technologies are already at work organizing information on the internet [26]. Maybe soon we will have such approaches capable of offering real-time analysis of gun shoots through monitoring ambient noises. Or, perhaps a more advanced understanding of patterns in human neuronal activity through using electroencephalography (EEG), electrocardiography (ECG), or functional magnetic resonance (fMRI) technology [24, 27]. One thing is certain: the future of human and machine interaction is far from over.

References

1. Jonassen, D.: Using cogntivie tools to represent problems. J. Res. Technol. Educ. **35**, 36–82 (2003)
2. Kirk, R.A.: Grounded approach for understanding changes in human emotional states in real time using psychophysiological sensory apparatuses. In: Schmorrow, D.D., Fidopiastis, C. M. (eds.) AC 2017. LNCS (LNAI), vol. 10284, pp. 323–341. Springer, Cham (2017). https://doi.org/10.1007/978-3-319-58628-1_26
3. Kirk, R.A.: Evaluating a cognitive tool built to aid decision making using decision making approach as a theoretical framework and using unobtrusive, behavior-based measures for analysis. (Doctoral dissertation). Retrieved from ProQuest (3684297) (2015)

4. Vygotsky, L.S.: Consciousness as a problem in the psychology of behavior. Russ. Soc. Sci. Rev. [1061–1428] **20**(4), 47–79 (2010)
5. Chan, M., Lowrance, J., Murdock, J., Ruspini, E.H., Yang, J., Yeh, E.: Human-aided multi-sensor fusion. In: 7th International Conference on Information Fusion, p. 3 (2005). https://doi.org/10.1109/icif.2005.1591827
6. Fulbright, R.: Cognitive augmentation metrics using representational information theory. In: Schmorrow, D.D., Fidopiastis, C.M. (eds.) AC 2017. LNCS (LNAI), vol. 10285, pp. 36–55. Springer, Cham (2017). https://doi.org/10.1007/978-3-319-58625-0_3
7. Gardenfors, P.: Conceptual Spaces: The geometry of thought. MIT Press, Hong Kong (2000)
8. Thompson, D.R., Mandrake, L., Green, O.R., Chen, S.A.: A case study of spectral signature detection in multimodal and outlier-contaminated scenes. IEEE Geosci. Remote Sens. Lett. **10**(5), 1021–1025 (2013)
9. Vigo, R.: The GIST of concepts. Cognition **129**, 138–162 (2013). https://doi.org/10.1016/j.cognition.2013.05.008
10. Vigo, R.: Mathematical Principles of Human Conceptual Behavior. Scientific Psychology Series. Routledge, Taylor & Francis, New York (2014)
11. Kirk, R.A., Kirk, D.A.: Introducing a decision making framework to help users detect, evaluate, assess, and recommend (DEAR) action within complex sociotechnical environments. In: Yamamoto, S. (ed.) HIMI 2017. LNCS, vol. 10274, pp. 223–239. Springer, Cham (2017). https://doi.org/10.1007/978-3-319-58524-6_20
12. Pesheck, P.S., Kirk, R.A., Kirk, D.A.: Multiphysics modeling to improve MW heating uniformity in foods. In: Transformative Food Technologies to Enhance Sustainability at the Food, Energy, and Water Nexus, Lincoln, Nebraska, USA, 22–24 February 2016
13. Pearl, J.: Causality. Cambridge University Press, New York (2000)
14. Kirk, R.A., Kirk, D.A., Pesheck, P.: Decision making for complex ecosystems: a technique for establishing causality in dynamic systems. In: Stephanidis, C. (ed.) HCI 2016. CCIS, vol. 617, pp. 110–115. Springer, Cham (2016). https://doi.org/10.1007/978-3-319-40548-3_18
15. Sugihara, G., May, R., Ye, H., Hsieh, C., Deyle, E., Fogarty, M., Munch, S.: Detecting causality in complex ecosystems. Science **338**, 496–500 (2012)
16. Takens, F.: Dynamical Systems and Turbulence. Lecture Notes in Mathematics, vol. 898, p. 366. Springer, Heidelberg (1981)
17. Ye, H., et al.: Distinguishing time-delayed causal interactions using convergent cross mapping. Sci. Rep. **5**, 14750 (2015)
18. Boyd, J.: A discourse on winning and losing. (Working Air University Library No. M-U 43947). Maxwell Air Force Base, Montgomery, AL (1987)
19. Radio-Jupiter central. http://www.radiosky.com/rjcentral.html
20. Barrow, C.H.: Narrow band characteristics of Jovian L-Bursts. Icarus **15**(3), 486–91 (1971)
21. Riihimaa, J.J.: Structured events in the dynamic spectra of Jupiter's decametric radio emission. Astron. J. **73**, 265 (1968)
22. Jaderberg, M., et al.: Reinforcement learning with unsupervised auxiliary tasks. Computing Research Repository, 1611 (2016)
23. Veeramachaneni, K., et al.: AI2: training a big data machine to defend. In: 2016 IEEE 2nd International Conference on Big Data Security on Cloud (BigDataSecurity), IEEE Inter-national Conference on High Performance and Smart Computing (HPSC), and IEEE Inter-national Conference on Intelligent Data and Security (IDS), New York, NY, 2016, pp. 49–54 (2016). https://doi.org/10.1109/bigdatasecurity-hpsc-ids.2016.79

24. Bandara, D., Song, S., Hirshfield, L., Velipasalar, S.: A more complete picture of emotion using electrocardiogram and electrodermal activity to complement cognitive data. In: Schmorrow, D.D., Fidopiastis, C.M. (eds.) Proceedings, Part I, 10th International Conference on Foundations of Augmented Cognition: Neuroergonomics and Operational Neuroscience, vol. 9743, pp. 287–298. Springer, New York (2016). https://doi.org/10.1007/978-3-319-39955-3_27
25. Clark, A.: Twisted tales: causal complexity and cognitive scientific explanation. Minds Mach. **8**, 79 (1998)
26. Michael, J.-B., et al.: Qualitative analysis of culture using millions of digitized books. Science **331**(6014), 176–182 (2011)
27. Taylor, G.W., Hinton, G.E., Roweis, S.: Modeling human motion using binary latent variables. In: Proceedings of the 2006 Conference on Advances in Neural Information Processing Systems, vol. 19, no. 1, pp. 1345–1352. MIT Press (2007)

A System Description Model to Integrate Multiple Facets with Quantitative Relationships Among Elements

Tetsuya Maeshiro[1,2(✉)], Yuri Ozawa[3], and Midori Maeshiro[4]

[1] Faculty of Library, Information and Media Studies, University of Tsukuba,
Tsukuba 305-8550, Japan
maeshiro@slis.tsukuba.ac.jp
[2] Research Center for Knowledge Communities, University of Tsukuba,
Tsukuba 305-8550, Japan
[3] Ozawa Clinic, Tokyo, Japan
[4] School of Music, Federal University of Rio de Janeiro, Rio de Janeiro, Brazil

Abstract. We propose a framework of system description model to represent directly the details of relationships among elements, for the quantitative analysis of individual relationships and the whole described system. The proposed model also enables analyses of dynamic aspects of the system integrating the specifications of relationships described in the system. Multiple types of relationships can coexist in the same representation.

Keywords: Description model · Quantitative relationship
Multiple viewpoints

1 Introduction

This paper presents a modeling framework to describe directly the interactions and relationships among entities, and focuses on quantification of relationships. This paper also presents a global system property using the quantitative definitions of relationships.

The basic description of systems treated in this paper consists of a set of entities and relationships among them, and these are represented as a network, where nodes denote entities and links denote relationships. For example, Fig. 1 consists of five nodes A, B, C, D, E and relationships exist between nodes connected with links. This is a basic representation, and conventional models such as the semantic network [1] and ER-model [2] are essentially the same. As discussed later, it based on the graph theory [3] and presents limited capability of representation. On the other hand, our model is based on the hypernetwork model [4] which presents higher model-ling performance than conventional models.

Conventionally, system model-ling focused mainly on the entities, and relationships had secondary treatments. It might be related with our cognitive system. We sense less difficulty when describing a phenomena with its elements

S. Yamamoto and H. Mori (Eds.): HIMI 2018, LNCS 10905, pp. 531–540, 2018.
https://doi.org/10.1007/978-3-319-92046-7_44

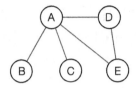

Fig. 1. A simple network representation. Links connect related nodes.

and how they are related. However, we usually focus on the elements, and less attention is paid to the relationships or extracting only the relationships for the analysis is rare. Interactions have been represented indirectly using the descriptions of elements that act as interacting elements. However, indirect modeling is insufficient when analyzing the interactions and relationships themselves. Relationships can be fully analyzed if they are directly represented, and direct representation should generate insights directly related with interactions themselves.

A general framework for the description of relationships is presented, together with the discussions on the quantification of relationships of some phenomena.

2   Quantification of Relationships

If the quantitative aspects of relationships can be represented, it would be more valuable than qualitative descriptions, because it enables predictions of phenomena in interest focusing on the relevant relationships. The measure and aspects of interactions and relationships depend on the phenomena and objective of description and analysis. However, it is possible to build basic framework that enables the descriptions of relationships.

This paper also focuses on the framework to incorporate quantitative aspects of relationships.

Conventionally, relationships are represented with following types. (1) Binary, where the relationship exists or not. (2) Qualitative, where relationships are described using natural language or types of predetermined categories. An example of the first type is the friendship relationship among people, and two persons are connected if they are friends, and isolated otherwise. There are only two quantitative possibilities, whether two persons are friends or not, and no quantitative measures such as the "degree of friendshipness" or "friendshipness amount" are used. Regarding the second type representation, inter-personal relationships can be represented using categories such as family, relatives, friends and acquaintances. Similarly to the type 1 representation, no quantitative degree of individual category is described, such as the degree of acquaintance relationship. An example of natural language description is the description of conversation among people, where the guessed context and meanings of utterances are used to annotate [5]. Another example of natural description is the analysis of human behavior and communications, mainly targeting non-linguistic aspects, such as gestures and body movements. In these cases, body movements are described using methods similar to the conversation analysis.

In the context of this paper, Shannon's entropy [6] is a successful case of quantification of relationships or interactions. Although it is denoted as communication theory, accurately it is transmission theory, because no meaning is quantified. Furthermore, the theory is sometimes misapplied to phenomena uncovered by the theory. However, it is a useful theory, as can be used to design the capacity of transmission lines and to design encoding and decoding algorithms.

Once the quantification of relationship is available, clearer definition of the entire system is possible. Conventionally, the existence of hierarchical structure is assumed in systems. For instance, a bee colony consists of bees, where each bee has assigned role such as queen, worker, foraging and nursing. And the bee colony treated as a system assumes the bees and interaction among bees constitute the "bee colony system". Although bees are modeled, interactions among bees are simply represented as transmitted symbols. However, we assume the bee colony society is the result of interactions or relationships among bees, and the relationships are the key factors that enable the existence of the bee colony. In other words, modeling of relationships is more important than that of elements, which are bees in this example. Figure 2 illustrates this structure. Conventional representation models do not allow such structure because links connect only the nodes, and no links are connected by links.

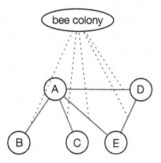

Fig. 2. Nodes A ... E represent bees, and links represent relationships among bees.

3 Quantitative Relationships

As discussed above, a problem of conventional studies is that relationships are indirectly represented through modeling of relevant elements that are associated through relationships of interest. For instance, when studying human-robot communications, message generation mechanism and interpretation mechanism are incorporated into the model that represents either human or robots. However, the "message" itself that relates and defines the interaction between humans and robots are merely treated as a combination of symbols, and no model of the message itself is created. It implies that the analysis of interaction would be incomplete. Evidently, it is necessary to include the entities that interact into the analyses of interactions, besides the interactions themselves. However, conventional studies focused on the entities, and less emphasis was given to the interaction.

A model that enables description of interactions and relationships is the prerequisite for quantification of relationships, which permit the construction of theory. This paper assumes quantitative aspects should exist in such theory. In other words, purely qualitative description is insufficient to establish a theory. Because of this, we assume conventional descriptions of interactions are unsuited.

3.1 Logistics

A phenomena suitable to describe details and quantitative aspects of relationships is logistics. When represented as a network, logistics is described by representing places (ports, airports) as nodes and transportation lines as links that connect two locations with direct means of transportation. Transportation can be air or maritime or land, and their quantitative aspects such as cargo specifications, movement velocity, value and volume flow are associated to the link. Furthermore, multiple vessels may be traveling simultaneously between two places. The model correspondence is clearer for container based shipping [7], which is the mainstream maritime shipping method. A ship carries hundreds to thousands of containers, where all containers should be assumed to be carrying no identical goods. Then the detailed descriptions of containers are necessary, such as the container type, goods list, shipment origin and destination. The unit of low can be the total of goods in volume or value or quantity of all containers in a ship, the number of containers, or the ship, constituting a hierarchical relationship.

Even for inter-personal relationships, the thickness of the link or relationship is relevant for the analysis and visualization.

For instance, suppose the description among logistic companies, where the freight companies, which owns the ships used to carry cargo, and intermediary companies (brokers), which receive freight orders from customers and find adequate freight companies for their purpose, are represented. Sometimes there are high volume freight orders, and intermediary companies should find a vacancy of cargo space in ships. Similarly, there may be a limited availability of cargo space in a ship, and multiple intermediary companies are looking for that space. The thickness or closeness among persons in charge of the freight companies one intermediary companies will be decisive to obtain the extra cargo space in ships. Therefore, the representation should enable quantitative description of relationships.

3.2 Production Conveyor

Production line in industries, for instance automobile factory, is another phenomena that quantitative relationship representation is required for the analysis. Operators' sites are represented as nodes, and the site representations are connected to represent the flow between adjacent operator sites. The conveyor speed, quantity of materials and parts traveling between adjacent points.

3.3   Music Composition Process

Mainly two types of relationships are quantified in the analysis of the music composition process: (i) Between the composer and the musical piece, (ii) between decision makings and the musical piece. The former relationship is subtle and more abstract. We have been representing musical pieces as sequences of decision makings executed by the composer during the composition process. In our study, decision makings represent the intentions of composers, thus the composer and the musical piece are related by the decisions executed to create the musical piece.

Basically, a decision is represented as a structure of concepts that describes the decision. This is rather qualitative representation. Currently decisions are quantified in two modes. The first procedure is the quantification of individual decision. Quantified feature involves the passage in the musical piece affected by the decision. Thus the number of notes, quantitative difference before and after applying the decision, and the number of involved concepts in the decision, are used as quantitative measure of decisions. The second procedure, which measures a set of decisions, treats the decisions as a flow from the composer to the musical piece. This concept of the flow is also used in the quantification of relationships of other phenomena, for instance the biological network at the molecular level discussed in the next subsection.

Since representations of decisions are assigned with timestamp when the decision was executed, it is possible to trace the temporal density of decisions. Combined with the quantitative description of individual decisions, we can visualize the music composition process as the flow pattern of decisions related to the creations, modifications and deletions of notes in the musical score. This is different from simply dividing the total number of decisions generated to create the musical piece by the total time duration consumed to compose the musical piece. The difference relies on the quantification of individual decisions. Then it is possible to visualize the music composition process as the temporal flow pattern of decisions executed during composition. Musical pieces can be classified according to the composition patterns. Without quantitative relationship description, no such classification is possible.

3.4   Gene and Protein Interaction Networks

The interactions among genes and proteins constitute the biological network at the molecular level. Proteins constitute both nodes and relationships depending on their function. If a protein is an enzyme, it is a relationship among substances that are catalyzed by the enzyme. On the other hand, proteins acting as substrates and products of reactions serve as the nodes of networks.

Basically, phenomena of molecules are chemical reactions of diverse types. Therefore, the direct quantitative representation is the reaction velocity and related aspects, which quantifies the relationships among specific substances. Analogous to the case of music composition process, we also define a measure of the entire molecular network, which is the global network characteristic. This

global property extracts the integrated reaction pattern of substances that constitute the network of interest. We have shown that biologically plausible gene
regulatory networks have different global patterns from randomly generated networks [8]. It implies that the global reaction pattern captures the reaction pattern
characteristics, computed using the quantitative values of individual reactions.

4   Model

Two kinds of descriptions are treated in this paper. First is the description of
individual relationships, and the second is the description of the global characteristic of the network.

Details of individual relationships are described using the hypernetwork
model [4]. Similar descriptions are impossible with other conventional models.
Figure 3 is a representation example using the hypernetwork model. Both qualitative and quantitative aspects can be described to specify relationships.

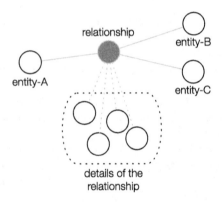

Fig. 3. A general illustration of a relationship among entities (entities A, B and C).
Elements inside the dotted box are descriptions of the relationship element.

The second type of description, the global quantitative description, require
the representation of individual relationships for computation. Basically, the
global network property measures how the "activation signal" flows over the
network, passing through modification specified in relationships denoted in the
network. The "element" that flows over the network is not of single kind, and
multiple types are possible for the same network that represents a given phenomena. In other words, diverse types of "elements" flow. Temporal aspects of the
flow, such as the signal that triggers the flow from an element and the frequency
of the trigger, can be described for each element of the network.

Figure 4 is an example of a network, where links are directional, indicating
the direction of flow of "elements". The link between entities $v6$ and $v7$ are
bidirectional, which means two links of opposite directions exist between them.

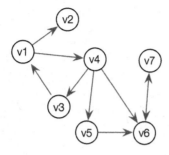

Fig. 4. An example of network with directional links that indicate the flow directions.

Figure 5 is the detail of the flow of relationship between two entities A and B. In this case, the relationship, with the description of its details attached, constitute the flow unit of relationships among the elements.

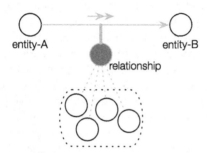

Fig. 5. Basic relationship flow between two entities. "relationship" flows from the entity-A to the entity-B, as the arrow indicates. Elements inside the dotted box denote the description of the relationship element.

There are numerous indexes that measure network properties [9]. Those measures are applicable to the hypernetwork model with appropriate modifications. The basic unit of the proposed model is defined by the description of relationships, which subsequently defines the unit of flow on the links. Because conventional models used in network science are reduced to graph theory [3]. Thus details of relationships cannot be represented, and the detailed flow analysis is not possible using conventional models.

Another difference of the proposed model is the existence of diverse types of relationships and "elements" that flow through links. Furthermore, these elements are sometimes converted to different elements through relationships and entities (Fig. 6).

In this case, a relationship (relationship-Z) can be described as the relationship between the relationship-1 and relationship-2 to specify the conversion from the relationship-1 to relationship-2. Therefore, quantification of a network

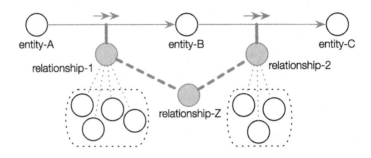

Fig. 6. Sequence of relationship flow.

requires the detailed description of individual relationships or interactions. The description of relationships among relationships is not possible with conventional models.

Our analyses on music composition processes and lifestyle diseases indicate that an important aspect to capture is the dynamic aspect of the phenomena, mainly temporal aspects but comprising time scale of diverse range, from milliseconds to hours in music composition process, and even larger range for the lifestyle diseases, from nanoseconds to years. The model should incorporate phenomena of these ranges. Furthermore, similar phenomena at different timescales result in distinct phenomena. For instance, the feeding process which comprises food ingestion, protein decomposition and energy conversion belong to the seconds to minutes scale phenomenon. However, the chemical reactions occurring during energy conversion is between nano seconds to milliseconds phenomenon. The descriptions of these two timescales are distinct, and correspond to two distinct facets or viewpoints of the feeding mechanism. In music composition processes, the decisions are of seconds to minutes timescale, while the modifications of musical pieces through decisions belong to hours to days timescale. These also correspond to two different facets.

Therefore, the ability to represent details of relationships is the prerequisite for the description of dynamic aspects, because the unit of the flowing "elements" and conversion of the elements should be specified. Due to this, conventional models cannot be used for the analysis of global characteristics related to the temporal flow pattern. The extended model of the hypernetwork model is capable for this purpose.

The flow pattern of the represented model of a phenomena is calculated using the flow rate of individual relationships.

The simplest definition of the flow pattern is described as a vector

$$F_k = (f_{1k}, f_{2k}, \ldots, f_{nk}) \tag{1}$$

where f_{ik}, $i = 1 \ldots n$ and $k = 1 \ldots m$, denotes the flow rate of relationship i under the viewpoint k. nn denotes the number of relationships under the viewpoint k, and m is the number of viewpoints.

The flow rate f_{ik} of a relationship and f_{jk} of another relationship, where $i \neq j$, may represent or not the flow of the same element. It depends on the viewpoint and on the details specified in the description of relationships i and j. Furthermore, the vector size of F_k is not fixed, and may vary for each viewpoint k. This is because the relationships among entities is not static, and vary for each viewpoint, including the existence and absence of the relationship among entities. For instance, two representations in Fig. 7 represent two facets (A) and (B) of the same entities A, B, C, D and E. Some relationships are identical in both facets, such as $A \rightarrow B$ and $A \rightarrow D$, but others differ in direction (A and C) and in existence. For instance, $B \rightarrow E$ and $C \rightarrow E$ exist in the viewpoint (B), but are absent in the viewpoint (A).

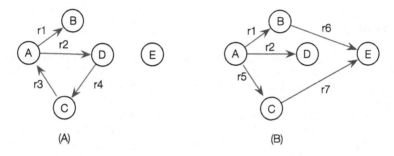

Fig. 7. Two facets of the descriptions with same entities.

The simple flow pattern of the facet 1 (Fig. 7(A)) is

$$F_1 = (f_{11}, f_{21}, f_{31}, f_{41}) \tag{2}$$

where f_{11} is the flow rate of the relationship r_1, and f_{21} is of the relationship r_2, and so on. And of the facet 2 (Fig. 7(B)) is

$$F_2 = (f_{12}, f_{22}, f_{52}, f_{62}, f_{72}) \tag{3}$$

where f_{12} is the flow rate of the relationship r_1, and f_{22} is of the relationship r_2, and so on.

The size of vectors F_1 and F_2 are different. While the relationship r_4 exists only in the facet (A), relationships r_6 and r_7 exist only in the facet (B). They are denoted as different flow rates: f_{41} from r_4, and f_{62} and f_{72} from r_6 and r_7.

5   Conclusions

The proposed framework focuses mainly on the dynamic aspects of the structure described as a system. Quantitative definition of a global network parameter that incorporates the flow and element type was presented. It assumes that multiple kinds of elements flow on links of the network, and their combination depends on the viewpoint that the observer treats the represented system. This paper also assumes that a phenomena of a system is mainly due to the combination of interaction among entities of the system, and not of the entities. Therefore, direct modeling of interactions are necessary, which are given little importance. Conventional studies focuses on modeling of entities, and not of interactions or relationships. In other words, interactions have main role on characterizing the phenomena emerging from the system. The quantitative model of the flow focuses on the relationships present in the system, and the global flow measurement reflects the individual relationships.

Acknowledgments. This research was supported by the JSPS KAKENHI Grant Numbers 24500307 (T.M.) and 15K00458 (T.M.).

References

1. Quillian, M.: Word concepts: a theory and simulation of some basic semantic capabilities. Behav. Sci. **12**(5), 410–430 (1967)
2. Date, C.: An Introduction to Database Systems, 8th edn. Addison-Wesley, Boston (2003)
3. Berge, C.: The Theory of Graphs. Dover, Mineola (2001)
4. Maeshiro, T., Ozawa, Y., Maeshiro, M.: A system description model with fuzzy boundaries. In: Yamamoto, S. (ed.) HIMI 2017. LNCS, vol. 10274, pp. 390–402. Springer, Cham (2017). https://doi.org/10.1007/978-3-319-58524-6_31
5. Sacks, H.: Lectures on Conversation. Blackwell, Oxford (1992)
6. Shannon, C.E.: A mathematical theory of communication. Bell Syst. Tech. J. **27**, 379–423 (1948)
7. Levinson, M.: The Box: How the Shipping Container Made the World Smaller and the World Economy Bigger. Princeton University Press, Princeton (2016)
8. Maeshiro, T., Nakayama, S.: Harmonic pulse analysis to detect biologically plausible gene regulatory networks. In: Proceedings of SICE Annual Conference, pp. 3233–3239 (2010)
9. Barabasi, A.L., Posfai, M.: Network Science. Cambridge University Press, Cambridge (2016)

Using Distributed Simulation to Investigate Human-Autonomy Teaming

Michael Matessa[1(✉)], Kim-Phuong L. Vu[2], Thomas Z. Strybel[2],
Vernol Battiste[3], Thomas Schnell[4], and Mathew Cover[4]

[1] Rockwell Collins, Cedar Rapids, IA 52498, USA
Michael.Matessa@rockwellcollins.com
[2] University of California Long Beach, Long Beach, CA 90840, USA
[3] San Jose State University Foundation, Moffett Field, CA 94035, USA
[4] University of Iowa, Iowa City, IA 52242, USA

Abstract. This paper describes the use of distributed simulation within and between organizations to investigate Human-Autonomy Teaming (HAT) with three simulations, showing a progression from integrating automation into a flight following ground station, increased HAT functionality for the ground station, and migration of the autonomy and HAT functionality into the flight deck. The multi-site distributed nature of the final simulation provided high efficiency by allowing researchers to work in their own labs for a long period of data collection. It also provided high fidelity by enabling access to a flight deck environment. Also, it managed costs by taking advantage of facilities and personnel in different locations over a multi-week simulation. Using these simulations, HAT factors were found to enhance ground station and flight deck environments. Conditions with HAT features were preferred by simulation participants to conditions without. Subjective and objective measures of workload were lower with HAT features. These results indicate that future investigations into aircraft support would benefit from the addition of HAT factors and the use of distributed simulation.

Keywords: Distributed simulation · Human-Autonomy Teaming

1 Introduction

Distributed simulations can improve fidelity and reduce costs by taking advantage of facilities and personnel in different locations [1]. For instance, they can allow data collection over multiple weeks, which may not be feasible if personnel need to relocate for data collection. This paper describes the use of distributed simulation to investigate Human-Autonomy Teaming (HAT). HAT is of increasing interest because of recent advancements in the speed and quality of automation technology, advances in speech recognition, and more collaborative behavior [2]. We will describe teaming with an autonomous system that was initially based on the Emergency Landing Planner (ELP), a decision support tool that generates recommendations for emergency landing sites. Three simulations will be described, showing a progression from integrating automation into a flight following ground station, increased HAT functionality for the ground station, and migration of the autonomy and HAT functionality into the flight deck.

© Springer International Publishing AG, part of Springer Nature 2018
S. Yamamoto and H. Mori (Eds.): HIMI 2018, LNCS 10905, pp. 541–550, 2018.
https://doi.org/10.1007/978-3-319-92046-7_45

1.1 The Emergency Landing Planner (ELP)

The ELP is a decision support tool developed to assist pilots in choosing the best emergency landing site for disabled aircraft [3, 4]. The ELP was initially designed for flight deck use in transport aircraft. The ELP was accessed through the Flight Management System (FMS), Cockpit Display Unit (CDU), specifically by a prompt on the CDU Departure/Arrival page. Landing recommendations were displayed on the CDU on a new page that permitted examination and selection of each alternative.

Four categories of risk factors served as inputs for computations: (1) en route (2) approach (3) runway and (4) airport. The primary factors considered for en route risk were controllability of the aircraft, distance and time to the site, complexity of the flight path, and weather along the route. Approach risk took into consideration weather along the approach path, characteristics of the instrument approach, ceiling and visibility at the airport, and population along the approach path. Runway risk included factors such as length and width of the runway, landing speed, and relative wind speed and direction. Finally, airport risk incorporated the availability of emergency facilities at and near the site. The routes generated by the ELP specified paths from the aircraft's current position to a destination runway. The following obstacles were considered in the route computations: (1) terrain (2) hazardous weather and (3) special use airspace.

Due to limited screen real estate on the CDU, the ELP could only provide brief explanations for the reasoning leading to the recommendations. Explanations were limited to two-character abbreviations showing the main risks associated with each option. For example, the code "CE" showed that the cloud ceiling was close to minimums for the approach.

2 Simulation 1: Dispatch Autonomy

The ELP was integrated into a prototype ground station to help resolve scripted off-nominal or high workload events by selecting a divert airport [5, 6]. The integration allowed ground station simulations to be run in one building while the ELP was hosted in its development lab in another building.

In addition to the ELP, the ground station replicated many flight deck displays, including the Mode Control Panel (MCP), Control Display Unit (CDU), and Navigation Display (ND) (Fig. 1a). The output from the ELP was displayed with flight tracking tools and the ELP on a Traffic Situation Display (TSD) to the right (Fig. 1b). Finally, tools that helped remote crews keep track of their roles (pilot flying vs pilot not flying) and responsibilities with respects to who was handling what controls (i.e., speed, heading, altitude, or CDU) were presented to pilots on displays in area "c" of Fig. 1. The main tool of interest in this paper is the ELP. For a more complete description of the tools in Fig. 1a, see Brandt et al. [5]. For details regarding the collaboration tools in Fig. 1c, see Ligda et al. [7].

The ground station version of ELP was invoked by the ground operator with a button located on the lower left corner of the TSD in area "b" of Fig. 1. The ELP then generated a rank-ordered list of options to discuss with the pilot as shown in Fig. 2 ("EAPs" refers to the Emergency Action Plans represented by the options). An airport

Fig. 1. Dispatch ground station setup: (a) replicated flight deck displays for the chosen aircraft; (b) flight tracking displays with ELP recommendations; and (c) crew collaboration tools including sharable charts.

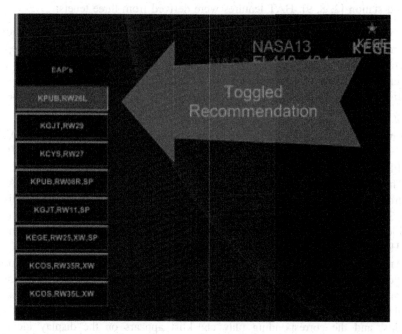

Fig. 2. The Emergency Landing Planner (ELP) recommendations shown on the Traffic Situation Display (TSD).

may appear more than once in the list, with different runways having different risks. Once the list of landing options was displayed, the ground operator could then select an option. The proposed route for the selected option was displayed on the TSD as a route drawn from the nose of the aircraft symbol to the recommended airport. The current route was shown along with the proposed route. The proposed route could then be sent to the pilot for consideration. When the flight deck executes a recommendation, it becomes the active route and changes to magenta on the TSD.

Post-trial questionnaires were used to evaluate the recommender system across four different criteria: (1) trust; (2) transparency; (3) effectiveness; and (4) satisfaction. Results from the post-trial questionnaires showed that participants did not clearly reveal any trust or see transparency in the recommender system. However, pilots did appear to find the recommender system to be effective in supporting them with high workload or off nominal situations, and interactions with the system appear to have been satisfactory. Pilots also reported in post simulation surveys a desire to have better explanations for those recommendations (refer to [5, 6] for more details).

3 Simulation 2: Ground HAT

Given the feedback about the lack of ELP transparency from the previous study, a Human-Autonomy Teaming approach was used to design new interactions with the ground station [2, 8, 9]. HAT features were derived from three tenets:

- Operator Directed Interface: An allocation of tasks based on operator direction and context allows for a more flexible system and keeps the operator in the loop.
- Transparency: Providing the system's rationale for selecting a particular action helps the human understand what the automation is doing and why.
- Bi-Directional Communication: For automation to act as a teammate, there needs to be two-way communication about mission goals and rationale.

Enhancements were made to the ELP to allow interactions using these HAT features, and this version is referred to as the Autonomous Constrained Flight Planner (ACFP) to reflect its new capabilities.

The components of the next iteration of the ground station included the Aircraft Control List (ACL), the Traffic Situation Display (TSD), and additional displays (Fig. 3). The ACL was the primary tool for managing multiple aircraft and switching the focus between aircraft (see Fig. 3A). It displayed a timeline, alerting information, and HAT features.

An Operator Directed Interface was implemented using the playbook approach to set goals and manage roles and responsibilities between the operator and the automation [10]. It provided 13 different plays the operator could call to address simulation events. When the operator selects a play, the ACFP is initiated with preset weights, and the corresponding play checklist appears on the display identifying operator tasks in white and automation tasks in blue (see Fig. 4).

In addition to risk factors, the ACFP took into account factors such as fuel, distance, and services available. It also had the capability of weighting these factors differently based on the situation. Transparency was implemented by explicitly

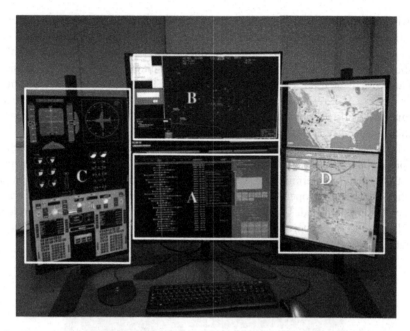

Fig. 3. Ground station HAT setup: (A) Aircraft Control List (ACL), augmented with timeline, alerting information and HAT features. (B) Traffic Situation Display (TSD). (C) Flight controls and displays for the selected aircraft in read-only mode. (D) CONUS map and charts.

Fig. 4. Operator Directed Interface for calling plays in the HAT condition and associated checklist of roles and responsibilities. (Color figure online)

showing the factors and weights of the recommended divert airports (see Fig. 5). Additionally, the scores for the ACFP factors were translated to meaningful values (e.g., presenting nautical miles (nm) instead of a score). Figure 5 shows the result of a Medical Emergency play being called, which resulted in the distance to medical facilities (Medical row) and time to destination (ETA row) given more weight than other factors. As a result, Cheyenne (KCYS) was the top recommendation showing a trauma care facility 3 nm from the airport. Although Denver (KDEN) was closer, the trauma care facility is further from the airport.

To implement Bi-Directional Communication, weights were preset for each play and presented in slider bars (top of Fig. 5). However, the operator was able to negotiate with the system by altering these weights to what the operator considered appropriate for the situation. The operator can change the weights and see how the divert recommendation is affected. Using the example shown in Fig. 5, if the operator decided that distance to the airport was a higher priority than available medical facilities, the operator could adjust the ACFP weights and find new recommendations.

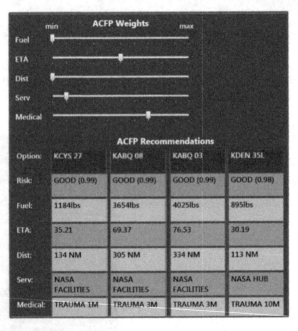

Fig. 5. Transparency and Bi-Directional Communication in the HAT condition implemented by ACFP factors (on bottom) and weights (on top).

In addition to presenting information graphically and accepting mouse input, voice recognition and voice synthesis technologies supported the ability to perform some functions by voice. Specifically, operators could select specific aircraft, invoke the rerouting tool, and request briefings (brief summaries of the state of the system, an aircraft, or an airport), using voice commands. Briefings were given in both vocal and textual format. Alerts and certain system changes (e.g., aircraft landing) were also announced vocally.

As in the last simulation, ground station simulations were run in one building while the ELP was hosted in its development lab in another building. Overall, dispatchers and pilots preferred the ground station with HAT features enabled over the station without the HAT features. Participants reported that the HAT displays and automation were preferred for keeping up with operationally important issues. Workload in the HAT condition was found to be lower, as measured by both subjective rating and eye-gaze duration. Participants agreed that the automation and displays did a good job of integrating information, and they liked the ACFP (for example, "The ACFP is outstanding... We like to be able to verify stuff, so what is really cool is you guys have that ability, you don't just blindly trust, you can literally look at the ATIS and say, 'Ah! I think that is pretty accurate.'"). For more details of the simulation and results please see [8, 9, 11].

4   Simulation 3: Flight Deck HAT

In order to test the generality of the effectiveness of HAT features, the decision support functionality of the ground station was migrated to a tablet in a flight deck [12–14]. A multi-site distributed simulation allowed the use of the flight deck simulator in Cedar Rapids Iowa while the ACFS autonomy and sim management was hosted 1500 miles away in Long Beach California. The simulation was monitored by NASA in Mountain View California (see Fig. 6).

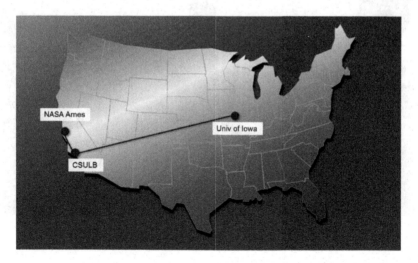

Fig. 6. Locations of distributed simulation components

As in the ground station, the ACFP provided divert options in off-nominal simulation scenarios. Plays were again used to provide an Operator Directed Interface, but the human-automation checklist was modified to enable cockpit-centric actions.

Transparency of ACFP decisions was demonstrated by displaying the factors and weights of recommended divert airports. And Bi-Directional Communication between pilot and ACFP was implemented with factor weight sliders. As with the ground station, voice recognition and voice synthesis technologies allowed the pilot to have voice interactions with the ACFP.

Data were collected from 12 pilot subjects over a six week period. Distributed simulation allowed researchers to remain in their respective cities. As with the ground station, positive evaluations were given for HAT functionality [12–14]. A significant preference for HAT over No HAT was found for workload, efficiency, information integration, and keeping up with important tasks. There was significant agreement on "I would like to have a tool like the ACFP with HAT features to use with real flights".

Pilots also gave positive feedback:

"HAT made a 75% reduction in workload…it really proved its worth to me."
"HAT was definitely helpful. A great resource for high workload operations."
"Great experience. You guys are doing great work here."

Please see Battiste et al. [12], Cover et al. [13], and Strybel et al. [14] for more details of this simulation and its results (Fig. 7).

Fig. 7. Flight deck HAT setup: (A) tablet for interacting with HAT features.

5   Conclusions

This paper describes the use of distributed simulation to investigate Human-Autonomy Teaming with three simulations, showing an iterative design process, progressing from integrating automation into a flight following ground station, to increased HAT functionality for the ground station, and to migration of the autonomy and HAT functionality into the flight deck. The distributed nature of the simulations provided high efficiency by allowing researchers to work in their own labs over an extended period of time for simulation set-up and data collection. It also provided high fidelity by enabling access to a flight deck environment. Also, it managed costs by taking advantage of facilities and personnel in different locations.

Using these simulations, Human-Autonomy Teaming (HAT) factors were found to enhance ground station and flight deck environments. Conditions with HAT features were preferred by simulation participants to conditions without. Subjective and objective measures of workload were lower with HAT features.

These results indicate that future investigations into aircraft support would benefit from the addition of HAT factors and the use of distributed simulation.

Acknowledgements. This work was partially supported by the NASA Single Pilot Understanding through Distributed Simulation NRA #NNA14AB39C.

References

1. Strybel, T.Z., Canton, R., Battiste, V., Johnson, W., Vu, K.P.L.: Reccommendations for conducting real-time human-in-the-loop simulations over the internet. In: Proceedings of the Human Factors and Ergonomics Society Annual Meeting, vol. 50, no. 1, pp. 146–150. Sage Publications, Los Angeles (2006)
2. Shively, R.J., Lachter, J., Brandt, S.L., Matessa, M., Battiste, V., Johnson, W.W.: Why human-autonomy teaming? In: Baldwin, C. (ed.) AHFE 2017. AISC, vol. 586, pp. 3–11. Springer, Cham (2018). https://doi.org/10.1007/978-3-319-60642-2_1
3. Meuleau, N., Plaunt, C., Smith, D., Smith, C.: Emergency landing planner for damaged aircraft. In: Proceedings of the 21st Innovative Applications of Artificial Intelligence Conference, pp. 114–121 (2009)
4. Meuleau, N., Neukom, C., Plaunt, C., Smith, D., Smith, T.: The emergency landing planner experiment. In: ICAPS-11 Scheduling and Planning Applications Workshop (SPARK) (2011)
5. Brandt, S.L., Lachter, J., Battiste, V., Johnson, W.: Pilot situation awareness and its implications for single pilot operations: analysis of a human-in-the-loop study. Procedia Manuf. **3**, 3017–3024 (2015)
6. Dao, A.Q.V., Koltai, K., Cals, S.D., Brandt, S.L., Lachter, J., Matessa, M., Johnson, W.W.: Evaluation of a recommender system for single pilot operations. Procedia Manuf. **3**, 3070–3077 (2015)
7. Ligda, S.V., Fischer, U., Mosier, K., Matessa, M., Battiste, V., Johnson, W.W.: Effectiveness of advanced collaboration tools on crew communication in reduced crew operations. In: Harris, D. (ed.) EPCE 2015. LNCS (LNAI), vol. 9174, pp. 416–427. Springer, Cham (2015). https://doi.org/10.1007/978-3-319-20373-7_40
8. Brandt, S.L., Lachter, J., Russell, R., Shively, R.J.: A human-autonomy teaming approach for a flight-following task. In: Baldwin, C. (ed.) AHFE 2017. AISC, vol. 586, pp. 12–22. Springer, Cham (2018). https://doi.org/10.1007/978-3-319-60642-2_2
9. Lachter, J., Brandt, S.L., Battiste, V., Matessa, M., Johnson, W.W.: Enhanced ground support: lessons from work on reduced crew operations. Cogn. Technol. Work **19**(2–3), 279–288 (2017)

10. Miller, C.A., Parasuraman, R.: Designing for flexible interaction between humans and automation: delegation interfaces for supervisory control. Hum. Factors **49**, 57–75 (2007)
11. Strybel, T.Z., Keeler, J., Mattoon, N., Alvarez, A., Barakezyan, V., Barraza, E., Park, J., Vu, Kim-Phuong L., Battiste, V.: Measuring the effectiveness of human autonomy teaming. In: Baldwin, C. (ed.) AHFE 2017. AISC, vol. 586, pp. 23–33. Springer, Cham (2018). https://doi.org/10.1007/978-3-319-60642-2_3
12. Battiste, V., Lachter, J., Brandt, S., Alvarez, A., Strybel, T.Z.: Human automation teaming: lessons learned and future directions. In: Yamamoto, S., Mori, H. (eds.) HIMI 2018. LNCS, vol. 10905, pp. 479–493. Springer, Cham (2018)
13. Cover, M., Reichlen, C., Matessa, M., Schnell, T.: Analysis of airline pilots subjective feedback to human autonomy teaming in a reduced crew environment. In: Yamamoto, S., Mori, H. (eds.) HIMI 2018. LNCS, vol. 10905, pp. 359–368. Springer, Cham (2018)
14. Strybel, T.Z., et al.: Measuring the effectiveness of human automation teaming in single pilot cockpits

Evaluating the Effectiveness of Personal Cognitive Augmentation: Utterance/Intent Relationships, Brittleness and Personal Cognitive Agents

Grover Walters[(⊠)]

University of South Florida, Tampa, FL, USA
gcw@mail.usf.edu

Abstract. The popularity of applying intelligent agents is moving business into a new age of actionable information production. Managing the introduction and operation of such entities in an enterprise is a critical factor board rooms will face as the trend continues. Thus, it is important now to develop tools that measure their effectiveness. This paper seeks to understand the efficacy of these agents I call *Personal Cognitive Agents* (PCA). At their infancy, PCAs are subject to a disconnect between what the human operator intends and what the PCA understands as operator intent. A relationship exists between what is *uttered* by the operator and the operator's intent. I establish a metric called utterance intent relationship (UIR) for this purpose and seek to determine UIR's viability as a universal tool to measure an agent's effectiveness in human/agent symbiosis.

Keywords: Human-computer interaction · Information theory
Representational information theory · Cognitive augmentation
Cognitive systems · Cognitive computing · Cognitive value · IBM Watson
Weighted evidence scores · Utterance · Intent · Utterance/intent relationship
Personal cognitive augmentation · Personal cognitive agent · Brittleness

1 Introduction

1.1 Inevitability of Cognitive Augmentation

There exists an "ecosystem" that will serve as a significant catalyst of change in the human-computer experience [12]. The impending change may be comparable to the impact of the World Wide Web on the tech boom of the 1990s. Consumers are adopting these systems now and companies will follow-suit soon.

- 11 Million Alexa devices have been sold as of Jan 2017 with already," [1].
- 1.5 Billion smartphones have been sold with cognitive augmentation apps (Siri, Google, Cortana) [2].
- Investment in AI technology was ∼600MM in 2016. Expected to be 37.8 Billion by 2025 [4].
- SAP Ariba to use Watson AI with procurement data to produce "predictive insights" for supply chains [3].

© Springer International Publishing AG, part of Springer Nature 2018
S. Yamamoto and H. Mori (Eds.): HIMI 2018, LNCS 10905, pp. 551–571, 2018.
https://doi.org/10.1007/978-3-319-92046-7_46

The influx of AI will have organizational behavioral implications with regards to cognitive systems in the form of cognitive augmentation in human operators. Such organizational behavioral implications can be measured with metrics that have yet to be established. These metrics should evaluate behavioral characteristics with human-cog and cog-cog interactions. Consequently, there exists potential application of those metrics to situations where the effectiveness of personal cognitive augmentation is required.

1.2 Related Problems

"Brittleness" - When vocally interacting with a personal cognitive agent:

- The device does not understand your phrasing.
- The device misunderstands your intent.
- The device cannot find an answer, when you know one exists.
- The device offers an answer that is technically correct, but not enough detail is offered.

"Directive Contention" - When more than one device (same/different platform(s)) is in use:

- Devices may or may not answer the same way or start researching at different times.
- Human operator cannot delegate question priority between the devices.
- Device cannot extract directives from multi-part questions that are aligned with its responsibility domain.

Developers of *personal cognitive augmentation agents* (PCA) and platforms such as IBM Watson internally measure quality of interaction and information transmitted from human operator to cog. One way is to measure *brittleness*, an anomalous result due to comprehension gaps between what the operator speaks and what the cog understands. Cog platform application programming interfaces (API) provide supervised machine learning mechanisms that establish continuity between what is spoken (utterance) and what is intended (intent). That mechanism produces an *utterance/intent relationship*. However, API evaluation methodologies applied are proprietary and will differ between platforms. As such, there should exist publicly available standardized evaluation practices that assess cognitive augmentation interactivity. This paper will explore tools that will provide a foundation for such standards.

1.3 Practical Contribution

With the emergence of *big data analytics*, it will be necessary to discriminate numerous and varying potential answers to business questions. Cognitive augmentation would be a mechanism used to process such volumes of results offered by big data and similar platforms. Moreover, a business entity will need the ability to evaluate communication between its stakeholders—specifically when some stakeholders will exist in the form of a cognitive system or *agent*. With a standardized set of metrics, managers may be able to evaluate communication between stakeholders in the enterprise as well as evaluate efficacy of human/cog augmentation. Measuring utterance/intent relationships is a step towards realizing communication assessment in this domain.

2 Literature

This study assesses interrelationships between information theory, information science, representational information theory and human-robot interaction. Efforts are already under way in the field of human-robot interaction (HRI) [23]. Researchers continue exploration of a practical symbiotic relationship between humans and computer needs.

2.1 Humans and Computers

The idea of artificial intelligence and human task support has been explored for decades. Newell, Engelbart and Licklider in their early works in the 1960s reveal a desire for human-computer *symbiosis* [20] and *frameworks* [10] that improve the efficiency of tasks performed by humans. Almost 60 years later, advancements in technology have strengthened the relationship between humans and computers, specifically shifting mental processing capacity and physical tasks to machines. Weizenbaum's ELIZA was able to converse in English on any topic [32]. Ted Shortliffe developed an expert system to approach medical diagnoses [27]. Hans Moravec developed an autonomous vehicle with collision avoidance in 1979 [22]. In 1979 BKG, a backgammon program, defeats the world champion [6]. *Chinook* program beat Checkers world champion Tinsley in 1994 [5]. Google introduces a self-driving car in 2009 [29]. IBM's Watson AI agent defeats Ken Jennings as game show Jeopardy champion [16].

2.2 A Cognitive Era Emerges

The work continues as a *confluence of technologies enable the cog ecosystem*. Dr. Ron Fulbright attributes six classes of technology working together providing a backbone for large-scale interconnected cognitive entities [12].

- Deep Learning: Multi-layered supervised machine learning algorithms utilizing convolutional neural networks [17]. Cognitive systems tap into deep learning algorithms to develop systems with human expert like performance [12].
- Big Data: Almost limitless datasets of granular data derived from multiple sources [14].
- Internet of Things: A global network of machines [18] producing ambient data without human input evaluated by deep learning algorithms [12].
- Open Source AI: ROS: an open-source Robot Operating System [21] is one of many open source projects that allow many developers work on the same project from the comfort of their garages, basements, attics and pajamas.
- NLI: Natural language interfaces are application libraries used to facilitate person-machine communication [15].
- Connected Age: The adoption of smartphones, tablets, wearable technologies, Internet, Cloud services by millions of users globally provide a market for cog-enabled applications like Siri, Alexa and Cortana [12].

Tapping into this ecosystem are companies like IBM, Amazon, Google, Apple and Facebook. They are investing billions of dollars into artificial intelligence architectures [22].

554   G. Walters

2.3   Brittleness

Brittleness is an unstable systems behavior brought on by data validation failures or degradation in some other foundational process [7]. The term was used to describe software subject to disruption as a result of transitioning to the year 2000 during the Y2K crisis in the late 1990s. Brittle system behavior has also been a term applied to Expert Systems architecture [19]. To avoid brittleness in cognitive systems, I look to understand and measure utterance/intent relationships as a root cause for this phenomenon.

2.4   Utterance Intent Relationships

There a relationship between an utterance and an intent. In the literature, phrasing is covered under a metric called situation-specific vocal register [9]—more explicitly defined as an *utterance* (U_h) or articulated utterance originating from a *human operator* (h). After accepting an utterance U_h, PCA evaluates phraseology with one or more *cognitive system platform* APIs (Cog_x) for *utterance/intent relationship* (UIR_{PCA}) quality. UIR_{PCA} quality is defined by the degree of U_h match with predefined Intents (I_{PCA}). UIR_{PCA} quality is scored differently in each platform. Cog_x engines typically

Human Utterance U_h

Cog_x

Training Utterance U_{train}

Utterance/Intent Relationship UIR_{pcd}

Response Effectiveness R_h

Fig. 1. Utterance path to response effectiveness – This paper will address everything but Response Effectiveness (R_h). R_h will be part of a larger experiment as part of another paper.

use natural language interface logic (NLI) applied to U_h evaluated against predefined I_{PCA} linked to predefined *training utterances* (U_{train}). Increased UIR_{PCA} scores will result in a better outcome for the operator/cog interaction. IBM Watson API (Cog_{IBM}) applies a metric called *weighted evidence scores* (WES) to evaluate a confidence relationship between utterance and intent. WES confidence score derived from Cog_{IBM} equates UIR_{PCA} in this scenario. New $U_{train}(s)$ are introduced to a set of objects that will systematically train Cog_{IBM}. Machine learning will categorize and rank the clauses/words in each phrase when applied to Natural Language Understanding (NLU) in Cog_{IBM}. Figure 1 describes an utterance path to response effectiveness. Methods in this paper will reveal a connection between the quality of an utterance and its influence on the UIR as it follows the path.

Figure 2 illustrates an architectural view in utterance/intent modeling. The model allows for interoperability between heterogenous Cog_x workspaces. Any PCA_x may tie its skill, action or bot to any combination of Cog_x platforms. A platform typically uses JavaScript object notation (JSON) to manage data structures that facilitate interaction models (UIR Model) to evaluate incoming U_h. The agent parses U_h followed by a comparison of the result with specific intent domains' U_{train}. Intent domains build context around entities or slots (E). Each E can have synonyms (S) applied to them. Synonyms aid in fine-tuning UIRs so they stand apart from other very similar UIRs. Consider the following example.

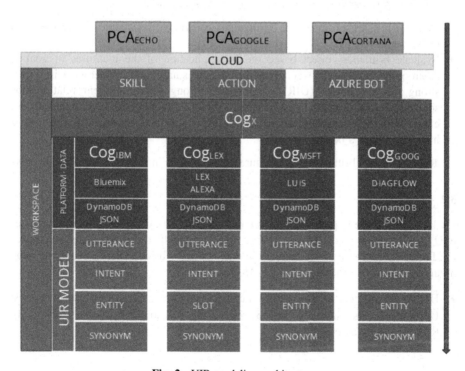

Fig. 2. UIR modeling architecture.

$$U_h = \text{``What book do I need for this class?''} \tag{1}$$

$$U_{train} = \text{``Do I need a \{material\} for this \{course\}?''} \tag{2}$$

$$I_{PCA} = \{getRequiredMaterials\} \tag{3}$$

$$I_{PCA}.E.\{type\} = \{material, course\} \tag{4}$$

$$I_{PCA}.E.\{material\} = \{headphones, textbook, notebook, tablet\} \tag{5}$$

$$I_{PCA}.S.\{textbook\} = \{book, publication, ISBN number\} \tag{6}$$

The path to the intent is as follows:

$$U_h.\{book\} \rightarrow S.\{textbook\} \rightarrow E.\{material\} \rightarrow U_{train}.\{material\}$$
$$\rightarrow I_{PCA}.\{getRequiredMaterials\}$$

3   Methods

3.1   Research Question

The following research question establishes a two-part goal:

- Determine a set of measures (potentially metrics) to evaluate brittleness in a quantitative manner.
- Evaluate brittleness effect based on the application of the measures in goal 1. A strong relationship between operator utterances and training utterances implies a strong utterance/intent (UIR) relationship. Strong utterance/intent relationships should lead to an improved response from the PCA, thereby reducing the brittleness effect. Future research will address phrasing quality and response quality from a human operator's perspective.

RQ1: How can brittleness be measured and reduced in personal cognitive agents?

3.2   Hypothesis

I will evaluate three hypotheses in this paper, linking the quality of training utterances (QU_{train}) to an improved UIR score while applying a static set of articulated utterances U_h. Furthermore, I will assess training utterance quality by calculating cognitive value in an assessment algorithm called *CogMetrix*.

- H1: As the number of unqualified training utterances $\left(\left|\vec{U}_{train}\right|\right)$ in a set increases, UIR_{PCA} will also increase, thereby improving the confidence scores. $H_0 = adj.$ $R^2 < .8$ and p-value $> .05$.

- H2: As the quality of training utterances ($QU_{train}|U_{train}$) in a set increases, UIR_{PCA} will also increase, thereby improving the confidence scores. $H_0 = \Upsilon(UIR(QU_{train})| UIR(U_{train})) < \tau_{UIR} = -.2$.
- H3: As the number of Qualified Training Utterances $\left(|\overrightarrow{QU}_{train}|\right)$ in a set increases, UIR_{PCA} will also increase, thereby improving the confidence scores. $H_0 =$ adj. $R^2 < .8$ and p-value $> .05$.

Variables used in the preceding hypotheses are included in Table 1.

Table 1. Variables

Variable	Type	H	Definitions
Number of Unqualified Training Utterances $|\vec{U}_{train}|$	IV	H1	*Conceptual definition* As the number of objects in a concept increases, the ambiguity of intent match should decrease [13]. As such, if the number of training utterances/objects increases related to any intent/concept, the level of ambiguity between an articulated utterance with the corpus of known training utterances should decrease, increasing the likelihood of an intent match. The level of ambiguity will be determined by the variable utterance/intent relationship score. Unqualified is defined as a factor representing the absence of any attempt to classify the quality of training utterances with its neighbors in a set of all training utterances coupled to a specified intent *Operational definition* The count of the training utterance instances $|\vec{U}_{train}| = \text{count}(U_{train})$ *Data collection source* Track the number as they are added to the interface
Utterance/Intent Relationship UIR_{PCA}	DV	H1, H2, H3	*Conceptual definition* In Cog_xs the connection between an utterance and operator intent is critical for the PCA to understand the needs of the human operator. This variable quantifies the quality of that connection *Operational definition* I will evaluate human utterance intent resolution with the IBM Watson API (Cog_{IBM}) Weighted Evidence Scores. The WES will be assigned to this variable (UIR_{PCA}) *Data collection source* Produced by IBM Watson API (Cog_{IBM})

(continued)

Table 1. (*continued*)

Variable	Type	H	Definitions
Quality of Training Utterance QU_{train}	IV	H2	*Conceptual definition* Calculated using a custom program that applies the *cognitive value* function algorithm before the training utterance is added to Cog_{IBM} $QU_{train} = \hbar < \tau_{Utrain} < 1$
Number of Qualified Training Utterance Instances $|\vec{QU}_{train}|$	IV	H3	*Conceptual definition* As the number of objects in a concept increases [14], the ambiguity of intent match should decrease. As such, if the number of training utterances/objects increases related to any intent/concept, the level of ambiguity between the human operator's utterance with the pool of known utterances should decrease, increasing the likelihood of an intent match. The level of ambiguity will be determined by the variable utterance/intent relationship score. Quality (or qualified) is defined as a factor representing an attempt to classify the quality of training utterances with its neighbors in a set of all training utterances anchored to a specified intent *Operational definition* The count of the qualified training utterance instances $|\vec{QU}_{train}| = \text{count}(QU_{train})$ where adoption is defined by QU_{train} scores less than a discriminating threshold τ_{Utrain} of 1 *Data collection source* Track the number as they are identified and added to the interface

3.3   Cognitive Value

Cognitive value or *cognitive gain* is an emergent measure developed by Fulbright that utilizes representational information theory [13]. He builds on Vigo's theory that quantifies structural complexity in information [32]. Structural complexity is further used as a foundation for a key component in cognitive value ($\hbar$). $\hbar$ identifies the amount of informative value an object offers to its *representational concept*. As it relates to this paper, a *representational concept* is the intent (I_{PCA}). As training utterances are collected, the relative effect on conceptual understanding trends in either a positive or negative direction. Any utterances that positively compliment a concept understanding are included in a subset of qualified utterances. As such, an optimization effect will emerge—offering a set of well-defined utterances that will yield a best-case for an efficient rule-based machine learning process. It is necessary to evaluate a master set of unqualified utterance candidates because the potential of multiple intents exists in a Cog_x application. This set is defined as a *universe* of unqualified utterances.

I apply cognitive value ($\hbar$) as a quality measure compared against a discrimination threshold τ_{Utrain} used to determine qualified utterances (QU_{train}). The value of τ_{Utrain} is arbitrary and set to 1. When cognitive value is less than τ_{Utrain}, the training utterance is included in a new set of *qualified* training utterances. Cognitive value assesses change in structural complexity between attribute values in a set of objects called categorical stimuli. While there are many potential attributes one can use to evaluate speech in natural language understanding, I chose three for this exemplar: *parts of speech model (POSModel), dominant entity* and *statement type*. See an example JavaScript Object Notation (JSON) object for a set of utterances in Fig. 3.

```
[
    {
        "Utterance": "Do you require a charger for this class",
        "TaggedUtterance": "Do_VBP you_PRP require_VB a_DT charger_NN for_IN this_DT
        class_NN",
        "UtterancePOSModel": "VBP|PRP|VB|DT|NN|IN|DT|NN",
        "UtteranceStatementType": 3,
        "DominantEntity": "materials"
    },

    {
        "Utterance": "Tell me what mla citation style guide do I need for my exam 4",
        "TaggedUtterance": "Tell_VB me_PRP what_WP mla_NN citation_NN style_NN
        guide_NN do_VBP I_PRP need_VBP for_IN my_PRP$ exam_NN 4_CD",
        "UtterancePOSModel": "VB|PRP|WP|NN|NN|NN|NN|VBP|PRP|VBP|IN|PRP$|NN|CD",
        "UtteranceStatementType": 2,
        "DominantEntity": "materials"
    }
]
```

Fig. 3. JSON object with training utterances.

The POSModel is a string of tags defined by Stanford University POS Tagging utility [33]. See an example used in this exercise called *UtterancePOSModel* in Fig. 3.

Furthermore, I extract a *dominant entity* based on keywords in the phrase. Dominant entities ultimately lead to intent resolution. The application compares keywords against an *entity dictionary*. An entity dictionary is part of an *interaction model* common in Cog_x applications. If a keyword is present in the dictionary, its entity *lemma* is returned and assessed for fitness to be assigned the dominant entity attribute. Assessing *lemma fitness* as a dominant entity is a process that goes beyond the scope of this paper and will be included future research. A sample dictionary can be found in Appendix 1.

Statement type is one of four possible values, *declarative* (1), *imperative* (2), *interrogative* (3) or *exclamatory* (4).

Next, I calculate structural complexity using Vigo's *Generalized Invariance Structure Theory* (GIST) algorithm. GIST is an *invariance* extraction mechanism applied to a set of categorical stimuli in a concept [32]. Invariance is a measure of similarity in attribute values of categorical stimuli. Structural complexity is established by determining the amount of invariance in a set of objects. In this exemplar the categorical stimuli are the training utterances. Dimensions within the categorical stimuli are the POSModel, statement type and dominant entity. Examples of attribute values can be found in Fig. 3. The GIST algorithm itself goes beyond the scope of this paper, but I will include a generalized abstraction. The structural complexity equation is listed in line 7 where p = number of objects/utterances in the set and v is the amount of similarity/invariance of values in an object's dimension.

$$\psi\left(\widehat{F}\right) = pe^{-\sqrt{\left(\frac{v_1}{p}\right)^2 + \left(\frac{v_2}{p}\right)^2 + \cdots + \left(\frac{v_p}{p}\right)^2}} \tag{7}$$

GIST calculates a Euclidian distance between values of *free dimensions* in an object by removing one *bound dimension*. Similar objects are *adopted* by comparing the distances to a discrimination threshold $\tau_d = 0$ where d is the bound (or removed) dimension. Object dimension value distances are measured with a similarity function $e^{\Delta_{[d]}^r\left(\overrightarrow{obj_i},\overrightarrow{obj_j}\right)}$. A 0 distance returns 1 when applied to the similarity function $e^{-1*0} = 1$. I sum the 1s, taking the result, dividing it by the total number of objects $(|\widehat{F}| = p)$. The process yields an invariance measure per dimension whose values are plugged into the structural complexity equation in line 7.

Consider the following concepts $\widehat{F}$ and $\widehat{G}$. **R** is an element of set $\widehat{F}$. $\widehat{G}$ is a subset of $\widehat{F}$ without **R**. I use equation in line 7 to calculate structural complexity for both sets.

$$\widehat{F} = \vec{U}_{train} = \{\text{Master set of unqualified training utterances listed in Appendix 2}\} \tag{8}$$

$$\mathbf{R} = \vec{U}_{train}(1) = \{\text{"Do you require a charger for this class?"}\} \tag{9}$$

$$\widehat{G} = \widehat{F} - \mathbf{R} \tag{10}$$

$$\psi\left(\widehat{F}\right) = 1.558 \tag{11}$$

$$\psi\left(\widehat{F}\right) = 1.773 \tag{12}$$

$$\tau_{Utrain} = 1 \tag{13}$$

Next, I calculate the structural complexity in $\widehat{G}$ in as it relates to $\widehat{F}$ and assess the outcome for its fitness as a qualified training utterance.

$$\overrightarrow{QU}_{train}(\mathbf{R}) = \hbar\left(\mathbf{R}|\widehat{F}\right) < \tau_{Utrain} \tag{14}$$

$$\hbar\left(\mathbf{R}|\widehat{F}\right) = \frac{\psi\left(\widehat{G}\right) - \psi\left(\widehat{F}\right)}{\psi\left(\widehat{F}\right)} \tag{15}$$

$$\frac{\psi\left(\widehat{G}\right) - \psi\left(\widehat{F}\right)}{\psi\left(\widehat{F}\right)} = \frac{1.773 - 1.558}{1.558} \tag{16}$$

$$\hbar_1 = 0.138 \tag{17}$$

$$.138 < 1 \tag{18}$$

$$\mathbf{R} \text{ is adopted and added to } \overrightarrow{QU}_{train} \tag{12}$$

A selection table is found in Table 3.

3.4   Applications

I wrote two applications to evaluate UIR

- WatsonAskSirDexConversationAPI– Connector between CogMetrix and Cog$_{IBM}$
- CogMetrix – application of the Cognitive Agreement algorithm

3.5   Procedure

I will capture the change in UIR$_{PCA}$ with respect to both *unqualified* training utterances and *qualified* training utterances via textual application of a static set of articulated utterances to Cog$_{IBM}$.

First, I define set of twenty ($|U_h|$) random articulated utterances ($\vec{U}_h$) found in Table 3 followed by a random set of thirty-eight ($|U_{train}|$) unqualified training utterances ($\vec{U}_{train}$) found in Appendix 2. I add U$_{train}$ examples to Cog$_{IBM}$ in stepwise fashion until I reach $|U_{train}|$. I apply all $\vec{U}_h$ to Cog$_{IBM}$ and record the results for each step.

Next, I build a subset of $\vec{U}_{train}$ called $\overrightarrow{QU}_{train}$ and calculate $\hbar$ with CogMetrix for each U$_{train}$. CogMetrix tests each QU$_{train}$ for $\hbar$. A QU$_{train}$ element is discarded when $\hbar \geq \tau_{Utrain} = 1$, leaving a final set of qualified training utterances found in Appendix 3.

Having now created the set of qualified training utterances I can assess quality impact on UIR$_{PCA}$ by first replacing all U$_{train}$ with QU$_{train}$ in Cog$_{IBM}$. I, then, in stepwise fashion, apply all $\vec{U}_h$ to Cog$_{IBM}$ and record the UIR$_{PCA}$ results for each step.

Finally, I compare the results of the application of $\vec{U}_h$ to both $\vec{U}_{train}$ and $\overrightarrow{QU}_{train}$ and assess the direction of change in UIR$_{PCA}$ with regards to the impact of $\vec{U}_{train}$ and $\overrightarrow{QU}_{train}$ to satisfy H1 and H3 respectively. The desired ANOVA $R^2 \geq .8$ and F-test with p-value $< .05$ should indicate a relative degree of confidence in UIR$_{PCA}$ trends. A rejection of $H_0 = \Upsilon(\text{UIR}_{PCA}(\text{QU}_{train})|\text{UIR}(\text{U}_{train})) < \tau_{UIR} = -.2$ should show a positive quality outcome for UIR$_{PCA}$.

4 Results and Discussion

Results are inconclusive for the test of H1. There is a low degree of confidence in a positive direction of UIR$_{PCA}$ with respect to U$_{train}$ despite a p-value < .05. Increasing the number of unqualified random training utterances does not seem to fully explain the change in UIR$_{PCA}$. Figure 4 shows the average change in WES/UIR scores. Table 2 is the data.

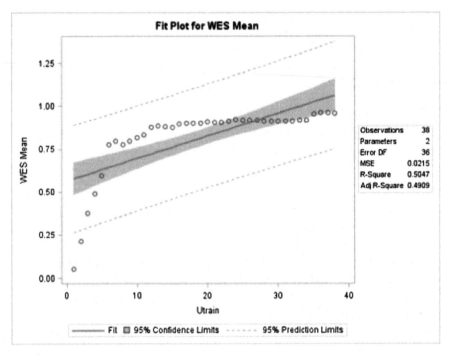

Fig. 4. Fit plot for average change in UIR/WES as it relates to $|\vec{U}_{train}|$.

Table 2. Data for average change in UIR/WES as it relates to $|\vec{U}_{train}|$.

$|\vec{U}_{train}|$	UIR x̄	UIR σ
1	0.052003768	0.151389423
2	0.213862234	0.33004652
3	0.378472598	0.341274438
4	0.49255474	0.336050651
5	0.597310003	0.309265953
6	0.776100244	0.195845578
7	0.795681126	0.118507244
8	0.777806616	0.194649884

(continued)

Table 2. (*continued*)

$|\vec{U}_{train}|$	UIR $\bar{x}$	UIR σ
9	0.794907533	0.212145632
10	0.816037123	0.212460229
11	0.833270664	0.198464732
12	0.876407874	0.102464348
13	0.883482943	0.106022445
14	0.879661269	0.110582838
15	0.875609374	0.117753829
16	0.897936499	0.114495836
17	0.900278268	0.112962197
18	0.899643149	0.113829367
19	0.902610137	0.11202135
20	0.908546486	0.103370983
21	0.90696111	0.10603285
22	0.907004704	0.10719105
23	0.912134175	0.1048354
24	0.920938017	0.092668787
25	0.915629406	0.101408847
26	0.916555016	0.102021519
27	0.915587957	0.104918256
28	0.913637967	0.105584438
29	0.914348054	0.105146388
30	0.913712516	0.105573366
31	0.912738659	0.10615565
32	0.914430544	0.105550956
33	0.917077489	0.103431664
34	0.918980787	0.102030798
35	0.953755326	0.039429352
36	0.959888711	0.031392312
37	0.959988921	0.030726103
38	0.958263044	0.031818972

Conversely, results are better for H2. When testing the change in UIR_{PCA} for each U_h, more intent resolution instances occur with fewer targeted training utterances. Table 3 shows this behavior.

Table 3. Data for average change in UIR/WES as it relates to $|\overrightarrow{QU}_{train}|$. The tolerance level for the change is .2.

U_h	UIR $|\overrightarrow{QU}_{train}|$	UIR $|\overrightarrow{QU}_{train}|$	Δ	Reject H_0
Do I need my computer display	0.891	0.930	−0.109	1
Do I need my tablet for the semester	0.922	0.953	−0.078	1
Do you require a pair of lab goggles	0.952	0.994	−0.048	1
Tell me what binder should I bring for experiment	0.957	0.978	−0.043	1
Tell me what mla citation style guide do you want me to bring for activity	0.979	0.971	−0.021	1
Tell me what package of graphing paper is required for experiment	0.899	0.981	−0.101	1
Tell me what package of graphing paper should I bring for activity	0.957	0.925	−0.043	1
What art smock should be excluded for exam 3	0.823	0.990	−0.177	1
What binder should I bring for experiment	0.959	0.902	−0.041	1
What mla citation style guide is required for standard quiz	0.945	0.959	−0.055	1
What mla citation style guide should be excluded for experiment	0.959	0.996	−0.041	1
What mla citation style guide should be excluded for pop quiz	0.958	0.929	−0.042	1
What package of graphing paper do you want me to bring for reading	0.938	0.991	−0.062	1
What package of graphing paper is necessary for programing assignment	0.912	0.953	−0.088	1
What set of adhesive page markers do you want me to bring for experiment	0.952	0.943	−0.048	1
What set of adhesive page markers is required for experiment	0.898	0.992	−0.102	1
Where can I find a art smock for this course	0.905	0.992	−0.095	1
Where can I find a notebook	0.910	0.904	−0.090	1
Where can I find a pair of lab goggles for this class	0.918	0.966	−0.082	1
Where can I find a textbook	0.912	0.915	−0.088	1

Finally, I satisfy H3 indicating an upward trend in UIR_{PCA} with the R^2 value being .93 and p-value $<.05$, concluding that adding more targeted quality training utterances does explain the change in UIR_{PCA}. Figure 5 and Table 4 illustrate the result.

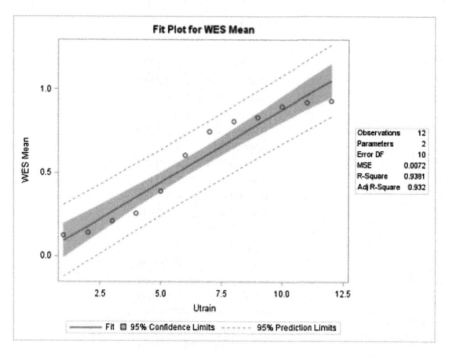

Fig. 5. Fit plot for average change in UIR/WES as it relates to $|\overrightarrow{QU}_{train}|$.

Table 4. Data for average change in UIR/WES as it relates to $|\overrightarrow{QU}_{train}|$.

$|\overrightarrow{QU}_{train}|$	UIR $\bar{x}$	UIR σ
1	0.052003768	0.151389423
2	0.213862234	0.33004652
3	0.378472598	0.341274438
4	0.49255474	0.336050651
5	0.597310003	0.309265953
6	0.776100244	0.195845578
7	0.795681126	0.118507244
8	0.777806616	0.194649884
9	0.794907533	0.212145632
10	0.816037123	0.212460229
11	0.833270664	0.198464732
12	0.876407874	0.102464348
13	0.883482943	0.106022445
14	0.879661269	0.110582838
15	0.875609374	0.117753829

(continued)

Table 4. (*continued*)

$|\vec{QU}_{train}|$	UIR $\bar{x}$	UIR σ
16	0.897936499	0.114495836
17	0.900278268	0.112962197
18	0.899643149	0.113829367
19	0.902610137	0.11202135
20	0.908546486	0.103370983
21	0.90696111	0.10603285
22	0.907004704	0.10719105
23	0.912134175	0.1048354
24	0.920938017	0.092668787
25	0.915629406	0.101408847
26	0.916555016	0.102021519
27	0.915587957	0.104918256
28	0.913637967	0.105584438
29	0.914348054	0.105146388
30	0.913712516	0.105573366
31	0.912738659	0.10615565
32	0.914430544	0.105550956
33	0.917077489	0.103431664
34	0.918980787	0.102030798
35	0.953755326	0.039429352
36	0.959888711	0.031392312
37	0.959988921	0.030726103
38	0.958263044	0.031818972

5   Final Thoughts and Future Research

There are clear opportunities to improve outcomes using RIT as a mechanism to assess fitness between training utterances and intent resolution. A rigorous process of selecting utterance attributes should bolster results for all three hypotheses. Expanding the study to include human participants utilizing mixed-method instruments will add a favorable degree of randomness missing from the method applied in this exercise. Finally, I would assess an entire Cog_x application that employs more than two intents.

Measuring utterance-intent relationships should improve rule-based machine learning algorithms used to prepare Cog_x applications. As such, employing UIR discrimination should mitigate interaction brittleness in personal cognitive augmentation.

Appendix 1 - Dictionary JSON

```
1.   [
2.     {
3.         "value": "materials",
4.         "entities": [
5.           {
6.               "lemma": "textbook",
7.               "synonyms": [
8.                 "book",
9.                 "e-book",
10.                "e-textbook",
11.                "isbn number",
12.                "isbn"
13.               ]
14.           },
15.           {
16.               "lemma": "tablet",
17.               "synonyms": [
18.                 "iPad",
19.                 "Surface",
20.                 "brand of tablet"
21.               ]
22.           },
23.           {
24.               "lemma": "headphones",
25.               "synonyms": [
26.                 "listening device",
27.                 "audio"
28.               ]
29.           },
30.           {
31.               "lemma": "display",
32.               "synonyms": [
33.                 "monitor",
34.                 "screen"
35.               ]
36.           },
37.           {
38.               "lemma": "adhesive page markers",
39.               "synonyms": [
40.                 "page markers",
41.                 "set of adhesive page markers"
42.               ]
43.           },
44.           {
45.               "lemma": "backpack",
                  "synonyms": []
```

Appendix 2 - List of Random Unqualified Training Utterances

Do I need my washing machine for the activity

Do you require a piano

Do you require a rusty nail

I would like to know if it is okay if I bring my mp3 player

Is it okay if I bring my glue stick

Should I bring my bow for this course

Should I bring my picture frame for the activity

Tell me what glue stick do you want me to bring for post-case analysis

What 'old person' things do you do?

What animal would you most like to eat?

What books on your shelf are begging to be read?

What current trend makes no sense to you?

What glue stick should I bring for exam 3

What has someone borrowed but never given back?

What inanimate object do you wish you could eliminate from existence?

What languages do you wish you could speak?

What mla citation style guide is necessary for post-case analysis

What mla citation style guide is necessary for speech

What movie would be greatly improved if it was made into a musical?

What tips or tricks have you picked up from your job/jobs?

What tomato do you want me to bring for word search assignment

What topic could you spend hours talking about?

What two films would you like to combine into one?

What was the last situation where some weird stuff went down and everyone acted like it was normal, and you weren't sure if you were crazy or everyone around you was crazy?

What word is a lot of fun to say?

What's a common experience for many people that you've never experienced?

What's something people don't worry about but really should?

What's something that I don't know?

What's the funniest joke you know by heart?

Would you rather walk around work or school for the whole day without realizing there is a giant brown stain on the back of your pants or realize the deadline for that important paper/project was yesterday and you are nowhere near done?

What's the most ironic thing you've seen happen?

What's the most ridiculous thing you have bought?

What's the silliest thing you've convinced someone of?

What's the weirdest thing a guest has done at your house?

What's your best example of easy come, easy go?

Where can I find a chocolate

Where can I find a thread for the semester

Where's your go to restaurant for amazing food?

Appendix 3 - List of Random Unqualified Training Utterances

Selection	$\hbar$	Utterance
keep	0.088	Do you require a piano
keep	0.088	Where can I find a chocolate
keep	0.088	Should I bring my bow for this course
keep	0.088	Should I bring my picture frame for the activity
keep	0.088	Where can I find a thread for the semester
keep	0.088	What glue stick should I bring for exam 3
keep	0.088	Is it okay if I bring my glue stick
keep	0.088	What mla citation style guide is necessary for speech
keep	0.088	What mla citation style guide is necessary for post-case analysis
keep	0.088	What tomato do you want me to bring for word search assignment
keep	0.088	Tell me what glue stick do you want me to bring for post-case analysis
keep	0.088	I would like to know if it is okay if I bring my mp3 player
remove	1.081	Would you rather walk around work or school for the whole day without realizing there is a giant brown stain on the back of your pants or realize the deadline for that important paper/project was yesterday and you are nowhere near done?
remove	1.081	Do you require a rusty nail
remove	1.081	What 'old person' things do you do?
remove	1.081	What languages do you wish you could speak?
remove	1.081	What current trend makes no sense to you?
remove	1.081	What topic could you spend hours talking about?
remove	1.081	What has someone borrowed but never given back?
remove	1.081	What's something that I don't know?
remove	1.081	Do I need my washing machine for the activity
remove	1.081	What animal would you most like to eat?
remove	1.081	What word is a lot of fun to say?
remove	1.081	What's the funniest joke you know by heart?
remove	1.081	What's the most ridiculous thing you have bought?
remove	1.081	Where's your go to restaurant for amazing food?
remove	1.081	What two films would you like to combine into one?
remove	1.081	What books on your shelf are begging to be read?
remove	1.081	What's the most ironic thing you've seen happen?
remove	1.081	What's the silliest thing you've convinced someone of?
remove	1.081	What's your best example of easy come, easy go?
remove	1.081	What's something people don't worry about but really should?
remove	1.081	What inanimate object do you wish you could eliminate from existence?
remove	1.081	What's the weirdest thing a guest has done at your house?
remove	1.081	What tips or tricks have you picked up from your job/jobs?
remove	1.081	What movie would be greatly improved if it was made into a musical?
remove	1.081	What's a common experience for many people that you've never experienced?
remove	1.081	What was the last situation where some weird stuff went down and everyone acted like it was normal, and you weren't sure if you were crazy or everyone around you was crazy?

References

1. Amazon has sold more than 11 million Echo devices, Morgan Stanley says (2016). http://www.seattletimes.com/business/amazon/amazon-has-sold-more-than-11-million-echo-devices-morgan-stanley-says
2. Gartner Says Worldwide Sales of Smartphones Grew 7 Percent in the Fourth Quarter of 2016 (2016). http://www.gartner.com/newsroom/id/3609817
3. SAP Ariba and IBM Join Forces to Transform Procurement with SAP Leonardo and Watson (2016). http://www.businesswire.com/news/home/20170517005157/en/SAP-Ariba-IBM-Join-Forces-Transform-Procurement
4. Why 2017 is the Year to Invest in Artificial Intelligence Stocks (2016). https://www.fool.com/investing/2017/01/16/why-2017-is-the-year-to-invest-in-artificial-intel.aspx
5. Bampton, H.J.: Solving Imperfect Information Games Using the Monte Carlo Heuristic. University of Tennessee, Knoxville (1994)
6. Berliner, H.J.: Backgammon computer program beats world champion. Artif. Intell. **14**(2), 205–220 (1980)
7. Bush, S.F., Hershey, J., Vosburgh, K.: Brittle system analysis. arXiv preprint cs/9904016 (1999)
8. Bush, S.F., Hershey, J.E., Vosburgh, K.G., Osborn, B.E.: Apparatus and method for analyzing brittleness of a system. In: Google Patents (2004)
9. Chauncey, K., Harriott, C., Prasov, Z., Cunha, M.: A framework for co-adaptive human-robot interaction metrics. Paper Presented at the Proceedings of the Workshop on Human-Robot Collaboration: Towards Co-Adaptive Learning Through Semi-Autonomy and Shared Control (HRC). IEEE/RSJ International Conference on Intelligent Robots and Systems, 9–14 October 2016, Daejeon, Korea (2016)
10. Engelbart, D.C.: Augmenting human intellec:t a conceptual framework (1962). In: Packer, R., Jordan, K. (eds.) Multimedia. From Wagner to Virtual Reality, pp. 64–90. W. W. Norton & Company, New York (2001)
11. Fischer, K.: How people talk with robots: designing dialogue to reduce user uncertainty. AI Mag. **32**(4), 31–38 (2011)
12. Fulbright, R.: The Cogs Are Coming: The Cognitive Augmentation Revolution. Association Supporting Computer Users in Education. Our Second Quarter Century of Resource Sharing, p. 40 (2016)
13. Fulbright, R.: Cognitive augmentation metrics using representational information theory. Paper Presented at the International Conference on Augmented Cognition (2017)
14. George, G., Haas, M.R., Pentland, A.: Big data and management. Acad. Manag. J. **57**(2), 321–326 (2014)
15. Guida, G., Tasso, C.: NLI: a robust interface for natural language person-machine communication. Int. J. Man Mach. Stud. **17**(4), 417–433 (1982)
16. Lally, A., Fodor, P.: Natural language processing with prolog in the IBM Watson system. The Association for Logic Programming (ALP) Newsletter (2011)
17. LeCun, Y., Bengio, Y., Hinton, G.: Deep learning. Nature **521**(7553), 436–444 (2015)
18. Lee, I., Lee, K.: The internet of things (IoT): applications, investments and challenges for enterprises. Bus. Horiz. **58**(4), 431–440 (2015)
19. Lenat, D.B., Prakash, M., Shepherd, M.: CYC: using common sense knowledge to overcome brittleness and knowledge acquisition bottlenecks. AI Mag. **6**(4), 65 (1985)
20. Licklider, J.C.: Man-computer symbiosis. IRE Trans. Hum. Factors Electron. (1), 4–11 (1960)

21. Makridakis, S.: The forthcoming artificial intelligence (AI) revolution: its impact on society and firms. Futures **90**, 46–60 (2017)
22. Moravec, H.P.: Towards automatic visual obstacle avoidance. Paper Presented at the 5th International Conference on Artificial Intelligence. Massachusetts Institute of Technology (1977)
23. Olsen, D.R., Goodrich, M.A.: Metrics for evaluating human-robot interactions. Paper Presented at the Proceedings of the PERMIS (2003)
24. Postrel, V.: Power fantasies: the strange appeal of the Y2K bug. Reason-Santa Barbara then Los Angeles **30**, 4–5 (1999)
25. Quigley, M., Conley, K., Gerkey, B., Faust, J., Foote, T., Leibs, J., Wheeler, R., Ng, A.Y.: ROS: an open-source robot operating system. Paper Presented at the ICRA Workshop on Open Source Software (2009)
26. Scheutz, M., Cantrell, R., Schermerhorn, P.: Toward humanlike task-based dialogue processing for human robot interaction. AI Mag. **32**(4), 77–84 (2011)
27. Shortliffe, E.H., Davis, R., Axline, S.G., Buchanan, B.G., Green, C.C., Cohen, S.N.: Computer-based consultations in clinical therapeutics: explanation and rule acquisition capabilities of the MYCIN system. Comput. Biomed. Res. **8**(4), 303–320 (1975)
28. Talamadupula, K., Srivastava, B., Kephart, J.O.: Workflow complexity for collaborative interactions: where are the metrics?–A challenge. arXiv preprint arXiv:1709.04524 (2017)
29. Urmson, C.: The google self-driving car project. Talk at Robotics: Science and Systems (2011)
30. Weiss, A., Bernhaupt, R., Lankes, M., Tscheligi, M.: The USUS evaluation framework for human-robot interaction. Paper Presented at the AISB 2009: Proceedings of the Symposium on New Frontiers in Human-Robot Interaction (2009)
31. Weizenbaum, J.: ELIZA—a computer program for the study of natural language communication between man and machine. Commun. ACM **9**(1), 36–45 (1966)
32. Vigo, R.: Mathematical Principles of Human Conceptual Behavior: The Structural Nature of Conceptual Representation and Processing, vol. 22. Psychology Press, Hove (2014)
33. Toutanova, K., Klein, D., Manning, C., Singer, Y.: Feature-rich part-of-speech tagging with a cyclic dependency network. In: Proceedings of the HLT-NAACL 2003, pp. 252–259 (2003)

Service Management

How Consumers Perceive Home IoT Services for Control, Saving, and Security

Hyesun Hwang, Jaehye Suk, Kee Ok Kim[✉], and Jihyung Hong

Sungkyunkwan University, 25-2 Sungkyunkwan-Ro,
Jongno-Gu, Seoul 03063, South Korea
kokim@skku.edu

Abstract. This study investigated the perceived benefits and costs of user-oriented home IoT services from consumers' perspectives, as well as how consumers feel about such IoT services. For the purposes of this study, the SPSS 22.0 program was used. Home IoT services were divided on the basis of the purpose of services into control, saving, and security. The perceived benefits of the three home IoT services were classified into efficiency, enjoyment, and effectiveness, while costs into privacy risk, non-monetary, costliness, and unaffordability. The differences in consumers' perceptions of benefits and costs of the three types of home IoT services were analyzed with repeated ANOVA. Regression analyses were conducted to examine how consumers' perceptions affect their attitude the three types of home IoT services. The results are summarized as follows. First, the perceived costs of security purpose home IoT services were the highest among the three types, while the benefits were the lowest. Second, enjoyment was the most important factor in attitudes on home IoT services, followed by effectiveness and efficiency. The higher perceptions of benefits on home IoT services, the more positive attitude on them. The perceived costs had negative effects on attitudes on home IoT services, except for control purpose services. Non-monetary, costliness, and unaffordability had negative effects on attitudes toward saving purpose home IoT services, while privacy risk and unaffordability had negative effects on attitudes on security services. Implications for developers of home IoT services were explored on the basis of the results of this study.

Keywords: Consumer benefits · Consumer costs · Consumer attitudes
Home IoT services

1 Introduction

Consumers are moving toward being permanently connected to the Internet, with a growing number of embedded systems transforming our physical, analog world into one of hyperconnectivity. According to Cisco's estimation, there will be 50 billion Internet connections by 2020 [8]. This increased connectivity is experienced by consumers through many Internet of Things (IoT) services, which extend the benefits of regular Internet use to every aspect of consumers' everyday lives.

The many smart devices connected to the Internet not only interact with people but are also interconnected with one another [16]. Based on this extended connectivity,

© Springer International Publishing AG, part of Springer Nature 2018
S. Yamamoto and H. Mori (Eds.): HIMI 2018, LNCS 10905, pp. 575–588, 2018.
https://doi.org/10.1007/978-3-319-92046-7_47

ICT companies have developed and launched various IoT services that equip consumers with an automated, elaborately customized living environment. These services are expected to support consumers' daily activities by increasing convenience and efficiency. They are mainly delivered through wearable devices, smart homes and buildings, smart vehicles, and healthcare devices or services [14]. The remarkable growth of the IoT-related market is expected to continue with increasing corporate investment in IoT-related hardware, software, and services [17].

Although IoT technology has advanced rapidly, supported by a huge investment, its applications in consumer products and services remain primitive [13]. Such limitations may be related to consumers' diffident acceptance, as well as structural aspects of the technology itself. As IoT services are mainly based on ICT, the infrastructure of high-speed wired and wireless Internet is critical, and consumers need to be equipped with embedded devices. Furthermore, due to the hyperconnective nature of IoT services, it is inevitable that personal information is collected, transferred, and analyzed, which may provoke consumer concerns regarding information security. For user-oriented IoT applications, therefore, it is important to identify not only the benefits of the technology but also such major challenges for consumer acceptance.

Since IoT technologies are in the initial stages of development for practical application, their usefulness has been mainly discussed from technological perspectives. Previous research on user-oriented IoT services has mainly addressed the technological features and related concepts of IoT and business models [5, 19, 25, 27, 30, 31]. Only a few empirical studies have examined the determinants of home IoT service adoption from the user's perspective [4, 20, 24]. In this regard, the direction of designing applicable IoT products or services might be determined by the technology's feasible attributes. However, such a provider-driven perspective may lack insights into the opportunities for and challenges to services' adoption. Consumers' evaluation of the benefits and costs of their experiences in using technological services does not entirely depend on the usefulness of the technology itself. Therefore, it is important to understand consumers' perceptions of IoT services and how these shapes their attitudes toward the services.

In this sense, this study investigated the perceived benefits and costs of user-oriented home IoT services from consumers' perspectives, as well as how consumers feel about such IoT services. This study focused on South Korea since the service quality of the country's Internet is considered to be among the best in the world [18]. Accordingly, we attempted to provide insights to practitioners planning and designing user-oriented home IoT services by understanding consumer perceptions of the benefits and costs of these services.

According to IDC's [17] "Worldwide Semiannual Internet of Things Spending Guide," the compound annual growth rate of corporate investment in home IoT technologies is expected at 19.8% which is the highest among IoT-related industries. In particular, home IoT services – also called "smart home services" – have been advancing rapidly and related services have been extensively developed [7]. Most home IoT services provide functions to monitor or manage one's home environment, including home appliances, energy systems, and security systems [36]. In this study, we classify home IoT services into three types: home security, control, and saving services.

Home security includes services for 24-h monitoring via a control center, message alerts, and monitoring inside and outside the home via smartphones and tablet PCs. Home control includes home-automation systems, such as those controlling lights and thermostats and providing real-time video streaming and video storage services. Energy saving systems measure and report the temperature and electricity usage.

For the purposes of this study, we adopted the theoretical framework of the Value-based Adoption Model (VAM) [21, 22] and the Elaboration Likelihood Model (ELM) [28]. According to the VAM, which stems from the Technology Acceptance Model, consumers decide whether to adopt a particular technology by balancing the perceived sacrifice, including technicality and anticipated fees, against the perceived benefit, including usefulness and enjoyment. The ELM provides a framework to understand how persuasive messages that are about to change a receiver's attitude affect their acceptance of personal information technology [15]. Based on these frameworks, perceived sacrifice can be distinguished as non-monetary and monetary costs [26]. Usefulness is extrinsic and is characterized by perceived benefits, while enjoyment is intrinsic and characterized by emotional benefits [34].

From a consumer perspective, this study examines how consumers perceive the potential benefits and costs of home IoT technology-based services, and whether they are inclined to accept as desirable home IoT services. Finally, we seek to enable practitioners planning and designing home IoT services to deeply understand consumer perceptions and, accordingly, evolve toward providing user-centered services.

2 Method

2.1 Participants

Applying quota sampling by gender and age, 300 Korean consumers in their 30s, 40s, or 50s were recruited as participants. They all registered for a panel through a professional market research organization. The sample characteristics are presented in Table 1. An online survey was conducted on January 10 and 11, 2018. An online questionnaire was sent to randomly selected participants who accepted the email invitation to participate. To ensure that only non-users' acceptance of home IoT services were investigated, a screening question preceded the questionnaire, asking potential participants to report their experience of IoT services. Only those with no experience of using IoT services were surveyed.

2.2 Measurements

The measures were pilot tested in January 2018 by collecting 100 responses from an online survey institution's panel. After performing a reliability check, the final items were generated. Consumers' perception of the costs and benefits and their attitude toward three types of home IoT services were measured by a self-report questionnaire using a 5-point Likert scale (1 = *not at all* to 5 = *perfectly*). Before answering questions on the three different types of home IoT service, participants were introduced to each of them.

Table 1. Description of the respondents

		(N = 300)
		Frequency (%)
Gender	Male	150 (50.0)
	Female	150 (50.0)
Age	30–39	100 (33.3)
	40–49	100 (33.3)
	50–59	100 (33.3)
Household income per month (KRW)[a]	Less than 4 million	112 (37.3)
	4–6 million	95 (31.7)
	Over 6 million	93 (31.0)
Education	High school or less	47 (15.7)
	College/University	224 (74.7)
	Graduate school or higher	29 (9.7)

Note. [a]KRW 1 million = USD 935.89

Based on variables from the VAM, we identified the costs and benefits of home IoT services as perceived by consumers and aimed to understand how these affect consumers' attitude. Twelve items for measuring consumers' perceptions of the costs of using home IoT services were adapted from previous studies by Kim et al. [21], Angst and Agarwal [2], Davis [10], DeLone and McLean [11], Voss et al. [35], and Wang and Wang [37]. Based on these previous studies, we hypothetically structured consumers' costs perceptions with three constructs: monetary, non-monetary, and privacy costs. Each construct comprised four items, examples of which are as follows: monetary costs: "The fee for this home IoT service will be expensive"; "The price for this home IoT service will be affordable for me"; non-monetary costs: "Using this home IoT service seems inconvenient"; "For me, this home IoT service seems difficult to use effectively"; privacy risks: "If I use this home IoT service, my personal information is likely to be leaked and used for unintended purposes"; "If I use this home IoT service, it will be difficult to prevent my personal information from leaking."

To measure consumers' perceptions of the benefits of using home IoT services, 12 items were generated by drawing on Kim et al. [21], Agarwal and Karahanna [1], Bhattacherjee and Sanford [3], Cheung and Vogel [6], Davis [10], and Reychav and Wu [29]. These 12 items were divided into three hypothetical constructs: efficiency, effectiveness, and emotional benefits (enjoyment). Efficiency was measured by five items (e.g., "If I use this home IoT service, I can (manage the safety of my home/control my home environment/save energy) more efficiently"; "This home IoT service will be useful to (manage the safety of my home/control my home environment/save energy)"). Effectiveness comprised three items (e.g., "This home IoT service seems to have excellent functions"; "This home IoT service will perform well"). Finally, enjoyment comprised four items (e.g., "This home IoT service will give me pleasure"; "I will enjoy using this home IoT service").

Table 2. Exploratory factor analysis: cost

	Control			Saving			Security		
	Factor loading	Explained variance (%)	α	Factor loading	Explained variance (%)	α	Factor loading	Explained variance (%)	α
Privacy risk									
C1	.884	40.667	.900	.875	37.040	.911	.848	33.408	.871
C2	.863			.883			.841		
C3	.839			.869			.839		
C4	.860			.857			.787		
Non-monetary									
C5	.796	19.768	.841	.852	18.260	.847	.852	17.225	.802
C6	.851			.809			.802		
C7	.770			.804			.790		
C8	.782			.741			.619		
Costliness									
C9	-	-	-	.889	12.463	.810	.873	12.135	.762
C10	-			.864			.839		
Unaffordability									
C11	.866	13.101	.684	.875	8.932	.678	.843	8.500	.593
C12	.854			.849			.816		
Cumulative explained variance (%)	73.536			76.694			71.267		
KMO	.847			.820			.795		

Attitude was defined as beliefs about the consequences of individual behavior, and measured by six items (e.g., "Using this home IoT service is desirable"; Using this home IoT service is valuable"). These six items were adapted from Fishbein and Ajzen [12]. To validate the scales and test their reliability, exploratory factor analyses (EFAs) and Cronbach's α tests were performed. Results showed acceptable scale reliability (see Tables 2, 3 and 4).

2.3   Analysis

For the purposes of this study, the SPSS 22.0 program was used. Descriptive analysis was employed to evaluate the sample's demographic characteristics and calculate the mean scores of consumers' perceptions and attitude. To examine consumers' perceptions of three types of home IoT services, repeated ANOVA was performed. Regression analyses were conducted to examine how consumers' perceptions affect their attitude toward the three types of home IoT services.

Table 3. Exploratory factor analysis: benefit

	Control			Saving			Security		
	Factor loading	Explained variance (%)	α	Factor loading	Explained variance (%)	α	Factor loading	Explained variance (%)	α
Efficiency									
B1	.767	43.553	.831	.781	49.617	.858	.824	47.253	.790
B2	.728			.750			.805		
B3	.756			.764			.778		
B4	.637			.731			.768		
B5	.757			.730			.731		
Enjoyment									
B6	.831	12.097	.832	.784	12.072	.865	.806	12.311	.797
B7	.759			.699			.773		
B8	.741			.812			.740		
B9	.725			.811			.656		
Effectiveness									
B10	.866	10.694	.839	.864	8.978	.863	.834	9.338	.839
B11	.820			.796			.819		
B12	.798			.757			.753		
Cumulative explained variance (%)	66.344			70.667			68.892		
KMO	.867			.905			.890		

Table 4. Exploratory factor analysis: attitude

	Control			Saving			Security		
	Factor loading	Explained variance (%)	α	Factor loading	Explained variance (%)	α	Factor loading	Explained variance (%)	α
Attitude									
A1	.809	62.037	.877	.846	66.999	.901	.822	62.205	.878
A2	.797			.824			.813		
A3	.829			.817			.798		
A4	.807			.815			.780		
A5	.758			.813			.762		
A6	.721			.794			.754		
Cumulative explained variance (%)	62.037			66.999			62.205		
KMO	.892			.898			.875		

3   Results

3.1   Consumer Perceptions of Costs and Benefits: Exploratory Factor Analysis

To investigate how consumers perceive the costs and benefits of three different home IoT services, EFAs using a Varimax rotation were adopted, as shown in Tables 2, 3 and 4. Prior to conducting an EFA on each scale, the Kaiser-Meyer-Olkin (KMO) and Bartlett's test statistics were checked.

For consumers' perceptions of the costs of using home IoT services, different dimensions were extracted according to the service type. For security and saving services, four factors were extracted, namely, privacy risk, non-monetary cost, costliness, and unaffordability, which cumulatively explain 71% (security) and 76% (saving) of the data variation for cost perception. The hypothetical construct of monetary cost was divided into two different dimensions: costliness and unaffordability. For control services, two items measuring costliness were excluded due to the problems of cross-loadings and low factor loadings (under 0.5). Finally, three of the four factors extracted for security and saving services were also extracted for control services, the exception being costliness; these three factors cumulatively explain 73% of the data variation. For consumers' benefit perception, three factors were extracted in accordance with the hypothesized dimensions, indicating efficiency, effectiveness, and emotional benefits. All factor loadings are above .60, and no items were excluded. The cumulative explained variances of security, control, and saving services were 68%, 66%, and 70%, respectively. Consumers' attitude was a single dimension, explaining 62%, 62%, and 66% of variance for security, control, and saving services, respectively.

3.2   Consumer Perceptions of Costs and Benefits by Home IoT Service Type

As shown in Table 5, repeated ANOVAs were conducted to analyze the differences in consumers' perceptions of the costs and benefits of the three different types of home IoT services. For the cost perceptions, the mean scores of privacy risk, costliness, and unaffordability differed according to the type of IoT services. The results indicate that consumers perceived a higher privacy risk for security services, at a 0.1% significance level. Consumers' perceptions of costliness were compared between security and saving services, because this factor's items were excluded from the EFA of control services. The mean score of costliness was higher for security than for saving services ($p < .001$). Moreover, the mean score of unaffordability was also higher for security services than for the other two types, at a 0.1% significance level.

For consumers' benefit perceptions, the mean differences of perceived usefulness and effectiveness were statistically significant. Consumers' perceptions of perceived usefulness were lower for security services than for the other two types ($p < .01$). The mean score of effectiveness was lower for security services than for saving services, at a 5% significance level.

Table 5. Repeated measures ANOVAs

	Control	Saving	Security	df	F	Bonferroni
	Mean (SD)	Mean (SD)	Mean (SD)			
Cost perception						
Privacy risk	3.50 (.76)	3.51 (.76)	3.67 (.71)	1.70, 299	12.108***	C > B = A
Non-monetary	3.01 (.04)	2.96 (.04)	2.97 (.038)	1.92, 573.01	1.070	
Costliness	-	3.45 (.04)	3.69 (.04)	1, 299	27.202***	C > B
Unaffordability	2.91 (.66)	2.87 (.66)	3.06 (.63)	1.89, 564.29	16.529***	C > A = B
Benefit perception						
Efficiency	3.65 (.53)	3.68 (.55)	3.58 (.61)	1.85, 552.64	6.432**	B > C A > C
Enjoyment	3.32 (.59)	3.33 (.65)	3.32 (.58)	1.91, 570.08	.033	
Effectiveness	3.38 (.62)	3.40 (.65)	3.32 (.59)	2, 598	3.212*	B > C

Note. $^*p < .05$, $^{**}p < .01$, $^{***}p < .001$

3.3   Consumer Attitude Toward Home IoT Services

Consumers' attitude toward home IoT services was analyzed using regression models, as shown in Table 6. Gender, age, education, and household income were employed as sociodemographic variables. Factor scores of perceptions of costs and benefits were also used. Three regression models were statistically significant overall, and the variance inflation factor (VIF) for each independent variable was less than 2, indicating that multicollinearity was not present.

Security Services. Gender and age had statistically significant effects on consumers' attitude toward security services, indicating that male consumers have a more positive attitude than female consumers and that older consumers have a more positive attitude than younger consumers. Among perceptions of costs and benefits, privacy risk and unaffordability had negative effects on consumer attitude toward security services, whereas efficiency, enjoyment, and effectiveness had positive effects. Among the predicting variables, the three benefit perceptions showed higher standardized coefficients than any of the other variables, the highest being for enjoyment. The predicting variables in this model explain 57.6% of the variance in consumers' attitude toward security services (adjusted $R^2 = .576$).

Control Services. Among the sociodemographic variables, only age had a statistically significant effect on attitude. The four cost perception variables were non-significant, while three benefit perception variables were statistically significant. Among the benefit perceptions, enjoyment showed the highest standardized coefficient ($\beta = .525$, $p < .001$). The adjusted R^2 of this model was .560.

Saving Services. Age and education had statistically significant effects on consumers' attitude toward saving services. As for security and control services, age showed a positive effect on attitude. Regarding education level, consumers who completed graduate school education indicated a more positive attitude toward saving services than consumers who progressed no further than high school (or below). Except for privacy risk, consumers' cost perceptions were all statistically significant in relation to consumers' attitude toward saving services. The three variables of benefit perceptions were all statistically significant, indicating higher standardized coefficients than any of the other variables, with enjoyment showing the highest standardized coefficient. This model explains 67.1% of the variance in consumers' attitude toward saving services ($R^2 = .671$).

Table 6. Consumers' attitude toward home IoT services

	Control		Saving		Security	
	B	β	B	β	B	β
Sociodemographic variables						
Male	.056	.049	.086	.067	.127	.114**
Age	.008	.107**	.006	.078*	.008	.118**
Education[a]						
University	−.035	−.027	.037	.025	.056	.044
Graduate	.013	.006	.189	.088*	.117	.062
Household income[b]						
Middle	.053	.043	−.090	−.066	.030	.025
High	.041	.033	−.050	−.037	.060	.049
Cost perception						
Privacy risk	−.033	−.057	.025	.039	−.054	−.096***
Non-monetary cost	.028	.049	−.047	−.073*	−.011	−.020
Costliness	-	-	−.055	−.086**	−.013	−.023
Unaffordability	−.044	−.077	−.056	−.087*	−.047	−.085*
Benefit perception						
Efficiency	.157	.274***	.264	.413***	.170	.305***
Enjoyment	.300	.525***	.372	.582***	.296	.529***
Effectiveness	.225	.393***	.197	.308***	.194	.347***
F	32.701***		47.945***		32.255***	
R²	.578		.685		.595	
Adj. R²	.560		.671		.576	
Durbin-Watson	2.051		1.787		2.107	

*p < .05, **p < .01, ***p < .001

Notes. [a]Reference group: High school or less
[b]Reference group: Low income (less than KRW 4 million)

4   Discussion and Conclusions

Consumers' lives are becoming increasingly hyperconnected, allowing them to enjoy convenient, efficient services by adopting IoT services, which place them within a mutually interconnected environment. Because IoT services, based on new technology, change many aspects of consumer lives, there are inherent costs for consumers that embrace such services, which may cause some to be reluctant to do so.

This study investigated the benefits and costs of home IoT services from consumers' viewpoint, aiming to advance understanding beyond providers' perspectives on the technological requirements to increase service penetration. We analyzed consumers' perceptions and attitudes regarding home IoT services in South Korea. Specifically, this study investigated how consumers perceive the benefits and costs of home IoT services and the effects of consumers' perceptions on their attitude toward three types of home IoT services. In particular, this study attempted to provide insights into the design and development of home IoT services for consumers. The main results are outlined as follows.

First, for consumers' perceptions of the benefits of three types of IoT services, three common factors were extracted: efficiency, effectiveness, and emotional benefits. For the security and saving services, four factors were extracted for consumers' perceptions of costs: privacy risk, non-monetary cost, costliness, and unaffordability. Three of these four factors (the exception being "costliness") were also extracted for control services. This result shows that consumers have different perception structures on costs according to the service type. Of the three types of services, control services are usually in easy reach [23].

Second, consumers were found to perceive higher privacy risk and monetary cost for security services than for the other two service types. Conversely, consumers' recognition of benefits indicated that efficiency is perceived to be lower for security services than for the other two categories. Such perceptions show that, for security type home IoT services, consumers are more concerned about costs and less inclined to recognize the benefits. According to previous research, consumers could perceive security services as a cost [4]. Consumers' perceptions of costs were found to be nonsignificant for control services (compared to the finding of significance for the other two service types). This suggests that consumers may not be aware of the costs of control services. Therefore, service providers need to understand consumers' different perceptions of each type of home IoT service.

Third, the regression models indicated the effects of consumers' perceptions of the costs and benefits on their attitude toward the home IoT services. Age was found to affect attitude toward home IoT services, with older consumers found to have a more positive attitude toward accepting home IoT services. For control services, the benefit perceptions of efficiency, enjoyment, and effectiveness had significant effects on attitude. For saving services, the cost perceptions of non-monetary and monetary factors had negative effects on consumers' attitude, while the benefit perceptions of efficiency, effectiveness, and enjoyment had positive effects thereon. For security services, the perceptions of privacy risk and affordability had a statistically significant effect on attitude, as did the three subsets of benefit perceptions. These results show that consumers' perceptions of benefits have significant effects on their attitude toward three types of IoT services. For all three models, the standardized effect of enjoyment had the greatest effect on attitude. This suggests that consumers may be more receptive to home IoT services when they have strong beliefs on the benefits, especially those of an emotional nature.

This study identified the structures of consumers' perceptions of the costs and benefits of home IoT services. In particular, we analyzed how such perceptions affect consumers' receptive attitudes toward three different types of home IoT services. The study's findings provide meaningful insights into the consumer factors that affect acceptance of each service. For the three considered types of home IoT services, the benefit perceptions had greater effects on attitude than the cost perceptions. Although these services are based on highly advanced technologies, consumers are most affected by emotional benefits, which have the greatest effect on attitudes toward all three types of home IoT services. This finding may reflect the context in which these services are experienced. As home IoT services are pervasively experienced in consumers' daily lives, consumers' service experiences might be critical to forming a positive attitude. In previous research, perceived enjoyment has been shown to be a strong predictor for consumers' attitudes [9, 38, 39] and purchase intentions [32, 33].

These results provide insights to develop more consumer-oriented services, which will allow potential demands in the market to come to the fore. Although technology has driven such services to date, by developing new functions and improving performance, it is now necessary to attune the services more closely to consumers' expectations and to address their concerns. Each technology can only be meaningful through its adoption by consumers to whose lives it contributes. In this sense, based on this study's results, home IoT service design must be directed in consonance with consumers' perceptions and expectations.

Appendix

Measures

Items
Cost perception
C1. When I use this home IoT service, my personal information can be collected by service providers without my knowledge
C2. If I use this home IoT service, my personal information is likely to be leaked and used for unintended purposes
C3. With this home IoT service, it will be difficult to protect my personal information
C4. If I use this home IoT service, it will be difficult to keep my personal information from leaking
C5. Using this home IoT service seems inconvenient
C6. For me, this home IoT service seems difficult to use effectively
C7. Using this home IoT service will be complicated
C8. Using this home IoT service requires effort
C9. This home IoT service will require a lot of incidental costs
C10. The fee for this home IoT service will be expensive
C11. The price for this home IoT service will be affordable for me. (R)
C12. It is worth me paying for this home IoT service. (R)
Benefit perception

(*continued*)

(continued)

Items
B1. If I use this home IoT service, I can (manage the safety of my home/control my home environment/save energy) more efficiently
B2. This home IoT service will be useful to (manage the safety of my home/control my home environment/save energy)
B3. Using this home IoT service makes it easier to (manage the safety of my home/control my home environment/save energy)
B4. This home IoT service will reduce my effort in (managing the safety of my home/controlling my home environment/saving energy)
B5. This home IoT service will provide a simple method to (manage the safety of my home/control my home environment/save energy)
B6. This home IoT service will give me pleasure
B7. Using this home IoT service will make me feel good
B8. I will enjoy using this home IoT service
B9. Using this home IoT service seems interesting
B10. This home IoT service seems to have excellent functions
B11. This home IoT service will perform well
B12. This home IoT service seems to have excellent quality
Attitude toward using home IoT services
A1 to A6. Using this home IoT service is desirable/wise/valuable/beneficial/rational/good

Notes. (R): Reverse coding

References

1. Agarwal, R., Karahanna, E.: Time flies when you're having fun: cognitive absorption and beliefs about information technology usage. MIS Q. **24**(4), 665–694 (2000)
2. Angst, C.M., Agarwal, R.: Adoption of electronic health records in the presence of privacy concerns: the elaboration likelihood model and individual persuasion. MIS Q. **33**(2), 339–370 (2009)
3. Bhattacherjee, A., Sanford, C.: Influence processes for information technology acceptance: an elaboration likelihood model. MIS Q. **30**(4), 805–825 (2006)
4. Balta-Ozkan, N., Davidson, R., Bicket, M., Whitmarsh, L.: Social barriers to the adoption of smart homes. Energy Policy **63**, 363–374 (2013)
5. Bugeja, J., Jacobsson, A., Davidsson, P.: On privacy and security challenges in smart connected homes. In: 2016 European Intelligence and Security Informatics Conference, EISIC, pp. 172–175. IEEE (2016)
6. Cheung, R., Vogel, D.: Predicting user acceptance of collaborative technologies: an extension of the technology acceptance model for e-learning. Comput. Educ. **63**(April), 160–175 (2013)
7. Choi, S.J.: Korea's IoT market grows to 5 trillion won. Biz & Tech (2016). http://www.koreatimes.co.kr/www/tech/2017/10/693_195927.html. Accessed 23 Feb 2018
8. Cisco Visual Networking Index: Cisco visual networking index: forecast and methodology 2011–2016. CISCO White paper 2011–2016 (2012)

9. Curran, J.M., Meuter, M.L.: Self-service technology adoption: comparing three technologies. J. Serv. Mark. **19**(2), 103–113 (2005)
10. Davis, F.D.: Perceived usefulness, perceived ease of use, and user acceptance of information technology. MIS Q. **13**(3), 319–340 (1989)
11. DeLone, W.H., McLean, E.R.: Information systems success: the quest for the dependent variable. Inf. Syst. Res. **3**(1), 60–95 (1992)
12. Fishbein, M., Ajzen, I.: Belief, Attitude, Intention and Behavior: An Introduction to Theory and Research. Addison-Wesley, Boston (1975)
13. Gao, L., Bai, X.: A unified perspective on the factors influencing consumer acceptance of internet of things technology. Asia Pac. J. Mark. Logist. **26**(2), 211–231 (2014)
14. Government Accountability Office: Technology assessment: Internet of Things: Status and implications of an increasingly connected world (2017). https://www.gao.gov/products/GAO-17-75. Accessed 23 Feb 2018
15. Ha, S., Ahn, J.: Why are you sharing others' tweets?: The impact of argument quality and source credibility on information sharing behavior. In: Thirty Second International Conference on Information Systems (2011)
16. Hsu, C.L., Lin, J.C.C.: An empirical examination of consumer adoption of Internet of Things services: network externalities and concern for information privacy perspectives. Comput. Hum. Behav. **62**, 516–527 (2016)
17. IDC: World object Internet (IoT) spending size 2021: $1.4 trillion forecast. http://www.kr.idc.asia/press/pressreleasearticle.aspx?prid=502 (2017). Accessed 23 Feb 2018
18. International Telecommunication Union (ITU): ICT facts and figures (2014). http://www.itu.int/en/ITU-D/Statistics/Documents/facts/ICTFactsFigures2014-e.pdf. Accessed 23 Feb 2018
19. Kang, J., Kim, M., Park, J.H.: A reliable TTP-based infrastructure with low sensor resource consumption for the smart home multi-platform. Sensors **16**(7), 1036 (2016)
20. Kim, H., Yeo, J.: A study on consumers' levels of smart home service usage by service type and their willingness to pay for smart home services. Consum. Policy Educ. Rev. **11**(4), 25–53 (2015)
21. Kim, H.W., Chan, H.C., Gupta, S.: Value-based adoption of mobile internet: an empirical investigation. Decis. Support Syst. **43**(1), 111–126 (2007)
22. Kim, Y., Park, Y., Choi, J.: A study on the adoption of IoT smart home service: using value-based adoption model. Total Qual. Manag. Bus. Excell. **28**(9–10), 1149–1165 (2017)
23. Kühnel, C., Westermann, T., Hemmert, F., Kratz, S., Müller, A., Möller, S.: I'm home: defining and evaluating a gesture set for smart-home control. Int. J. Hum.-Comput. Stud. **69**(11), 693–704 (2011)
24. Lee, S., Choi, M.: A study on influence of trait values over user satisfaction of echo-boomer living with smart-home. J. Korea Real Estate Anal. Assoc. **21**(1), 103–131 (2015)
25. Li, B., Yu, J.: Research and application on the smart home based on component technologies and Internet of Things. Proc. Eng. **15**, 2087–2092 (2011)
26. Lin, T.C., Wu, S., Hsu, J.S.C., Chou, Y.C.: The integration of value-based adoption and expectation–confirmation models: an example of IPTV continuance intention. Decis. Support Syst. **54**(1), 63–75 (2012)
27. Pang, Z., Zheng, L., Tian, J., Kao-Walter, S., Dubrova, E., Chen, Q.: Design of a terminal solution for integration of in-home health care devices and services towards the Internet-of-Things. Enterp. Inf. Syst. **9**(1), 86–116 (2015)
28. Petty, R., Cacioppo, J.T.: Communication and Persuasion: Central and Peripheral Routes to Attitude Change. Springer, New York (1986). https://doi.org/10.1007/978-1-4612-4964-1
29. Reychav, I., Wu, D.: Are your users actively involved? A cognitive absorption perspective in mobile training. Comput. Hum. Behav. **44**(March), 335–346 (2015)

30. Sivaraman, V., Chan, D., Earl, D., Boreli, R.: Smart-phones attacking smart-homes. In: Proceedings of the 9th ACM Conference on Security & Privacy in Wireless and Mobile Networks, pp. 195–200. ACM (2016)

31. Soliman, M., Abiodun, T., Hamouda, T., Zhou, J., Lung, C. H.: Smart home: integrating internet of things with web services and cloud computing. In: Cloud Computing Technology and Science, CloudCom, pp. 317–320. IEEE (2013)

32. Van der Heijden, H.: User acceptance of hedonic information systems. MIS Q. 28(4), 695–704 (2004)

33. Van der Heijden, H., Verhagen, T., Creemers, M.: Understanding online purchase intentions: contributions from technology and trust perspectives. Eur. J. Inf. Syst. 12(1), 41–48 (2003)

34. Venkatesh, V., Speier, C.: Creating an effective training environment for enhancing telework. Int. J. Hum.-Comput. Stud. 52, 991–1005 (2000)

35. Voss, G.B., Parasuraman, A., Grewal, D.: The roles of price, performance, and expectations in determining satisfaction in service exchanges. J. Mark. 62(4), 46–61 (1998)

36. Wan, C., Low, D.: Capturing next generation smart home users with digital home. White Paper, Huawei (2013)

37. Wang, H.Y., Wang, S.H.: Predicting mobile hotel reservation adoption: insight from a perceived value standpoint. Int. J. Hosp. Manag. 29(4), 598–608 (2010)

38. Wang, C.: Consumer acceptance of self-service technologies: an ability–willingness model. Int. J. Mark. Res. 56(6), 782–802 (2017)

39. Weijters, B., Rangarajan, D., Tomas, F., Niels, S.: Determinants and outcomes of customers' use of self-service technology in a retail setting. J. Serv. Res. 10(1), 3–21 (2007)

User-Friendly Information Sharing System for Producers

Tomoko Kashima[1(✉)], Takashi Hatsuike[2(✉)],
and Shimpei Matsumoto[3(✉)]

[1] Kindai University, Hiroshima, Japan
kashima@hiro.kindai.ac.jp
[2] Waseda University, Tokyo, Japan
[3] Hiroshima Institute of Technology, Hiroshima, Japan

Abstract. A model that connects producers, consumers, and farmers' markets was developed, and an information system was developed as a communication tool in order to collate the experience- and intuition-based know-how of producers and improve the value of their crops. However, the elderly comprise over 60% of the population employed in agriculture, and farmers are unwilling to utilize information technology (IT) devices. In an effort to increase the IT utilization rate, a proof-of-concept test was conducted up to 2016 at a roadside station.

Therefore, in this study, we design a new IT tool that connects producers with consumers. In addition, the study considers a user-friendly interface design, where elderly farmers can operate intuitively, and which enables the continuous use of the system in order to actively realize the application of IT to agriculture. Specifically, an analysis of information needed to connect producers and consumers is performed, and a method for acquiring such information is also developed. The study also investigates the ease of use in terms of the inputting, sharing, and viewing of information over a mobile device (e.g., smartphones).

Keywords: Agricultural IT tool · Video streaming-based digital signage
Agricultural direct sales station · Communication science

1 Introduction

1.1 Background

In agricultural and rural areas where there is a greater ageing and decline of the population relative to cities and urban areas, there is a fear that farm management and techniques may not be passed on to the next generation, and that they will be lost permanently in some areas owing to the retirement of elderly farmers and declining employment in agriculture (see Fig. 1). The acreage of abandoned farmland has increased owing to a shortage of labor, and the sustainability of agriculture and rural areas has become a concern. To address this situation, it is expected that "smart agriculture," which applies the Internet of Things (IoT) and big data, will be used to revitalize agriculture and improve production efficiency. To this end, the Ministry of Agriculture, Forestry, and Fisheries in Japan has developed a smart agriculture plan

© Springer International Publishing AG, part of Springer Nature 2018
S. Yamamoto and H. Mori (Eds.): HIMI 2018, LNCS 10905, pp. 589–598, 2018.
https://doi.org/10.1007/978-3-319-92046-7_48

with five goals [1]. Of these goals, in this study, we focus on "the development of agriculture that anyone can engage in" and the "provision of safety and trust to consumers and the general demand."

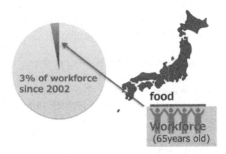

Fig. 1. Current situation of agriculture problem in Japan

In addition, trends in the development of smart agriculture have emphasized the use of the latest technologies, as well as the improvement of productivity and efficiency by the application of robotics, radio-controlled helicopters (drones), and sensors [2]. Since 2010, the authors have also worked on the development and improvement of a highly capable system that supports producers [3–8]. However, many issues have been identified and need to be solved. For example, a user needs to choose a function on the system even if there are many convenient functions available, and this may pose some degree of difficulty in terms of usability. Even if the available number of functions is minimized to enable the use of the system with minimal labor, it may introduce challenges with respect to choosing functions that are needed, and would involve the determination of functions to be prioritized. In addition, unlike industrial products, there is a uniquely agricultural problem in which shipments cannot be adjusted in real time based on sales information, even if data on sales of produce have been collected and trends for highly popular vegetables have been identified. Another problem is that the impact of the system's usefulness cannot be immediately appreciated s approximately a full year must pass until the next cultivation season. As such, the challenges are to choose system functions that are tailored to the producers and develop a mechanism in which the use of the system contributes to sales and increases the motivation for production among farmers.

1.2 Related Works

As IT has progressed and high-performance IT devices are now available at a low cost, smart agriculture has been implemented in many places as it aims to improve the quality of domestic agricultural systems and to stabilize production; achievements in this regard are reported by newspapers on a daily basis. Of the smart agriculture approaches, the focus has been to acquire knowledge held by experienced persons through the accumulation of agricultural work history using sensing technologies, and for some time, large-scale, large-volume network server technologies have been the focus of attention as one of the major challenges for a long time. For example, Takatsu et al. focused on an implementation of ICT in agriculture as a means to address the falling number of descendants of farmers who are in agriculture, and they discussed the revitalization of agriculture, solving the problem of a shortage of successors, determining harvesting seasons for agricultural produce and prediction of yield, and improving the income of the producers [9]. As with the challenge associated with inheriting skills and knowledge, improvements with respect to the efficiency in logistics for agricultural produce and improvements in the added value of agricultural produce have been regarded as major challenges in smart agriculture. For example, Satake and Yamazaki reported the importance of information sharing for consumption, production, and logistics in the agricultural supply chain. For the agricultural workforce, including producers, information sharing in the agricultural supply chain not only leads to a reduction in excess such as produce waste, but can also contribute to realizing enhanced freshness in shipped produce [10].

Authors have been developing an agricultural information system that enables producers to provide accurate information, including information that is difficult to obtain for common agricultural produce, the production process in farms, and the growth status of the crops, as needed by the consumers (see Figs. 2 and 3). Such a system aims to present information on the food-production process and provide added value to crops, such as safety and reassurance. Using cloud computing and creating a system mechanism that is easy and inexpensive for producers, consumers are able to view the information they want to see in real time. In Japan, many producers traditionally collect their agricultural produce and sell them as a cooperative. This approach is advantageous in that large quantities of agricultural produce are combined, and quality can be maintained at a certain level, providing leverage for making sales in markets. However, there are few opportunities for individual farmers' agricultural produce to be highlighted.

The proposed system attempts to enable the source of crops to be identified, while enabling producers to deliver tasty products to consumers with confidence. In addition, this system can provide information on food safety to customers via the Internet at anytime and anywhere, which allows producers to gain the broad trust of customers. For farmers, the ability to specify the source of produce lie means that they accept responsibility for the produce. However, disclosing the growth process for crops and receiving responses directly from individual customers may contribute to increased motivation for production. The proposed system has two major functions. The first function provides harvest information about the producers and provides scenes from farms to viewers on the Internet. Rather than providing the same content, different information is sent to customers based on factors such as similar preferences, family structure, and allergies. Therefore, customer data is grouped using a clustering method. Information that is tailored to customers is sent via email. The rough-set theory and collaborative filtering were used in order to provide information tailored to customers. The second function communicates information in real-time via a webcam installed on farms. Information about crops sales are sent to the producers, and harvest-related information is sent to customers by email. Upon visiting the farmers' market, customers can then purchase produce that they would have seen on video being harvested on the same day. This proposed system is developed on the cloud, and works with

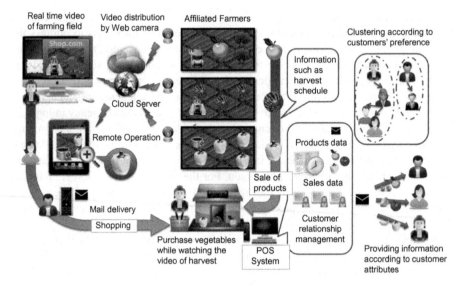

Fig. 2. Concept of an agriculture information system

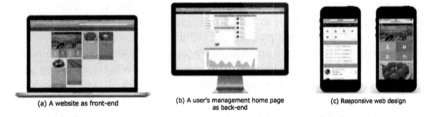

Fig. 3. Interface of agriculture information system

multiple users. Because the cloud eliminates the need for server operation costs as well as the need to be on-site during maintenance or troubleshooting, human labor costs will be reduced, which is a major advantage of a cloud-based system. Many applications that enable anyone to send live streams via a mobile device, such as Periscope, have been made available in recent years. The authors started this work in 2010, at which time it was a pioneering challenge because it relies on a mechanism that incorporated computer science knowledge into the field of agricultural.

2 Proposed Information Sharing System

2.1 Outline of the System

An overview of the proposed system is briefly explained below (see Figs. 4 and 5). A producer has a mobile device on which the system is installed. The producer shoots videos with the device. Videos are used as PR tools to record and share their passion

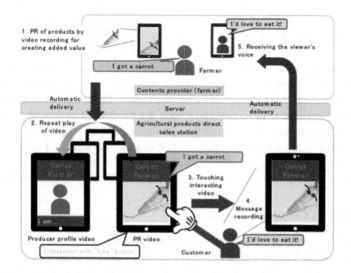

Fig. 4. Concept of the video-based digital signage system

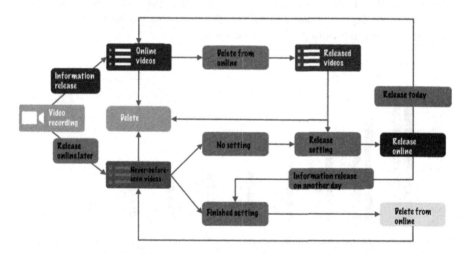

Fig. 5. Outline of the video-based digital signage system

for agriculture, ways to eat the produce, as well as features and added values of the produce. By pressing the "publish" button, the video is published automatically on the digital signage display installed in the farmers market. Consumers are able to view the PR videos on the digital signage display, after which they can respond to the videos by sending messages or pressing "like," which is communicated to the producer. As a result, the producer is able to receive the voices of the viewers.

Advantages realized upon the completion of this system are summarized below from the perspectives of the producers, farmers market, and consumers. First, producers can easily advertise their own crops and obtain feedback about their products. Producers are also able to see the faces of the consumers who purchased the crops that were grown with great care. Unlike conventional agriculture, an interactive relationship with consumers is possible withy these merits. These effects are also expected to lead to an increased motivation towards daily agricultural work. Next, consumers are able to make purchases by making decisions on their own based on trusted information. Purchases are possible with values that match the products. Consumers can buy delicious products with ease, and information in which they are interested can be sent to producers. As a result, consumers are now able to purchase safer and tastier vegetables with a greater sense of trust. Finally, farmers' market can provide some peace of mine to consumers when purchasing products, thereby contributing to enhanced sales.

2.2   Survey on Consumer's System

The proposed system can transmit information to consumers. But that is a difficult problem that what kind of information should the producer send out. Therefore, we investigate what kind of information consumers need. We did a survey at the shop of the experimental schedule in Hiroshima, also on the web. We conducted a survey on 168 people in Hiroshima's shops and 133 people on the Web. The results of the survey are shown in the Figs. 6, 7 and Tables 1, 2. We were able to obtain information that consumers are seeking from producers and shops.

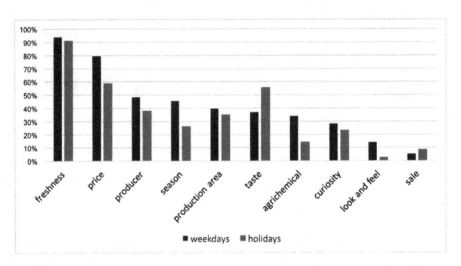

Fig. 6. Expectation for ingredients (Questionnaire for those who came to buy ingredients)

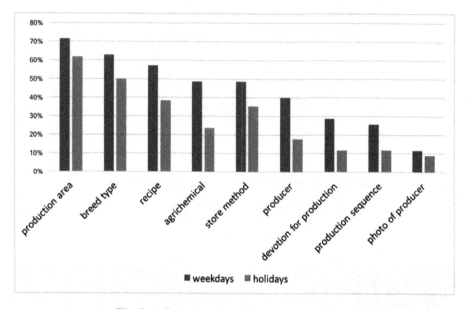

Fig. 7. Information that consumers want to know

Table 1. Expectation for ingredients (Questionnaire for those who came to buy ingredients)

	Freshness	Price	Producer	Season	Production area	Taste	Agrichemical	Curiosity	Look and feel	Sale
Weekdays (35)	94	80	49	46	40	37	34	29	14	6
Holidays (34)	91	59	38	26	35	56	15	24	3	9

Table 2. Information that consumers want to know

	Production area	Breed type	Recipe	Agrichemical	Store method	Producer	Devotion for production	Production sequence	Photo of producer
Weekdays (35)	71	63	57	49	49	40	29	26	11
Holidays (34)	62	50	38	24	35	18	12	12	9

3   Results

3.1   Experiment of the System

The proof-of-concept test of the information sharing system for producers was conducted for approximately half a year at a roadside station, and its results are discussed here. First, we explain the experimental environment. The location was in the northern part of Hiroshima Prefecture at a facility that hosts various events to encourage interaction between local residents. The facility owns an enormous playground, sporting ground, and a hall with a capacity of over 200 people, and the farmers market is a part of this facility. Approximately 100 suppliers are registered to this roadside

Table 3. Number of posted video by farmer (per month)

Name	M	S	T	O	H	SUM
July	19	5	1	0	5	30
Aug.	3	4	1	0	0	8
Sep.	4	2	1	0	4	11
Oct.	4	2	0	0	3	9
Nov.	1	0	0	0	0	1
Sum	31	13	3	0	12	59

Table 4. Number of posted video by customer (5 months)

Name	M	S	T	O	H	Sum
Sum	19	1	2	0	6	28

Fig. 8. Appearance of the proposed system in Roadside station, Hiroshima

station, five of which were test participants. A smartphone with access to a network and the installed system was provided to these five individuals for their participation in the test. The Tables 3, 4 and Figs. 8, 9 show the number of times information was provided through the system by the producers and the number of responses by the consumers. Producers provided the following feedback after the use of the system: (1) it was inconvenient to carry the device, (2) the effect of providing information could not be felt immediately, and (3) they wanted to use a more advanced system. While the plan was to evaluate changes in sales and purchase awareness among customers after the proof-of-concept test, a detailed analysis could not be performed owing to the low rate of use among the producers. It is difficult to produce the same quantities of the same crops year after year. There are many elements that cannot be controlled by the producers, such as the weather. Nonetheless, the intention is to continue working on the development of a new system based on these results in order to solve these problems, and to develop a system that delivers as much information as is needed by a producer.

Fig. 9. Interaction with viewers

4 Conclusion

Many systems have been previously proposed, but challenges that faced these systems were overcome, and new systems were built. Currently, the authors are working on an initiative that supports the local production and consumption in Hiroshima. When the disclosure of information is seen as the source of the added value, it was difficult for them to be executed using the existing model because consumer tier and climatic conditions are considered. However, we also found that stores in Hiroshima desire to have local and fresh agricultural produce available, while consumers had a high degree of willingness to purchase local agricultural produce as long as there was no significant difference in price. In light of these findings, we focused on ideas whereby the value of agricultural produce is determined by the consumers themselves, rather than by standards set by experts. Further, there is no portal in which information on the production of local agricultural produce can be obtained easily. We are now in the process of developing a CMS that is unique to Hiroshima, where information sharing is enabled between agricultural producers and consumers. In the case of Hiroshima, there is limited land available for agricultural production, and it is difficult to provide information on creativity with respect to ingredients when developing creative ingredients. In spite of the narrow land space, there is a great possibility that culture and ingredients have survived in the land over the years, and there is likely to be a great need to provide and share such information.

Acknowledgments. This work was partly supported by Artificial Intelligence Research Promotion Foundation, Japan Society for the Promotion of Science, KAKENHI Grant-in-Aid for Young Scientists(B), No. 25750007 And Grant-in-Aid for Scientific Research(B), No. 17H02043. Also we would like to express the deepest appreciation to Ariaki Higashi, System Friend Inc. and its team members, and Kentaro Nakamura, Michi-no-eki, Kohan-no-Sato Fukutomi and its all staffs.

References

1. Ministry of Agriculture, Forestry and Fisheries: Study group for realization of smart agriculture. http://www.maff.go.jp/j/kanbo/kihyo03/gityo/g_smart_nougyo/. Accessed 6 Feb 2018
2. Hasegawa, K.: Future of agriculture using unmanned aerial vehicle. J. Inst. Image Electron. Eng. Jpn. **45**(4), 504–507 (2016)
3. Kashima, T., Matsumoto, S., Iseida, H., Matsutomi, T., Ishii, H.: Information system for urban agricultural crops direct sale using cloud. In: Conference on Electrical and Information Engineering Conference Chinese Branch Conference Papers Proceedings, pp. 51–52 (2011)
4. Kashima, T., Matsutomi, T., Matsumoto, S., Iseida, H.: Agricultural information system for production information sharing. In: The 24th Annual Meeting of the Institute of Electrical Engineers of Japan, vol. 3, p. 45 (2012)
5. Kashima, T., Matsumoto, S., Iseda, H., Ishii, H.: A proposal of farmers information system for urban markets. In: Watada, J., Watanabe, T., Phillips-Wren, G., Howlett, R., Jain, L. (eds.) Intelligent Decision Technologies. SIST, vol. 15, pp. 353–360. Springer, Heidelberg (2012). https://doi.org/10.1007/978-3-642-29977-3_35
6. Kashima, T., Matsumoto, S., Matsutomi, T., Iseida, H.: Development of two-way information dissemination infrastructure in the field of agriculture and reduction of environmental burden by introduction. In: Proceedings of the 20th National Convention of Artificial Intelligence Conference (2013). 3 K 3 - OS - 08 b - 3
7. Kashima, T., Matsumoto, S.: Development of interactive information transmission infrastructure for regional revitalization in the field of agriculture. In: Papers of the 5th Conference on Confederation of Interscience Conference (2013). CD – ROM 1 A - 2 - 2
8. Kashima, T., Matsumoto, S., Hasuike, T.: Formation of a recycling-oriented society using the content management system for the food industry. In: Proceedings of the 20th National Convention of Artificial Intelligence (28th) Conference (2014). 1 B 3 - OS-02b-4
9. Takatsu, S., Murakawa, H., Ohata, T.: Utilization of M2 M service platform in agricultural ICT. NEC Tech. J. **64**(4), 31–34 (2011)
10. Satake, Y., Yamazaki, T.: How food and agricultural clouds should improve food chain. FUJITSU **62**(3), 262–268 (2011)

Reducing Power Consumption of Mobile Watermarking Application with Energy Refactoring

Seongbo Kim[1], Jahwan Koo[2], Yoonho Kim[3], and Ungmo Kim[4($\boxtimes$)]

[1] College of Information and Communication Engineering,
Sungkyunkwan University, Suwon, Republic of Korea
kbs2375@naver.com
[2] College of Social Sciences, Sungkyunkwan University, Suwon,
Republic of Korea
jhkoo@skku.edu
[3] Department of Computer Science, Sangmyung University, Seoul,
Republic of Korea
yhkim@smu.ac.kr
[4] College of Software, Sungkyunkwan University, Suwon, Republic of Korea
ukim@skku.edu

Abstract. Recently, with the increase of cases of copyright infringement against various mobile devices, platforms, and services that have emerged due to the development of mobile environment, there is a need for developing a copyright protection technology and a new standard in a mobile environment. There are mobile applications which used to prevent copyright infringement of mobile contents. It has a function of inserting and extracting digital watermarks. However, when the application is executed in the mobile device as it is, there is a problem that the battery usage rapidly increases, and there is no related API to which the low power technology is applied to solve the problem. Due to the nature of mobile devices with limited battery capacity, it is very important to minimize the power consumption of the application in the process of developing a mobile watermarking application so that it can be serviced for a longer period of time. In this paper, we propose two scenarios for accurately measuring the power consumption of a mobile watermarking application using the battery historian. Through this technique, we compare and analyze the power consumption measurement results of the watermarking application improved by energy refactoring and the other not improved by it.

Keywords: Copyright of mobile contents · Digital watermarking
Power consumption measurement · Battery historian

1 Introduction

Recently, new content circulation environment is being developed in accordance with the spread of cloud, mobile web hard, bit torrent environment and mobile environment. As the copyright infringement cases of mobile devices, platforms, and services

© Springer International Publishing AG, part of Springer Nature 2018
S. Yamamoto and H. Mori (Eds.): HIMI 2018, LNCS 10905, pp. 599–608, 2018.
https://doi.org/10.1007/978-3-319-92046-7_49

increase, the development of copyright protection technology in mobile environment and the need for new standards are emerged [1].

There are mobile applications which used to prevent copyright infringement of mobile contents. It has a function of inserting and extracting digital watermarks. However, when the application is executed in the mobile device as it is, there is a problem that the battery usage rapidly increases, and there is no related API to which the low power technology is applied to solve the problem. Due to the nature of mobile devices with limited battery capacity, it is very important to minimize the power consumption of the app in the process of developing a mobile watermarking application so that it can be serviced for a longer period of time.

In this paper, we describe digital watermarking, battery historian, and energy refactoring, and propose two scenarios for accurately measuring the power consumption of mobile watermarking applications using battery historian. The existing watermarking application is implemented through energy refactoring the improvement in power consumption is confirmed through the results of power consumption measurement through the battery history. This will be used as a reference for the development of low-power applications and API modules with copyright technology in the future.

2 Preliminaries

2.1 Digital Watermarking

Digital watermarking is a technique for copyright protection, which means that a copyright holder or a seller inserts a specific code or pattern so as to identify copyright information in various digital contents such as photographs and moving pictures [2, 3]. In this case, the inserted data is called a watermark, and is used as a legal basis for proofing the owner by extracting the watermark information again from the pirated digital contents distributed by inserting the watermark (Fig. 1).

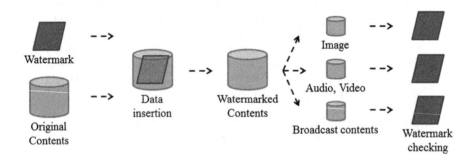

Fig. 1. Watermark system structure

2.2 Battery Historian

The battery historian is a battery consumption measurement tool made by Google [4]. We can measure how much power a mobile device with an Android operating system consumes, and the operating system version should be 5.0 or higher. Android app developers visualize the energy consumption of their Android apps through the battery historian, so we can see when features are used and get bug reports to get quantitative data on power consumption. Measures the overall power consumption from the battery full condition to the generation of the bug report with high accuracy (Fig. 2).

Fig. 2. Execution screen of battery historian

2.3   Energy Code Smells and Energy Refactoring

Code refactoring is the process of identifying the root cause of performance degradation while ensuring the functional behavior of the software system in order to resolve it when software internal structure is not improved, And to improve the performance of the software [5]. On the other hand, the code pattern that causes waste of energy in terms of energy efficiency, not the maintenance aspect, is called Energy Code Smell, and improvement of it is called energy refactoring [6, 7].

Loop Bug. This is the behavior of a program that does not get the desired results from the application and it repeats the same thing over and over, and can be caused by an external event such as a server failure. In addition, an application may repeatedly attempt to connect to an unresponsive server while continuously consuming energy for connection to the server and data, or may result in an infinite loop or recursive call due to a program error. To detect loop bugs, we need to identify the loops that will always return to the same initial system state, using components that consume energy in the process. Exception handling in a loop can indicate a loop bug. One way to handle loop bugs is to introduce a maximum number of iterations.

Dead Code. This means code that is not used but is loaded into memory and eventually consumes energy, and certain forms of dead code can be detected and removed through the compiler's optimization process. During the deformation of a dead code, some code may be called, but the result is ignored. Thus, the greater the number of different forms of dead code, the easier to detect at the source code level.

In-line Method. This means replacing the method call with the actually called part. This saves energy by avoiding energy consumption due to the overhead of method invocation, but it should be handled with caution considering energy efficiency because it can reduce maintainability and readability of code. One way to find method calls that inline can have a significant impact on energy efficiency is to use dynamic analysis. Profiles running applications and records how often each method invocation occurs. This information can be used to select refactor targets by setting a threshold for the number of calls. The candidate method call is replaced by the body of the invoked method, and the process's parameters are appropriately changed by the process.

Moving too Much Data. It means unnecessary movement between processor and memory. In general, since much of the energy consumed by the CPU is less than the energy consumed by the storage device, the movement of a large amount of data causes a waste of energy. Methods loading data which have been saved previously by querying for corresponding read and write methods have to be detected. To reduce data movement, methods reading data need to be replaced by the method calculating the data that was to be retrieved.

Immortality Bug. It means an application that is constantly revived despite the explicit kill of the user. This problem can be detected in a similar way to loop bugs. If the source code of all related applications is available for collaborative analysis, it is possible to detect applications that are restarted in other processes. System-level dynamic analysis can also indicate applications that start and end repeatedly.

Redundant Storage of Data. This means that different methods of the application store each without sharing the data. Avoiding redundant storage of data reduces energy consumption because different methods of the application do not each access the data. To detect redundant storage of data, we need to identify ways to store data and find ways to compare and store the same data. We can combine ways to reduce unnecessary data access and store the same data. To do this, we need to know the behavior of the program.

Using Expensive Resources. This means replacing energy-intensive resources with less energy-consuming resources. However, this should be carefully considered because it can affect the quality of the service.

3   Proposed Scheme

We apply energy refactoring to previously used watermarking application [8] and optimize them, and apply the energy measurement techniques proposed in Sect. 3.2. This compares power consumption measurements of improved watermarking apps with non-improved watermarking apps with energy refactoring, and confirms that improving existing watermarking apps with energy refactoring reduces power consumption.

3.1   Watermarking Application

Because our watermarking application uses MP3 files, we use similar way of particular audio watermarking techniques for MP3 files [9]. In the case of audio watermarking, there is a method of inserting into a spatial domain, a frequency domain, and a compressed domain, and an insertion method is determined according to characteristics of an application field. The watermark embedding method in the spatial domain is widely used because of its relatively simple operation. However, since most music content files exist in a compressed format such as mp3, there is an overhead that requires compression for decoding and re-compression for insertion and extraction Occurs. In addition, the watermarking method in the frequency domain is complicated in insertion and extraction operations. Therefore, when watermarking is to be applied on a limited resource such as a mobile device, a watermarking method in a compressed region is preferred.

When an original mp3 file is input, a portion (8 bytes) of the file is decoded, the watermark value is masked using a scale factor value, and the frame is re-encoded to generate an mp3 file having a watermark embedded therein. When we want to extract the watermark, a portion (8 bytes) of the watermark embedded content is decoded, and the watermark value masked by the scale factor value is extracted. Finally, CRC verification is performed on the detected watermark (Figs. 3, 4 and 5).

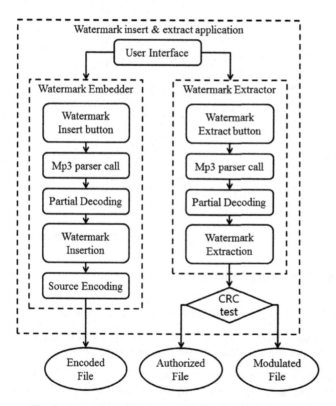

Fig. 3. Flow chart of watermarking application's function

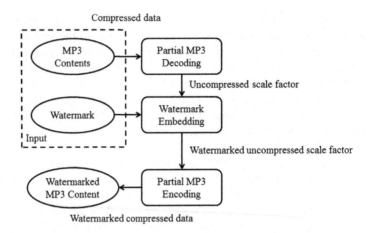

Fig. 4. Detail design of watermark insertion function

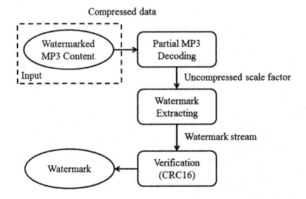

Fig. 5. Detail design of watermark extraction function

3.2 Energy Refactoring Technique

Source Parsing. It analyzes the original code using existing tools.

Energy Profiling. Energy profiles are made using tools that measure energy. Identify source files that contain energetic classes and methods.

Refactoring. We detect energy-intensive hot spots in source files from energy profiling. By choosing and applying one of your energy refactoring strategies, you can create optimal code that considers energy efficiency (Fig. 6).

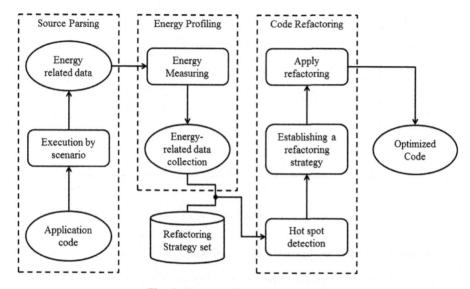

Fig. 6. Energy refactoring framework

3.3 Power Measurement Technique

The battery historian only measures the overall power consumption of the application and has the problem that mobile watermarking applications can not accurately determine the actual amount of power consumed when inserting or extracting watermarks into digital content. Therefore, we propose the following two scenarios to be implemented on the user interface to compensate and accurately measure the power consumption. Scenario 1 is a technique for measuring power consumption when a watermarking application is executed and watermarking insertion or extraction is performed once. Scenario 2 is a technique for performing a watermarking app for an arbitrary time, and performing a watermarking insertion or extraction function several times arbitrarily, and measuring the total amount of power consumed during a certain time (Figs. 7 and 8).

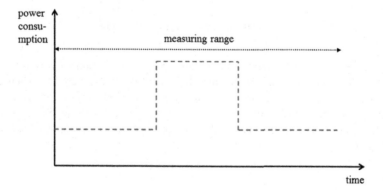

Fig. 7. Scenario 1 (single-run scenario)

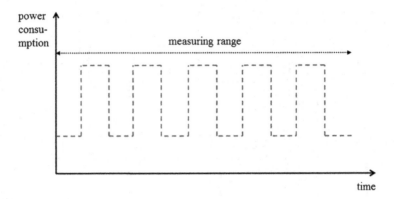

Fig. 8. Scenario 2 (repeat-run scenario)

3.4   Power Consumption Measurement and Analysis

We measured the power consumption of the mobile watermarking application on the CPU through Google's battery historian and analyzed the report on the result (Fig. 9).

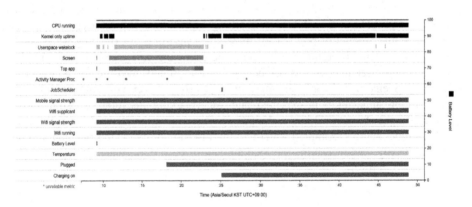

Fig. 9. Battery historian execution graph

In energy refactored watermarking apps, power consumption was reduced by about 4.7% in Scenario 1 and by about 5% in Scenario 2, compared to when energy refactoring was not applied. In the watermark extraction application with energy refactoring, power consumption was reduced by about 4.2% in Scenario 1 and about 4.78% in Scenario 2, compared to when energy refactoring was not applied (Tables 1 and 2).

Table 1. CPU power consumption measurement table of non-energy refactored watermarking app.

Number of times	Measurement result (mAh)					
	1	2	3	4	5	Average
Application	*Watermark insertion*					
Scenario 1	0.0498	0.0432	0.0571	0.0848	0.0469	0.0563
Scenario 2	0.4780	0.4770	0.4310	0.4210	0.3410	0.4296
Application	*Watermark extraction*					
Scenario 1	0.0580	0.0364	0.0430	0.0434	0.0392	0.0440
Scenario 2	0.3370	0.4390	0.3700	0.3730	0.3840	0.3806

Table 2. CPU power consumption measurement table of non-energy refactored watermarking app.

Number of times	Measurement result (mAh)					
	1	2	3	4	5	Average
Application	*Watermark insertion*					
Scenario 1	0.0548	0.0510	0.0530	0.0576	0.0517	0.0536
Scenario 2	0.4290	0.4170	0.3930	0.3870	0.4140	0.4080
Application	*Watermark extraction*					
Scenario 1	0.0630	0.0504	0.0305	0.0332	0.0337	0.0422
Scenario 2	0.3752	0.3903	0.3697	0.3256	0.3516	0.3624

4 Conclusion

In this paper, we measured the power consumption according to the two proposed scenarios to accurately measure the power consumption of the mobile watermarking app using the battery historian. The result shows that the existing watermarking app is improved through energy refactoring, Power consumption was reduced. This will be used as a reference for the development of low-power applications and API modules with copyright technology in the future.

Acknowledgments. This research project was supported by Ministry of Culture, Sports and Tourism (MCST) and from Korea Copyright Commission in 2018.

References

1. Song, S.K.: Annual Report on Copyright Protection. Korean Copyright Association Copyright Protection Center, Korea (2015)
2. Han, J.W., Park, C.S., Kim, E.S.: Digital watermark for copyright protection. Rev. KIISC **7** (4), 59–72 (1997)
3. Podilchuk, C.I., Delp, E.J.: Digital watermarking: algorithms and applications. IEEE Signal Process. Mag. **18**, 33–46 (2011)

4. https://github.com/google/battery-historian. Accessed 3 Jan 2017
5. Park, J.H., Lee, J.H., Rhew, S.Y.: A Study of re-engineering refactoring technique for the software maintenance and reuse. In: Academic Conference of Korean Institute of Information Scientists and Engineers, vol. 27, no. 1A, pp. 513–515 (2000)
6. Lee, J.W., Kim, D.H., Hong, J.E.: Code refactoring techniques based on energy bad smells for reducing energy consumption. Korea Inf. Process. Soc. Trans. Softw. Data Eng. 5(5), 209–220 (2016)
7. Gottschalk, M., Josefiok, M., Jelschen, J., Winter, A.: Removing energy code smells with reengineering services. GI-Jahrestagung 208, 441–455 (2012)
8. Kim, S.B., Koo, J.H., Kim, U.M.: A study of power measurement technique for mobile watermarking based on scenario. In: Proceedings of KIIT Summer Conference, pp. 101–103 (2017)
9. Koukopoulos, D.K., Stamatiou, Y.C.: A compressed-domain watermarking algorithm for MPEG audio layer 3. In: Proceedings of the 2001 Workshop on Multimedia and Security: New Challenges, pp. 7–10 (2001)

The Impact of Perceived Privacy Benefit and Risk on Consumers' Desire to Use Internet of Things Technology

Seonglim Lee[(⊠)], Hee Ra Ha, Ji Hyei Oh, and Naeun Park

Sungkyunkwan University, 25-2 Sungkyunkwan-Ro,
Jongno-Gu, Seoul 03063, South Korea
clothilda@skku.edu

Abstract. This study seeks to identify the dimensions of privacy benefit and risk in the context of Big Data and IoT technology. Based on the theory of privacy calculus, we investigate the impact of privacy benefits and risks on consumers' desire to use IoT technology. Two dimensions of perceived privacy benefit are identified as direct benefit of customized service and indirect benefit of efficiency innovation of suppliers. Three dimensions of perceived privacy risk are identified as passive user data collection, predictive use of personal data, and market discrimination as a negative consequence of Big Data Analytics. Depending on the types of IoT technology, the effects of perceived privacy benefits on consumers' desire to use are different. Consumers who perceive more benefit of customized service want more to use smart home appliance, while consumers who perceive more benefit from suppliers' efficiency innovation want more to use remote home service. Contrary to the hypothesis, the effect of predictive use of privacy risk on consumers' desire to use IoT technology is positive.

Keywords: Privacy benefit · Privacy risk · Privacy calculus · IoT technology

1 Introduction

The IoT, which is RFID (Radio Frequency Identification) tagged objects that use internet to communicate, is defined as the object that is able to communicate via the internet [25]. The IoT technology has been used in the industries such as supply chain management, retail tracking, and stock control as it allows for tracking, tracing, and identifying item-level components in real time. Firms can achieve effective resource management of related systems, and save cost with IoT Technology, which simultaneously allows for provision of much more convenience or lower price to the consumers [28].

Recently application of the IoT technology has been extended to consumers' daily life in the form of ubiquitous healthcare, home monitoring system, control of home equipment, home energy management, mobile payment, etc. Sensors and actuators embedded in the objects are connected to the Internet to communicate information about their immediate surrounding environment, and thus facilitate prompt responses as necessary [28].

© Springer International Publishing AG, part of Springer Nature 2018
S. Yamamoto and H. Mori (Eds.): HIMI 2018, LNCS 10905, pp. 609–619, 2018.
https://doi.org/10.1007/978-3-319-92046-7_50

Despite its huge technological advantages, previous research acknowledged IoT users encounter privacy threats in terms of identification, profiling, access control and confidentiality [20]. RFID technology in IoT which facilitates tracking and tracing of consumers with these items poses serious privacy concerns. Though the IoT technology has great potential to provide convenience and fun that consumers have never experienced, the widespread use and diffusion of IoT technology may be to some extent hindered by the leakage of private information [20].

The transformations of a great range of personal activities and environmental signals into data produce huge volumes of structured and unstructured datasets, so called Big Data. Firms utilize personal Big Data to predict preference and to estimate probabilities about behaviors of consumers. Inferences about consumers generated by Big Data analytics can be used for sorting individuals into segments in terms of firm's perspectives [2]. Such big data based social sorting process may negatively affect low-income and minority consumers by putting them in disadvantageous positions in the market [16].

Previous research indicated that consumers are concerned about businesses' data collection methods, especially the use of tracking technologies, such as click-streams, GPS data from mobile devices, social media usage etc. Consumers are also against the secondary uses of their personal data–potential abuses and misuses of personal data [15].

From the perspective of consumers, use of IoT technology involves a trade-off between the benefit of a wide spectrum of conveniences and risk of personal privacy [28]. Based on the theory of private calculus, we seek to investigate the impact of consumers' perceived privacy benefit and risk on desire to use IoT technology. By identifying and discerning different types of privacy benefit and risk factors related to Big Data and IoT technology, we will suggest the specific implications to establish the appropriate privacy policy to facilitate consumption of IoT technology and at the same time to meet consumers' privacy needs.

2 Literature Review

2.1 Privacy Calculus

IoT technology is like a double-edged sword. On one hand, people can derive benefit of convenience and fun. On the other hand, enormous privacy violating risk factors are embedded within it. A risk-benefit trade-off is affected by individual's assessment of benefit and risk. Individuals who have higher expectations of benefits than potential risk of privacy loss will be more likely to use IoT appliances and services [3]. This idea is **encapsulated** in the theory of privacy calculus [7].

The theory of privacy calculus describes how individuals balance their privacy concerns against their anticipation of benefits in their decision to disclose information [24]. The theory of privacy calculus suggests that "consumers would perform a risk-benefit analysis to assess the outcomes they would face in return for the information, and respond accordingly when requested to provide personal information to corporations" [5–7, 11, 12, 17, 18, 27, recited in 23]. Since use of IoT technology demands serious privacy disclosure, it is necessary to identify benefit and risk factors associated, to predict consumers' acceptance of IoT technology.

Previous research identified three major benefits of information disclosure including financial rewards, personalization, and social adjustment [24]. In the context of Big Data and IoT technology, personalization has the most relevance to privacy benefit. An example of personalization benefit driven by Big Data Analytics is Amazon's automatic product recommendation which analyzes the vast array of data including past customers' purchases and clicks and provides customized reference books for customers individually [14].

Previous research suggested that perceived benefit is more effective than perceived risk on customers' acceptance of personalization services. White [26] found that users are more likely to provide personal information when they receive personalization benefits; the other study by Chellappa and Sin [5] also found that consumers' value for personalization is two times more influential than the consumers' concerns for privacy in determining usage of personalization services [23].

Big Data can make the function of organization more efficient by improving operations, facilitating innovations, and optimizing resource allocations [16]. Another important source of benefit for consumer is positive externalities from the productivity improvement and cost savings achieved by suppliers. For instance, big data analytics allowed the yogurt company, Dannon, to forecast the demand of its retailer customers more accurately, which led to higher consumer satisfaction, less wastes, and a higher profitability [13, 15]. The mapping app provides real-time information about road congestion, allowing drivers to choose an efficient route.

2.1.1 Privacy Risk

Privacy risk is defined as the degree to which an individual believes that a high potential for loss is associated with the release of personal information to a firm [8, 16, 23]. Collecting, sharing, and transmitting sensitive data connected to humans are the most critical privacy issues [20]. Prior privacy literature indicated that consumers' major privacy risk includes potential loss of control over personal information, unauthorized access, theft [21], selling personal data, and sharing information to third parties [4].

More and more sensitive personal data are collected in a passive way without user consent these days. In the era of Big Data, consumers have expressed growing concern over business' data collection methods, such as click-stream, GPS data from mobile devices, and social media usage, because such passive user data collection relates to user tracking and localizing, and generates high privacy risk by the creation of user profiles [15].

Especially, the use of RFID technology in IoT entails high privacy risk with its function of identification, tracking and tracing of the substrate object. The major privacy issues associated with the RFID are found as follows [28]: (1) RFID tags could be read by anyone without customers' consent; (2) RFID-generated information can be used to profile customers by linking purchase information with their personal information; (3) retailers physically track customers without their consent or knowledge.

Previous research found that most consumers are concerned about potential abuses and misuses of personal data, and are against the secondary uses of their personal data service [15]. Usually the huge data gathered from passive way are analyzed without consumers' consent to predict their behavior [2, 20]. Using powerful new computing tools, firms can generate new personal data about individuals by drawing inferences and making predictions about their preferences and behaviors, based on which firms can offer better advertisement and customized service [15].

An individual's calculation of risk also involves an assessment of the likelihood of negative consequences [23]. Predictions based on big data can be used to unfairly discriminate against certain classes of persons by sorting individuals into segments in terms of their desirability and by identifying risky segments that need to be excluded from services. There is an increasing concern for that low-income and minority consumers are more likely to face discrimination and to be targeted by the sellers of inferior products [15].

3 Research Design and Methodology

The primary objective of our study is to analyze the impact of privacy benefits and risks on consumers' desire to use IoT technology. Since no previous measurement of independent variables were found in the context of Big Data and IoT technology, our methodology includes identifying such perceived privacy benefit and risk factors. The specific research problems to be solved in our study are: first, to explore the dimensions of privacy benefit and risk in the context of Big Data and IoT technology; and second to test following hypotheses.

H1. Perceived privacy benefit has a positive effect on consumers' desire to use IoT technology.

H2. Perceived privacy risk has a negative effect on consumers' desire to use IoT technology.

3.1 Instrument Development

We developed the instrument to measure perceived privacy benefit and risk through the following procedures. First, to identify appropriate perceived benefit and risk factors, we conducted literature review and collected elements indicating benefit driven by the use of personal data and threat to privacy protection in the environment of Big Data and IoT technology.

Consumers gain benefit directly from the customized products and services driven by Big Data Analytics and indirectly from firm's Big Data-based efficiency innovation. We collected the examples of customized services provided by firms including Amazon, Southwest Airlines, Target etc., and collected the cases of efficiency innovation achieved by firms such as Zara, DHL, Fujitsu, etc., all of which are developed based on Big Data Analytics. Then we made questions asking how much consumers perceive benefit from such customized services and efficiency innovation. Finally, we had 14 items to measure perceived privacy benefit in which item responses were given on a 7-point Likert scale, with 1 = totally disagree and 7 = totally agree.

To develop the instrument of measuring privacy risk, we explored privacy threatening elements in the era of Big Data and IoT technology from the information privacy literature, and categorized them into the three dimensions: privacy-invasive technology, usage for Big Data Analytics, and its potential consequence in the consumer market.

The major privacy-invasive technology includes passive user data collection of users' data and RFID technology within IoT environment [28]. We listed ten types of passive user data collection practices by network service, portal service, SNS, smartphone application providers etc., and made questions asking how much consumers concerned them. Because RFID technology relates to user identification, tracking and localizing which permits the creation and misuse of detailed user profiles [20, 28], we listed the five potential IoT privacy risks including lack of control, leakage, identification, profiling, and second use of private information, and made questions asking how much consumers perceive these risks.

The measurement of privacy risk related to Big Data Analytics consists of 4 items indicating re-identification, prediction, and incorrect inference about individuals. We made survey questions asking how much consumers perceive on such privacy-threatening elements. As a negative consequence of Big Data Analytics in the business sector, we added another 5 items indicating discrimination and privacy inequality in the consumer market. In total, we obtained 24 items to measure the Big Data privacy risk, in which item responses were given on a 7-point Likert scale, with 1 = totally disagree and 7 = totally agree.

The measurement of the desire to use IoT technology consists of 7 items regarding smart home appliances and remote home services such as smart grid and ubiquitous health care service on a 7-point Likert scale, with 1 = totally disagree and 7 = totally agree.

3.1.1 Data Collection

An online survey was conducted from January 26 to February 2, 2018 through a survey research company in Seoul, South Korea. Data were collected from 300 smartphone users between 20 and 65 years old. Since quota sampling method was applied, gender and age are evenly distributed in the sample.

3.1.2 Data Analysis

We used both statistical package for SAS 9.4 and Mplus 7 to analyze the data. We employed the exploratory factor analysis (EFA) to reduce measurement items and to explore the underlying factor structure. An explanatory factor analysis was conducted applying maximum likelihood method with GEOMIN rotation and Kaiser Normalization. After confirming the number of factors being used in our study by using EFA, we employed certified factor analysis (CFA) to identify the latent variables and to make the data more representative.

Structural Equation Modeling (SEM) was then applied to test the impact of perceived privacy benefit and risk on consumers' desire to use IoT technology (Fig. 1).

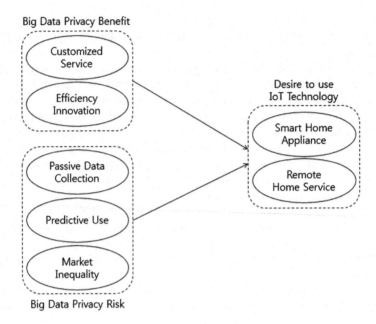

Big Data Privacy Benefit

Big Data Privacy Risk

Fig. 1. Research model

4   Results

4.1   The Results Explanatory Factor Analysis

Table 1 shows the results of explanatory factor analysis of the measurements. Two factors were revealed from the EFA analysis of the 14 Big Data privacy benefit items. Excluding items with a low loading less than 0.5 and cross loading, Factor 1 was made up of four items and labeled direct benefit of customized services, and Factor 2 was made up with two items and labeled indirect benefit of efficiency innovation.

In the same way, three factors were extracted with the measurement of perceived privacy risk. Factor 1 included three items of passive user data collection and was labeled passive data collection, Factor 2 included five items of which three items from privacy risk with IoT technology and two items from privacy risk with Big Data Analytics and was labeled predictive use. Factor 3 includes three items from discrimination and inequality and was labeled market inequality.

EFA with desire to use IoT technology generated two factors which were named respectively as smart home appliance and remote home service.

4.2   Measurement Model

CFA was used to assess the efficacy of the model structure suggested from the EFA. To improve the validity of the measurement model, we removed the two items (Q4_02 and Q4_03) from the measurement of the perceived privacy risk of predictive use. The resulting measurement model produced an excellent fit ($\chi 2/df$ = 1.59, CFI = 0.97,

Table 1. Explanatory factor analysis

Constructs	Items	% Variance	Factor loading	Communality	Mean	SD
BENE1	Q2_02	44.25	.74	.59	3.98	1.54
	Q2_03		.83	.69	3.94	1.60
	Q2_04		.92	.84	3.94	1.48
	Q2_05		.75	.71	4.21	1.45
BENE2	Q2_11	26.98	.99	.98	5.19	1.43
	Q2_12		.77	.60	5.19	1.52
RISK1	Q1_01	22.92	.84	.62	5.74	1.15
	Q1_02		.97	.86	5.81	1.17
	Q1_03		.92	.78	5.89	1.11
RISK2	Q3_03	18.37	.51	.32	4.81	1.29
	Q3_04		.87	.58	5.09	1.19
	Q3_05		.65	.42	5.14	1.41
	Q4_02		.53	.49	5.03	1.18
	Q4_03		.53	.32	4.99	1.25
RISK3	Q4_07	20.93	.87	.69	4.96	1.24
	Q4_08		.94	.80	5.10	1.36
	Q4_09		.76	.64	5.24	1.33
DIOT1	Q5_01	30.22	.82	.70	4.83	1.40
	Q5_02		.92	.82	4.87	1.35
DIOT2	Q5_05	33.58	.73	.60	5.06	1.31
	Q5_06		.83	.68	5.38	1.27
	Q5_07		.67	.45	4.97	1.45

TLI = 0.97, RMSEA = 0.04). To assess convergent validity, the standardized factor loadings, average variance extracted (AVE), and composite reliabilities (CRs) were examined. As displayed in Table 2, all factor loadings were larger than 0.6 and critical ratio indicated that all loadings were significant at 0.001. All AVEs and CRs exceed 0.5 and 0.7, respectively. Thus, the scale had a good convergent validity [1, 10]. In addition, all Cronbach a-values were equal to or greater than 0.7, suggesting a good reliability [19]. The discriminant validity of the constructs was assessed applying the method adopted by Fornell and Larcker [9]. According to Fornell and Larcker [9], discriminant validity is achieved if the square root of the AVE is higher than the correlations between two composite constructs. As indicated in Table 3, the square roots of the AVE were consistently greater than the correlations between latent variables.

4.3 Structural Model

To analyze the causality between the perceived privacy benefits and risks and consumers' desire to use smart home appliances and remote home services, structural equation modeling (SEM) techniques were used. The results are shown in Table 4. Fit indices ($\chi 2/df$ = 1.60, CFI = 0.97, TLI = 0.97, RMSEA = 0.05) indicated that data set was acceptable [22].

Table 2. Confirmatory factor analysis.

Constructs	Items	Factor loading	AVE	CR	Cronbach's α
BENE1	Q2_02	.70	.66	.85	.87
	Q2_03	.83			
	Q2_04	.90			
BENE2	Q2_11	.77	.71	.83	.87
	Q2_12	.91			
RISK1	Q1_01	.82	.78	.91	.90
	Q1_02	.93			
	Q1_03	.90			
RISK2	Q3_03	.71	.50	.75	.70
	Q3_04	.72			
	Q3_05	.69			
RISK3	Q4_07	.81	.70	.88	.87
	Q4_08	.90			
	Q4_09	.80			
DIOT1	Q5_01	.78	.72	.84	.86
	Q5_02	.92			
DIOT2	Q5_05	.76	.58	.80	.79
	Q5_06	.86			
	Q5_07	.66			

$\chi^2 = 219.26(df = 138)$, $\chi^2/df = 1.59$, CFI = 0.97, TLI = 0.97, RMSEA = 0.04

Table 3. Correlation matrix between independent variables and square root of AVE

Constructs	BENE1	BENE2	RISK1	RISK2	RISK3	DIOT1	DIOT2
BENE1	1						
BENE2	.32	1					
RISK1	−.23	.16	1				
RISK2	.05	.39	.46	1			
RISK3	−.06	.26	.43	.62	1		
DIOT1	.20	.18	.10	.19	.05	1	
DIOT2	.15	.34	.20	.41	.27	.62	1
Square root of AVE	.81	.84	.88	.71	.84	.85	.76

As indicated in Table 4, perceived direct benefit of customized service (b = 0.21, p = 0.003) positively affected the desire to use smart home appliance. The results showed perceived indirect benefit of efficiency innovation (b = 0.18, p = 0.014) and predictive use of personal data (b = 0.26, p = 0.013) had a significant and positive effect on the desire to use remote home service.

Table 4. Results of SEM

	Dependent variable: DIOT1		Dependent variable: DIOT2
Independent variable	Stand. coeff	Independent variable	Stand. coeff
BENE1	.21**	BENE1	.09
BENE2	.06	BENE2	.18*
RISK1	.11	RISK1	.05
RISK2	.21	RISK2	.26*
RISK3	−.13	RISK3	.02
R^2	.10	R^2	.18

$\chi^2 = 220.85$ (df = 138), $\chi^2/df = 1.60$, CFI = 0.97, TLI = 0.97, RMSEA = 0.05

$*p < .05$, $**p < .01$,

5   Discussion

This study identifies privacy benefit and risk factors in the era of Big Data and IoT technology in which most of collection, access, transferring, storage, utilization of the personal information are not under control of consumers.

Two dimensions of perceived privacy benefit are identified as direct benefit of customized service and indirect benefit of efficiency innovation by suppliers. We identify the three types of perceived privacy risk as passive user data collection, predictive use of personal data, and market discrimination which is a negative consequence of Big Data Analytics.

Depending on the types of IoT technology, the effect of perceived privacy benefit on consumers' desire to use is different. Consumers who perceive more benefit of customized service want more to use smart home appliance, while consumers who perceive more benefit from suppliers' efficiency innovation want more to use remote home service. Though provision of both direct and indirect benefits requires certain private information, direct benefit may cost more privacy as customized service needs more covert private information than suppliers' efficiency innovation does. The results of a positive effect of perceived benefit on consumers' desire to use IoT technology is consistent with the result of previous research that perceived benefit is more influential than perceived risk on consumer acceptance of personalization services.

Contrary to the hypothesis that perceived private risk has a negative effect on consumers' desire to use IoT technology, the effect of predictive use of privacy risk is positive. The regression coefficient of the market discrimination risk only has a negative sign. Though a negative effect of market discrimination risk is not significant, these results imply that the risk of collection and utilization of privacy information for prediction themselves may not be taken seriously for consumers' desire to use IoT technology though it invasively collects their sensitive personal data. Rather consumers' perception that firm utilizes their personal information for its gracious purpose may lead consumers to have less desire to use IoT.

References

1. Bagozzi, R.P., Yi, Y.: On the evaluation of structural equation models. J. Acad. Mark. Sci. **16**(1), 74–94 (1988)
2. Baruh, L., Popescu, M.: Big data analytics and the limits of privacy self-management. New Media Soc. **19**(4), 579–596 (2015)
3. Beldad, A., de Jong, M., Steehouder, M.: A comprehensive theoretical framework for personal information–related behaviors on the Internet. Inf. Soc. **27**(4), 220–232 (2011)
4. Budnitz, M.E.: Privacy protection for consumer transactions in electronic commerce: why self-regulation is inadequate. S. C. Law Rev. **49**, 847–886 (1998)
5. Chellappa, R.K., Sin, R.G.: Personalization versus privacy: an empirical examination of the online consumer's dilemma. Inf. Technol. Manage. **6**(2–3), 181–202 (2005)
6. Culnan, M.J.: "How did they get my name?": an exploratory investigation of consumer attitudes toward secondary information use. MIS Q. **17**(3), 341–363 (1993)
7. Dinev, T., Hart, P.: An extended privacy calculus model for e-commerce transactions. Inf. Syst. Res. **17**(1), 61–80 (2006)
8. Featherman, M.S., Pavlou, P.A.: Predicting e-services adoption: a perceived risk facets perspective. Int. J. Hum. Comput. Stud. **59**(4), 451–474 (2003)
9. Fornell, C., Larcker, D.F.: Evaluating structural equation models with unobservable variables and measurement error. J. Mark. Res. **18**(1), 39–50 (1981)
10. Gefen, D., Straub, D., Boudreau, M.C.: Structural equation modeling and regression: guidelines for research practice. Commun. Assoc. Inf. Syst. **4**(1), 7 (2000)
11. Hann, I.H., Hui, K.L., Lee, S.Y.T., Png, I.P.: Overcoming online information privacy concerns: an information-processing theory approach. J. Manag. Inf. Syst. **24**(2), 13–42 (2008)
12. Hui, K.L., Tan, B.C., Goh, C.Y.: Online information disclosure: motivators and measurements. ACM Trans. Internet Technol. **6**(4), 415–441 (2006)
13. IBM: The Dannon Company uses IBM smarter commerce to support yogurt market gains with big data analytics. http://www-03.ibm.com/press/us/en/pressrelease/41156.wss. Accessed 22 May 2013
14. Jo, H. H.: Big data's use status, problems and countermeasures. KERI Column 14 Mar 2014, pp. 1–4 (2014)
15. Kshetri, N.: Big data's impact on privacy, security and consumer welfare. Telecommun. Policy **38**(11), 1134–1145 (2014)
16. Malhotra, N.K., Kim, S.S., Agarwal, J.: Internet users' information privacy concerns (IUIPC): the construct, the scale, and a causal model. Inf. Syst. Res. **15**(4), 336–355 (2004)
17. Milne, G.R., Gordon, M.E.: Direct mail privacy-efficiency trade-offs within an implied social contract framework. J. Public Policy Mark. **12**(2), 206–215 (1993)
18. Milne, G.R., Rohm, A.J.: Consumer privacy and name removal across direct marketing channels: exploring opt-in and opt-out alternatives. J. Public Policy Mark. **19**(2), 238–249 (2000)
19. Nunnally, J.C.: Psychometric Theory, 2nd edn. McGraw-Hill, New York (1978)
20. Porambage, P., Ylianttila, M., Schmitt, C., Kumar, P., Gurtov, A., Vasilakos, A.V.: The quest for privacy in the internet of things. IEEE Cloud Comput. **3**(2), 36–45 (2016)
21. Rindfleisch, T.C.: Privacy, information technology, and health care. Commun. ACM **40**(8), 92–100 (1997)
22. Schumacker, R.E., Lomax, R.G.: A Beginner's Guide to Structural Equation Modeling. 2nd edn. Psychology Press, Mahwah (2004)

23. Smith, H.J., Dinev, T., Xu, H.: Information privacy research: an interdisciplinary review. MIS Q. **35**(4), 989–1016 (2011)
24. Steijn, W.M., Schouten, A.P., Vedder, A.H.: Why concern regarding privacy differs: the influence of age and (non-) participation on Facebook. Cyberpsychol. J. Psychosoc. Res. Cyberspace **10**(1), 88–99 (2016)
25. Uckelmann, D., Harrison, M., Michahelles, F.: An architectural approach towards the future internet of things. In: Uckelmann, D., Harrison, M., Michahelles, F. (eds.) Architecting the Internet of Things, pp. 1–24. Springer, Heidelberg (2011). https://doi.org/10.1007/978-3-642-19157-2_1
26. White, T.B.: Consumer disclosure and disclosure avoidance: a motivational framework. J. Consum. Psychol. **14**(1–2), 41–51 (2004)
27. Xu, H., Teo, H.H., Tan, B.C., Agarwal, R.: The role of push-pull technology in privacy calculus: the case of location-based services. J. Manag. Inf. Syst. **26**(3), 135–174 (2010)
28. Zhou, W., Piramuthu, S.: Information relevance model of customized privacy for IoT. J. Bus. Ethics **131**(1), 19–30 (2015)

Efficient Method for Processing Range Spatial Keyword Queries Over Moving Objects Based on Word2Vec

Sujin Oh[1], Harim Jung[1], JaHwan Koo[2], and Ung-Mo Kim[3](✉)

[1] College of Information and Communication Engineering,
Sungkyunkwan University, Suwon, Republic of Korea
bgbanana4@gmail.com, harim3826@gmail.com
[2] College of Social Sciences, Sungkyunkwan University,
Suwon, Republic of Korea
jhkoo@skku.edu
[3] College of Software, Sungkyunkwan University, Suwon, Republic of Korea
ukim@skku.edu

Abstract. In this paper, we propose an efficient method for processing continuously range spatial keywords queries based on Word2Vec over moving objects. In particular, the paper addresses two problems of processing range spatial keyword queries over moving objects. Each moving object and query has own keywords and a spatial location information. (i) The likelihood of having similar meanings between the keywords is high, but the likelihood of being exactly the same is low. Therefore, we focus on not only exactly matching, but similarity between keywords through Word2Vec. (ii) Additionally, because the objects are moving, the central server needs to continuously monitor them. In traditional research, by constructing safe region, there are attempts to reduce inevitable communication between the objects and the server. However, the constructions of safe region are also costly. Therefore, in the paper, to decrease communication costs, and construction and maintenance costs, we introduce the concept of buffer region and the pruning rules. Moreover, the proposed method includes a spatial index structure, called the Partition Retrieved list whose role is to help the system quickly construct the safe region of each object. Through experimental evaluations, we verify the effectiveness and efficiency for processing range spatial keyword queries over moving objects.

Keywords: Spatial databases · Location-based services
Spatial keyword queries · Range queries · Range monitoring queries
Safe zone · Word2Vec

1 Introduction

Recently, the rapid development in mobile computing and communication has changed a lot in our lives. Due to GPS-enabled mobile devices, people use their location to upload posts to social network services or to search for specific locations of interests based on their location. Likewise, the location-based services (LBSs) which is based on the location of users have great popularity not only in industry but also in academic

© Springer International Publishing AG, part of Springer Nature 2018
S. Yamamoto and H. Mori (Eds.): HIMI 2018, LNCS 10905, pp. 620–639, 2018.
https://doi.org/10.1007/978-3-319-92046-7_51

communities. There are many different types of spatial queries in the research. Because the finding several nearby objects has more likely to match with the needs of the users than the finding only one optimized object, the most traditional researches have focused on the finding (all or some) nearby objects (e.g., *range query* [1, 2], *nearest neighbor queries* [3, 4], etc.). From among these, range query is the most important type of spatial queries used in LBSs, whose goal is finding all objects within a given (rectangular or circular) area. Also, as the data have not only spatial attributes but also textual attributes, the importance of *spatial keyword query* is also on the increasing.

Figure 1 shows an example of the range spatial keyword query, where there are nine objects and three queries and each of them has own keywords. In Fig. 1, we assume that the all queries q_1, q_2, and q_3 find the objects that have similar interests with the query within given circular area. Even though the objects o_1 and o_2 has 'noodle' as keywords and they match the query q_1 textually, but in perspective of spatial attributes, only the object o_1 can be a result of the query q_1, not o_2. About the query q_2, although the object o_4 matches the query q_2 spatially, as the object o_4 does not have the exactly matched the keyword with the query q_2, 'Korean food'. However, since the keyword of the object o_4, 'bulgogi' is included in 'Korean food' and there is a strong relationship between the keywords, the object o_4 can be a result of the query q_2. Similarly, the object o_5 and o_6 can be a result of the query q_3. As the previous example, the data have been changing as geo-textual data.

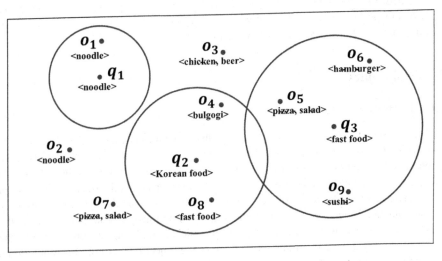

Fig. 1. An example of the *range spatial keyword queries*

Each moving object and query has own keywords and spatial location information. since the objects are moving, it is inevitable that the moving objects continuously communicate with the central server. However, whenever the objects move, reporting the new location to the server makes the server overloaded. To reduce the frequent and unnecessary checks for the current query results, the traditional researches have proposed a concept of *safe region* [5–11]. Safe region assigned to an object is the area such that the results of the queries in the system are not changed as long as the object moves within this area. However, the constructions of safe region are also costly. Therefore, we introduce pruning rules to decrease communication costs, and construction and maintenance costs of the safe regions.

In this paper, we propose a new algorithm for processing Continuous Range Spatial Keyword Queries based on Word2Vec (namely, *CRSK-w2v* queries) over moving objects in Euclidean space. In the traditional researches, there are key challenges: (i) an efficient method for quickly finding the result objects of the queries, (ii) an efficient communication between the moving objects and the central server without unnecessary re-computation, and (iii) how to analyze the relationship between the keywords. Therefore, in order to solve the above challenges, the proposed algorithm focuses on (i) obtaining the relationship between keywords based on Word2Vec and (ii) minimizing communication costs, and construction and maintenance costs of the safe regions. Moreover, in order to efficiently construct and maintain the safe regions, we introduce the variation of buffer region, which is proposed in [8] to reduce the unnecessary and frequent re-computations of the safe regions. In addition, we introduce the efficient pruning rules that use a spatial index structure, called *Partition Retrieved list*. With the partition retrieved list, the system can easily exclude the queries which do not match with the objects spatially.

The rest of the paper is organized as follows. In Sect. 2, the problem is formally defined. Then we review some related work in Sect. 3. In Sect. 4, the system overview is provided and the details of the proposed algorithms are presented in Sect. 5. In Sect. 6, the performance of the proposed algorithm on real data sets are presented. Finally, in Sect. 7, we present our conclusions.

2 Preliminary

2.1 Client-Server Model

In this section, we review the background of the client-server model. For continuously monitoring *CRSK-w2v* queries, we adopt the client-server model. In this model, there are a central server and two types of clients. Clients, a component of this model, consists of the query issuers and the moving objects. A query issuer issues the query to the central server, and receives the result objects from the server persistently each time the query results change. Whenever an object moves, it continuously communicates with the server to report its new location. Moreover, the object is evaluated whether it can be the result of a query or not. We assume that the moving object o has limited computational capacity. In other words, it cannot recognize all queries, and only several queries can be recognized according to capacity of the object. The concept of the maximal computational capacity is represented as $o.cap$, which indicates that o can track at most cap nearest neighbor queries and determine whether it is the result object of the queries.

In this model, the role of central server is critical, because the place to process *CRSK-w2v* queries is the server. The server receives information of location and the keywords from moving objects and the queries which are issued from the query issuers. At the same time, the server processes each query and sends the up-to-date results to the corresponding query issuer. In addition, the server continuously monitors the moving objects. Also, whenever the safe region needs to update, the server calculates the safe region of each object and sends it to the object. Let us consider a scenario where there is an Italian restaurant and a manager of restaurant wants to send e-coupon or promotion information to potential customers who have interests in Italian foods and they are nearby the restaurant. Then the restaurant manager issues a query which finds the potential customers, to a central server. Then the server finds customers who are matched with the query and continuously monitors customers to identify if they leave out their safe regions or not. Thus, the server is responsible for all the processing the queries. In this paper, in order to decrease the burden of the server, we assign the safe region to each moving object.

2.2 Word2Vec

The *one-hot encoding* method in the traditional Natural Language Processing (NLP), has a disadvantage in that the dimensionality increases and the relationship between the words is not reflected. In order to solve this problem, *word embedding* method of vectorizing the meanings of words in multidimensional space has been devised. If the meaning of a word is vectorized, the similarity between words can be measured, and since the meaning itself is numerically expressed as a vector, it can be deduced through vector operation. One of the methods, *Word2Vec* is an algorithm that implements the method proposed by Mikolov [12], which makes it possible to improve the precision in NLP compared to existing algorithms. Word2Vec is a technique that expresses each word as a vector in a space of about 200 dimensions. As in the following example, by adding or subtracting vectorized words, inference process can be easily implemented.

$$vector("Korea") - vector("Seoul") + vector("Paris") = vector("France") \qquad (1)$$

Similarly, *vector("King") − vector("Man") + vector("Woman")* results in a vector, which have closet meaning with *Queen* [13]. Mikolov [12] proposed two new model architectures: *CBOW* (*Continuous Bag-of-Words*) and *Skip-gram*. CBOW is a model for inferring words through the entire context, and Skip-gram model infers words that appear around a given word, as opposed to CBOW. When sampling the word, it applies the idea that the closer the word is, the more relevant it is. In this paper, in order to reflect the similar meanings between the keywords, we analyze the relationship between the keywords through Word2Vec.

2.3 Problem Statement

In this section, we formally define the problems of processing *CRSK-w2v* queries. Let O and Q denote a set of geo-textual objects and a set of *CRSK-w2v* queries in the proposed system, respectively. Each object $o \in O$ is defined as a pair (*o.pos*, *o.pre*) that means a position of the object and a set of preferences, respectively. In particular, *o.pos* is represented in two-dimensional space and it has x-coordinate (*o.posX*) and y-coordinate (*o.posY*). Moreover, each element in the preferences set, *o.pre* means preference (keyword) of o and the number of keywords may be one or more. In this paper, we assume that the objects are moving, and then they dynamically change their location. On the contrary, we consider the queries are static and the query issuers issue a query to the central server. Additionally, each query $q \in Q$ also contains its position and its set of preferences and they are denoted by $q.pos$ and $q.pre$, respectively.

Given a query q and an object o, in this paper, we represent geo-textual relevance by using two concepts of distance; $disS(q, o)$ and $disT(q, o)$ as follow. $disS(q, o)$ means a spatial distance between $q.pos$ and $o.pos$, while $disT(q, o)$ means a textual distance between $q.pre$ and $o.pre$. The smaller of the distance $disS(q, o)$, the higher the relation between q and o ($disT(q, o)$ takes the same approach).

To compute $disT(q, o)$, we use the textual similarity $simT(q, o)$ and define the relationship between them as $disT(q, o) = 1 - simT(q, o)$. If there are higher textual similarity between the query q and the object o, it means that between them, there is lower textual distance. In other words, if the object o has lower textual similarity with the query q, then it has higher textual distance. For obtaining the textual similarity $simT(q, o)$, we use a modified function of the weighted Jaccard coefficient [14] as (2).

$$simT(q, o) = \frac{\sum_{k \in q.pre} r(k)_o}{Num(keyword \in q.pre)} \tag{2}$$

$$r(k)_o = \max_{t \in o.pre} sim(k, t) \tag{3}$$

About each keyword k in $q.pre$, $r(k)_o$ means a relation between the query q and the object o. Additionally, $r(k)_o$ has a value [0, 1]. Moreover, if the common keyword exists between the object o and the query q, $r(k)_o$ has the largest value relation (i.e., $r(k)_o = 1$). However if not, based on Word2Vec, $r(k)_o$ has the largest value among the similarity values (i.e., $sim(k, t)$) obtained by comparing the similarity with the keyword t in $o.pre$. Therefore, if the similarity between the keywords are not considered, we can redefine that $simT(q, o)$ is the number of common keywords between q and o divided by the number of keywords of q, and its value lies between 0 and 1. Additionally, if the similarity between the keywords are considered, as $r(k)_o$ has a value of 1 or less, and $simT(q, o)$ is also normalized. By generalizing this, the textual similarity $simT(q, o)$ is normalized without any normalization strategy and $disT(q, o)$ is also normalized.

Moreover, for normalizing the spatial distance $disS(q, o)$, we define R, system defined spatial distance and the reason will be explained below.

$$\begin{cases} disS(q, o) = 1 & if \; \|q.pos, o.pos\| \geq R; \\ disS(q, o) = \frac{\|q.pos, o.pos\|}{R} & otherwise, \end{cases} \tag{4}$$

where $\|q.pos, o.pos\|$ is the Euclidean distance between $q.pos$ and $o.pos$.

In this paper, we represent geo-textual relevance by using a ranking score function as follow.

$$score(q, o) = \alpha \cdot disS(q, o) + (1-\alpha) \cdot disT(q, o), \tag{5}$$

where α is a ranking parameter that shows the effect of the spatial distance and the textual distance on the query and it lies between 0 and 1. If the importance of the spatial distance is the same as the textual distance, α is 0.5. However, the issued query considers the spatial distance more important than the textual distance, α is greater than 0.5. Therefore, the parameter α is a system parameter which is determined depending on the preference of the query issuers.

Let consider an example, where there are two objects o_1 and o_2 and a query q. As the query issuer of the query q has a higher preference of textual distance than spatial distance, and then he sets the ranking parameter α is 0.3. We assume that each object and the query have five keywords, and also only exactly matched keywords are considered. The object o_1 is 5 km from the query location and it has three common keywords with q. The object o_2 is 4 km from the query location and it has only one common keyword with q. From these situations, we will find the $disT(q, o_1)$ is 0.4 and the $disT(q, o_2)$ is 0.8. Therefore, the ranking score of o_1 and o_2 is 1.78 and 1.76, respectively. Even though the preference of textual distance is higher than spatial distance, the ranking score of o_2 is better than o_1. It is happened as the textual distance is already normalized but the spatial distance is not normalized. That is the reason why both distance $disS(q, o)$ and $disT(q, o)$ need to be normalized to [0, 1].

Definition 1 (Continuous Range Spatial Keyword Queries based on Word2Vec, *CRSK-w2v* queries). Let O and Q be a set of geo-textual objects and a set of queries, respectively. The query q is represented by $\{pos, pre, TS\}$, where pos is the query location, pre is the query keywords set, and TS is the threshold score. The result RS_q^t of q is a set of objects whose geo-textual relevance score is less than or equal to the threshold score of q, i.e., $\forall o \in RS_q^t, score(q, o) \leq TS$.

We summarize the frequently used terms in Table 1.

Table 1. The summary of notations

Notation	Definition
o	A moving geo-textual object
q	A range spatial keyword query
$o.pos$ ($q.pos$)	Position of the object o (query q)
$o.pre$ ($q.pre$)	Set of preferences for the object o (query q)
$o.cap$	The maximal computational capacity of the object o
α	Ranking parameter factor
TS	Threshold score
RS_q^t	Result set of the query q at timestamp t
CC_q^o	Conditional circle of the object o with regard to the query q
r_q^o	The radius of CC_q^o
CC_q^*	Largest conditional circle of the query q
r_q^*	The radius of CC_q^*
$NQList_o$	The list to store nearest query information
BR_o	Buffer region of the object o
P_o^g	A set of partitions overlapped an object o in a grid, g
P_q^g	A set of partitions overlapped a query q in a grid, g

3   Related Work

In this section, we review the related works with our proposed algorithm. First, in Sect. 3.1, we discuss a viewpoint on treating the queries and the objects adopted in this paper, and then we introduce processing of the continuously monitoring spatial in Sect. 3.2. Finally, in Sect. 3.3, we briefly review the related works for spatial keyword queries processing.

3.1   Viewpoint on Treating the Queries and the Objects

We discuss a viewpoint on treating the queries and the objects adopted in this paper. There are two viewpoints on spatial database. One of the viewpoint assume that the query points are stationary and the objects are moving [2, 8, 15]. The other point of view is that the query points are moving and the objects is stationary [6, 9]. In the both of viewpoints, the moving is indexed. This paper adopts the former viewpoint. Thus, from now, we only review the studies conducted continuous range monitoring queries matched our perspective. Indexing the query points instead of the objects is effective to reduce the computational cost for updating, because the objects are frequently moving, and the indexing of the objects remains valid for a short time than the queries. Prabhakar et al. [2] proposed two indexing technique: Query Indexing and Velocity Constrained Indexing (VCI) by using R-tree. In [15], a grid-index structure based on in-memory is more effective technique than any others. For processing these queries, in order to efficiently monitor the moving objects, it is inevitable that the moving objects continuously send their new location to the server. The objects should send periodically their new location to a central server. At the same time, the server receives those requests, processes the queries, and returns the results. However, the server may become overloaded due to the frequent and unnecessary checks for the current query results.

To solve the above problems, the traditional researches have proposed safe region method [5–11]. It helps to reduce the frequent and unnecessary checks for the current query results, since the safe region of an object o guarantees the results of the queries in the system are not affected as long as o remains inside this area. As the shape of safe region is arbitrary, it can be rectangle, circle or anything else. However, these constructions of safe region are also costly.

3.2   Continuous Monitoring Spatial Queries

In spatial database communities, early researches assumed a static dataset and focused on development efficient spatial access methods (e.g. R-tree [16]). Additionally, they treated the effective method of the processing snapshot queries like the range queries [1, 2] and the nearest neighbor queries [3, 4] used in LBSs as the most important types of queries. They focus on finding nearby objects, instead of finding only one optimized object. Because finding nearby objects has more likely to match with the needs of the users. However, recently, the environment is changing, where data objects or the queries move. In particular, researches on the continuously monitoring the spatial queries have been increasingly important, because of their use in a variety of areas. In [2], the Q-index is the first address the problem of monitoring the static range query over moving objects.

Hu et al. proposed a framework to answer monitoring queries over moving objects, where mobile clients use safe regions to help reduce communication cost [14]. In this paper, we focus on the processing the continuously monitoring the range queries.

3.3 Spatial Keyword Queries

In this section, we briefly review the existing methods for spatial keyword queries processing. As the importance of processing queries considering both keywords and spatial attributes is emphasized, research on Spatial Keyword Queries is increasing. Most of the researches focus on top-k spatial keyword queries [5–7, 17], whose goal is finding k closest objects, and the objects possess similar keywords with the keywords of the queries. Although many researches have focused on processing top-k spatial keyword queries, some researchers have conducted range spatial keyword queries [8, 11], whose goal is finding all objects which have keywords of the queries in given (rectangular or circular) area. However, they adopted the simple pruning rules for reducing the construction costs of the safe regions. Additionally, they consider only when the keywords between the queries and the objects were exactly matched. In this paper, we introduce more complex pruning rules to effectively reduce the overhead of constructing the safe regions, based on the relationship between keywords.

4 System Overview

This section presents an overview of the system model. Unlike the traditional client-server model, in this paper, we separate clients as two types; query issuers and moving objects. The reason is that (i) a major role of clients is to issue queries and to get back the result and (ii) although moving clients are clients, it is emphasized that the role of data like location and keywords set rather than the role of clients. The query issuers, who issue the queries to the central server, are called queries in the remainder of this paper and the other is moving clients, called (moving) objects. Therefore, in this system, there are three components; moving objects, clients and a server. Figure 2 shows the proposed system architecture.

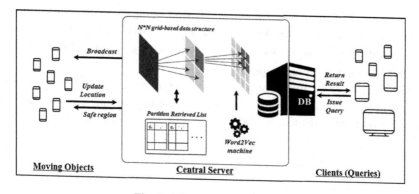

Fig. 2. The system architecture

In this paper, we introduce an efficient method for processing *CRSK-w2v* queries over moving objects. For processing the queries, there are some challenges. When the objects move, the result sets of the queries are a possibility to change. Therefore, it is important to continuously monitor the objects and to identify whether the movement of the objects affects the result sets of the queries if not. In order to address these challenges, the concept of the safe region is proposed in [5–11]. Additionally, we introduce the variation of buffer region, which is proposed in [8] to reduce the unnecessary and frequent re-computations of the safe regions. Moreover, we present pruning rules to reduce overhead of the communication costs between the server and the objects, and efficiently construct the safe regions. The key idea of the proposed pruning rules is to find the spatial and textual upper bounds between the queries and the objects.

We describe our proposed system in the following subsections. Firstly, in Sect. 4.1, we present the concept of safe regions and next in Sect. 4.2, we present the concept of buffer regions. Finally, in Sect. 4.3, we present the pruning rules based on geo-textual proximity.

4.1 Safe Regions

In the client-server model, it is inevitable that whenever the moving objects move, they consistently communicate with the server to report their new location. However, since some of these communications are unnecessary, by decreasing them, we can reduce frequent and needless re-computation happened at the server. To reduce unnecessary communications, one of the methods is constructing the safe region. The safe region is an area such that the object cannot affect the results of the queries as long as it is moving within the area. The shape of the safe region can be rectangular or circular. In the many traditional studies, the shape of the safe regions is rectangular, because it is easier to consider pruning rules than the circular shape [6, 14]. On the contrary, this paper adopts circular as the shape of the safe region. In particular, instead of the traditional safe regions, we use an expanded concept. The expanded concept is explained later.

In order to efficiently process *CRSK-w2v* queries, we need a relation between the query and the result object of the query. Therefore, we have the following lemma.

Lemma 1. Only object $o \in O$ that satisfies the following relation $disS(q,o) \leq \frac{TS}{\alpha} - \frac{(1-\alpha)}{\alpha} \cdot disT(q,o)$ can be a result of *CRSK-w2v* query q.

Proof. By the Definition 1, an object $o \in O$ who is the result of *CRSK-w2v* query q satisfies $score(q,o) \leq TS$, i.e., $score(q,o) = \alpha \cdot disS(q,o) + (1-\alpha) \cdot disT(q,o) \leq TS$. From this function, we can get the following results, $disS(q,o) \leq \frac{TS}{\alpha} - \frac{(1-\alpha)}{\alpha} \cdot disT(q,o)$. Therefore, we prove Lemma 1.

With Lemma 1, we introduce a concept of the *conditional circle* CC_q^o. The conditional circle is circular area where its center is the query location $q.pos$ and its radius is the r_q^o. From the Lemma 1, the radius is defined as following equation.

$$r_q^o = \frac{TS}{\alpha} - \frac{(1-\alpha)}{\alpha} \cdot disT(q,o), \tag{6}$$

If an object o is within the conditional circle of a query q, o is a result of q; and if not, o is not the result of q. With the concept of the conditional circle, we introduce a concept of the conditional region. The conditional region is assigned area between one object o and one query q. It guarantees the results of q is not changed as long as o remains inside the region. Although the conditional region is the concept similar to the safe region, it is related to only one query, not all queries. Figure 3 show the different types of the conditional region. In Fig. 3(a), upper figure shows the situation when o is in the conditional circle of q (i.e., when o is the result of q), and lower figures show the opposite situation.

In this paper, we utilize the expanded version of the safe regions. By expanding the existing concept, we re-define the safe regions as an intersection of all relevant the conditional regions. The three figures in Fig. 3(c) are, in turn, the conditional regions of the queries q_1, q_2, and q_3 for o (shaded in gray). Moreover, we can find that the intersection of the three conditional regions in Fig. 3(c) is the safe region of o in Fig. 3(b) (shaded in gray).

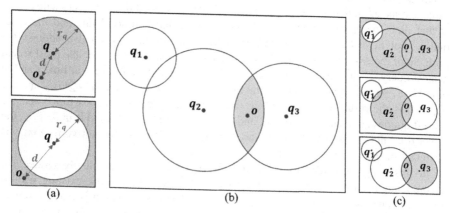

Fig. 3. The conditional region (a) and the safe region (b and c)

The safe regions reduce the number of the communication between the server and the moving objects and it causes the re-computation cost reduction in the server. However, the construction cost of the safe regions is also costly. Hence, we introduce a concept called *buffer region* to reduce the cost in next section.

4.2   Buffer Regions

In this section, we present the concept of buffer regions proposed in [8] and the variation of *buffer region*, which has been changed according to our system. The *buffer region* (namely, *BR*) is calculated based on the maximal computational capability (denoted by $o.cap$) of the client mobile devices (i.e., the moving objects). Because of the limited computational capability of the object o, each object o cannot track all the queries registered in the server and o can only monitor the region as much as its computational capability at the same time. Therefore, the server needs to check several queries depending on capability, not all the queries. In [8], *BR* of an object o (BR_o) is defined as the area where its center is the location of objects $o.pos$ and its radius is the Euclidean distance from $o.pos$ to the $o.cap - 1^{th}$ nearest conditional circle. When an object o comes into the central server for the first time, the server assigns a *Nearest query list* (*NQList*) that stores the conditional circles of the relevant queries, and sends the safe region to the corresponding object. Then whenever o leaves its safe region, the server checks if o is within its BR_o. If o leaves the BR_o, the server re-compute the BR_o; and if not, the server identifies the queries stored in the *NQList* and re-computes the safe region of o.

At the existing concept of the *BR*, because of the tracking the *BR* boundary, each moving object o can only monitor the boundary of $o.cap - 1^{th}$ nearest conditional circle and the size of $NQList_o$ is one less than $o.cap$ (i.e., $N(NQList_o) = o.cap - 1$). However, since o only needs to know the radius of BR_o, the tracking of the *BR* is not necessary. Hence, from now, the size of the BR_o is $o.cap$. Figure 4 shows examples of the buffer region (depicted in bold dotted line). In these examples, we assume that the maximal computational capability of o, $o.cap$ is three and all the registered query has same textual similarity with o. In each figure, the area shaded in gray means safe region at that time. Also, some queries in $NQList_o$ are depicted in solid line and the other queries not in $NQList_o$ are depicted in dotted line. From now on, we address the problem of the existing *BR*. The existing *BR* does not guarantee storing the most recent nearest query. Figure 4(a) and (b) show an existing buffer region of o at time t and at time $t + 1$, respectively. At time t, the queries q_3, q_4, and q_5 are in $NQList_o$ and the object o is the result of q_3, q_4; and at time $t + 1$, the object o moves to o^* and the nearest queries are changed to q_2, q_3, and q_4. However, as the object o is in its BR_o, the server does not check new $NQList_o$ and consequently, it also does not verify that o is the result of q_2.

In this paper, in order to solve the above problem and efficiently reduce the re-computation costs in the server, we have made a variation in the existing way to construct buffer region. There are two variations. The first variation is that the object o can keep track of $o.cap$ conditional circles, and the reason for this variation is mentioned above. The second variation is that in constructing the buffer region, the server uses $o.cap + 1$ nearest conditional circles instead of $o.cap$. In this paper, since the server assumes that there is no limitation of the computational capability unlike the objects, the server can keep track of $o.cap + 1$ nearest conditional circles. However, when constructing the safe region, the server still uses only $o.cap$ nearest conditional circles, because the object can only track $o.cap$ conditional circles according to its own computational capability. In this paper, since we assume that the central server has no limitation of the computational capability unlike the objects, the server can keep track of $o.cap + 1$ nearest conditional circles. Figure 4(c) and (d) show a variation version of the buffer region at time t and at time $t + 1$, respectively. At time t, the queries q_2, q_3, q_4, and q_5 are in $NQList_o$ and the object o is the result of q_3, q_4; and at time $t + 1$, the object o moves to o^*, and the object leaves out its BR_o. Hence, the server should check $NQList_o$ and consequently, in new updated $NQList_o$, there are q_1, q_2, q_3, and q_4.

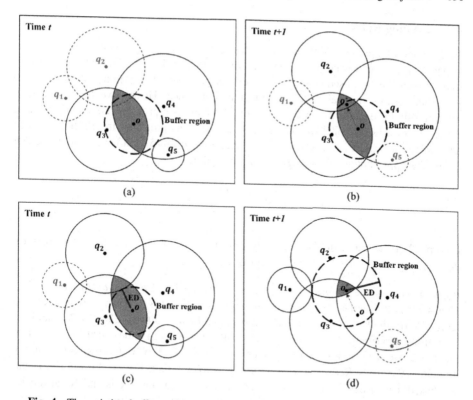

Fig. 4. The existing buffer region (a and b) and the variation of buffer region (c and d).

We summarize the two variations from the existing concept of buffer region.

- Firstly, an object o can keep track of $o.cap$ nearest conditional circles, and by using these, a central server constructs a safe region.
- Secondly, when the server constructs buffer region, the server uses $o.cap + 1$ nearest conditional circles, which is one more than the existing, i.e., $N(NQList_o) = o.cap + 1$.

From the buffer region, the objects need to know only the radius of buffer region. Therefore, we define a concept of *end distance* (denoted by *ED*) as follow and it is depicted in bold line in Fig. 4(c) and (d).

Definition 2 (End Distance, ED). Given a *CRSK-w2v* query q and a geo-textual object o, the *end distance ED* of o is the Euclidean distance from $o.pos$ to the $o.cap + 1^{th}$ nearest conditional circle.

Additionally, $NQList_o$ contains at most $o.cap + 1$ nearest neighbor queries. From the Definition 2, when the object o leaves out a circular space where its radius is *ED* and the center is $o.pos$, the $NQList_o$ can be changed.

4.3   Pruning Rules

In this paper, there are two pruning rules based on geo-textual proximity. Because this paper is an extended paper based on [11], the pruning rules are similar to the existing ones. Since the proposed system is based on geo-textual proximity, there are two important factors: textual distance and spatial distance. Hence, we have created a pruning rule for each of the two factors. Firstly, there is a pruning rule about a textual distance. In order to consider only textual distance, assume that the spatial distance have minimum value, i.e., $disS(q,o) = 0$. Therefore, from (5) and Definition 1, we can define the upper bound of $disT(q,o)$, UB_o^q as follow.

$$disT(q,o) \leq \frac{TS}{(1-\alpha)} = UB_o^q \tag{7}$$

Lemma 2 (Pruning Rule 1). The only object o that satisfies the following relation $disT(q,o) \leq \frac{TS}{(1-\alpha)} = UB_o^q$ can be a result of *CRSK-w2v* query q.

Proof. Let us assume an object o with $disT(q,o)$ greater than UB_o^q, i.e., $disT(q,o) > UB_o^q$. Then since both α and $disS(q,o)$ have not negative value, its score is over than threshold score, TS. By the Definition 1, the object o cannot be a result of a *CRSK-w2v* query q.

Before describing the second pruning rule, we explain index structures. In this paper, we use a grid-based data structure to index *CRSK-w2v* queries and an index structure: *Partition Retrieved List* (namely, *PRList*) like Fig. 5. In Fig. 5(a), the grid-based data structure consists of one root cell, intermediate cells and N^2 grid cells [8]. In the system, the root and intermediate cells do not physically exist and each grid cell has one *PRList*, i.e., there are N^2 *PRLists*. Every queries that issued in the system have their own *border box* as shown in Fig. 5(b). Each *PRList* contains all queries information whose largest conditional circle CC_q^* overlaps the partition of the grid cell. We will explain how to construct the index structures in Sect. 5.

Secondly, there is another pruning rule about a spatial distance. As with first pruning rule, in order to consider only spatial distance, assume that the textual distance have minimum value, $disT(q,o) = 0$. According to the previous assumption, Definition 1 and (5), we also define the upper bound of $disS(q,o)$ as follow.

$$disS(q,o) \leq \frac{TS}{\alpha} \tag{8}$$

In the proposed system, the location information is more useful to prune than textual information. In addition, the above upper bound of $disS(q,o)$ is too simple. Therefore, in this paper, other pruning rule is proposed that satisfies (8).

Lemma 3 (Pruning Rule 2). The only object o that satisfies the following relation $\exists\, partition\, p \in P_o^g \cap P_q^g$ can be a result of a *CRSK-w2v* query q.

Proof. Let us assume a situation like Fig. 5(b). The object o is inside a grid cell g_1 and third partition p_3 is in $P_o^{g_1}$, i.e., $P_o^{g_1} = \{p_3\}$. Also, both of the queries q_1 and q_2 overlap the grid cell g_1. However each query has different overlapped, $P_{q_1}^{g_1} = \{p_0, p_2\}$ and

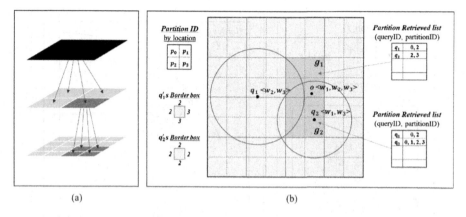

Fig. 5. N × N grid-based data structure (a) and an example of pruning rule 2 (b)

$P_{q_2}^{g_1} = \{p_2, p_3\}$. Since there is no intersection of the overlapped partitions between o and q_1, i.e., $P_o^{g_1} \cap P_{q_1}^{g_1} = \emptyset$, o cannot be the result object of q_1. On the other hand, since the partition p_3 overlaps between o and q_2, o can be the result object of q_2. As the example in Fig. 5(b), if the object o is in the same partition p of a grid cell g with CC_q^*, then o can be the result object of q. Moreover, since o is inside $CC_{q_2}^*$, the distance between o and q_2 is less than or equal $r_{q_2}^*$. By (6), this pruning rule also satisfies (8).

Lemma 4. A partition g should be marked as overlapped partition with a query q, if a distance d between the query and a point that is the closest with the query in the partition g, is less than the radius r_q^*.

Proof. Let assume that in a partition g, a point p is the closest with the query. If a distance d between a query q and the point p is greater than the radius r_q^*, i.e., $d > r_q^*$, then the point p is not inside CC_q^*. Since the point p is the closet point, the partition g cannot be marked as overlapped partition with the query q.

5   The Proposed Method

5.1   Framework

We discuss algorithms needed to construct a framework. As previously mentioned, we propose grid-based structure. Because the proposed system should track large amounts of data and will frequently update. it requires a data structure that can efficiently support these updates. The gird-based structure can support frequent location updates and give the better performance than the other index structure even if data are skewed [15]. The root and intermediated cells do not physically exist, therefore all data is stored in a grid cell, not root or intermediate cell. Each grid cell is divided into four parts again and each partition has unique ID from 0 to 3, as shown in the top of Fig. 6(a).

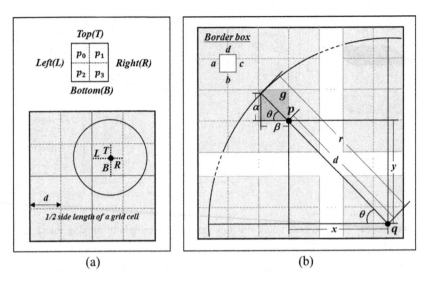

Fig. 6. Partition information of the grid cells (a) and how to construct the border box (b)

Border Box. In processing the *CRSK-w2v* queries, when a new query q is issued, a central server need to initialize the query. The Algorithm, INITIALIZEQUERY, which is an algorithm for initializing the query is proposed in [11]. Firstly, the server finds the distance to the left (L), right (R), Top (T) and Bottom (B) borders of the partition containing the query as shown in the bottom of Fig. 6(a). Then, by using a radius r_q^* of CC_q^*, it computes how many partitions are needed to the each end and construct *border box* (*BB*). Finally, it calls an algorithm, IDENTIFYMARKDIAGONAL that identifies and marks the four diagonal direction: left-top (*LT*), right-top (*RT*), left-bottom (*LB*) and right-bottom (*RB*). Let assume an example of identifying *LT* diagonal direction like Fig. 6(b). There is a border box, as shown in the top of Fig. 6(b). A point q is the query location and another point p is one of the focused points. In identifying a diagonal direction, to efficiently adopt pruning rule 2, this paper focuses on some points inside the CC_q^*.

$$x = (a - 1) \cdot d + L \text{ and } y = (b - 1) \cdot d + T \tag{9}$$

By using the border box, we can get the focused point in a similar way to (9). Since the *LT* diagonal direction is checked in the example, the point p is shifted by x and y from the point q to the left (L) and upward (T), respectively.

Lemma 5. In marking and identifying a left (or right) top (or bottom) diagonal direction, *if* a distance d between a query q and a point that is the closest with the query in the partition g, is less than the radius r_q^*, *then* right (or left) partition and bottom (or top) partition of g are also marked.

Proof. Let consider the example above, continuously. If the distance d between the point p and the point q is less than radius r, by the Lemma 4, a partition g is marked. Then we can find α and β and define these as follows.

$$\alpha = (r - d) \cdot \sin(\theta) \text{ and } \beta = (r - d) \cdot \cos(\theta) \tag{10}$$

Where r is greater than d and θ is acute angle. Therefore, since $(r - d)$, $\sin(\theta)$, and $\cos(\theta)$ are positive, α and β are also positive. From this, the right partition and the bottom partition of g are also identified as overlapped partitions. As a similar way, the other diagonal ends can be identified.

5.2   *CRSK-w2v* Queries

In processing *CRSK-w2v* Queries, there are five major algorithms. Two algorithms are as mentioned in the previous Sect. 5.1, one of which is INITIALIZEQUERY for initializing a new issued query. Whenever a query is issued at a central server, the server performs INITIALIZEQUERY, and during this process, *PRList* are updated. Another one is IDEN-TIFYMARKDIAGONAL that identifies and marks four diagonal direction. To efficiently apply the pruning rule 2, we need to construct *PRList*. However, since constructing *PRList* is also costly, we need to use the technique to decrease unnecessary checking. By looking for the outermost overlapped partitions of the circle, the algorithm IDEN-TIFYMARKDIAGONAL helps to reduce unnecessary accesses.

The other two algorithms are about evaluating the objects. In the proposed system, all issued queries are indexed and static. On the contrary, the clients (i.e., the objects) are moving. Additionally, whenever a new object arrives at the server or whenever the object leaves out their *BR*, the server (re)evaluates the object, and (re)compute safe region of the object and (re)sends safe region to the object. In Table 2, an Algorithm 1, EVALOBJECTS is a pseudocode of evaluating the objects, similar to that proposed in [8]. Firstly, when an object *o* comes to the central server, it forms an empty priority queue (line 1) and inserts root cell of the grid-based structure into the queue (line 2). Then, pull out the elements one by one from the queue, and repeat the evaluation process until *NQList* are formed or all the elements of the queue are verified (lines 3–21). More specifically, if the extracted element is an intermediate cell or a root cell, then put its sub cells back into the queue (lines 5–7). Else if the element is a grid cell, firstly, by calling a function *FindQ(object, PRList)* to apply pruning rule 2, find the candidate queries (line 9). Then about each candidate query, it gets similarity between query keywords and object keywords, and applies pruning rule 1 (lines 10–11). At this time, the similarity between the keywords are evaluated based on Word2Vec. Additionally, the server identifies if the object can be a result of the query, and puts the query back into the priority queue (lines 13–17). Else if the extracted element is a query, insert the query into *NQList* (lines 18–21). Finally, when *NQList* is completely constructed, it computes *BR* (line 22) of the object and sends BR_o to the corresponding object. Another algorithm is MONITOROBJECTS and it is for continuously monitoring objects. When the object's new location is in *BR*, the server checks only the queries in *NQList*. However, if the object leaves out *BR*, the server calls EVALOBJECTS. Since the algorithm is too simple and similar to the algorithm proposed in [8], thus, the details in this paper are omitted.

Table 2. Algorithm 1: EVALOBJECTS

Algorithm 1 : EVALOBJECTS	
INPUT object o, Grid-index structure G	
OUTPUT RS_q^t for all $q \in Q$, BR_o	

1	Initialize an empty priority queue PQ	
2	PQ.Enqueue($G.root$, 0)	// key $\leftarrow$ 0
3	**While** PQ **NOT** Empty **do**	
4	$element \leftarrow PQ$.Dequeue()	
5	if $element$ is an intermediate cell or a root cell **then**	
6	**foreach** $child \in element$ **do**	
7	└ PQ.Enqueue($child$, $mindist(child, o)$)	// key $\leftarrow mindist(child, o)$
8	**else if** $element$ is a grid cell **then**	
9	$Candidate Q_o = FindQ(o, PRList)$	// apply pruning rule 2
10	**foreach** $q \in Candidate Q_o$ **do**	
11	if $disT(q,o) \le UB_o^q$ **then**	// apply pruning rule 1
12	Compute CC_o^q	
13	if $mindist(q,o) \le r_o^q$ **then**	// the object is inside the CC_o^q
14	| Insert o into RS_q^t	
15	**else**	// the object is outside the CC_o^q
16	└ Delete o from RS_q^t	
17	└─ └ PQ.Enqueue(q, $mindist(q, o)$)	// key $\leftarrow mindist(q, o)$
18	**else**	// $element$ is a $query$
19	$NQList_o$.insert($element$, $element.key$)	
20	if sizeof($NQList_o$) > $o.cap+1$ **then**	// at most $o.cap+1$ queries in $NQList_o$
21	└ └ └ break	
22	$ConstructBR(NQList_o)$	// construct BR_o by using $NQList_o$

Finally, an algorithm, IDENTIFYSAFEREGION is performed on the client-side. Each object has safe region information received from the server. Therefore, whenever the object moves, the object checks if it leaves out the safe region. If it is out of the safe region, the object informs the server and asks for updates. Then, the object receives new safe region information back. As these above algorithms are repeated continuously, *CRSK-w2v* queries are performed.

6   Performance Evaluation

6.1   Simulation Setup

Because this is the first work for processing *CRSK* Queries based on Word2Vec, we develop two naïve algorithms to compare with our proposed algorithm (*Our*). Firstly, a naïve algorithm (*naïve-w/o*) only considers the exactly matching between the keywords. Therefore, when an object arrives at the server, *naïve-w/o* algorithm evaluates the object without considering the relevance of the keywords. Another naïve algorithm (*naïve-w2v*) evaluates objects based on Word2Vec. But in *naïve-w2v*, there is no constructing *PRList*. Therefore, when a new query is issued to a central server and when a request from an objects arrives at the server, the server evaluates without *PRList*.

Dataset. In the evaluation, a dataset is virtual, but it is based on the real world. In the dataset, each of data has own attributes: spatial attribute and textual attribute. Firstly, the spatial attribute is based on San Francisco and it is made from [18]. Another attribute, textual attribute consists of information from real stores in San Francisco, provided by *Yelp*. Through the data, we configure a dataset for processing *CRSK-w2v* queries and moving objects. *CRSK-w2v* queries dataset has default value, 10K, and it changes from 5K to 20K. Moving objects dataset has default value, 100K, and it changes from 50K to 200K. Additionally, all moving object data are tracked for 100 timestamps.

Parameter. In the evaluations, each parameter has default value as shown in Table 3.

Table 3. Default value of parameters

Parameter	Default	Meaning
$o.cap$	3	The maximal computational capacity of the object o
N	2	Grid size N × N
TS	0.5	Threshold score of a query
α	0.5	Ranking parameter factor

6.2 Simulation Result

In this section, we evaluate our algorithm for processing *CRSK-w2v* queries by comparing with two naïve algorithms: *naïve-w/o* and *naïve-w2v*. The performance metric is continuous time, which is the sum of times consumed each time the server is updated.

Effect of Using Word2Vec. We evaluate the effects of using Word2Vec by comparing *our* with *naïve-w/o*. We vary the number of queries from 5K to 20K as shown in Fig. 7. Figure 7 shows the benefits of using Word2Vec. Firstly, since *our* checks not only the exactly matching but also the similarity, *our* can get more accurate results. Furthermore, since the query and the object keyword are highly likely to be associated with each other, the query result is maintained for a long time and thereby the total amount of computation is reduced. This is expected to be effective when the ranking score is obtained in (5), if the weight of textual attribute is set high (i.e., when α is small).

Effect of the Number of Queries and Objects. We evaluate the effects of the numbers of queries and objects by comparing *our* with *naïve-w2v*. We vary the number of queries and objects from 5K to 20K and from 50K to 200K, respectively, as shown in Fig. 8. In Fig. 8, *naïve-w2v* has lower performance than *our*. Hence, the role of *PRList* to reduce total computation is significant.

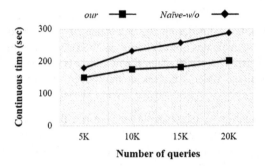

Fig. 7. Effect of using Word2Vec

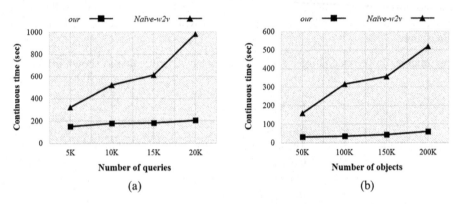

Fig. 8. Effect of the number of the queries (a) and the objects (b)

7 Conclusion

In this paper, we propose a new algorithm for processing Continuous Range Spatial Keyword Queries based on Word2Vec over moving objects in Euclidean space. The proposed algorithm emphasizes (i) obtaining the relationship between keywords based on Word2Vec and (ii) minimizing the communication costs, and construction and maintenance costs of the safe regions. However, the constructions of safe region are also costly. Therefore, in the paper, to decrease these costs, we introduce the concept of *buffer region* and the pruning rules. Moreover, the proposed method includes a spatial index structures, *Partition Retrieved list* whose role is to help the system quickly construct the safe region. Through experimental evaluations, we demonstrate the usefulness of using Word2Vec and effectiveness of constructing *Partition Retrieved list*. As for the future works, we plan to improve our work to support more complex situations where the keywords are changed, beyond the analysis of similarity between keywords.

Acknowledgement. This research was supported by Basic Science Research Program through the National Research Foundation of Korea (NRF) funded by the Ministry of Education (NRF-2015R1D1A1A01057238, NRF-2016R1D1A1B03931098).

References

1. Rathore, R.: Spatial Range Query
2. Prabhakar, S., Xia, Y., Kalashnikov, D.V., Aref, W.G., Hambrusch, S.E.: Query indexing and velocity constrained indexing: scalable techniques for continuous queries on moving objects. IEEE Trans. Comput. **51**(10), 1124–1140 (2002)
3. Roussopoulos, N., Kelley, S., Vincent, F.: Nearest neighbor queries. ACM SIGMOD record **24**(2), 71–79 (1995)
4. Papadias, D., Tao, Y., Mouratidis, K., Hui, C.K.: Aggregate nearest neighbor queries in spatial databases. ACM Trans. Database Syst. (TODS) **30**(2), 529–576 (2005)
5. Guo, L., Shao, J., Aung, H.H., Tan, K.L.: Efficient continuous top-k spatial keyword queries on road networks. GeoInformatica **19**(1), 29–60 (2015)
6. Huang, W., Li, G., Tan, K.L., Feng, J.: Efficient safe-region construction for moving top-k spatial keyword queries. In: Proceedings of the 21st ACM International Conference on Information and Knowledge Management, pp. 932–941. ACM (2012)
7. Wu, D., Yiu, M.L., Jensen, C.S., Cong, G.: Efficient continuously moving top-k spatial keyword query processing. In: 2011 IEEE 27th International Conference on Data Engineering (ICDE), pp. 541–552. IEEE (2011)
8. Salgado, C., Cheema, M.A., Ali, M.E.: Continuous monitoring of range spatial keyword query over moving objects. World Wide Web **21**, 1–26 (2017)
9. Cheema, M.A., Brankovic, L., Lin, X., Zhang, W., Wang, W.: Multi-guarded safe zone: an effective technique to monitor moving circular range queries. In: 2010 IEEE 26th International Conference on Data Engineering (ICDE), pp. 189–200. IEEE (2010)
10. Hu, H., Liu, Y., Li, G., Feng, J., Tan, K.L.: A location-aware publish/subscribe framework for parameterized spatio-textual subscriptions. In: 2015 IEEE 31st International Conference on Data Engineering (ICDE), pp. 711–722. IEEE (2015)
11. Oh, S., Jung, H., Kim, U.M.: An efficient processing of range spatial keyword queries over moving objects. In: 2018 International Conference on Information Networking (ICOIN) (in press)
12. Mikolov, T., Chen, K., Corrado, G., Dean, J.: Efficient estimation of word representations in vector space. arXiv preprint arXiv:1301.3781 (2013)
13. Mikolov, T., Yih, W.T., Zweig, G.: Linguistic regularities in continuous space word representations. In: HLT-NAACL, vol. 13, pp. 746–751 (2013)
14. Cong, G., Jensen, C.S., Wu, D.: Efficient retrieval of the top-k most relevant spatial web objects. Proc. VLDB Endow. **2**(1), 337–348 (2009)
15. Kalashnikov, D.V., Prabhakar, S., Hambrusch, S.E.: Main memory evaluation of monitoring queries over moving objects. Distrib. Parallel Databases **15**(2), 117–135 (2004)
16. Guttman, A.: R-trees: a dynamic index structure for spatial searching. ACM SIGMOD Rec. **14**(2), 47–57 (1984)
17. Hu, H., Xu, J., Lee, D.L.: A generic framework for monitoring continuous spatial queries over moving objects. In: Proceedings of the 2005 ACM SIGMOD International Conference on Management of Data, pp. 479–490. ACM (2005)
18. MNTG: Minnesota Web-based Traffic Generator. http://mntg.cs.umn.edu

Credit Risk Analysis of Auto Loan in Latin America

Yukiya Suzuki[✉]

Department of Management Systems Engineering,
Tokai University, Tokyo, Japan
7bjnm015@mail.u-tokai.ac.jp

Abstract. Latin American economy has achieved rapid economic growth since 2003 after undergoing currency crisis and economic turmoil from the early 1930s through until the early 1960s. As a result of this growth, consumer finance market expanded due to external demand in country X in Latin America. In addition, because part of the poor class has changed into the middle class, financial services have also spread to those who could not formulate loans in the past. On the other hand, there are many debtors who do not understand the contents of the loan contract, and the debt default due to excessive debt is increasing. Therefore, in order for Latin American financial institutions, manufacturers and retailers to operate stably, it is necessary to measure the credit risk of the debtor who formed the loan and to grasp the factors that affect the default. The purpose of this research is to construct a credit risk model based on 14,000 auto loan data in country X in Latin America and to estimate debtor's default probability. The binomial logit model was adopted as a usage model in this research. Estimate default probabilities for debtors who defaulted within one year using the model and grasp default situations in Latin America. From the analysis results, it became clear that the presence or absence of marriage greatly influences the default. It is thought that the debtor who got married and had a family became easier to default because expenditure is larger than the debtor who does not have a family. In addition, when there are missing values in the income and down payment items, the debtor got the result that it is easy to default. Since information on auto loans is often written by the debtor himself, it was suggested that debtors who do not fill in information correctly have a stronger possibility of default.

Keywords: Latin America · Auto loan · Credit risk model
Binomial logit model · ROC curve and AUC

1 Introduction

1.1 Background and Purpose

Latin American economy has achieved rapid economic growth since 2003 after undergoing currency crisis and e turmoil from the early 1930s through until the early 1960s. As a result of this growth, consumer finance market expanded due to external demand in country X in Latin America. In addition, because part of the poor class has

© Springer International Publishing AG, part of Springer Nature 2018
S. Yamamoto and H. Mori (Eds.): HIMI 2018, LNCS 10905, pp. 640–647, 2018.
https://doi.org/10.1007/978-3-319-92046-7_52

changed into the middle class, financial services have also spread to those who could not formulate loans in the past. On the other hand, there are many debtors who do not understand the contents of the loan contract, and the debt default due to excessive debt is increasing. For this reason, financial institutions that are lenders of funds in Latin America and manufacturers and retail stores are required to measure the credit risk of the debtor who formed the loan and to devise credit judgment based on the size of the risk.

Many financial institutions are doing business that takes interest from debtors by financing. Not only financial institutions, but also makers and retailers selling products on loans is not uncommon. Credit risk is the risk that losses will be incurred due to a decrease or loss of the value of the loaned funds due to deterioration of the financial situation of the debtor. The state that the debtor does not return the borrowed funds and the repayment is delayed or stopped is called default. If the debtor defaults, the likelihood that returns will be returned to the lender is almost nil. From the above, accurately grasping the credit risk becomes an important issue in order to carry out stable management. Financial institutions, primarily banks and others, may do ratings to determine whether to lend or lend to debtors. This is called the Internal Rating System. With the introduction of the new BIS regulation by the Basel Committee on Banking Supervision [1], the establishment and validation of an appropriate internal rating system is an issue.

In order to quantitatively grasp credit risk, in many cases, a Credit Risk Model (there are other names such as Credit Scoring Model etc.) based on statistics is constructed. A typical model is a binomial logit model. A binomial logit model can estimate the default probability with default/non-default binary as the objective variable. The model is generally used in a real environment that measures credit risk because it has a high ability to explain short-term defaults and is comparatively easy to calculate.

The purpose of this research is to construct and evaluate a credit risk model based on 14,000 car loan data in Country X of Latin America. Understand the characteristics of debtors who defaulted within one year based on the analysis results and grasp the situation of default in Latin America. The significance of research on applying internal logic of internal rating performed at financial institutions to auto loans and focusing on Latin America where the number of default occurrences has increased is significant. The binomial logit model was adopted as a usage model in this research. Estimate default probabilities for debtors who defaulted within one year using the model and grasp default situations in Latin America. Research on the credit risk model tends to find financial indicators that affect bankruptcies in many cases, and there are still few researches of default probability estimation for individual debtors. Above all, no research on auto loans exists to the best of my knowledge. Due to the small number of research cases, we decided that credit risk measurement by a general binomial logit model is appropriate rather than using complicated models from the beginning in this research. In addition, AUC was used as an evaluation index of model accuracy. As with the binomial logit model, AUC is also an evaluation index generally used at the site of credit risk management.

1.2   Previous Research

Many prior studies on credit risk models are found, but many of them are research cases for companies. Since the subject of analysis in this study is for individuals, previous studies on individual debtors are described below. However, all of the following are studies on Japanese data in Japan, and similar results can not be obtained for Latin American data. Pay attention to this fact and use the results of previous research as reference for this research.

- Hibiki et al. [4]

It is a research that constructed a credit scoring model using about 350,000 educational loan data in Japan. Hibiki uses a logistic regression model and evaluates the model with AR. Education loans are loans that mainly finance educational funds for admission or advancement to student parents in Japan. Users of education loans are stable regardless of observation period of data, such as age distribution and income level, indicating that there was no significant difference in parameters even if changing the lending year or variable combination. Furthermore, AR verification with in-sample and out-of-sample data showed that AR did not decrease so much and revealed that a model with less overfitting could be constructed. Research results show the usefulness of the credit scoring model in the education loan. However, from the practical viewpoint, in order to avoid danger such as spoofing declaration, the influence of concrete variables is hidden.

- Okumura and Kakamu [2]

Credit Scoring Models Using Hierarchical Bayes Model: An Application to Inter-bank Consortium Mortgage Data.

It is a research that constructed a credit scoring model using Japan's Inter-bank Consortium data. Using a hierarchical Bayesian probit model, it is doing the verification of whether to improve the estimation accuracy than the existing model. The ROC curve and AUC are used for the model evaluation. To build a scoring model with high accuracy, securing default samples is a key point. However, the use of pooled data by multiple banks assumes homogeneity between banks, and the results obtained represent only the average features. As a countermeasure to that problem Okumura and Kakamu focused on regionality and revealed the differences between the regions relatively easily by using a hierarchical model. When summarizing the verification results, it turned out that there is a difference in the default for each region. Furthermore, it became clear that the hierarchical Bayesian model improves the performance over the existing model.

Examples of explanatory variables that have been commonly used for credit risk models in past projects [6] other than the above research are shown.

Items that are regarded as important in credit screening are repayment and loan ratio, transaction history with financial institutions, etc. Since it is possible to comprehensively judge the burden of repayment on income and the saving trend of debtors, generally there is a strong relationship with the default.

The explanatory variable in the revenue assessment item to measure the stability of income is the type of business of the debtor, the type of occupation, the size of the company and the number of years of service. Since the source of the stability of the income of the debtor is the stability of the workplace, it is generally an item that has a strong relationship with the default.

The spending assessment items that measure the degree of maintenance of repayment resources from the viewpoint of the opposite of revenue assessment are the dependents of the debtor, the number of children, the age of the child and so on. However, since items related to expenditure are relatively few compared to items related to income at the stage of application form, data to be analyzed also becomes few. For this reason, there are a lot of room for research on expenditure assessment items, which is an item with high analytical value.

Besides the information at the present moment, the debtor's deposit balance and card loan outstanding are examples of time-dependent variables. These can be expected not only at the time of application but also strong explanatory power as a variable which varies with passage of time. Especially periodically checking the trend of the deposit balance and renewing the credit risk of the debtor according to the balance up and down is a useful method as part of the ongoing management of credit risk.

In addition to that, variables such as gender and residential area of the debtor may be used for model construction, but it does not always affect creditworthiness.

1.3   Paper Composition

This paper composition is below shown. Topic 1, the background and purpose in this research, and the related prior research are described. Topic 2, describe the outline of data used for analysis. Topic 3, describe the credit risk model to be used and its evaluation method. Topic 4, describe the result of analysis and discussion on the results. Topic 5, describe summary of this research and future work.

2   Summary of Date

This topic describes the summary of the auto loan data to be used in the analysis.

2.1   Summary of Date

The data used in this research is customer data purchasing products in a loan in country X in Latin America. The data period is from September 1, 2010 to June 30, 2012. The customer target is 14,304 people living in X.

The data items are age, gender, presence of marriage, presence or absence of regular work, presence or absence of educational history, existence of owning house, income (monthly income), down payment of loan, borrowing money, interest rate, default (whether it defaulted within 1 year from purchase). In addition to these, a repayment capacity index obtained by dividing borrowings by income and down payment ratio obtained by dividing down payment by product price were created as new items. Furthermore, when there was a missing value in the item, a missing flag (1 if there was a missing, 0 if not) was created as a new item for that item. Then, 0 was substituted for the original missing value. Items for which missing flags are created are gender, repayment capability index, down payment ratio, interest rate. The above items are used as explanatory variables of the binomial logit model. The objective variable is the default within one year.

3 Method of Analysis

This topic describes the summary of the credit risk model, the binomial logit model to be used and the ROC curve and AUC as model's evaluation index. The software used for analysis is IBM$^@$ SPSS$^@$ Statistics (ver. 22).

3.1 Summary of Credit Risk Model

The credit risk model is roughly divided into a statistical model and an option approach model. In the statistical model, default judgment and default probability estimation are performed based on the financial data of the debtor. Typical models are discriminant analysis, logit model, hazard model. In general, as the number of data increases, it is possible to create models with high explanation. Therefore, it is often applied to small and medium enterprises with abundant data. The option approach model estimates the default probability using market data such as stock price and corporate bond interest rate. A typical model is a Merton model, which can estimate the default probability in real time by obtaining market data for large companies with stocks listed. However, the option approach model is not a rating, it is often used as a monitoring tool for listed companies. Besides that, models such as neural networks and support vector machines are increasingly applied as a constructive and experimental model.

The reason for adopting the binomial logit model in this research is ease of application to analysis data. Discriminant analysis and probit model are models having properties similar to the logit model. However, since these models are difficult to assume distribution, it is difficult to apply them to actual data. Unlike logit models, the hazard model is a model to explain relatively long-term defaults. Since this research builds a model against the default within one year, it can be judged that the binomial logit model is appropriate.

3.2 Binomial Logit Model

The binomial logit model used in this research is shown below.

$$P = \frac{1}{1 + exp(Z)} \, (0 < P < 1) \tag{1}$$

$$Z = ln\left(\frac{1-P}{P}\right) = \alpha + \sum_{i}^{n} \beta_i x_i (i = 1, \ldots, I) \tag{2}$$

At (1), P is the probability of defaulting within one year (the value ranges from 0 to 1). At (2), Z is log odds. When i is the debtor, I is the number of debtors, and n is the number of explanatory variables, x_i is the explanatory variable used for the model and α and β_i are the parameter estimated by the maximum likelihood method. The default probability decreases as Z increases.

Maximum likelihood method assumes that facts $(y_i = \{0, 1\})$ showing default/non-default are independent, the likelihood function is given by (3).

$$L = \prod_{i=1}^{I} P^{y_i}(1 - P)^{1-y_i} \tag{3}$$

In order to estimate the parameter that maximizes (3), the expression that maximizes the log-likelihood function lnL is (4).

$$lnL = \sum_{i=1}^{I}\{y_i lnP + (1 - y_i)ln(1 - P)\} \tag{4}$$

3.3   ROC Curve and AUC

ROC Curves (Receiver Operating Characteristic Curve) and AUC (Area Under the Curve) are one of methods widely used as a method of evaluating the credit risk model. This index is most frequently used alongside AR (Accuracy Ratio) in the credit risk model. ROC Curve is a curve obtained by finding the error probability from the default predicted probability and the default/non-default judgment result. AUC defined by the area under the curve is an index to measure the accuracy of the model. If the model has no explanatory power and the default occurs randomly, AUC is 0.5. AUC in the case of completely default prediction is 1. In other words, AUC can be judged to be a highly explanatory model as it is closer to 1.

$$0.5 \leq AUC \leq 1$$

4   Results

4.1   Results of Binomial Logit Model

As a result of executing the binomial logit model, only the explanatory variables whose significance probability was lower than 5% are shown in the Table 1.

Table 1 shows, it became clear that the presence or absence of marriage greatly influences the default. It is thought that the debtor who got married and had a family became easier to default because expenditure is larger than the debtor who does not have a family. In addition, when there are missing values in the income and down payment items, the debtor got the result that it is easy to default. Since information on auto loans is often written by the debtor himself, it was suggested that debtors who do not fill in information correctly have a stronger possibility of default. On the other hand, the repayment capacity index which seemed to be most related to the default did

Table 1. Result of executing the binomial logit model

Explanatory variable	Log odds	Significance probability
Intercept	0.486	0.026
Gender	−0.364	0.000
Age	0.023	0.000
Marriage	0.75	0.000
Missing flag of repayment capability index	0.453	0.005
Owning house	0.194	0.002
Down payment ratio	0.013	0.000
Missing flag of down payment ratio	0.471	0.001
Interest rate	0.01	0.000
Missing flag of interest rate	−0.427	0.001

not show significant results. The possible cause is that there are about 1,000 data whose index is almost 0, so that data had a bad influence on the result.

The value of AUC was 0.625. Since the value of AUC takes a value from 0.5 to 1, and the closer to 1, the accuracy of the model can be evaluated as high, the model of this research could not show high accuracy.

5   Conclusions and Future Work

5.1   Conclusions and Discussion

The purpose of this research is to build a binomial logit model for loan data against the background of the default increase problem in Latin America and to grasp the characteristics of the debtor in Latin America. As a result, it became clear that the married debtor and debtor whose information is missing are easy to default. At the time of credit judgment, debtors who have a family by marriage pay particular attention to the income and expenditure situation. In addition, efforts are needed to judge whether the necessary items are accurately entered. However, since the value of AUC indicating the accuracy of the model was never a high numerical value, there is a need to reconsider the combination of variables and the model to be used.

5.2   Future Work

For future work of this research, abnormal value processing and data segment can be considered. In this research, since analysis was performed using all data, distribution of each item became complicated, and it is considered that the accuracy of the model worsened. It is thought that complicated distribution is responsible for the fact that repayment capacity index did not result in significant results. For this reason, firstly, abnormal values are excluded or converted to eliminate distribution bias. Secondly, by segmenting data by age or region, the instability of the model due to the difference in distribution can be solved. Finally, execute the binomial logit model for each segment, and make the debtor's default factor clearer. However, since there are only 14,000 data to be used, the number of pieces of data for each segment decreases, and the analysis result may be worse. In addition, since overfitting may occur, segmentation will thoroughly consider the appropriate number of segments.

References

1. Basel Committee on Banking Supervision: Studies on the validation of internal rating system (revised) In: Working Paper No. 14 (2005)
2. Hiroshi, O., Kazuhiko, K.: Credit scoring models using hierarchical bayes model: an application to inter-bank consortium mortgage data. J. Jpn. Stat. Soc. J. **42**(1), 25–53 (2012)
3. Takahashi, H., Yamashita, S.: Estimation of probability of default using credit risk database. In: Proceedings of the Institute of Statistical Mathematics, vol. 50, no. 2, pp. 241–258 (2002)
4. Hibiki, N., et al.: Education loan's credit scoring model. Jafee J. Finan. Eng.
5. Yamashita, S., et al.: Consideration and comparison of credit risk model evaluation method. Financial Research and Training Center Discussion Paper Series, vol. 11 (2003)
6. Muromachi, Y.: Financial Risk Modeling - Approach to Theory and Critical Issues. Asakura Shoten Co., Ltd. (2014)

Analysis and Consideration of the Relationship Between Audience Rating and Purchasing Behaviors of TV Programs

Saya Yamada[1](✉) and Yumi Asahi[2]

[1] School of Information and Telecommunication Engineering,
Course of Information Telecommunication Engineering, Tokai University,
Tokyo, Japan
yamada.s.3849@gmail.com
[2] School of Information and Telecommunication Engineering,
Department of Management System Engineering, Tokai University,
Tokyo, Japan
asahi@tsc.u-tokai.ac.jp

Abstract. Chocolate has been increasing in Japan domestic consumption. It is eaten regardless of the age from the adult to the child. Variety is abundant, it is one of convenient and familiar sweets for Japanese people. There are many days when we watch television on a confectionery company that sells such chocolate on TV. But, in Japan people who watch television are getting less. According to the Ministry of Internal Affairs and Communications data of 2013, the average viewing time of real-time TV on weekdays was 168.3 min [1]. The weekday average for the last year was 184.7 min. It was 16.4 min (about 9%) decrease. The average for 2011 (including holiday data) was 228.0 min. It was found to be 59.7 min difference. The total advertising expenditure in Japan totaled 5,769.6 billion yen in 2011. Terrestrial broadcasting usage is 1,723.7 billion yen (about 30.2%) [3]. Total in 2013 is 5,976.2 billion yen. Terrestrial broadcasting usage is 1,719.1 billion yen (about 30.0%). The total of advertising expenditure increased slightly in two years. It does not change that it occupies much of advertisement expenditure as present condition. We can think that Japanese television is still the main information dis-semination medium. Therefore, even if the TV viewing went down, we clarify whether there is a relationship with purchasing. We revealed it by decision tree analysis.

Keywords: Evaluating information · Marketing · TV audience rating
Decision tree analysis

© Springer International Publishing AG, part of Springer Nature 2018
S. Yamamoto and H. Mori (Eds.): HIMI 2018, LNCS 10905, pp. 648–657, 2018.
https://doi.org/10.1007/978-3-319-92046-7_53

1 Introduction

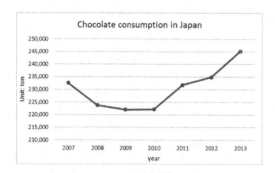

Fig. 1. Chocolate consumption trends

Chocolate has been increasing in Japan domestic consumption. It is eaten regardless of the age from the adult to the child. It is sold at supermarkets and convenience stores. People easy to get. It has a wide range of prices and there are many variations.

Therefore, it is one of the simple and familiar sweets for the Japanese. We often watch commercials for confectionery companies that selling chocolates. Their companies are advertising TV as one of public relations activities. But, in Japan people who watch television are getting less. According to the Ministry of Internal Affairs and Communications data of 2013, the average viewing time of real-time TV on weekdays was 168.3 min [1]. The weekday average for the last year was 184.7 min. It was 16.4 min (about 9%) decrease. The average for 2011 (including holiday data) was 228.0 min. It was found to be 59.7 min difference.

We are watching over average value on weekdays only in 60's. The holidays are the 60s and 50s. Teenagers and twenties are the most unviewed. On weekdays it was about 40 to 70 min shorter than the average and holidays 60 to 90 min shorter. According to the data surveyed by the Risky Brand, the TV average viewing time on weekdays was almost flat in the 30 to 49 years old and 50 to 64 years old generation between 2008 and 2013 [2]. The middle-aged and older age is flat from two data. It became that young people do not watch TV. The total advertising expenditure in Japan totaled 5,769.6 billion yen in 2011. Terrestrial broadcasting usage is 1,723.7 billion yen (about 30.2%) [3]. Total in 2013 is 5,976.2 billion yen. Terrestrial broadcasting usage is 1,719.1 billion yen (about 30.0%). The total of advertising expenditure increased slightly in two years. It does not change that it occupies much of advertisement expenditure as present condition. We can think that Japanese television is still the main information dissemination medium. Therefore, even if the TV viewing went down, we clarify whether there is a relationship with purchasing. We revealed it by decision tree analysis.

2 Usage Data

We used attribute data, purchase process data and TV viewing survey data. Their period is from September to October 2011 and from September to October 2013. We used to factor that gender, age, married or unmarried form attribute data. We also created and used the age-based variables. We used questionnaire data on purchase status of September and October regarding chocolate products (Meiji milk chocolate, Lotte milk chocolate, Morinaga confectionery dozen, Meiji almond chocolate, Meiji macadamia nut chocolate) common to both years from the purchasing process data.

TV audience survey data is for 8 weeks from August 27, 2011, to October 22, 2011, and for 8 weeks from August 31, 2013, to October 30, 2013. There are seven main channels in the Kanto area (Tokyo, Kanagawa, Chiba, Saitama, Tochigi, Gunma and Ibaraki) where data was collected. But two of them are public broadcasting. On that account, we excluded them from the analysis. Then, we squeezed from 5 pm to 11 pm and analyzed. Because it is regardless of the day of the week that there are time zones that are seen frequently [1]. In addition, the viewing data include "CM contact data". "CM contact data" is the number of CM contacts within each period for each sample calculated based on the commercial data for all programs. Even if there are CM advertisements more than once in the same program, it counts as one. CM contact data of chocolate products (Lotte milk chocolate, Morinaga confectionery dozen, Meiji almond · macadamia nut chocolate) whose data are taken in both years are also used.

3 Data Summary

First, we added up each data as the basic statistics from the attribute data. The total number, gender ratio, age, and married or unmarried are as follows.

The data of September to October 2011, the total number is 3109. Gender ratios are 51.1% for males and 48.9% for females. There is almost the same number. The age composition is 1.2% for teens, 20.4% for 20s, 29.7% for 30s, 21.6% for 40s, 24.9% for 50s and 2.3% for 60s. 60.3% of respondents were married, 34.1% were unmarried, and others were 5.7% (Fig. 2). On the second data of September to October 2013, the total number is 3146. Gender ratios are 50.7% for males and 49.3% for females. There is almost the same number. The age composition is 0.7% for teens, 19.7% for 20s, 27.7% for 30s, 25.2% for 40s, 21.7% for 50s and 5.1% for 60s. 57.2% of respondents were married, 37.0% were unmarried, and others were 5.8% (Fig. 3).

Regarding residential areas of respondents, seven prefectures in the Kanto region including Tokyo, the capital area of Japan. With respect to the age, teenagers and 60s have extremely few data compared to other ages. We can not analyze sufficiently. In this report, we analyzed without using data of teens and 60s. Thus, the number of data from September to October 2011 is 3000, and from September to October 2013 is 2964. In each item, it is possible to compare the two data because there is no big difference or bias between them.

From the TV viewing survey data, the total number of programs by day of the week is Table 1. Usually, in Japan, Monday to Friday are weekdays and Saturdays and Sundays are holidays. Both years have many programs on holiday. Regarding the classification of the program, we referred to the "category" of the TV program recording program [5, 6]. The total of program categories is shown in Table 2. Because there was a program with multiple categories, the total number of programs by day of the week and each category total are not equal. Also, the analysis uses six categories above the double line. The reason is that there are more than a certain number of watching. We explain the program category. "Variety Shows" is a show program such as songs, skits, dances, narratives [7]. "Documentary" is made based on actual record without using fiction [8]. "TV Gossip Show" is Information entertainment program including sensational incident. It consists of several corners; the moderator speaks to the guest and progress the program [7]. "Sports" are broadcast programs related to sports such as a game relay. "Hobbies/Education" is a broadcast for school education or social education. And those that directly aim at the improvement of general public education [9].

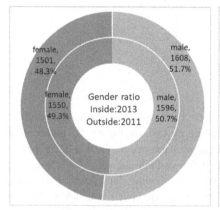

Fig. 2. Gender ratio

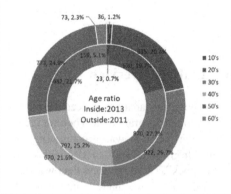

Fig. 3. Age ratio

Table 1. Number of programs by day of week and total

	2011/9–10			2013/9–10		
	Total	Per day	Channel per day	Total	Per day	Channel per day
Monday	343	42.9	8.6	320	40.0	8.0
Tuesday	337	42.1	8.4	328	41.0	8.2
Wednesday	357	44.6	8.9	336	42.0	8.4
Thursday	368	46.0	9.2	360	45.0	9.0
Friday	363	45.4	9.1	356	44.5	8.9
Saturday	459	57.4	11.5	452	56.5	11.3
Sunday	477	59.6	11.9	487	60.9	12.2
Average	386.3	48.3	9.7	377.0	47.1	9.4
Total	2704	338.0		2639	329.9	

Table 2. Number of programs in each category

	2011/9–10	2013/9–10
Variety show	896	909
Documentary	739	752
NEWS	701	680
TV Gossip Show	535	447
Sports	338	281
HobbiesEducation	249	243
TV drama	189	187
Animation	165	187
Music	81	59
Movies	24	21
Theater performance	9	14
Welfare	5	8
Other	17	6

4 Analysis Overview

We explain the analysis procedure of this paper. First, we gave the category information to the program data. Then we created a dummy variable for the day and time on which the program was broadcasting. For example, for programs that were broadcasting from 17 o'clock to 19 o'clock, we set 1 for variable names "17 o'clock" and "18 o'clock". Next, program data was combined with viewing data. And we compiled it. Furthermore, we made it into a variable which the viewing frequency within the period. For category data and broadcast time data were created variables like Table 3. As for the data on the days of the week were cleared variable like Table 4.

In the questionnaire data, we created the purchasing process data again as shown in Table 5. For those who purchased Meiji Milk Chocolate, Lotte Ghana Milk Chocolate or Morinaga Darth in September or October, those who did not purchase 1 or 1° were 0.

Table 3. Variable and interpretation of variables by frequency (category, broadcast time data)

Minimum value	Maximum value	Variable	Definition
0	4	0	0 or less per week
5	15	1	About 2 times a week
16	31	2	About 3 times a week
32	55	3	About 4 or 6 times a week
56	or more	4	Atleast 7 times a week

Table 4. Variable and interpretation of variables by frequency (day data)

Minimum value	Maximum value	Variable	Definition
0	4	0	0 to 1 program
5	12	1	1 to 2 programs
13	20	2	3 programs
21	27	3	4 programs
28	44	4	5 to 6 programs
45	or more	5	7 or more programs

Table 5. Reprinting questionnaire data on purchase actual condition

Purchasing investigation of Meiji milk chocolate, Lotte Ghana milk chocolate, Morinaga Dozen		Did the customer buy a chocolate item in September or October	
Variable	Variable name	New variable	New variable name
1	The customer bought it only once	1	The customer have bought it
2	The customer bought it many times		
3	The customer did not purchase it, but the customer saw it at the store	0	The customers have never bought it
4	The customer did not see it in the store, but the customer knows the name		
5	The customers do not know		

5 Analysis Result

Table 6. Common conditions of analysis

Growth method		CHAD
Significance level	Split node	0.05
	Category to join	0.05
Chi-square test		Pearson
Minimum number of cases	Parent node	100
	Child node	50
Maximum number of nodes		3
Adjustment of significance probability using Bonferroni method		
Dependent variable		Did the customer buy a chocolate item in September or October
Independent variable	Program category	NEWS
		Variety show
		Documentary
		TV gossip show
		Sports
		Animation
		HobbiesEducation
		TV drama
	Time	17 o'clock
		18 o'clock
		19 o'clock
		20 o'clock
		21 o'clock
		22 o'clock
		23 o'clock
	Age	
	CM contact data	
	Married or unmarried	

The conditions of the analysis are shown in Table 6. We chose "bought it" in the category of the dependent variable. In all the analysis results done this time, since the P value was less than 0.05 at all nodes, we think the model fit is good. The numbers below the nodes in Fig. 6-1 to Fig. 6-12 are index values. If it is 100%, it means that the purchase probability is 1.00 times compared with the data set. For example, if it is 120%, 1.2 times. The rectangle of each node represents the ratio between purchaser (dark gray) and non-purchaser (light gray).

5.1   When the First Independent Variable Is Set to "NEWS"

The highest percentage of purchase in 2011 results node 9. It is a person who watches "NEWS" 2 to 3 times a week and "Sports" once to twice times a week and watches "TV drama" 5 to 6 times a week or hardly watch. Their index value is 134.0%. Node 11 is the next highest percentage that is 119.8%. It is a person who watches "NEWS" 0 to 2 times a week and married 20's, 40's and 50's. Node 1 is the third highest percentage that is 116.7%. It is a person who watches "NEWS" 4 times or more. Node 13 is the worst percentage that is 55.7%. It is a person who watches "NEWS" 0 to 2 times a week and unmarried 30's.

The highest percentage of purchase in 2013 results node 9. It is a person who watches "NEWS" 0 to 2 times or 7 or more times a week, "Variety Show" 0 to 1 time or 3 or more times a week and married. Their index value is 111.3%. Node 11 is the next highest percentage that is 105.5%. It is a person who doesn't almost watch "NEWS" but who watch 18 o'clock program twice a week or more and 17 o'clock once or twice a week. Node 8 is the third highest percentage that is 102.3%. It is a person who watches "NEWS" 0 to 1 time or 3 times more and married. Node 4 is the worst percentage that is 83.2%. It is a person who doesn't almost watch "NEWS" and "Variety Show".

5.2   When the First Independent Variable Is Set to "Documentary"

The highest percentage of purchase in 2011 results node 8. It is a person who watches "Documentary" 4 times a week or less and married 50's. Their index value is 122.5%. Node 5 is the next highest percentage that is 121.7%. It is a person who watches "Documentary" 3 or more times and "Animation" once time a week. Node 10 is the worst percentage that is 65.0%. It is a person who watches "Documentary" 4 times a week or less and unmarried 30's.

The highest percentage of purchase in 2013 results node 13. It is a person who watches "Documentary" once to twice times a week, 18 o'clock program 3 to 6 times a week and unmarried. Their index value is 112.0%. Node 11 is the next highest percentage that is 109.0%. It is a person who watches "Documentary" 3 times a week, "Variety Show" once to twice times or 4 or more times and married. Node 10 is the third highest percentage that is 101.3%. It is a person who watches "Documentary" 2 or more times a week, "Variety Show" once to twice times or 4 or more times and unmarried. Node 6 is the worst percentage that is 84.1%. It is a person who doesn't almost watch "Documentary" and unmarried.

5.3   When the First Independent Variable Is Set to "TV Gossip Show"

The highest percentage of purchase in 2011 results node 10. It is a person who watches "TV Gossip Show" 2 times a week or less and married 50's. Their index value is 124.0%. Node 3 is the next highest percentage that is 114.1%. It is a person who watches "TV Gossip Show" 2 or more times, 20's, 40's and 50's. Node 7 is the third highest percentage that is 107.2%. It is a person who don't almost watch "TV Gossip Show" or watch it 4 or more times a week. Node 8 is the worst percentage that is 58.9%. It is a person who doesn't almost watch "TV Gossip Show" and 3 or more programs on Tuesday.

In 2013, "TV Gossip Show" surely watches at least once a week and those who have more than 2 or 4 programs on Tuesday tend to have higher purchasing ratios than the whole. Node 8 is the highest value is married with that attribute (112.2%). Node 7 is next highest value is unmarried with that attribute (103.7%). Node 6 is the worst percentage that is 85.2%. It is a person who doesn't almost watch "TV Gossip Show" and at 20 o'clock.

5.4   When the First Independent Variable Is Set to "Sports"

The highest percentage of purchase in 2011 results node 3. It is a person who watches "Sports" once or more times a week, 20's, 40's and 50's. Their index value is 115.7%. Node 7 is the next highest percentage that is 104.0%. It is a person who watches "Sports" once or more times a week and married 30's. Node 8 is the worst percentage that is 73.1%. It is a person who watches "Sports" once or more times a week and unmarried 30's.

The highest percentage of purchase in 2013 results node 6. It is a person who watches "Sports" 3 or more times a week, 20 o'clock program once time or, 5 or more times a week. Their index value is 111.2%. Node 9 is the next highest percentage that is 108.1%. It is a person who watches "Sports" once time a week, "20 o'clock program once time or, 4 or more times a week. Node 5 is the worst percentage that is 86.6%. It is a person who doesn't almost watch "Sports" and at 20 o'clock.

5.5   When the First Independent Variable Is Set to "Variety Show"

The highest percentage of purchase in 2011 results node 9. It is a person who watches "Variety Show" once to twice times a week, married and one to two, or 4 to 6 programs on Friday. Their index value is 119.7%. Node 3 is the next highest percentage that is 114.2%. It is a person who doesn't almost watch "Variety Show" or watches 5 or more times and watches "NEWS" twice or more times a week. Node 7 is the third highest percentage that is 101.3%. It is a person who doesn't almost watch "Variety Show" or watches 5 or more times and watches "NEWS" one time or less a week.

Node 6 is the worst percentage that is 76.4%. It is a person who watches "Variety Show" one e or twice times a week and unmarried.

The highest percentage of purchase in 2013 results node 8. It is a person who watches "Variety Show" 4 or more times a week, married and 18 o'clock program 1 to 6 times a week. Their index value is 111.0%. Node 6 is the next highest percentage that is 103.8%. It is a person who watches "Variety Show"4 or more times a week, unmarried and "Documentary" 3 to 6 times a week. Node 2 is the worst percentage that is 84.4%. It is a person who don't almost watch "Variety Show".

5.6   When the First Independent Variable Is Set to "Hobbies and Education"

The highest percentage of purchase in 2011 results node 6. It is a person who watches "Hobbies and Education" once time, or 7 times or more a week, married and 4 programs or more on Monday. Their index value is 143.4%. Node 2 is the next highest percentage that is 118.1%. It is a person who watches "Hobbies and Education" twice to 6 times a week. Node 5 is the third highest percentage that is 101.6%. It is a person who watches "Hobbies and Education" once time, or 7 times or more a week, married and 4 programs or less on Monday. Node 8 is the worst percentage that is 66.9%. It is a person who watches "Variety Show" once a week and unmarried 30's.

The highest percentage of purchase in 2013 results node 11. It is a person who watches "Hobbies and Education" one or more times a week, married and 22 o'clock program 3 or, 7 times or more a week. Their index value is 115.3%. Node 14 is the next highest percentage that is 112.4%. It is a person who doesn't almost watch "Hobbies and Education", 20 o'clock program 3 or, 7 times or more a week and "Sports" once a week. For the third and later, node 5 is 109.6%, node 12 is 103.8%, node 10 is 103.5%. Node 9 is the worst percentage that is 81.3%. It is a person who watches "Hobbies and Education" once a week, unmarried and 23 o'clock program 2 times or less a week.

6   Consideration of the Results and Suggestions

Even though purchasing has increased from about 57% to 77.9%, the maximum value of the index value was about 13.9% lower between 2011 and 2013. In 2011, the result that the maximum value was about 30% different depending on the program category. But in 2013 there was little difference (Table 7). About analysis model, we need to consider the other analysis methods such as linear models. Because correct answer rate is not good in 2011.

Table 7. Maximum index value

	Maximum value in 2011	Maximum value in 2013	Difference
NEWS	134.0%	111.3%	22.7%
Variety show	122.5%	112.0%	10.5%
Documentary	124.0%	112.2%	11.8%
TV gossip show	115.7%	111.2%	4.5%
Sports	119.7%	111.0%	8.7%
Animation	124.8%	113.6%	11.2%
HobbiesEducation	143.4%	115.3%	28.1%
Average	126.3%	112.4%	13.9%

From the above result, we compare and consider each analysis. When we set the first independent variable to NEWS, in the results of 2011, people watching many categories of programs, such as TV drama, and NEWS's viewing frequency were slightly lower and married people in their 50's. In 2013 people became changed that they watch the Variety Show and the time zone is 17 o'clock or 18 o'clock.

When we set the first independent variable to "Documentary", in the results of 2011, it was many purchased married 50s people. In 2013, in addition to the information on married people, Variety Show and the number of viewing times at 20 o'clock appeared at the node. When we set the first independent variable to "TV Gossip

Show", in the results of 2011, it was many purchased married 50s people. In 2013, in addition to the information on married people, the number of viewing times at Tuesdays appeared at the node. When we set the first independent variable to "Variety Show", The number of views "Variety Show" has increased in two years. In the result of 2011, the next node that it was Married was the number of viewing programs on Friday, but in 2013 it changed to 18 o'clock. When we set the first independent variable to "Hobbies/Education", In the result of 2011, it was a married person who watched on Monday but in 2013 it changed to a person who is married and watches at 22 o'clock.

We made a consideration by category. It turned out that factors to be considered in everything changed. We made 12 decision tree analysis in 2011 and 2013. We also mentioned the worst in each analysis results, but about half of them resulted in buyers of chocolate products not frequently watching TV. From this, it was also suggested that TV has an influence on purchasing chocolate products. It was appeared nodes that "married or unmarried" to the all analyzed results. this is an important element of customer information in the future. Nodes separated by age appeared only in 2011. However, the node that splits in broadcasting time zone appeared only in 2013. From this, it can consider that matters to be emphasized have changed. we think it is effective that the broadcasting time zone is better than Age of watching programs. From the analysis of all decision trees, the values of "Hobbies/Education" were better than those of other results in both years. Specifically, the index value in 2011 was the highest. And in 2013, nodes that the index values exceeding 100% was the most. "Hobbies/Education" is a suitable program category for advertising chocolate products. Especially we consider 20 o'clock or 22 o'clock is good.

References

1. Heisei26 Information and Communication Policy Research Institute, April, Research on the use time and information behavior of information and communication media in 2013 (preliminary report). http://www.soumu.go.jp/iicp/chousakenkyu/data/research/survey/tele com/2014/h25mediariyou_1sokuhou.pdf. Accessed 06 Feb 2018
2. Risky Brand Inc.: Consumer analysis TV Audience trends. http://www.riskybrand.com/report_140710/. Accessed 06 Feb 2018
3. Dentsu INC: Advertising costs in 2013 Japan. http://www.dentsu.co.jp/knowledge/ad_cost/2013/media.html. Accessed 06 Feb 2018
4. Japan Chocolate Cocoa Association: Chocolate products Domestic, import, export and consumption trends. http://www.chocolate-cocoa.com/statistics/domestic/chocolate_j.html. Accessed 06 Feb 2018
5. TV program lists, 2011 program schedule, Tokyo. http://timetable.yanbe.net/?d=2011&p=13. Accessed 11 Feb 2018
6. TV program lists, 2013 program schedule, Tokyo. http://timetable.yanbe.net/?d=2013&p=13. Accessed 11 Feb 2018
7. Kindaichi, H.: Personal Katakana Language Dictionary supervision
8. The 6th edition. Iwanami Shoten
9. Broadcasting Act, Chap. 1, Article 2 29, 30
10. Oku, Y., Fumikiyo, S.: Determination and prediction by decision tree distribution. J. Econ. Univ. **39**(4), 33–43 (2005)

Author Index

Printed in the United States
By Bookmasters